Practical Reliability Engineering

Fourth Edition

'The concept of chance enters into the very first steps of scientific activity, by virtue of the fact that no observation is absolutely correct. I think chance is a more fundamental concept than causality, for whether in a concrete case a cause–effect relationship exists can only be judged by applying the laws of chance to the observations.'

Max Born,
Natural Philosophy of Cause and Chance

'A statistical relationship, however strong and however suggestive, can never establish a causal connection. Our ideas on causation must come from outside statistics, ultimately from some theory.'

Kendall & Stuart,
The Advanced Theory of Statistics

'Reliability is, after all, engineering in its most practical form.'

James R. Schlesinger
Former US Secretary of State for Defense

Practical
Reliability
Engineering

Fourth Edition

PATRICK D. T. O'CONNOR

with

DAVID NEWTON
DN Consultancy, UK

RICHARD BROMLEY
RGB Services Ltd, UK

JOHN WILEY & SONS, LTD

Other Wiley Editorial Offices

John Wiley & Sons Inc., 605 Third Avenue, New York, NY 10158-0012, USA

Jossey-Bass, 989 Market Street, San Francisco, CA 94103-1741, USA

Wiley-VCH Verlag GmbH, Pappelallee 3, D-69469 Weinheim, Germany

John Wiley & Sons Australia Ltd, 33 Park Road, Milton, Queensland 4064, Australia

John Wiley & Sons (Asia) Pte Ltd, 2 Clementi Loop #02-01, Jin Xing Distripark, Singapore 129809

John Wiley & Sons Canada Ltd, 22 Worcester Road, Etobicoke, Ontario, Canada M9W 1L1

British Library Cataloguing in Publication Data

A catalogue record for this book is available from the British Library

ISBN 0 470 84462 0 (HB) 0 470 84463 9 (PB)
Printed and bound by CPI Antony Rowe, Eastbourne, East Sussex
Typeset in 10/12pt Times by Kolam Information Services Private Limited, Pondicherry, India

This book is printed on acid-free paper responsibly manufactured from sustainable
forestry in which at least two trees are planted for each one used for paper production.

To Ina

Contents

Preface to the First Edition

This book is designed to provide an introduction to reliability engineering and management, both for students and for practising engineers and managers. The emphasis throughout is on practical applications, and the mathematical concepts described are accordingly limited to those necessary for solution of the types of problems covered. Practical approaches to problem-solving, such as the use of probability plotting techniques and computer programs, are stressed throughout. More advanced texts are cited for further reading on the mathematical and statistical aspects. The references given in the Bibliographies are limited to those considered to provide a direct continuation of the chapter material, with the emphasis on practical applications. Tables and charts are provided to complement the analytical methods described, and numerous worked examples are included.

The book describes and comments on the usage of the major national and government standards and specifications covering reliability engineering and management in the USA and the UK. It is considered that this is an important aspect of the practical approach, since so much engineering development work is now governed by such documents. The effects of current engineering, commercial and legislative developments, such as microelectronics, software-based systems, consumerism and product liability, are covered in some detail.

The requirements of the examination syllabi of the American Society for Quality Control, and the Institute of Quality Assurance (UK) in reliability engineering are covered, so the book will be suitable for use in courses leading to these qualifications. The emphasis on practical approaches to engineering and management, the comprehensive coverage of standards and specifications, and the overall layout of the book should make it equally as suitable as a general up to date reference for use in industry and in government agencies.

PATRICK O'CONNOR
1981

Preface to the Second Edition

I have received much helpful criticism of the first edition of my book since it appeared in 1981. Whilst the reviews have generally not been unfavourable, critics have pointed out that, despite the title, the book was not quite practical enough in some areas. I have also come to realize this through my own work, particularly on the application of mathematical modelling and statistics to reliability problems. Consequently, much of the revision for the second edition has been to add to what I consider to be the practical aspects of management and engineering for reliability.

I have added to the sections on reliability prediction, demonstration and measurement, to explain and to stress the fundamental and considerable uncertainty associated with attempts to quantify and forecast a property of engineered products which is inherently non-deterministic. I believe that when people involved in reliability work manage to unshackle themselves from the tyranny of the 'numbers game' the way is cleared for the practical engineering and management approaches that are the only ways to achieve the highly reliable products demanded by the markets of today. I have not removed the descriptions of the methods for quantifying reliability, since I believe that, when these are applied with commonsense and understanding of their inherent limitations, they can help us to solve reliability problems and to design and make better products.

I have added three new chapters, all related to the practical aspects.

The first edition described how to analyse test data, but included little on how to test. I have therefore written a new chapter on reliability testing, covering environmental and stress testing and the integration of reliability and other development testing. I am indebted to Wayne Tustin for suggesting this and for his help and advice on this subject.

The quality of manufacture is obviously fundamental to achieving high reliability. This point was made in the first edition, but was not developed. I have added a complete chapter on quality assurance (QA), as well as new material on integrated management of reliability and QA programmes.

Maintenance also affects reliability, so I have added a new chapter on maintenance and maintainability, with the emphasis on how they affect reliability, how reliability affects maintenance planning and how both affect availability.

I have also added new material on the important topic of reliability analysis for repairable systems. Harry Ascher, of the US Naval Research Laboratory, has pointed

xv

out that the reliability literature, including the first edition of my book, has almost totally ignored this aspect, leading to confusion and analytical errors. How many reliability engineers and teachers know that Weibull analysis of repairable system reliability data can be quite misleading except under special, unrealistic conditions? Thanks to Harry Ascher, I know now, and I have tried to explain this in the new edition.

I have also brought other parts of the book up to date, particularly the sections on electronic and software reliability.

The third reprint of the first edition included many corrections, and more corrections are made in this edition.

I am extremely grateful to all those who have pointed out errors and have helped me to correct them. Paul Baird of Hewlett Packard, Palo Alto, was particularly generous. Colleagues at British Aerospace, particularly Brian Collett, Norman Harris, Chris Gilders and Gene Morgan, as well as many others, also provided help, advice and inspiration.

Finally, my thanks go to my wife Ina for much patience, support and typing.

PATRICK O'CONNOR
1985

Preface to the Third Edition

The new industrial revolution has been driven mainly by the continuing improvements in quality and productivity in nearly all industrial sectors. The key to success in every case has been the complete integration of the processes that influence quality and reliability, in product specification, design, test, manufacture, and support. The other essential has been the understanding and control of variation, in the many ways in which it can affect product performance, cost and reliability. Teachers such as W. E. Deming and G. Taguchi have continued to grow in stature and following as these imperatives become increasingly the survival kit of modern industry.

I tried to stress these factors in the second edition, but I have now given them greater prominence. I have emphasized the use of statistical experimentation for preventing problems, not just for solving them, and the topic is now described as a design and development activity. I have added to the chapter on production quality assurance, to include process improvement methods and more information on process control techniques. These chapters, and the chapter on management, have all been enlarged to emphasize the integration of engineering effort to identify, minimize and reduce variation and its effects. The important work of Taguchi and Shainin is described, for the first time in this book. Chris Gray gave me much valuable help in describing the Taguchi method.

I have updated several chapters, particularly those on electronic systems reliability. I have also added a new chapter on reliability of mechanical components and systems. I would like to thank Professor Dennis Carter for his advice on this chapter.

I have taken the opportunity to restructure the book, to reflect better the main sequence of engineering development, whilst stressing the importance of an integrated, iterative approach.

I have once more been helped by many people who have contributed kind criticisms of the earlier edition, and I have tried to take these into account. I also would like to record with thanks my continuing debt to Norman Harris for his contributions to bridging the gap between engineering and statistics, and for helping me to express his ideas.

Finally, my heartfelt thanks go to my wife and boys for their forebearance, patience, and support. Having an author at home must place severe demands on love and tolerance.

PATRICK O'CONNOR
1990

Preface to the Third Edition
Revised

This revised edition has been produced in response to numerous suggestions that the book would be of greater value to students and teachers if it included exercise questions. David Newton and Richard Bromley have therefore teamed up with me to produce exercises appropriate to each chapter of the book.

The exercises cover nearly all of the types of questions that occur in the reliability examinations set by the UK Institute of Quality Assurance (IQA) and by the American Society for Quality Control (ASQC). The ASQC examination questions are of the multiple-choice type, which is not the format used here, but this should make no difference to the value of the exercises in preparing for the ASQC examination.

A solutions manual is available to teachers, free of charge, by writing to John Wiley and Sons Ltd in Chichester.

I would like to thank David Newton and Richard Bromley for their enthusiastic support in preparing this revised edition.

PATRICK O'CONNOR
1995

Preface to the Fourth Edition

It is over ten years since the last major revision and update to my book. Inevitably in that time there have been developments in engineering technology and in reliability methods. In this new edition I have tried to include all of the important changes that affect reliability engineering and management today. In keeping with the original aims of the book, I have emphasised those with practical implications.

The main changes and additions include:

—Updated and more detailed descriptions of how engineering products fail (Chapters 1, 8 and 9).
—More detailed description of the nature of variation in engineering (Chapter 2).
—Descriptions of the Petri net and M(t) methods (Chapters 6 and 12).
—More detailed description of the particular aspects of software in engineering systems, and updated descriptions of design, analysis and test methods (Chapter 10).
—Expanded descriptions of accelerated test methods for development and manufacturing (Chapters 11 and 13).
—Updated and expanded descriptions of test methods for electronics and acceptance sampling (Chapter 13).
—More detailed descriptions of management aspects, including standards, "six sigma", and supplier management (Chapter 15).
—Updated references to standards, and updated and expanded bibliographies.

Some of the new material is adapted from my book "Test Engineering", with permission from the publisher.

The questions and the answers manual (available separately from the publisher) have been augmented to cover the new material.

An Internet homepage has been created for the book, at www.pat-oconnor.co.uk/ practicalreliability.htm. The homepage includes listings of suppliers of reliability engineering related services and software.

I would like to express my gratitude to Prof. S.K. Yang for his kind assistance with the description of the Petri net method, Dr. Gregg Hobbs for his teaching and help on HALT/HASS testing, Prof. Jörgen Möltoft for helping with the description of the M(t) method, and Jim McLinn for providing additional material, questions and answers on aspects of accelerated testing and data analysis. I also thank all who have provided suggestions and pointed out errors. Last but certainly not least I thank my wife, Ina, again.

PATRICK O'CONNOR
2001

Acknowledgements

I am deeply indebted to many people for their help and constructive criticism. In particular, Dr R. A. Evans provided invaluable comments, corrections and suggestions for improvement. My colleague Mr Norman Harris provided much valuable assistance and several new insights on the material of Chapters 2 and 3. Professor A. D. S. Carter of the Royal Military College of Science gave very helpful advice on the analysis methods in Chapter 4, and I am grateful for his permission to reproduce material from his published papers on load–strength interference which appear in the Bibliography at the end of that chapter. Dr Bev Littlewood of the City University, London, was very generous with suggestions and criticisms on Chapter 10 on software reliability. Mr Kenneth Blemel also provided many suggestions for improvement and corrections. I am also indebted to the many others who encouraged and helped me, either in discussions, with comments on the text, or whose published work provided me with insights and guidance.

I acknowledge with thanks the permission I have been given to reproduce the following copyright material: Table 2.4, from W. Volk, *Applied Statistics for Engineers*, McGraw-Hill, New York (1958); Fig. 2.16, from *Industrial Quality Control* (December 1965); Fig. 2.17, from H. F. Dodge and H. G. Romig, *Sampling Inspection Tables*, Bell Telephone Laboratories (1944); Figs 4.4 to 4.7, from A. D. S. Carter, 'Reliability reviewed', *Proc. Inst. Mech. Eng.*, **193** (4) (1979); Table 6.1, adapted from R. H. Myers, K. L. Wong and H. M. Gordy, *Reliability Engineering for Electronic Systems*, Wiley, New York (1964); Figs 9.5 and 9.6, and Table 9.4, from R. I. Anderson, RDH 376: *Reliability Design Handbook*, Reliability Analysis Center, IIT Research Institute (1976); Appendix 4 from A. Hald, *Statistical Tables and Formulas*, Wiley, New York (1952); probability and hazard plotting papers from Technical and Engineering Aids to Management Inc. and H. W. Peel and Co. Ltd.

Finally, I would like to thank Miss Linda Rendle for her speedy and efficient typing of the manuscript, and the management of British Aerospace Dynamics Group, Stevenage Division, for encouragement and assistance in writing the book.

Notation and Definitions

LIST OF SYMBOLS

α	Producer's risk (in sample or sequential testing, s-probability that good lot or item will be rejected). Also Duane slope, also s-significance, s-confidence.
β	Consumer's risk (in sample or sequential testing, s-probability that bad lot or item will be accepted). Also shape parameter (slope) of Weibull distribution.
γ	Location parameter (minimum life) of Weibull distribution.
μ	Location parameter for a continuous distribution. Equal to mean value for s-normal distribution.
η	Characteristic life for Weibull distribution.
σ	Standard deviation (scale parameter) of s-normal distribution.
ν	Degrees of freedom.
θ	Mean time between failures (MTBF) for constant failure rate.
λ	Failure rate.
$\Phi(z)$	Cumulative value of standard s-normal variate, z.
$B(\cdot)$	B-life of Weibull distribution, e.g. B_{10} life.
C	s-confidence.
C_{p}, C_{p_k}	Process capability.
d	Design ratio = design MTBF/low limit MTBF.
$E(\cdot)$	s-expected value. The s-expected value of x is denoted $\hat{x}$.
$f(\cdot)$	Function of. For a distribution, the probability density function (p.d.f.).
$F(\cdot)$	Cumulative distribution function (c.d.f.).
F	Variance ratio.
$h(\cdot)$	Hazard function.
$H(\cdot)$	Cumulative hazard function.
L	Load.
n	Sample size.
N	Population size.
$P(\cdot)$	s-probability.
$P(\cdot \mid \cdot)$	Conditional s-probability.
r	s-correlation coefficient. Also rank value for median ranks.

R(·)   Reliability function.
S   Standard error. Also strength.
t   Time.
Var(·)   Variance (σ^2).
$\bar{x}$ Mean value of distributed variable, x.
z   Standard s-normal variate.

SELECTED DEFINITIONS

Reliability

Reliability. The ability of an item to perform a required function under stated conditions for a stated period of time (BS 4778).

(Note: the term reliability can also be denoted as a probability or as a success ratio.)

Redundancy. The existence of more than one means for accomplishing a given function. Each means of accomplishing the function need not necessarily be identical (MIL-STD-721B).

Redundancy, active. That redundancy wherein all redundant items are operating simultaneously rather than being switched on when needed (MIL-STD-721B).

Redundancy, standby. That redundancy wherein the alternative means of performing the function is inoperative until needed and is switched on upon failure of the primary means of performing the function (MIL-STD-721B).

Failure

Failure. The termination of the ability of an item to perform a required function (BS 4778).

Observed failure rate. For a stated period in the life of an item, the ratio of the total number of failures in a sample to the cumulative observed time on that sample. The observed failure rate is to be associated with particular and stated time intervals (or summation of intervals) in the life of the item, and under stated conditions (BS 4778).

Observed mean time between failures (MTBF). For a stated period in the life of an item, the mean value of the length of time between consecutive failures computed as the ratio of the cumulative observed time to the number of failures under stated conditions (BS 4778).

Observed mean time to failure (MTTF) (for non- repairable items). For a stated period in the life of an item, the ratio of the cumulative time for a sample to the total number of failures in the sample during the period under stated conditions (BS 4778).

Observed B-Percentile life. The length of observed time at which stated proportion (*B*%) of a sample of items has failed (BS 4778).

1

Introduction to Reliability Engineering

WHAT IS RELIABILITY ENGINEERING?

No one disputes the need for articles to be reliable. The average consumer is acutely aware of the problem of less than perfect reliability in domestic products such as TV sets and automobiles. Organizations such as airlines, the military and public utilities are aware of the costs of unreliability. Manufacturers often suffer high costs of failure under warranty. Argument and misunderstanding begin when we try to quantify reliability values, or try to assign financial or other benefit values to levels of reliability.

The simplest, purely producer-oriented or inspectors' view of reliability is that in which a product is assessed against a specification or set of attributes, and when passed is delivered to the customer. The customer, having accepted the product, accepts that it might fail at some future time. This simple approach is often coupled with a warranty, or the customer may have some protection in law, so that he may claim redress for failures occurring within a stated or reasonable time. However, this approach provides no measure of quality over a period of time, particularly outside a warranty period. Even within a warranty period, the customer usually has no grounds for further action if the product fails once, twice or several times, provided that the manufacturer repairs the product as promised each time. If it fails often, the manufacturer will suffer high warranty costs, and the customers will suffer inconvenience. Outside the warranty period, only the customer suffers. In any case, the manufacturer will also probably incur a loss of reputation, possibly affecting future business.

We therefore come to the need for a time-based concept of quality. The inspectors' concept is not time-dependent. The product either passes a given test or it fails. On the other hand, reliability is usually concerned with failures in the time domain. This distinction marks the difference between traditional quality control and reliability engineering.

Whether failures occur or not and their times to occurrence, can seldom be forecast accurately. Reliability is therefore an aspect of engineering uncertainty. Whether an item will work for a particular period is a question which can be answered as a probability. This results in the usual engineering definition of reliability as:

1

The probability that an item will perform a required function without failure under stated conditions for a stated period of time.

Reliability can also be expressed as the number of failures over a period.

Durability is a particular aspect of reliability, related to the ability of an item to withstand the effects of time (or of distance travelled, operating cycles, etc.) dependent mechanisms such as fatigue, wear, corrosion, electrical parameter change, etc. Durability is usually expressed as a minimum time before the occurrence of *wearout* failures.

The objectives of reliability engineering, in the order of priority, are:

1. To apply engineering knowledge and specialist techniques to prevent or to reduce the likelihood or frequency of failures.
2. To identify and correct the causes of failures that do occur, despite the efforts to prevent them.
3. To determine ways of coping with failures that do occur, if their causes have not been corrected.
4. To apply methods for estimating the likely reliability of new designs, and for analysing reliability data.

The reason for the priority emphasis is that it is by far the most effective way of working, in terms of minimizing costs and generating reliable products.

The primary skills that are required, therefore, are the ability to understand and anticipate the possible causes of failures, and knowledge of how to prevent them. It is also necessary to have knowledge of the methods that can be used for analysing designs and data. The primary skills are nothing more than good engineering knowledge and experience, so reliability engineering is the first and foremost the application of good engineering, in the widest sense, during design, development, manufacture and service.

Mathematical and statistical methods can be used for quantifying reliability (prediction, measurement) and for analysing reliability data. The basic methods are described in Chapter 2, to provide an introduction for some of the applications described subsequently. However, because of the high levels of uncertainty involved these can seldom be applied with the kind of precision and credibility that engineers are accustomed to when dealing with most other problems. In practice the uncertainty is often in orders of magnitude. Therefore the role of mathematical and statistical methods in reliability engineering is limited, and appreciation of the uncertainty is important in order to minimise the chances of performing inappropriate analysis and of generating misleading results. Mathematical and statistical methods can make valuable contributions in appropriate circumstances, but practical engineering must take precedence in determining the causes of problems and their solutions. Unfortunately not all reliability training, literature and practice reflect this reality.

Over-riding all of these aspects, though, is the management of the reliability engineering effort. Since reliability (and very often also safety) is such a critical parameter of most modern engineering products, and since failures are generated primarily by the people involved (designers, test engineers, manufacturing, suppliers, maintainers, users), it can be maximised only by an integrated effort that encompasses training, teamwork, discipline, and application of the most appropriate methods. Reliability engineering "specialists" cannot make this happen. They can provide support, training and tools, but only managers can organise, motivate, lead and provide the resources. Reliability engineering is, ultimately, effective management of engineering.

WHY TEACH RELIABILITY ENGINEERING?

Engineering education is traditionally concerned with teaching how manufactured products work. The ways in which products fail, the effects of failure and aspects of design, manufacture, maintenance and use which affect the likelihood of failure are not usually taught, mainly because it is necessary to understand how a product works before considering ways in which it might fail. For many products the tendency to approach the failed state is analogous to entropy. The engineer's tasks are to design and maintain the product so that the failed state is deferred. In these tasks he faces the problems inherent in the variability of engineering materials, processes and applications. Engineering education is basically deterministic, and does not usually pay sufficient attention to variability. Yet variability and chance play a vital role in determining the reliability of most products. Basic parameters like mass, dimensions, friction coefficients, strengths and stresses are never absolute, but are in practice subject to variability due to process and materials variations, human factors and applications. Some parameters also vary with time. Understanding the laws of chance and the causes and effects of variability is therefore necessary for the creation of reliable products and for the solution of problems of unreliability.

However, there are practical problems in applying statistical knowledge to engineering problems. These problems have probably deterred engineers in the past from using statistical methods, and texts on reliability engineering and mathematics have generally stressed the theoretical aspects without providing guidance on their practical application. To be helpful a theoretical basis must be credible, and statistical methods which work well for insurance actuaries, market researchers or agricultural experimenters may not work as well for engineers. This is not because the theory is wrong, but because engineers usually have to cope with much greater degrees of uncertainty, mainly due to human factors in production and use.

Some highly reliable products are produced by design and manufacturing teams who practise the traditional virtues of reliance on experience and maintenance of high quality. They do not see reliability engineering as a subject requiring specialist consideration, and a book such as this would teach them little that they did not already practise in creating their reliable products. Engineers and managers might therefore regard a specialist reliability discipline with scepticism. However, many pressures now challenge the effectiveness of the traditional approaches. Competition, the pressure of schedules and deadlines, the cost of failures, the rapid evolution of new materials, methods and complex systems, the need to reduce product costs, and safety considerations all increase the risks of product development. Figure 1.1 shows the pressures that lead to the overall perception of risk. Reliability engineering has developed in response to the need to control these risks.

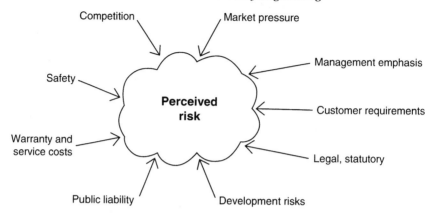

Figure 1.1   Perception of risk

Later chapters will show how reliability engineering methods can be applied to design, development and management to control the level of risk. The extent to which the methods are applicable must be decided for each project and for each design area. They must not replace normal good practice, such as safe design for components subject to cyclic loading, or application guidelines for electronic components. They should be used to supplement good practice. However, there are times when new risks are being taken, and the normal rules and guidelines are inadequate or do not apply. Sometimes we take risks unwittingly, when we assume that we can extrapolate safely from our present knowledge. Designers and managers are often overoptimistic or are reluctant to point out risks about which they are unsure.

It is for these reasons that an understanding of reliability engineering principles and methods is now an essential ingredient of modern engineering.

WHY DO ENGINEERING ITEMS FAIL?

There are many reasons why a product might fail. *Knowing, as far as is practicable, the potential causes of failures is fundamental to preventing them.* It is rarely practicable to anticipate all of the causes, so it is also necessary to take account of the uncertainty involved. The reliability engineering effort, during design, development and in manufacture and service should address all of the anticipated and possibly unanticipated causes of failure, to ensure that their occurrence is prevented or minimized.

The main reasons why failures occur are:

1. The design might be *inherently incapable*. It might be too weak, consume too much power, suffer resonance at the wrong frequency, etc. The list of possible reasons is endless, and every design problem presents the potential for errors, omissions, and oversights. The more complex the design or difficult the problems to be overcome, the greater is this potential.

2. The item might be *overstressed* in some way. If the stress applied exceeds the strength then failure will occur. An electronic component will fail if the applied electrical stress (voltage, current) exceeds the ability to withstand it, and a mechanical strut will buckle if the compression stress applied exceeds the buckling strength. Overstress failures such as these do happen, but fortunately not very often, since designers provide margins of safety. Electronic component specifications state the maximum rated conditions of application, and circuit designers take care that these rated values are not exceeded in service. In most cases they will in fact do what they can to ensure that the in-service worst case stresses remain below the rated stress values: this is called 'de-rating'. Mechanical designers work in the same way: they know the properties of the materials being used (e.g. ultimate tensile strength) and they ensure that there is an adequate margin between the strength of the component and the maximum applied stress. However, it might not be possible to provide protection against every possible stress application. If a 110 V appliance is connected to a 240 V supply it will probably fail, and a tyre will burst if sufficiently over-inflated.

3. Failures might be caused by *variation*. In the situations described above the values of strength and load are fixed and known. If the known strength always exceeds the known load, as shown in Fig. 1.2, then failure will not occur. However, in most cases, there will be some uncertainty about both. The actual strength values of any population of components will vary: there will be some that are relatively strong, others that are relatively weak, but most will be of nearly average strength. Also, the loads applied will be variable. Figure 1.3 shows this type of situation. As before, failure will not occur so long as the applied load does not exceed the strength. However, if there is an overlap between the distributions of load and strength, and a load value in the high tail of the load distribution is applied to an item in the weak tail of the strength distribution so that there is overlap or *interference* between the distributions (Fig. 1.4), then failure will occur. We will discuss load and strength interference in more detail in Chapter 4.

4. Failures can be caused by *wearout*. We will use this term to include any mechanism or process that causes an item that is sufficiently strong at the start of its life to

Figure 1.2 Load–strength—discrete values

Figure 1.3   Load–strength—distributed values

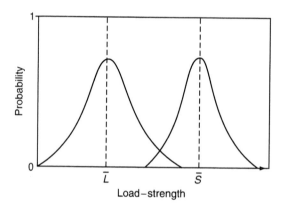

Figure 1.4   Load–strength—interfering distributions

become weaker with age. Well-known examples of such processes are materi-alfatigue, wear between surfaces in moving contact, corrosion, insulation deteri-oration, and the wearout mechanisms of light bulbs and fluorescent tubes. Figure 1.5 illustrates this kind of situation. Initially the strength is adequate to withstand the applied loads, but as weakening occurs over time the strength decreases. In every case the average value falls and the spread of the strength distribution widens. This is a major reason why it is so difficult to provide accurate predic-tions of the lives of such items.

5. Failures can be caused by other time-dependent mechanisms. Battery run-down, creep caused by simultaneous high temperature and tensile stress, as in turbine discs and fine solder joints, and progressive drift of electronic component param-eter values are examples of such mechanisms.

6. Failures can be caused by *sneaks*. A sneak is a condition in which the system does not work properly even though every part does. For example, an electronic system might be designed in such a way that under certain conditions incorrect operation occurs. The fatal fire in the Apollo spacecraft crew capsule was caused in

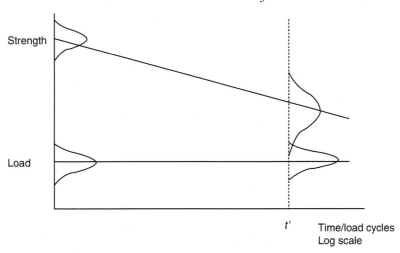

Strength

Load

t' Time/load cycles
Log scale

Figure 1.5   Time-dependent load and strength variation

this way: the circuit design ensured that an electrical short circuit would occur when a particular sequence was performed by the crew. Sneaks can also occur in software designs.

7. Failures can be caused by *errors*, such as incorrect specifications, designs or software coding, by faulty assembly or test, by inadequate or incorrect maintenance, or by incorrect use. The actual failure mechanisms that result might include most of the list above.

8. There are many other potential causes of failure. Gears might be noisy, oil seals might leak, display screens might flicker, operating instructions might be wrong or ambiguous, electronic systems might suffer from electromagnetic interference, etc.

Failures have many different causes and effects, and there are also different perceptions of what kinds of events might be classified as failures. The burning O-ring seals on the Space Shuttle booster rockets were not classed as failures, until the ill-fated launch of Challenger. We also know that all failures, in principle and almost always in practice, can be prevented.

PROBABILISTIC RELIABILITY

The concept of reliability as a probability means that any attempt to quantify it must involve the use of statistical methods. An understanding of statistics as applicable to reliability engineering is therefore a necessary basis for progress, except for the special cases when reliability is perfect (we know the item will never fail) or it is zero (the item will never work). For practical purposes a hammer might be 100 per cent reliable if used for driving tacks, but would have a zero reliability if used to stop a train. If used to break rocks, its reliability for a given period might have some value between 0 and 100 per cent. In engineering we try to ensure 100 per cent reliability, but our experience tells us that we do not always succeed. Therefore reliability statistics are usually concerned with probability values which are very high (or very low: the probability that a failure does occur, which is 1 – reliability). Quantifying such numbers brings increased uncertainty, since we need correspondingly more information. Other sources of uncertainty are introduced because reliability is often about people who make and people who use the product, and because of the widely varying environments in which typical products might operate.

Further uncertainty, often of a subjective nature, is introduced when engineers begin to discuss failures. Should a failure be counted if it was due to an error that is hoped will not be repeated? If design action is taken to reduce the risk of one type of failure, how can we quantify our trust in the designer's success? Was the machine under test typical of the population of machines?

Reliability is quantified in other ways. We can specify a reliability as the mean number of failures in a given time (failure rate), or as the *mean time between failures* (MTBF) for items which are repaired and returned to use, or as the *mean time to failure* (MTTF) for items which are not repaired. For repaired items, it is often assumed that failures occur at a constant rate, in which case the failure rate $\lambda = (\text{MTBF})^{-1}$. However, this is only a special case, valuable because it is often true and because it is easy to understand.

The application and interpretation of statistics in reliability are less straightforward than in, say, public opinion polls or measurement of human variations such as IQ or height. In these applications, most interest is centred around the behaviour of the larger part of the population or sample, variation is not very large and data are plentiful. In reliability we are concerned with the behaviour in the extreme tails of distributions and possibly unlikely combinations of load and strength, where variability is often hard to quantify and data are expensive.

Further difficulties arise in application of statistical theory to reliability engineering, owing to the fact that variation is often a function of time or of time-related factors such as operating cycles, diurnal or seasonal cycles, maintenance periods, etc. Engineering, unlike most fields of knowledge, is primarily concerned with change, hopefully, but not always, for the better. Therefore the reliability data from any past situation cannot be used to make credible forecasts of the future behaviour, without taking into account non-statistical factors such as design changes, maintainer training, and even imponderables such as unforeseeable production or service problems. The statistician working in reliability engineering needs to be aware of these realities.

Chapter 2 provides the statistical basis of reliability engineering, but it must always be remembered that quality and reliability data contain many sources of uncertainty and variability which cannot be rigorously quantified. It is also important to appreciate that failures and their causes are by no means always clear-cut and unambiguous. They are often open to interpretation and argument. They also differ in terms of importance (cost, safety, other effects). Therefore we must be careful not to apply conventional scientific, deterministic thinking to the interpretation of failures. For example, a mere count of total reported failures of a product is seldom useful or revealing. It tells us nothing about causes or consequences, and therefore nothing about how to improve the situation. This contrasts with a statement of a physical attribute such as weight or power consumption, which is usually unambiguous and complete. Nevertheless, it is necessary to derive values for decision-making, so the mathematics are essential. The important point is that the reliability engineer or manager is not, like an insurance actuary, a powerless observer of his statistics. Statistical derivations of reliability are not a guarantee of results, and these results can be significantly affected by actions taken by quality and reliability engineers and managers.

REPAIRABLE AND NON-REPAIRABLE ITEMS

It is important to distinguish between repairable and non-repairable items when predicting or measuring reliability.

For a non-repairable item such as a light bulb, a transistor, a rocket motor or an unmanned spacecraft, reliability is the survival probability over the item's expected life, or for a period during its life, *when only one failure can occur*. During the item's life the instantaneous probability of the first and only failure is called the *hazard rate*. Life values such as the mean life or *mean time to failure* (MTTF), or the expected life by which a certain percentage might have failed (say 10 per cent.), are other reliability characteristics that can be used. Note that non-repairable items may be individual parts (light bulbs, transistors) or systems comprised of many parts (spacecraft, microprocessors). When a part fails in a non-repairable system, the system fails (usually) and system reliability is, therefore, a function of the time to the first part failure.

For items which are repaired when they fail, reliability is the probability that failure will not occur in the period of interest, when *more than one failure can occur*. It can also be expressed as the *failure rate* or the *rate of occurrence of failures* (ROCOF). However, the failure rate expresses the instantaneous probability of failure per unit time, when several failures can occur in a time continuum. It is unfortunate that the expression *failure rate* has been applied loosely in the past to non-repairable items. What is really meant is that, in a repairable system which contains a part type, the part will contribute by that amount to the system failure rate. The part, being non-repairable, cannot have a failure rate. This distinction does not always have practical significance, but it can, as will be shown later. The expression ROCOF is sometimes used to stress this point.

Repairable system reliability can also be characterized by the *mean time between failures* (MTBF), but only under the particular condition of a constant failure rate. We are also concerned with the *availability* of repairable items, since repair takes time. Availability is affected by the rate of occurrence of failures (failure rate) and by maintenance time. Maintenance can be corrective (i.e. repair) or preventive (to reduce the likelihood of failure, e.g. lubrication). We therefore need to understand the relationship between reliability and maintenance, and how both reliability and maintainability can affect availability.

Sometimes an item may be considered as both repairable and non-repairable. For example, a missile is a repairable system whilst it is in store and subjected to scheduled tests, but it becomes a non-repairable system when it is launched. Reliability analysis of such systems must take account of these separate states. Repairability might also be determined by other considerations. For example, whether a

TV electronics board is treated as a repairable item or not will depend upon the cost of repair. An engine or vehicle will be treated as repairable only up to a certain age.

Repairable system reliability data analysis is covered in Chapter 12 and availability and maintainability in Chapter 14.

THE PATTERN OF FAILURES WITH TIME
(NON-REPAIRABLE ITEMS)

There are three basic ways in which the pattern of failures can change with time. The hazard rate may be decreasing, increasing or constant. We can tell much about the causes of failure and about the reliability of the item by appreciating the way the hazard rate behaves in time.

A constant hazard rate is characteristic of failures which are caused by the application of loads in excess of the design strength, at a constant average rate. For example, overstress failures due to accidental or transient circuit overload, or maintenance-induced failures of mechanical equipment, typically occur randomly and at a generally constant rate.

Material fatigue brought about by strength deterioration due to cyclic loading is a failure mode which does not occur for a finite time, and then exhibits an increasing probability of occurrence.

Decreasing hazard rates are observed in items which become less likely to fail as their survival time increases. This is often observed in electronic equipment and parts. 'Burn-in' of electronic parts is a good example of the way in which knowledge of a decreasing hazard rate is used to generate an improvement in reliability. The parts are operated under failure-provoking stress conditions for a time before delivery. As substandard parts fail and are rejected the hazard rate decreases and the surviving population is more reliable.

The combined effect generates the so-called *bathtub curve* (Figure 1.6). This shows an initial decreasing hazard rate or *infant mortality* period, an intermediate *useful life* period and a final *wearout* period. Death is a good analogy to failure of a non-repairable system, and the bathtub curve model is similar to actuarial statistical models.

THE PATTERN OF FAILURES WITH TIME (REPAIRABLE ITEMS)

The failure rates (or ROCOF) of repairable items can also vary with time, and important implications can be derived from these trends.

A constant failure rate (CFR) is indicative of externally induced failures, as in the constant hazard rate situation for non-repairable items. A CFR is also typical of complex systems subject to repair and overhaul, where different parts exhibit different patterns of failure with time and parts have different ages since repair or replacement. Repairable systems can show a decreasing failure rate (DFR) when reliability is improved by progressive repair, as defective parts which fail relatively

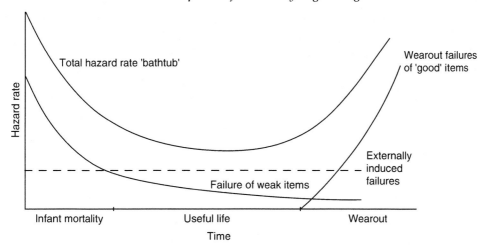

Figure 1.6 The 'bathtub' curve

early are replaced by good parts. 'Burn in' is applied to electronic systems, as well as to parts, for this purpose.

An increasing failure rate (IFR) occurs in repairable systems when wearout failure modes of parts begin to predominate.

The pattern of failures with time of repairable systems can also be illustrated by use of the bathtub curve (Fig. 1.6), but with the failure rate (ROCOF) plotted against age instead of the hazard rate.

The statistical treatment of failure data is covered in Chapters 2 and 3.

THE DEVELOPMENT OF RELIABILITY ENGINEERING

Reliability engineering, as a separate engineering discipline, originated in the United States during the 1950s. The increasing complexity of military electronic systems was generating failure rates which resulted in greatly reduced availability and increased costs. Solid state electronics technology offered long term hope, but conversely miniaturization was to lead to proportionately greater complexity, which offset the reliability improvements expected. The gathering pace of electronic device technology meant that the developers of new military systems were making increasing use of large numbers of new component types, involving new manufacturing processes, with the inevitable consequences of low reliability. The users of such equipment were also finding that the problems of diagnosing and repairing the new complex equipment were seriously affecting its availability for use, and the costs of spares, training and other logistics support were becoming excessive. Against this background the US Department of Defense and the electronics industry jointly set up the Advisory Group on Reliability of Electronic Equipment (AGREE) in 1952. The AGREE report concluded that, to break out of the spiral of increasing development and ownership costs due to low reliability, disciplines must be laid down as integral activities in the development cycle for electronic equipment. The report laid particular stress on the need for new equipments to be tested for several thousand hours in high stress cyclical environments including high and low temperatures, vibration and switching, in order to discover the majority of weak areas in a design at an early enough stage to enable them to be corrected before production commenced. Until that time, environmental tests of tens of hours duration had been considered adequate to prove the suitability of a design. The report also recommended that formal demonstrations of reliability, in terms of statistical confidence that a specified MTBF had been exceeded, be instituted as a condition for acceptance of equipment by the procuring agency. A large part of the report was devoted to providing detailed test plans for various levels of statistical confidence and environmental conditions.

The AGREE report was accepted by the Department of Defense, and AGREE testing quickly became a standard procedure. Companies which invested in the expensive environmental test equipment necessary soon found that they could attain levels of reliability far higher than by traditional methods. It was evident that designers, particularly those working at the fringes of advanced technology, could not be expected to produce highly reliable equipment without it being subjected to a test regime which would show up weaknesses. Complex systems and the components used in them included too many variables and interactions for the human designer to cope with infallibly, and even the most careful design re views and disciplines could not provide sufficient protection. Consequently it was necessary to make the product speak for itself, by causing it to fail, and then to eliminate the weaknesses that caused the failures. The Department of Defense (DOD) reissued the AGREE report on testing as US Military Standard (MIL-STD) 781, *Reliability Qualification and Production Approval Tests*.

Meanwhile the revolution in electronic device technology continued, led by integrated microcircuitry. Increased emphasis was now placed on improving the quality of devices fitted to production equipments. Screening techniques, in which devices are temperature cycled, vibrated, centrifuged, operated at electrical overstress and otherwise abused, were introduced in place of the traditional sampling techniques. With component populations on even single printed circuit boards becoming so large, sampling no longer provided sufficient protection against the production of defective equipment. These techniques were formalized in military standards covering the full range of electronic components. Compon-ents produced to these standards were called 'Hi-rel' components. Specifications and test systems for electronic components, based upon the US Military Standards, have been developed in the United Kingdom and in continental Europe, and internationally through the International Electrotechnical Commission (IEC).

Engineering reliability effort in the United States developed quickly, and the AGREE and reliability programme concepts were adopted by NASA and many other major suppliers and purchasers of high technology equipment. In 1965 the DOD issued MIL-STD-785–*Reliability Programs for Systems and Equipment*. This docu-ment made mandatory the integration of a programme of reliability engineering activities with the traditional engineering activities of design, development and production, as it was by then realized that such an integrated programme was the only way to ensure that potential reliability problems would be detected and eliminated at the earliest, and therefore the cheapest, stage in the development cycle. Much written work appeared on the cost-benefit of higher reliability, to show that effort and resources expended during early development and during production testing, plus the imposition of demonstrations of specified levels of reliability to MIL-STD-781, led to reductions in in-service costs which more than repaid the reliability programme expenditure. The concept of life cycle costs (LCC), or whole life costs, was introduced.

In the United Kingdom, Defence Standard 00–40, *The Management of Reliability and Maintainability* was issued in 1981. The British Standards Institution issued BS 5760—*Guide on Reliability of Systems, Equipments and Components*.

Starting in the early 1980s, the reliability of new Japanese industrial and commercial products took Western competitors by surprise. Items such as automobiles, electronic components and systems, and machine tools achieved levels of reliability far in excess of previous experience. These products were also less expensive and often boasted superior features and performance. The 'Japanese quality revolution' had been driven by the lessons taught by American teachers brought in to help Japan's post-war recovery. The two that stand out were J. R. Juran and W. Edwards Deming (Reference 9), who taught the principles of 'total quality management' (TQM) and continuous improvement. Japanese pioneers, particularly K. Ishikawa, also contributed. These ideas were all firmly rooted in the teaching of the American writer on management, Peter Drucker (Reference 10), that people work most effectively when they are given the knowledge and authority to identify and implement improvements, rather than being expected to work to methods dictated by 'management'.

These ideas proved to be much more effective than those prescribed in documents such as the standards mentioned above. They led to great increases in productivity and quality, and thus in market penetration, as Drucker had predicted. Many Western companies followed the new path that had been blazed and also made great improvements. However, the results of the Japanese head start are still apparent today in showrooms, shops, offices and factories.

Today we expect very high reliability in all of the products we use. Modern electronic components and systems seldom fail, and complex systems such as domestic equipment, automobiles and aircraft are far more reliable than the much simpler systems of twenty or thirty years ago. These improvements have been generated primarily by the application of the methods of TQM and continuous improvement. Many of the methods that had been standardized in the West have been shown to be less effective.

Another aspect of reliability thinking that has developed is the application of statistical methods. Since reliability can be expressed as a probability, and is affected by variation, in principle these methods are applicable. They form the basis of most teaching and literature on the subject. However, variation in engineering is usually of such an uncertain nature that refined mathematical and quantitative techniques can be inappropriate and misleading. This aspect will be discussed in later chapters.

Quality control (QC) of manufacturing processes obviously makes an essential contribution to the reliability of a product. In the chapters on manufacturing (Chapter 13) and reliability management (Chapter 15), the need to consider QC as an integral part of an overall reliability programme will be emphasized.

COURSES, CONFERENCES AND LITERATURE

Reliability engineering and management are now taught in engineering courses at a large number of universities, colleges and polytechnics, and on specialist short courses.

Conferences on general and specific reliability engineering and management topics have been held regularly in the United States since the 1960s and in Europe and elsewhere since the 1970s. The best known is the annual US Reliability and Maintainability Symposium (RAMS), sponsored by most of the important engineering associations and institutions in the United States. These conference proceedings contain much useful information and are often cited.

Journals on reliability have also appeared; some are referenced at the end of this chapter. Several books have been published on the subjects of reliability engineering and management; some of these are referenced at the end of other chapters.

Much of the reliability literature has tended to emphasize the mathematical and analytical aspects of the subject, with the result that reliability engineering is often considered by designers and others to be a rather esoteric subject. This is unfortunate, since it creates barriers to communication. More recently there has been a trend to emphasize the more practical aspects and to integrate reliability work into the overall management and engineering process. These aspects are covered in later chapters.

ORGANIZATIONS INVOLVED IN RELIABILITY WORK

Several organizations have been created to develop policies and methods in reliability engineering and to undertake research and training. A summary of these is given in the book's homepage (see Preface to the Fourth Edition, page xxi).

RELIABILITY AS AN EFFECTIVENESS PARAMETER

With the increasing cost and complexity of many modern systems, the importance of reliability as an effectiveness parameter, which should be specified and paid for, has become apparent. For example, a radar station, a process plant or an airliner must be available when required, and the cost of non-availability, particularly if it is un-scheduled, can be very high. In the weapons field, if an anti-aircraft missile has a less than 100 per cent probability of functioning correctly throughout its engagement sequence, operational planners must consider deploying the appropriate extra quantity to provide the required level of defence. The Apollo project second stage rocket was powered by six rocket motors; any five would have provided sufficient impulse, but an additional motor was specified to cater for a possible failure of one. As it happened there were no failures, and every launch utilized an 'unnecessary' motor. These considerations apply equally to less complex systems, e.g. vending and copying machines, even if the failure costs are less dramatic in absolute terms.

As an effectiveness parameter, reliability can be 'traded off' against other parameters. Reliability generally affects availability, and in this context maintainability is also relevant. Reliability and maintainability are often related to availability by the formula:

$$\text{Availability} = \frac{\text{MTBF}}{\text{MTBF} + \text{MTTR}}$$

where MTTR is the mean time to repair. This is the simplest steady-state situation. It is clear that availability improvements can be achieved by improving either MTBF or MTTR. For example, automatic built-in test equipment can greatly reduce diagnostic times for electronic equipment, at a cost of a slight reduction in overall reliability and an increase in unit costs. Many other parameters can be considered in trade-offs, such as weight, redundancy, cost of materials, parts and processes, or reduction in performance.

The greatest difficulty in estimating relationships for reliability trade-offs derives from the fact that, whereas it is possible to estimate quite accurately such factors as the cost and weight penalties of built-in test equipment, the cost of materials and components, or the worth of a measurable performance parameter, the effect on reliability cannot generally be forecast accurately, and reliability measurements can at best be made only within statistical limits imposed by the amount of data available. Even for quite simple systems, quantifying the separate contributions to a measured change in reliability due to different causes is a very uncertain exercise. Consequently, while there is considerable literature on the effectiveness of various reliability methods, few generally applicable estimating relationships are available. Selection of trade-offs must therefore be very much a matter of experience of similar projects in the knowledge that wide margins of error can exist.

RELIABILITY PROGRAMME ACTIVITIES

What, then, are the actions that managers and engineers can take to influence reliability? One obvious activity already mentioned is quality control, the whole range of functions designed to ensure that delivered products are compliant with the design. For many products, QC is sufficient to ensure high reliability, and we would not expect a match factory or a company mass-producing simple diecastings to employ reliability staff. In such cases the designs are simple and well proven, the environments in which the products will operate are well understood and the very occasional failure has no significant financial or operational effect. QC, together with craftsmanship, can provide adequate assurance for simple products or when the risks are known to be very low. Risks are low when safety margins can be made very large, as in most structural engineering. Reliability engineering disciplines may justifiably be absent in many types of product development and manufacture. QC disciplines are, however, essential elements of any integrated reliability programme.

A formal reliability programme is necessary whenever the risks or costs of failure are not low. We have already seen how reliability engineering developed as a result of the high costs of unreliability of military equipment, and later in commercial application. Risks of failure usually increase in proportion to the number of components in a system, so reliability programmes are required for any product whose complexity leads to an appreciable risk.

An effective reliability programme should be based on the conventional wisdom of responsibility and authority being vested in one person. Let us call him or her the reliability programme manager. The responsibility must relate to a defined object-ive, which may be a maximum warranty cost figure, an MTBF to be demonstrated or a requirement that failure will not occur. Having an objective and the authority, how does the reliability programme manager set about his or her task, faced as he or she is with a responsibility based on uncertainties? This question will be answered in detail in subsequent chapters, but a brief outline is given below.

The reliability programme must begin at the earliest, conceptual phase of the project. It is at this stage that fundamental decisions are made, which can signifi-cantly affect reliability. These are decisions related to the risks involved in the specification (design, complexity, producibility, etc.), development time-scale, re-sources applied to evaluation and test, skills available, and other factors.

The shorter the development time-scale, the more important is this need, particu-larly if there will be few opportunities for an iterative approach. The activities appropriate to this phase are an involvement in the assessment of these trade-offs and the generation of reliability objectives. The reliability staff can perform these functions effectively only if they are competent to contribute to the give-and-take inherent in the trade-off negotiations, which may be conducted between designers, production staff, marketing staff, customer representatives and finance staff.

As development proceeds from initial study to detail design, the reliability risks are controlled by a formal, documented approach to the review of design and to the imposition of design rules relating to components, materials and process selection, de-rating policy, tolerancing, etc. The objectives at this stage are to ensure that known good practices are applied, that deviations are detected and corrected, and that areas of uncertainty are highlighted for further action. The programme con-tinues through the initial hardware manufacturing and test stages, by planning and executing tests to generate confidence in the design and by collecting, analysing and acting upon test data. During production, QC activities ensure that the proven design is repeated, and further testing may be applied to eliminate weak items and to maintain confidence. The data collection, analysis and action process con-tinues through the production and in-use phases. Throughout the product life cycle, therefore, the reliability is assessed, first by initial predictions based upon past experience in order to determine feasibility and to set objectives, then by refining the predictions as detail design proceeds and subsequently by recording perform-ance during the test, production and in-use phases. This performance is fed back to generate corrective action, and to provide data and guidelines for future products.

The elements of a reliability programme are outlined in documents such as US MIL-STD-785, UK Defence Standard 00–40 and British Standard 5760 (References 1–3). The activities are described fully in subsequent chapters.

RELIABILITY ECONOMICS AND MANAGEMENT

Obviously the reliability programme activities described can be expensive. Figure 1.7 is a commonly-described representation of the theoretical cost–benefit relation ship of effort expended on reliability (or production quality) activities. However, despite its intuitive appeal and frequent presentation in textbooks and teaching on quality and reliability, this picture is misleading. Closer thought easily uncovers the

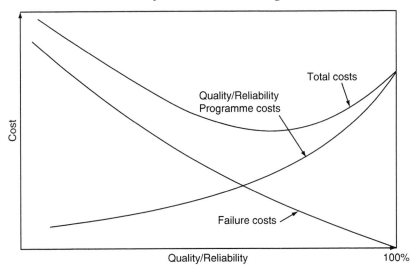

Figure 1.7 Reliability and life cycle costs (traditional view)

error in the picture. Since less than perfect reliability is the result of failures, all of which have causes, we should ask 'what is the cost of preventing or correcting the cause, compared with the cost of doing nothing?' When each potential or actual cause is analysed in this way, it is nearly always apparent that total costs continue to reduce indefinitely as reliability is improved. In other words, all effort on an effective reliability programme represents an investment, usually with a large payback over a short period. The only problem is that it is not easy to quantify the effects of given reliability programme activities, such as a certain amount of testing, on achieved reliability. However, experience shows clearly that the more realistic picture is as shown in Fig. 1.8. It was W. E. Deming (Reference 9) who first explained this relationship in his teaching on production quality. The truth has been clearly demonstrated by the success of the companies that have wholeheartedly adopted this teaching.

Achieving reliable designs and products requires a totally integrated approach, including design, training, test, production, as well as the reliability programme activities. In such an approach it is difficult to separately identify and cost those activities that are specifically devoted to reliability, as opposed to say performance or cost. The integrated engineering approach places high requirements for judgement and engineering knowledge on project managers and team members. Reliability specialists must play their parts as members of the team.

There are three kinds of engineering product, from the perspective of failure prevention:

1. *Intrinsically reliable components*, which are those that have high margins between their strength and the stresses that could cause failure, and which do not wear out within their practicable lifetimes. Such items include nearly all electronic components (if properly applied), nearly all mechanical non-moving components, and all correct software.

Figure 1.8   Reliability and life cycle costs (modern view)

2. *Intrinsically unreliable components*, which are those with low design margins or which wear out, such as badly applied components, light bulbs, turbine blades, parts that move in contact with others, like gears, bearings and power drive belts, etc.
3. *Systems* which include many components and interfaces, like cars, dishwashers, aircraft, etc, so that there are many possibilities for failures to occur, particularly across interfaces (e.g. inadequate electrical overstress protection, vibration nodes at weak points, electromagnetic interference, software that contains errors, etc.).

It is the task of design engineers to ensure that all components are correctly applied, that margins are adequate (particularly in relation to the possible extreme values of strength and stress, which are often variable), that wearout failure modes are prevented during the expected life (by safe life design, maintenance, etc.), and that system interfaces cannot lead to failure (due to interactions, tolerance mismatches, etc.). Because achieving all this on any modern engineering product is a task that challenges the capabilities of the very best engineering teams, it is almost certain that aspects of the initial design will fall short of the 'intrinsically reliable' criterion. Therefore we must submit the design to analyses and tests in order to show not only that it works, but also to show up the features that might lead to failures. When we find out what these are we must redesign and re-test, until the final design is considered to meet the criterion.

Then the product has to be manufactured. In principle, every one should be identical and correctly made. Of course this is not achievable, because of the inherent variability of all manufacturing processes, whether performed by humans or by machines. It is the task of the manufacturing people to understand and control variation, and to implement inspections and tests that will identify non-conforming product.

For many engineering products the quality of operation and maintenance also influence reliability.

The essential points that arise from this brief and obvious discussion of failures are that:

1. Failures are caused primarily by people (designers, suppliers, assemblers, users, maintainers). Therefore the achievement of reliability is essentially a management task, to ensure that the right people, skills, teams and other resources are applied to prevent the creation of failures.
2. Reliability (and quality) are not separate specialist functions that can effectively ensure the prevention of failures. They are the results of effective working by all involved.
3. There is no fundamental limit to the extent to which failures can be prevented. We can design and build for ever-increasing reliability.

Deming explained how, in the context of manufacturing quality, there is no point at which further improvement leads to higher costs. This is, of course, even more powerfully true when considered over the whole product life cycle, so that efforts to ensure that designs are intrinsically reliable, by good design and effective development testing, generate even higher pay-offs than improvements in production quality. The *'kaizen'* (continuous improvement) principle is even more effective when applied to up-front engineering.

The creation of reliable products is, therefore, primarily a management task. Guidance on reliability programme management and costs is covered in Chapter 15.

BIBLIOGRAPHY

1. US MIL-STD-785: *Reliability Programs for Systems and Equipment.* National Technical Information Service, Springfield, Virginia (suspended in 1976).
2. UK Defence Standard 00–40: *The Management of Reliability and Maintainability.* HMSO.
3. British Standard, BS 5760: *Reliability of Systems, Equipments and Components.* British Standards Institution, London.
4. US MIL-STD-721: *Definitions of Effectiveness Terms for Reliability, Maintainability, Human Factors and Safety.* Available from the National Technical Information Service, Springfield, Virginia.
5. British Standard, BS 4778: *Glossary of Terms Used in Quality Assurance* (including reliability and maintainability). British Standards Institution, London.
6. *Quality and Reliability Engineering International.* Wiley (published quarterly).
7. *IEEE Trans. Reliab.* IEEE (published quarterly).
8. *Proceedings of the US Reliability and Maintainability Symposia.* American Society for Quality and IEEE (published annually).
9. W. E. Deming, *Out of the Crisis.* MIT University Press, 1986 (originally published under the title *Quality, Productivity and Competitive Position*).
10. P. F. Drucker, *The Practice of Management.* Heinemann (1995).

QUESTIONS

1. Define (a) failure rate, and (b) hazard rate. Explain their application to the reliability of components and repairable systems. Discuss the plausibility of the 'bathtub curve' in both contexts.

2. (a) Explain the theory of component failures derived from the interaction of stress (or load) and strength distributions. Explain how this theory relates to the behaviour of the component hazard function.

   (b) Discuss the validity of the 'bathtub curve' when used to describe the failure characteristics of non-repairable components.

3. What are the main objectives of a reliability engineering team working on an engineering development project? Describe the important skills and experience that should be available within the team.

4. Briefly list the most common basic causes of failures of engineering products.

5. It is sometimes claimed that increasing quality and reliability beyond levels that have been achieved in the past is likely to be uneconomic, due to the costs of the actions that would be necessary. Present the argument against this belief. Illustrate it with an example from your own experience.

2

Reliability Mathematics

INTRODUCTION

The methods used to quantify reliability are the mathematics of probability and statistics. In reliability work we are dealing with uncertainty. In fact this is the case in much of modern engineering, and the probabilistic as opposed to the deterministic approach to engineering problems is becoming more widely applied. As an example, data may show that a certain type of power supply fails at a constant average rate of once per 10^7 h. If we build 1000 such units, and we operate them for 100 h, we cannot say with certainty whether any will fail in that time. We can, however, make a statement about the *probability* of failure. We can go further and state that, within specified statistical *confidence limits*, the probability of failure lies between certain values above and below this probability. If a sample of such equipments is tested, we obtain data which are called *statistics*.

Reliability statistics can be broadly divided into the treatment of *discrete functions*, *continuous functions* and *point processes*. For example, a switch may either work or not work when selected or a pressure vessel may pass or fail a test—these situations are described by discrete functions. In reliability we are often concerned with two-state discrete systems, since equipment is in either an operational or a failed state. Continuous functions describe those situations which are governed by a continuous variable, such as time or distance travelled. The electronic equipment mentioned above would have a reliability function in this class. The distinction between discrete and continuous functions is one of how the problem is treated, and not necessarily of the physics or mechanics of the situation. For example, whether or not a pressure vessel fails a test may be a function of its age, and its reliability could therefore be treated as a continuous function. The statistics of point processes are used in relation to repairable systems, when more than one failure can occur in a time continuum. The choice of method will depend upon the problem and on the type of data available.

Concepts such as statistical confidence are often written as *s-confidence* to distinguish them from the non-statistical meanings of the terms. This convention is adopted in the remainder of this book.

VARIATION

Reliability is influenced by variability, in parameter values such as resistance of resistors, material properties, or dimensions of parts. Variation is inherent in all manufacturing processes, and designers must understand the nature and extent of possible variation in parts and processes used. They must know how to measure and control this variation, so that the effects on performance and reliability are minimized.

Variation also exists in the environments that engineered products must withstand. Temperature, mechanical stress, vibration spectra, and many other varying factors must be considered.

Statistical methods provide the means for analysing, understanding and controlling variation. They can help us to create designs and develop processes which are intrinsically reliable in the anticipated environments over their expected useful lives.

Of course, it is not necessary to apply statistical methods to understand every engineering problem, since many are purely deterministic or easily solved using past experience or information available in sources such as databooks, specifications, design guides, and in known physical relationships such as Ohm's law. However, there are also many situations in which appropriate use of statistical techniques can be very effective in optimizing designs and processes, and for solving quality and reliability problems.

A cautionary note

Whilst statistical methods can be very powerful, economic and effective in reliability engineering applications, they must be used in the knowledge that variation in engineering is in important ways different from variation in most natural processes, or in repetitive engineering processes such as repeated, in-control machining or diffusion processes. Such processes are usually:

—Constant in time, in terms of the nature (average, spread, etc.) of the variation.
—Distributed in a particular way, describable by a mathematical function known as the *s*-normal distribution (which will be described later in this chapter).

In fact, these conditions often do not apply in engineering. For example:

—A component supplier might make a small change in a process, which results in a large change (better or worse) in reliability. Therefore past data cannot be used to forecast future reliability, using purely statistical methods. The change might be deliberate or accidental, known or unknown.
—Components might be selected according to criteria such as dimension or other measured parameter. This can invalidate the *s*-normal distribution assumption on which much of the statistical method is based. This might or might not be important in assessing the results.
—A process or parameter might vary in time, continuously or cyclically, so that statistics derived at one time might not be relevant at others.

—Variation is often deterministic by nature, for example spring deflection as a function of force, and it would not always be appropriate to apply statistical techniques to this sort of situation.

—Variation in engineering can arise from factors that defy mathematical treatment. For example, a thermostat might fail, causing a process to vary in a different way to that determined by earlier measurements, or an operator or test technician might make a mistake.

—Variation can be catastrophic, not only continuous. For example, a parameter such as a voltage level may vary over a range, but could also go to zero.

These points highlight the fact that variation in engineering is caused to a large extent by people, as designers, makers, operators and maintainers. The behaviour and performance of people is not as amenable to mathematical analysis and forecasting as is, say, the response of a plant to fertilizer or even weather patterns to ocean temperatures. Therefore the human element must always be considered, and statistical analysis must not be relied on without appropriate allowance being made for the effects of factors such as motivation, training, management, and the many other factors that can influence reliability.

Finally, it is most important to bear in mind, in any application of statistical methods to problems in science and engineering, that ultimately all cause and effect relationships have explanations, in scientific theory, engineering design, process or human behaviour, etc. Statistical techniques can be very useful in helping us to understand and control engineering situations. However, they do not by themselves provide explanations. We must always seek to understand causes of variation, since only then can we really be in control. *See the quotations on the flyleaf, and think about them.*

PROBABILITY CONCEPTS

Any event has a probability of occurrence, which can be in the range 0–1. A zero probability means that the event will not occur; a probability of 1 means that it will occur. A coin has a 0.5 (even) probability of landing heads, and a die has a 1/6 probability of giving any one of the six numbers. Such events are *s-independent*, i.e. the coin and the die logically have no memory, so whatever has been thrown in the past cannot affect the probability of the next throw. No 'system' can beat the statistics of these situations; waiting for a run of blacks at roulette and then betting on reds only appears to work because the gamblers who won this way talk about it, whilst those who lost do not.

With coins, dice and roulette wheels we can predict the probability of the outcome from the nominal nature of the system. A coin has two sides, a die six faces, a roulette wheel equal numbers of reds and blacks. Assuming that the coin, die and wheel are fair, these outcomes are also *unbiased*, i.e. they are all equally probable. In other words, they occur *randomly*.

With many systems, such as the sampling of items from a production batch, the probabilities can only be determined from the statistics of previous experience.

We can define probability in two ways:

1. If an event can occur in N equally likely ways, and if the event with attribute A can happen in n of these ways, then the probability of A occurring is

$$P(A) = \frac{n}{N}$$

2. If, in an experiment, an event with attribute A occurs n times out of N experiments, then as N becomes large, the probability of event A approaches n/N, i.e.

$$P(A) = \lim_{n \to \infty} \left(\frac{n}{N} \right)$$

The first definition covers the cases described earlier, i.e. equally likely s-independent events such as rolling dice. The second definition covers typical cases in quality control and reliability. If we test 100 items and find that 30 are defective, we may feel justified in saying that the probability of finding a defective item in our next test is 0.30, or 30 per cent.

However, we must be careful in making this type of assertion. The probability of 0.30 of finding a defective item in our next test may be considered as our *degree of belief*, limited by the size of the sample, in this outcome. This leads to a third, subjective, definition of probability. If, in our tests of 100 items, seven of the defectives had occurred in a particular batch of ten and we had taken corrective action to improve the process so that such a problem batch was less likely to occur in future, we might assign some lower probability to the next item being defective. This subjective approach is quite valid, and is very often necessary in quality control and reliability work. Whilst it is important to have an understanding of the rules of probability, there are usually so many variables which can affect the properties of manufactured items that we must always keep an open mind about statistically derived values. We must ensure that the sample from which statistics have been derived represents the new sample, or the overall population, about which we plan to make an assertion based upon our sample statistics.

A sample represents a population if all the members of the population have an equal chance of being sampled. This can be achieved if the sample is selected so that this condition is fulfilled. Of course in engineering this is not always practicable; e.g. in reliability engineering we often need to make an assertion about items that have not yet been produced, based upon statistics from prototypes.

To the extent that the sample is not representative, we will alter our assertions. Of course, subjective assertions can lead to argument, and it might be necessary to perform additional tests to obtain more data to use in support of our assertions. If we do perform more tests, we need to have a method of interpreting the new data in relation to the previous data: we will cover this aspect later.

The assertions we can make based on sample statistics can be made with a degree of s-confidence which depends upon the size of the sample. If we had decided to test ten items after introducing a change to the process, and found one defective, we might be tempted to assert that we have improved the process, from 30 per cent defectives being produced to only 10 per cent. However, since the sample is now much smaller, we cannot make this assertion with as high s-confidence as when we

Batch

Figure 2.1 Samples with defectives (black squares)

used a sample of 100. In fact, the true probability of any item being defective might still be 30 per cent, i.e. the population might still contain 30 per cent defectives.

Figure 2.1 shows the situation as it might have occurred, over the first 100 tests. The black squares indicate defectives, of which there are 30 in our batch of 100. If these are randomly distributed, it is possible to pick a sample batch of ten which contains fewer (or more) than three defectives. In fact, the smaller the sample, the greater will be the sample-to-sample variation about the population average, and the s-confidence associated with any estimate of the population average will be accordingly lower. The derivation of s-confidence limits is covered later in this chapter.

RULES OF PROBABILITY

In order to utilize the statistical methods used in reliability engineering, it is necessary to understand the basic notation and rules of probability. These are:

1. The probability of obtaining an outcome A is denoted by $P(A)$, and so on for other outcomes.
2. The *joint* probability that A *and* B occur is denoted by $P(AB)$.
3. The probability that A *or* B occurs is denoted by $P(A + B)$.
4. The *conditional* probability of obtaining outcome A, *given that B has occurred*, is denoted by $P(A|B)$.
5. The probability of the complement, i.e. of A *not* occurring, is $P(\bar{A}) = 1 - P(A)$.
6. If (and only if) events A and B are *s-independent*, then

$$P(A|B) = P(A|\bar{B}) = P(A)$$

and

$$P(B|A) = P(B|\bar{A}) = P(B) \tag{2.1}$$

i.e. $P(A)$ is unrelated to whether or not B occurs, and vice versa.

7. The joint probability of the occurrence of two s-independent events A and B is the product of the individual probabilities:

$$P(AB) = P(A)P(B) \tag{2.2}$$

This is also called the *product rule* or *series rule*. It can be extended to cover any number of s-independent events. For example, in rolling a die, the probability of obtaining any given sequence of numbers in three throws is

$$\frac{1}{6} \times \frac{1}{6} \times \frac{1}{6} = \frac{1}{216}$$

8. If events A and B are s-*dependent*, then

$$P(AB) = P(A)P(B|A) = P(B)P(A|B) \tag{2.3}$$

i.e. the probability of A occurring times the probability of B occurring given that A has already occurred, or vice versa.

If $P(A) \neq 0$, Eqn (2.3) can be rearranged to

$$P(B|A) = \frac{P(AB)}{P(A)} \tag{2.4}$$

9. The probability of any one of two events A or B occurring is

$$P(A + B) = P(A) + P(B) - P(AB) \tag{2.5}$$

10. The probability of A or B occurring, if A and B are s-independent, is

$$P(A + B) = P(A) + P(B) - P(A)P(B) \tag{2.6}$$

The derivation of this equation can be shown by considering the system shown in Fig. 2.2, in which either A or B, or A and B, must work for the system to work. If we denote the system success probability as P_s, then the failure probability, $P_f = 1 - P_s$. The system failure probability is the joint probability of A and B failing, i.e.

$$P_f = [1 - P(A)] [1 - P(B)]$$
$$= 1 - P(A) - P(B) + P(A)P(B)$$
$$P_s = 1 - P_f = P(A + B) = P(A) + P(B) - P(A)P(B)$$

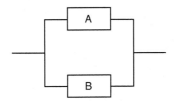

Figure 2.2 Dual redundant system

11. If events A and B are *mutually exclusive*, i.e. A and B cannot occur simultaneously, then

$$P(AB) = 0$$

and

$$P(A + B) = P(A) + P(B) \qquad (2.7)$$

12. If multiple, mutually exclusive probabilities of outcomes B_i jointly give a probability of outcome A, then

$$P(A) = \sum_i P(AB_i) = \sum_i P(A|B_i)P(B_i) \qquad (2.8)$$

13. Rearranging Eqn (2.3)

$$P(AB) = P(A)P(B|A) = P(B)P(A|B)$$

we obtain

$$P(A|B) = \frac{P(A)P(B|A)}{P(B)} \qquad (2.9)$$

This is a simple form of *Bayes' theorem*. A more general expression is

$$P(A|B) = \frac{P(A)P(B|A)}{\sum_i P(B|E_i)P(E_i)} \qquad (2.10)$$

where E_i is the *i*th event.

Example 2.1

The reliability of a missile is 0.85. If a salvo of two missiles is fired, what is the probability of at least one hit? (Assume *s*-independence of missile hits.)

Let A be the event 'first missile hits' and B the event 'second missile hits'. Then

$$P(A) = P(B) = 0.85$$

$$P(\bar{A}) = P(\bar{B}) = 0.15$$

There are four possible, mutually exclusive outcomes, $AB, A\bar{B}, AB, \bar{A}\bar{B}$. The probability of both missing, from Eqn (2.2), is

$$P(\bar{A})P(\bar{B}) = P(\bar{A}\,\bar{B})$$

$$= 0.15^2 = 0.0225$$

Therefore the probability of at least one hit is

$$P_s = 1 - 0.0225 = 0.9775$$

We can derive the same result by using Eqn (2.6):

$$P(A + B) = P(A) + P(B) - P(A)P(B)$$

$$= 0.85 + 0.85 - 0.85^2 = 0.9775$$

Another way of deriving this result is by using the *sequence tree diagram*:

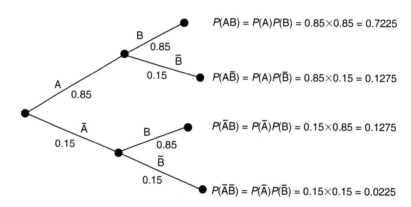

The probability of a hit is then derived by summing the products of each path which leads to at least one hit. We can do this since the events defined by each path are mutually exclusive.

$$P(AB) + P(A\bar{B}) + P(\bar{A}B) = 0.9775$$

(Note that the sum of all the probabilities is unity.)

Example 2.2

In Example 2.1 the missile hits are not s-independent, but are s-dependent, so that if the first missile fails the probability that the second will also fail is 0.2. However, if the first missile hits, the hit probability of the second missile is unchanged at 0.85. What is the probability of at least one hit?

$$P(A) = 0.85$$

$$P(B|A) = 0.85$$

$$P(\bar{B}|A) = 0.15$$

$$P(\bar{B}|\bar{A}) = 0.2$$

$$P(B|\bar{A}) = 0.8$$

The probability of at least one hit is

$$P(AB) + P(\bar{A}B) + P(\bar{B}A)$$

Since A, B and $A\bar{B}$ are s-independent,

$$P(AB) = P(A)P(B)$$
$$= 0.85 \times 0.85 = 0.7225$$

and

$$P(A\bar{B}) = P(A)P(\bar{B})$$
$$= 0.85 \times 0.15 = 0.1275$$

Since $\bar{A}$ and B are s-dependent, from Eqn (2.3),

$$P(\bar{A}B) = P(\bar{A})P(B|\bar{A})$$
$$= 0.15 \times 0.8 = 0.12$$

and the sum of these probabilities is 0.97.
  This result can also be derived by using a sequence tree diagram:

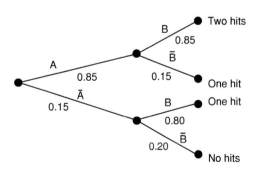

As in Example 2.1, the probability of at least one hit is calculated by adding the products of each path leading to at least one hit, i.e.

$$P(A)P(B) + P(A)P(\bar{B}) + P(\bar{A})P(B)$$
$$= (0.85 \times 0.85) + (0.85 \times 0.15) + (0.15 \times 0.80) = 0.97$$

Example 2.3

In the circuit shown, the probability of any switch being closed is 0.8 and all events are s-independent. (a) What is the probability that a circuit will exist? (b) Given that a circuit exists, what is the probability that switches a and b are closed?

Let the events that a, b, c and d are closed be A, B, C and D. Let X denote the event that the circuit exists.

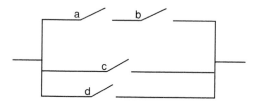

(a) $X = AB + (C + D)$

$$P(X) = P(AB) + P(C + D) - P(AB)P(C + D)$$
$$P(AB) = P(A)P(B)$$
$$= 0.8 \times 0.8 = 0.64$$
$$P(C + D) = P(C) + P(D) - P(C)P(D)$$
$$= 0.8 + 0.8 - 0.64 = 0.96$$

Therefore

$$P(X) = 0.64 + 0.96 - (0.96 \times 0.64) = 0.9856$$

(b) From Eqn (2.4),

$$P(AB|X) = \frac{P(ABX)}{P(X)}$$

A and B jointly give X. Therefore, from Eqn (2.8),

$$P(ABX) = P(AB)$$

So

$$P(AB|X) = \frac{P(AB)}{P(X)} = \frac{P(A)P(B)}{P(X)}$$

$$= \frac{0.8 \times 0.8}{0.9856} = 0.6494$$

Example 2.4

A test set has a 98 per cent probability of correctly classifying a faulty item as defective and a 4 per cent probability of classifying a good item as defective. If in a batch of items tested 3 per cent are actually defective, what is the probability that when an item is classified as defective, it is truly defective?

Let D represent the event that an item is defective and C represent the event that an item is classified defective. Then

$$P(D) = 0.03$$
$$P(C|D) = 0.98$$
$$P(C|\bar{D}) = 0.04$$

We need to determine $P(D|C)$. Using Eqn (2.10),

$$P(D|C) = \frac{P(D)P(C|D)}{P(C|D)P(D) + P(C|\bar{D})P(\bar{D})}$$

$$= \frac{(0.03)(0.98)}{(0.98)(0.03) + (0.04)(0.97)} = 0.43$$

This indicates the importance of a test equipment having a high probability of correctly classifying good items as well as bad items.

CONTINUOUS VARIATION

The differences, or *variations*, of parameters in engineering applications (machined dimensions, material strengths, transistor gains, resistor values, temperatures, etc.) are conventionally described in two ways. The first, and the simplest, is to state maximum and minimum values, or tolerances. This provides no information on the nature, or shape, of the actual distribution of values. However, in many practical cases, this is sufficient information for the creation of manufacturable, reliable designs.

The second approach is to describe the nature of the variation, using data derived from measurements. In this section we will describe the methods of statistics in relation to describing and controlling variation in engineering.

If we plot measured values which can vary about an average (e.g. the diameter of machined parts or the gain of a transistor) as a histogram, for a given sample we may obtain a representation such as Fig. 2.3(a).

In this case 30 items have been measured and the frequencies of occurrence of the measured values are as shown. The values range from 2 to 9, with most items having values between 5 and 7. Another random sample of 30 from the same population will usually generate a different histogram, but the general shape is likely to be similar, e.g. Fig. 2.3(b). If we plot a single histogram showing the combined data of

Figure 2.3 (a) Frequency histogram of a random sample, (b) frequency histogram of another random sample from the same population, (c) data of many samples shown with measurement intervals of 0.5

many such samples, but this time show the values in measurement intervals of 0.5, we get Fig. 2.3(c). Note that now we have used a percentage frequency scale. We now have a rather better picture of the distribution of values, as we have more information from the larger sample. If we proceed to measure a large number and we further reduce the measurement interval, the histogram tends to a curve which describes the population *probability density function* (p.d.f.) or simply the *distribution* of values. Figure 2.4 shows a general *unimodal* probability distribution, f(x) being the probability density of occurrence, related to the variable x. The value of x at which the distribution peaks is called the *mode*. Multimodal distributions are encountered in reliability work as well as unimodal distributions. However, we will deal only with the statistics of unimodal distributions in this book, since multimodal distributions are usually generated by the combined effects of separate unimodal distributions.

The area under the curve is equal to unity, since it describes the total probability of all possible values of x, as we have defined a probability which is a certainty as being a probability of one. Therefore

$$\int_{-\infty}^{\infty} f(x)\, dx = 1 \tag{2.11}$$

The probability of a value falling between any two values x_1 and x_2 is the area bounded by this interval, i.e.

$$P(x_1 < x < x_2) = \int_{x_1}^{x_2} f(x)\, dx \tag{2.12}$$

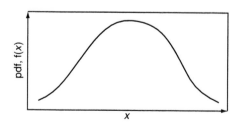

Figure 2.4 Continuous probability distribution

To describe a p.d.f. we normally consider four aspects:
1. The *central tendency*, about which the distribution is grouped.
2. The *spread*, indicating the extent of variation about the central tendency.
3. The *skewness*, indicating the lack of symmetry about the central tendency.
4. The *kurtosis*, indicating the 'peakedness' of the p.d.f.

Measures of central tendency

For a sample containing n items the sample *mean* is denoted by $\bar{x}$:

$$\bar{x} = \sum_{i=1}^{n} \frac{x_i}{n} \tag{2.13}$$

The sample mean can be used to *estimate* the population mean, which is the average of all possible outcomes. For a continuous distribution, the mean is derived by extending this idea to cover the range $-\infty$ to $+\infty$.

The mean of a distribution is usually denoted by μ. The mean is also referred to as the *location parameter, average value* or *s-expected value*, $E(x)$. In reliability work θ is used to denote the mean in certain special cases (see page 39).

$$\mu = \int_{-\infty}^{\infty} xf(x)\,\mathrm{d}x \tag{2.14}$$

This is analogous to the centre of gravity of the p.d.f. The *estimate* of a population mean from sample data is denoted by $\hat{\mu}$.

Other measures of central tendency are the *median*, which is the mid-point of the distribution, i.e. the point at which half the measured values fall to either side, and the *mode*, which is the value (or values) at which the distribution peaks. The relationship between the mean, median and mode for a right-skewed distribution is shown in Fig. 2.5. For a symmetrical distribution, the three values are the same, whilst for a left-skewed distribution the order of values is reversed.

Spread of a distribution

The spread, or dispersion, i.e. the extent to which the values which make up the distribution vary, is measured by its *variance*. For a sample size n the variance, Var (x) or $E(x - \bar{x})^2$, is given by

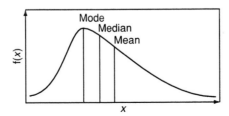

Figure 2.5 Measures of central tendency

$$\mathrm{Var}(x) = E(x - \bar{x})^2 = \frac{\sum_{i=1}^{n}(x_i - \bar{x})^2}{n}$$

(2.15)

Where sample variance is used to estimate the population variance, we use $(n - 1)$ in the denominator of Eqn (2.15) instead of n, as it can be shown to provide a better estimate. The *estimate of population variance* from a sample is denoted $\hat{\sigma}^2$ where

$$\hat{\sigma}^2 = \sum_{i=1}^{n} \frac{(x - \bar{x})^2}{n - 1}$$

(2.16)

The *population* variance σ^2, for a finite population N, is given by

$$\sigma^2 = \frac{\sum_{i=1}^{N}(x_i - \mu)^2}{N}$$

(2.17)

For a continuous distribution.

$$\sigma^2 = \int_{-\infty}^{\infty} (x - \mu)^2 \, f(x) \, dx$$

(2.18)

σ is called the *standard deviation* (SD) and is frequently used in practice instead of the variance. It is also referred to as the *scale parameter*. σ^2 is the second moment about the mean and is analogous to a radius of gyration.

Skewness and kurtosis

The third and fourth moments about the mean give the skewness and kurtosis of the distribution. Since we will not make use of these parameters in this book, the reader is referred to more advanced statistical texts for their derivation (e.g. Reference 8). A positively skewed distribution is skewed to the right, and vice versa. A symmetrical distribution has zero skew.

The cumulative distribution function

The cumulative distribution function (c.d.f.), $F(x)$, gives the probability that a measured value will fall between $-\infty$ and x:

$$F(x) = \int_{-\infty}^{x} f(x) \, dx$$

(2.19)

Figure 2.6 shows the typical ogive form of the c.d.f. with $F(x) \to 1$ as $x \to \infty$.

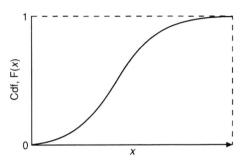

Figure 2.6 Typical cumulative distribution function (c.d.f.)

Reliability and hazard functions

In reliability engineering we are concerned with the probability that an item will survive for a stated interval (e.g. time, cycles, distance, etc.), i.e. that there is no failure in the interval (0 to x). This is the *s-reliability*, and it is given by the reliability function R (x). From this definition, it follows that

$$R(x) = 1 - F(x) = \int_x^\infty f(x)\ dx = 1 - \int_{-\infty}^x f(x)\ dx \qquad (2.20)$$

The *hazard function* or *hazard rate* h(x) is the conditional probability of failure in the interval x to $(x + dx)$, given that there was no failure by x:

$$h(x) = \frac{f(x)}{R(x)} = \frac{f(x)}{1 - F(x)} \qquad (2.21)$$

The *cumulative hazard function* H(x) is given by

$$H(x) = \int_{-\infty}^x h(x)\ dx = \int_{-\infty}^x \frac{f(x)}{1 - F(x)}\,dx \qquad (2.22)$$

Range of definite integrals

In engineering we do not usually encounter measured values below zero and the lower limit of the definite integral is then 0.

CONTINUOUS DISTRIBUTION FUNCTIONS

The normal (or Gaussian) distribution

By far the most widely used 'model' of the nature of variation is the mathematical function known as the *normal (or Gaussian) distribution*.

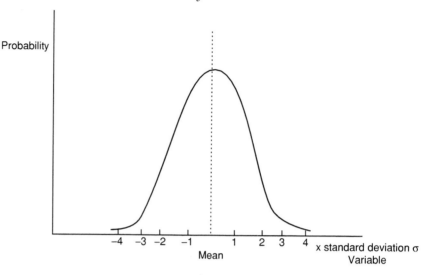

Figure 2.7 The s-normal (Gaussian) distribution

The s-normal p.d.f. is given by

$$f(x) = \frac{1}{\sigma(2\pi)^{1/2}} \exp\left[-\frac{1}{2}\left(\frac{x-\mu}{\sigma}\right)^2\right]$$

(2.23)

where μ is the location parameter, equal to the mean. The mode and the median are coincident with the mean, as the p.d.f. is symmetrical. σ is the scale parameter, equal to the SD.

A population which conforms to the s-normal distribution has variations which are symmetrically disposed about the mean (Figure 2.7) (i.e. the skewness is zero). Since the tails of the s-normal distribution are symmetrical, a given spread includes equal values in the left-hand and right-hand tails.

For s-normally distributed variables, about 68 per cent of the population fall between ±1 SD. About 95 per cent fall between ±2 SD, and about 99.7 per cent between ±3 SD.

An important reason for the wide applicability of the s-normal distribution is the fact that, when a value is subject to many additive sources of variation, irrespective of how these variations are distributed, the resulting composite distribution can be shown to approach the s-normal distribution. This is known as the *central limit theorem*. It justifies the use of the s-normal distribution in many applications, including engineering, particularly in quality control. The s-normal distribution is a close fit to most QC and some reliability observations, such as the sizes of machined parts and the lives of items subject to wearout failures, as well as to natural phenomena such as the heights of adults and strengths of materials.

Appendix 1 gives values for $\Phi(z)$, the *standardized normal* c.d.f., i.e. $\mu = 0$ and $\sigma = 1$. z represents the number of SDs displacement from the mean. Any normal distribution can be evaluated from the standardized normal distribution by calculating the standardized normal variate z, where

$$z = \frac{x - \mu}{\sigma}$$

and finding the appropriate value of $\Phi(z)$.

Example 2.5

The life of an incandescent lamp is s-normally distributed, with mean 1200 h and SD 200 h. What is the probability that a lamp will last (a) at least 800 h? (b) at least 1600 h?
(a) $z = (x - \mu)/\sigma$, i.e. the distance of x from μ expressed as a number of SDs. Then

$$z = \frac{800 - 1200}{200} = -2\,\text{SD}$$

Appendix 1 shows that the probability of a value not exceeding 2 SD is 0.977. Figure 2.8(a) shows this graphically, on the p.d.f. (the shaded area).
(b) The probability of a lamp surviving more than 1600 h is derived similarly:

$$z = \frac{1600 - 1200}{200} = 2\,\text{SD}$$

This represents the area under the p.d.f. curve beyond the $+2$ SD point. (Fig. 2.8(a)) or $1-$ (area under the curve to the left of $+2$ SD) on the c.d.f. (Fig. 2.8(b)). Therefore the probability of surviving beyond 1600 h is $(1 - 0.977) = 0.023$.

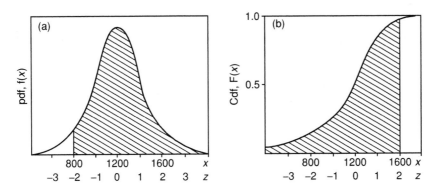

Figure 2.8   (a) The p.d.f. $f(x)$ versus x; (b) the c.d.f. $F(x)$ versus x (see Example 2.5)

The lognormal distribution

The lognormal distribution is a more versatile distribution than the normal as it has a range of shapes, and therefore is often a better fit to reliability data, such as for populations with wearout characteristics. Also, it does not have the normal distribution's disadvantage of extending below zero to $-\infty$. The lognormal p.d.f. is

$$f(x) = \begin{cases} \dfrac{1}{\sigma x (2\pi)^{1/2}} \exp\left[-\dfrac{1}{2} \left(\dfrac{\ln x - \mu}{\sigma} \right)^2 \right] & (\text{for } x \geq 0) \\ 0 & (\text{for } x < 0) \end{cases} \qquad (2.24)$$

In other words, it is the s-normal distribution with ln x as the variate. The mean and SD of the lognormal distribution are given by

$$\text{Mean} = \exp\left(\mu + \frac{\sigma^2}{2} \right)$$

$$\text{SD} = [\exp(2\mu + 2\sigma^2) - \exp(2\mu + \sigma^2)]^{1/2}$$

where μ and σ are the mean and SD of the ln data.

When $\mu \gg \sigma$, the lognormal distribution approximates to the s-normal distribution. The s-normal and lognormal distributions describe reliability situations in which the hazard rate increases from $x = 0$ to a maximum and then decreases.

The exponential distribution

The exponential distribution describes the situation wherein the hazard rate is constant. A Poisson process (page 71) generates a constant hazard rate. The p.d.f. is

$$f(x) = \begin{cases} a \exp(-ax) & (\text{for } x \geq 0) \\ 0 & (\text{for } x < 0) \end{cases} \qquad (2.25)$$

This is an important distribution in reliability work, as it has the same central limiting relationship to life statistics as the normal distribution has to non-life statistics. It describes the constant hazard rate situation. As the hazard rate is often a function of time, we will denote the independent variable by t instead of x. The constant hazard rate is denoted by λ. The mean life, or mean time to failure (MTTF), is $1/\lambda$. The p.d.f. is then written as

$$f(t) = \lambda \exp(-\lambda t) \qquad (2.26)$$

The probability of no failures occurring before time t is obtained by integrating Eqn (2.26) between 0 and t and subtracting from 1:

$$R(t) = 1 - \int_0^t f(t) \, dt = \exp(-\lambda t) \qquad (2.27)$$

$R(t)$ is the *reliability function* (or survival probability). For example, the reliability of an item with an MTTF of 500 h over a 24 h period is

$$R(24) = \exp\left(\frac{-24}{500} \right) = 0.953$$

Values of $\exp(-x)$ are tabulated in Appendix 2. Note that for small values of $\lambda t (< 0.1)$, the first two terms of the negative exponential series $(1 - \lambda t + \ldots)$ can be used to give a good approximation for $R(t)$.

$R(t)$ can also be derived as follows. If there are N items surviving at time t, then in a small period of time δt the number failing is δN. Then

$$\frac{\delta N}{\delta t} = -\lambda N$$

i.e. constant proportion to number surviving

$$\frac{\delta N}{N} = -\lambda \delta t$$

$$\ln N = -\lambda t - \ln C$$

$$N = C \exp(-\lambda t)$$

If N_0 is the number of items at $t = 0$ then $C = N_0$:

$$N = N_0 \exp(-\lambda t)$$

$$R(t) = \frac{N}{N_0} = \exp(-\lambda t)$$

For items which are repaired, λ is called the *failure rate*, and $1/\lambda$ is called the *mean time between failures* (MTBF) (θ). Note that 63.2 per cent of items will have failed by $t = \theta$.

If times to failure are exponentially distributed, the probability of x failures is Poisson-distributed (Page 55). For example, if the MTBF is 100 h, the probability of having more than 15 failures in 1000 h is derived as:

$$\text{Expected number of failures} = \frac{1000}{100} = 10$$

Referring to Fig. 2.17, for $\lambda(\mu) = 10$, $x = 15$:

$$P = 0.965$$

This is the probability of having up to 15 failures. Therefore the probability of more than 15 failures is

$$1 - 0.965 = 0.035$$

The gamma distribution

The gamma distribution describes, in reliability terms, the situation when partial failures can exist, i.e. when a given number of partial failure events must occur before an item fails, or the time to the ath failure when time to failure is exponentially distributed. The p.d.f. is

$$f(x) = \begin{cases} \dfrac{\lambda}{\Gamma(a)} (\lambda x)^{a-1} \exp(-\lambda x) & \text{(for } x \geq 0) \\ 0 & \text{(for } x < 0) \end{cases}$$

$$\mu = \frac{a}{\lambda} \tag{2.28}$$

$$\sigma = \frac{a^{1/2}}{\lambda}$$

where λ is the failure rate (complete failures) and a the number of partial failures per complete failure, or events to generate a failure. $\Gamma(a)$ is the *gamma function*:

$$\Gamma(a) = \int_0^\infty x^{a-1} \exp(-x) \, dx \tag{2.29}$$

When $(a-1)$ is a positive integer, $\Gamma(a) = (a-1)!$ This is the case in the partial failure situation. The exponential distribution is a special case of the gamma distribution, when $a = 1$, i.e.

$$f(x) = \lambda \exp(-\lambda x)$$

The gamma distribution can also be used to describe a decreasing or increasing hazard rate. When $a < 1$, $h(x)$ will decrease whilst for $a > 1$, $h(x)$ increases.

The χ^2 distribution

The χ^2 (chi-square) distribution is a special case of the gamma distribution, where $\lambda = \frac{1}{2}$, and $\nu = a/2$, where ν is called the number of *degrees of freedom* and must be a positive integer. This permits the use of the χ^2 distribution for evaluating reliability situations, since the number of failures, or events to failure, will always be positive integers. The χ^2 distribution is really a family of distributions, which range in shape from that of the exponential to that of the s-normal distribution. Each distribution is identified by the degrees of freedom.

In statistical theory, the χ^2 distribution is very important, as it is the distribution of the sums of squares of n, s-independent, s-normal variates. This allows it to be used for statistical testing, goodness-of-fit tests and evaluating s-confidence. These applications are covered later. The c.d.f. for the χ^2 distribution is tabulated for a range of degrees of freedom in Appendix 3.

The Weibull distribution

The Weibull distribution has the great advantage in reliability work that by adjusting the distribution parameters it can be made to fit many life distributions. The Weibull p.d.f. is (in terms of time t)

$$f(t) = \begin{cases} \dfrac{\beta}{\eta^{\beta}} t^{\beta-1} \exp\left[-\left(\dfrac{t}{\eta}\right)^{\beta}\right] & \text{(for } t \geq 0) \\ 0 & \text{(for } t < 0) \end{cases}$$

(2.30)

The corresponding reliability function is

$$R(t) = \exp\left[-\left(\frac{t}{\eta}\right)^{\beta}\right]$$

(2.31)

The hazard rate is

$$\frac{\beta}{\eta^{\beta}} t^{\beta-1}$$

β is the *shape parameter* and η is the *scale parameter*, or *characteristic life*—it is the life at which 63.2 per cent of the population will have failed.

When $\beta = 1$, the exponential reliability function (constant hazard rate) results, with

$$\eta = \text{mean life } (1/\lambda).$$

When $\beta < 1$, we get a *decreasing* hazard rate reliability function.
When $\beta > 1$, we get an *increasing* hazard rate reliability function.

When $\beta = 3.5$, for example, the distribution approximates to the normal distribution. Thus the Weibull distribution can be used to model a wide range of life distributions characteristic of engineered products.

So far we have dealt with the two-parameter Weibull distribution. If, however, failures do not start at $t = 0$, but only after a finite time γ, then the Weibull reliability function takes the form

$$R(t) = \exp\left[-\left(\frac{t-\gamma}{\eta}\right)^{\beta}\right]$$

(2.32)

i.e. a three-parameter distribution. γ is called the *failure free time, location parameter* or *minimum life*. It is sometimes denoted as t_0.

Another parameter that is used is the B-life, which is the time by which 3% of the population can be expected to fail. It is expressed as B_n, where n is the proportion failing.

The extreme value distributions

In reliability work we are often concerned not with the distribution of variables which describe the bulk of the population but only with the extreme values which

Table 2.1   Sample data taken randomly from a common population

Sample	Data							
1	30	31	$\bar{4}1$	<u>29</u>	39	36	38	30
2	31	34	<u>23</u>	27	29	32	$\bar{3}5$	35
3	<u>26</u>	33	$\bar{3}5$	32	34	29	30	34
4	<u>27</u>	33	30	31	31	36	28	$\bar{4}0$
5	<u>18</u>	$\bar{3}9$	25	32	31	34	27	37
6	<u>22</u>	36	$\bar{4}2$	27	33	27	31	31
7	39	35	32	39	32	<u>27</u>	28	32
8	33	34	32	30	34	$\bar{3}5$	33	<u>28</u>
9	32	32	$\bar{3}7$	25	33	35	35	<u>19</u>
10	28	32	36	37	<u>17</u>	31	$\bar{4}2$	32
11	26	<u>22</u>	32	23	33	$\bar{3}6$	36	31
12	36	31	$\bar{4}5$	<u>24</u>	30	27	24	27

can lead to failure. For example, the mechanical properties of a semiconductor wire bond are such that under normal operating conditions good wire bonds will not fracture or overheat. However, extreme high values of electrical load or extreme low values of bond strength can result in failure. In other words, we are concerned with the implications of the tails of the distributions in load–strength interactions. However, we often cannot assume that, because a measured value appears to be, say, *s*-normally distributed, that this distribution necessarily is a good model for the extremes. Also, few measurements are likely to have been made at these extremes. Extreme value statistics are capable of describing these situations asymptotically.

Extreme value statistics are derived by considering the lowest or highest values in each of a series of equal samples. For example, consider the sample data in Table 2.1, taken randomly from a common population. The overall data can be plotted as shown in Fig. 2.9 as $f(x)$. However, if we plot separately the lowest values and the highest values in each sample, they will appear as $g_L(x)$ and $g_H(x) \cdot g_L(x)$ is the extreme value distribution of the lowest extreme whilst $g_H(x)$ is the extreme value distribution of the highest extreme in each sample. For many distributions the distribution of the extremes will be one of three types:

Type I   —also known as the *extreme value* or *Gumbel distribution*.
Type II —also known as the *log extreme value distribution*.
Type III —for the lowest extreme values. This is the Weibull distribution.

Extreme value type I

The type I extreme value distributions for maximum and minimum values are the limiting models for the right and left tails of the exponential types of distribution, where this is defined as any distribution whose cumulative probability approaches unity at a rate which is equal to or greater than that for the exponential distribution. This includes most reliability distributions, such as the normal, lognormal and exponential distributions.

The probability density functions for maximum and minimum values, respectively, are

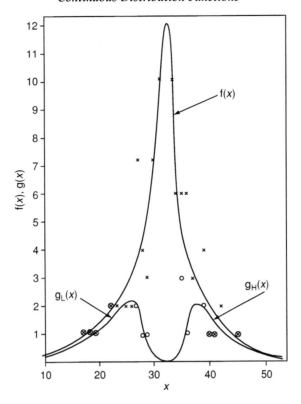

Figure 2.9   Extreme value distributions

$$f(x) = \frac{1}{\sigma}\exp\left\{-\frac{1}{\sigma}(x-\mu) - \exp\left[-\frac{1}{\sigma}(x-\mu)\right]\right\} \qquad (2.33)$$

$$f(x) = \frac{1}{\sigma}\exp\left\{\frac{1}{\sigma}(x-\mu) - \exp\left[\frac{1}{\sigma}(x-\mu)\right]\right\} \qquad (2.34)$$

The *reduced variate* is given by

$$y = \frac{x-\mu}{\sigma}$$

Substituting in Eqns (2.33) and (2.34), we can derive the c.d.f. in terms of the reduced variate y.

For maximum values:

$$F(y) = \int_{-\infty}^{y} \exp\{-[x + \exp(-x)]\}dx = \exp[-\exp(-y)] \qquad (2.35)$$

For minimum values:

$$F(y) = 1 - \exp[-\exp(y)] \tag{2.36}$$

The distribution of maximum values is right-skewed and the distribution of minimum values is left-skewed. The hazard function of maximum values approaches unity with increasing x, whilst that for minimum values increases exponentially.

For maximum values:

$$\mu_{e\mu_{max}} = \mu + 0.577\sigma$$

For minimum values:

$$\mu_{e\mu_{min}} = \mu - 0.577\sigma$$

The standard deviation $\sigma_{e\mu}$ is 1.283σ in both cases.

Extreme value type II

The extreme type II distribution does not play an important role in reliability work. If the logarithms of the variables are extreme value distributed, then the variable is described by the extreme value type II distribution. Thus its relationship to the type I extreme value distribution is analogous to that of the lognormal to the normal distribution.

Extreme value type III

The type III extreme value distribution for minimum values is the limiting modelfor the left-hand tail for distributions which are bounded at the left. In fact, the Weibull distribution is the type III extreme value distribution for minimum values, and although it was initially derived empirically, its use for describing the strength distribution of materials has been justified using extreme value theory.

The extreme value distributions related to load and strength

The type I extreme value distribution for maximum values is often an appropriate model for the occurrence of load events, when these are not bounded to the right, i.e. when there is no limiting value.

It is well known that engineering materials possess strengths well below their theoretical capacity, mainly due to the existence of imperfections which give rise to non-uniform stresses under load. In fact, the strength will be related to the effect of the imperfection which creates the greatest reduction in strength, and hence the extreme value distribution for minimum values suggests itself as an appropriate model for strength.

The strength, and hence the time to failure, of many types of product can be considered to be dependent upon imperfections whose extent is bounded, since only very small imperfections will escape detection by inspection or process control, justifying use of a type III (Weibull) model. On the other hand, a type I model might be more representative of the strength of an item which is mass-produced and not 100 per cent inspected, or in which defects can exist whose extent is not bounded, but which are not detected, e.g. a long wire, whose strength will then be a function of length.

For a system consisting of many components in series, where the system hazard rate is decreasing from $t = 0$ (i.e. bounded) a type III (Weibull) distribution will be a good model for the system time to failure.

SUMMARY OF CONTINUOUS STATISTICAL DISTRIBUTIONS

Figure 2.10 is a summary of the continuous distributions described above.

VARIATION IN ENGINEERING

Every practical engineering design must take account of the effects of the variation inherent in parameters, environments, and processes. Variation and its effects can be considered in three categories:

1. *Deterministic*, or *causal*, which is the case when the relationship between a parameter and its effect is known, and we can use theoretical or empirical formulae, e.g., we can use Ohm's law to calculate the effect of resistance change on the performance of a voltage divider. No statistical methods are required. The effects of variation are calculated by inserting the expected range of values into the formulae.
2. *Functional*, which includes relationships such as the effect of a change of operating procedure, human mistakes, calibration errors, etc. There are no theoretical formulae. In principle these can be allowed for, but often are not, and the cause and effect relationships are not always easy to identify or quantify.
3. *Random*. These are the effects of the inherent variability of processes and operating conditions. They can be considered to be the variations that are left unexplained when all deterministic and functional causes have been eliminated. For example, a machining process that is in control will nevertheless produce parts with some variation in dimensions, and random voltage fluctuations can occur on power supplies due to interference. Note that the random variations have causes. However, it is not always possible or practicable to predict how and when the cause will arise. The statistical models described above can be used to describe random variations, subject to the limitations discussed later.

Figure 2.10 Shapes of common failure distributions, reliability and hazard rate functions (shown in relation to t)

Variation can also be *progressive*, for example due to wear, material fatigue, change of lubricating properties, or electrical parameter drift.

Is the variation s-normal?

The central limit theorem, and the convenient properties of the *s*-normal distribution, are the reasons why this particular function is taught as the basis of nearly all statistics of continuous variation. It is a common practice, in most applications, to assume that the variation being analysed is *s*-normal, then to determine the mean and SD of the *s*-normal distribution that best fits the data.

However, at this point we must stress an important limitation of assuming that the *s*-normal distribution describes the real variation of any process. The *s*-normal p.d.f. has values between $+\infty$ and $-\infty$. Of course a machined component dimension cannot vary like this. The machine cannot add material to the component, so the dimension of the stock (which of course will vary, but not by much) will set an upper limit. The nature of the machining process, using gauges or other practical limiting features, will set a lower limit. Therefore the variation of the machined dimension would more realistically look something like Fig. 2.11. Only the central part might be approximately *s*-normal, and the distribution will have been *curtailed*. In fact all variables, whether naturally occurring or resulting from engineering or other processes, are curtailed in some way, so the *s*-normal distribution, while being mathematically convenient, is actually misleading when used to make inferences well beyond the range of actual measurements, such as the probability of meeting an adult who is 1 m tall.

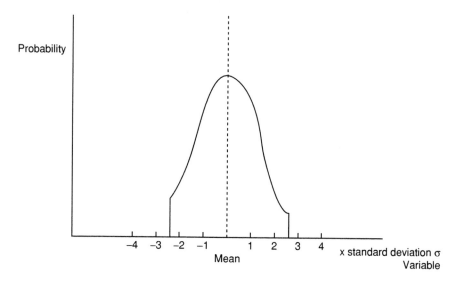

Figure 2.11   Curtailed *s*-normal distribution

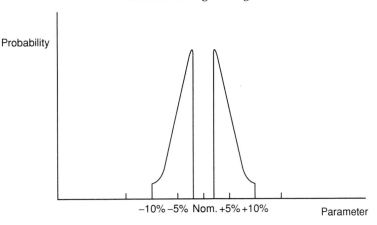

Figure 2.12 Effect of selection

There are other ways in which variation in engineering might not be *s*-normal. These are:

—There might be other kinds of selection process. For example, when electronic components such as resistors, microprocessors, etc. are manufactured, they are all tested at the end of the production process. They are then 'binned' according to the measured values. Typically, resistors that fall within ±2 per cent of the nominal resistance value are classified as precision resistors, and are labelled, binned and sold as such. Those that fall outside these limits, but within ±10 per cent become non-precision resistors, and are sold at a lower price. Those that fall outside ±10 per cent are scrapped. Because those sold as ±10 per cent will not include any that are ±2 per cent, the distribution of values is as shown in Fig. 2.12. Similarly, microprocessors are sold as, say, 200 MHz, 400 MHz, 800 MHz, etc. operating speeds depending on the maximum speed at which they function correctly on test, having all been produced on the same process. The different maximum operating speeds are the result of the variations inherent in the process of manufacturing millions of transistors and capacitors and their interconnections, on each chip on each wafer. The technology sets the upper limit for the design and the process, and the selection criteria the lower limits. Of course, the process will also produce a proportion that will not meet other aspects of the specification, or that will not work at all.

—The variation might be unsymmetrical, or *skewed*, as shown in Fig. 2.13. Distribution functions such as the lognormal and the Weibull can be used to model unsymmetrical variation. However, it is still important to remember that these mathematical models will still represent only approximations to the true variations, and the further into the tails that we apply them the greater will be the scope for uncertainty and error.

—The variation might be multimodal (Fig. 2.14), rather than unimodal as represented by distribution functions like the *s*-normal, lognormal and Weibull functions. For example, a process might be centred on one value, then an adjustment

Figure 2.13   Skewed distribution

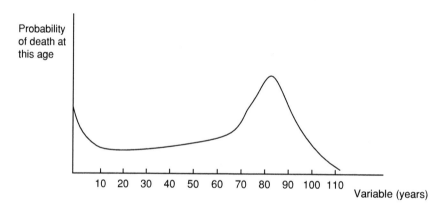

Figure 2.14   Bi-modal distribution

moves this nominal value. In a population of manufactured components this might result in a total variation that has two peaks, or a bi-modal distribution. A component might be subjected to a pattern of stress cycles that vary over a range in typical applications, and a further stress under particular conditions, for example resonance, lightning strike, etc.

Variation of engineering parameters is, to a large extent, the result of human performance. Factors such as measurements, calibrations, accept/reject criteria, control of processes, etc. are subject to human capabilities, judgements, and errors. People do not behave s-normally.

Walter Shewhart, in 1931 (Reference 3) was the first to explain the nature of variation in manufacturing processes. Figure 2.15 illustrates four very different kinds of variation, which, however all have the same means and SDs. These show

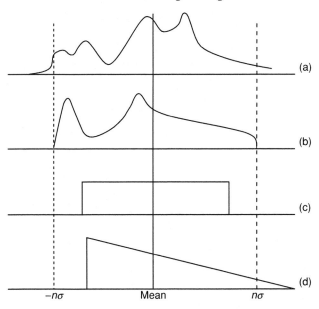

Figure 2.15 Four distributions with the same means and SDs (after W. A. Shewhart)

clearly how misleading it can be to assume that any variation is *s*-normal and then to make assertions about the population based upon the assumption.

Effects and causes

In engineering (and in many other applications) we are really much more concerned with the effects of variation than with the properties and parameters. If, for example, the output of a process varied as in Fig. 2.15(c), and the '±nσ' lines denoted the allowable tolerance, 100 per cent would be in tolerance. If, however, the process behaved as in (a) or (d), a proportion would be outside tolerance (only at the high end in the case of (d)). Variation can have other effects. A smaller diameter on a shaft might lead to higher oil loss or reduced fatigue life. Higher temperature operation might make an electronic circuit shut down. A higher proportion of fast processors on a production batch would result in higher profit. We must therefore first identify the effects of variation (they are often starkly apparent), and determine whether and to what extent the effects can be reduced. This is not simply a matter of 'reducing SD'.

The effects of variation can be reduced in two ways:

1. We can compensate for the variation, by methods such as gauging or 'select on test' (this curtails the original variation), by providing temperature compensating devices, etc.
2. We can reduce the variation.

In both cases we must understand the cause, or causes, of the variation. Shewhart categorized manufacturing variation into *'assignable'* and *'non-assignable'* causes.

(These are also referred to as *'special causes'* and *'common causes'*.) Assignable variation is any whose cause can be practically and economically identified and reduced: deterministic and functional variation fall into this category. Non-assignable variation is that which remains when all of the assignable variation has been removed. A process in this state is *'in control'*, and will have minimal, random, variation. Note that these are practical criteria, with no strict mathematical basis. Shewhart developed the methods of *statistical process control* (SPC) around this thinking, with the emphasis on using the data and charting methods to identify and reduce assignable variation, and to keep processes in control. SPC methods are described in detail in Chapter 13.

Tails

People like life insurance actuaries, clothes manufacturers, and pure scientists are interested in averages and SDs: the behaviour of the bulk of the data. Since most of the sample data, in any situation, will represent this behaviour, they can make credible assertions about these population parameters. However, the further we try to extend the assertions into the tails, the less credible they become, particularly when the assertions are taken beyond any of the data. In engineering we are usually more concerned about the behaviour of variation at the extremes, than that near the average. We are concerned by high stresses, high and low temperatures, slow processors, weak components, etc. In other words, it is the tails of the distributions that concern us. We often have only small samples to measure or test. Therefore, using conventional mathematical statistics to attempt to understand the nature, causes and effects of variation at the extremes can be misleading. However, these situations can be analysed using extreme value methods (page 41).

Conclusions

These are the aspects that matter in engineering, and they transcend the kind of basic statistical theory that is generally taught and applied. Later teachers, particularly W. E. Deming (Chapter 1, Reference 9) and Genichi Taguchi (Chapter 5) extended the ideas by demonstrating how reducing variation reduces costs and increases productivity, and by emphasizing the management implications.

Despite all of these reasons why conventional statistical methods can be misleading if used to describe and deal with variation in engineering, they are widely taught and used, and their limitations are hardly considered. Examples are:

—Most textbooks and teaching on SPC emphasize the use of the normal distribution as the basis for charting and decision-making. They emphasize the mathematical aspects, such as probabilities of producing parts outside arbitrary 2σ or 3σ limits, and pay little attention to the practical aspects discussed above.
—Typical design rules for mechanical components in critical stress application conditions, such as aircraft and civil engineering structural components, require that there must be a factor of safety (say 2) between the maximum expected stress and the lower 3σ value of the expected strength. This approach is really quite

arbitrary, and oversimplifies the true nature of variations such as strength and loads, as described above. Why, for example, select 3σ? If the strength of the component were truly normally distributed, about 0.1 per cent of components would be weaker than the 3-sigma value. If few components are made and used, the probability of one failing would be very low. However, if many are made and used, the probability of a failure among the larger population would increase proportionately. If the component is used in a very critical application, such as an aircraft engine suspension bolt, this probability might be considered too high to be tolerable. Of course there are often other factors that must be considered, such as weight, cost, and the consequences of failure. The criteria applied to design of a domestic machine might sensibly be less conservative than for a commercial aircraft application.

—The so-called '*six sigma*' approach to achieving high quality is based on the idea that, if any process is controlled in such a way that only operations that exceed plus or minus 6σ of the underlying distribution will be unacceptable, then fewer than one per million operations will fail. The exact quantity is based on arbitrary and generally unrealistic assumptions about the distribution functions, as described above. (The 'six sigma' approach entails other features, such as the use of a wide range of statistical and other methods to identify and reduce variations of all kinds, and the training and deployment of specialists. It is described in Chapter 15.)

DISCRETE VARIATION

The binomial distribution

The binomial distribution describes a situation in which there are only two outcomes, such as pass or fail, and the probability remains the same for all trials. (Trials which give such results are called *Bernoulli trials*.) Therefore, it is obviously very useful in QA and reliability work. The p.d.f. for the binomial distribution is

$$f(x) = \frac{n!}{x!(n-x)!} p^x q^{(n-x)} \tag{2.37}$$

$\dfrac{n!}{x!(n-x)!}$ may be written $\dbinom{n}{x}$

This is the probability of obtaining x good items and $(n-x)$ bad items, in a sample of n items, when the probability of selecting a good item is p and of selecting a bad item is q. The mean of the binomial distribution (from Eqn 2.13) is given by

$$\mu = np \tag{2.38}$$

and the SD from (Eqn 2.17)

$$\sigma = (npq)^{1/2} \tag{2.39}$$

The binomial distribution can only have values at points where x is an integer. The c.d.f. of the binomial distribution (i.e. the probability of obtaining r or fewer successes in n trials) is given by

$$F(r) = \sum_{x=0}^{r} \binom{n}{x} p^x q^{(n-x)} \tag{2.40}$$

Example 2.6

A frequent application of the cumulative binomial distribution is in QC acceptance sampling. For example, if the acceptance criterion for a production line is that not more than 4 defectives may be found in a sample of 20, we can determine the probability of acceptance of a lot if the production process yields 10 per cent defectives.

From Eqn (2.40),

$$F(4) = \sum_{x=0}^{4} \binom{20}{x} 0.1^x 0.9^{(20-x)}$$

$$= 0.957$$

When approximate values will suffice the nomogram in Fig. 2.16 can be used.

Example 2.7

An aircraft landing gear has 4 tyres. Experience shows that tyre bursts occur on average on 1 landing in 1200. Assuming that tyre bursts occur s-independently of one another, and that a safe landing can be made if not more than 2 tyres burst, what is the probability of an unsafe landing?

If n is the number of tyres and p is the probability of a tyre bursting,

$$n = 4$$

$$p = \frac{1}{1200} = 0.00083$$

$$q = (1 - p) = 0.99917$$

The probability of a safe landing is the probability that not more than 2 tyres burst.

$$F(2) = \binom{4}{2}(0.000\ 83)^2(0.999\ 17)^2 + \binom{4}{1}(0.000\ 83)^1(0.999\ 17)^3$$

$$+ \binom{4}{0}(0.000\ 83)^0(0.999\ 17)^4$$

$$= 0.000\ 004\ 159\ 7 + 0.003\ 325\ 006\ 9 + 0.996\ 670\ 831$$

$$= 0.999\ 999\ 997\ 7$$

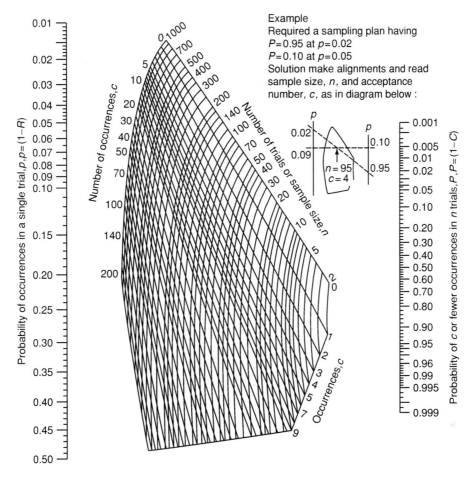

Figure 2.16 Nomogram of the cumulative binomial distribution

Therefore the probability of an unsafe landing is

$$1 - 0.999\ 999\ 997\ 7 = 2.3 \times 10^{-9}$$

The Poisson distribution

If events are Poisson-distributed they occur at a *constant average rate*, with only one of the two outcomes countable, e.g. the number of failures in a given time or defects in a length of wire:

$$f(x) = \frac{\mu^x}{x!}\exp(-\mu) \quad (x = 0,\ 1,\ 2,\ \ldots) \tag{2.41}$$

where μ is the mean rate of occurrence. The Poisson distribution can also be considered as an extension of the binomial distribution, in which n is considered infinite.

Figure 2.17 provides a chart for the cumulative Poisson distribution. For example, if we need to know the probability of not more than three failures occurring in 1000 h of operation of a system, when the mean rate of failures is 1 per 1000 h, we can read the answer from Fig. 2.17 as 0.98 ($\mu = 1/1000$, $x = 3$).

Since the Poisson distribution can represent the limiting case of the binomial distribution it gives a good approximation to the binomial distribution, when p or q are small and n is large. This is useful in sampling work where the proportion of defectives is low (i.e. $p < 0.1$).

The Poisson approximation is

$$f(x) = \frac{(np)^x}{x!} \exp(-np)$$

$$\left[\mu = np; \sigma = (np)^{1/2} = \mu^{1/2} \right]$$

(2.42)

This approximation allows us to use Poisson tables or charts in appropriate cases and also simplifies calculations.

Example 2.8

If the probability of an item failing is 0.001, what is the probability of 3 failing out of a population of 2000?

The binomial solution is

$$\binom{2000}{3} 0.999^{1997} 0.001^3 = 0.1805$$

This is rather more tedious to evaluate than the Poisson approximation. The Poisson approximation is evaluated as follows:

$$\mu = np$$

$$= 2000 \times 0.001 = 2$$

$$P(x = 3) = \frac{2^3}{3!} \exp(-2) = 0.1804$$

As the s-normal distribution represents a limiting case of the binomial and Poisson distributions, it can be used to provide a good approximation to these distributions. For example, it can be used when $0.1 > p > 0.9$ and n is large. Then

$$\mu = np$$

$$\sigma = (npq)^{1/2}$$

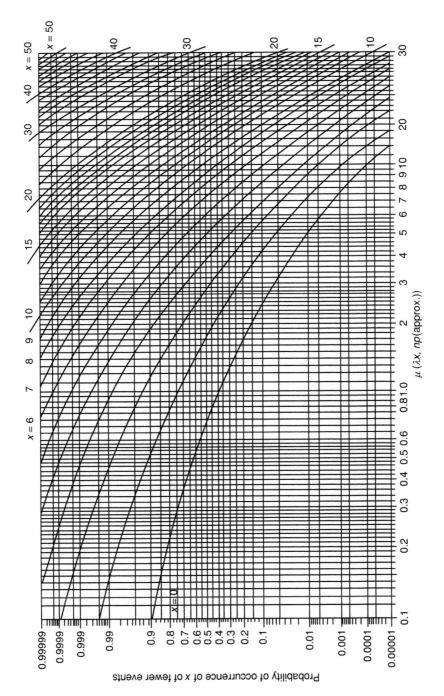

Figure 2.17 Curves showing probabilities of Poison distribution

Example 2.9

What is the probability of having not more than 20 failures if $n = 100$, $p = 0.14$?
  Using the binomial distribution,

$$P_{20} = 0.9640$$

Using the *s*-normal approximation

$$\mu = np = 14$$

$$\sigma = (npq)^{1/2} = 3.470$$

$$z = \frac{20 - 14}{3.47} = 1.73$$

Referring to Appendix 1, $P_{20} = 0.9582$.
  As $p \rightarrow 0.5$, the approximation improves, and we can then use it with smaller values of n. Typically, if $p = 0.4$, we can use the approximation with $n = 50$.

STATISTICAL CONFIDENCE

Earlier in this chapter we mentioned the problem of statistical confidence. *s*-confidence is the exact fraction of times the *confidence interval* will include the true value, if the experiment is repeated many times. The confidence interval is the interval between the *upper* and *lower s-confidence limits*. *s*-confidence intervals are used in making an assertion about a population given data from a sample. Clearly, the larger the sample the greater will be our intuitive confidence that the estimate of the population parameter will be close to the true value. *s*-confidence and engineering confidence must not be confused; *s*-confidence takes no account of engineering or process changes which might make sample data unrepresentative. Derived *s*-confidence values must always be interpreted in the light of engineering knowledge, which might serve to increase or decrease our engineering confidence.

s-confidence limits on continuous variables

If the population value x follows a *s*-normal distribution, it can be shown that the means, $\bar{x}$, of samples drawn from it are also *s*-normally distributed, with variance σ^2/n (SD $= \sigma/n^{1/2}$). The SD of the sample means is also called the *standard error of the estimate*, and is denoted $S_{\bar{x}}$.

  If x is not *s*-normally distributed, provided that n is large (> 30), $\bar{x}$ will tend to a *s*-normal distribution. If the distribution of x is not excessively skewed (and is unimodal) the *s*-normal approximation for $\bar{x}$ at values of n as small as 6 or 7 may be acceptable.

  These results are derived from the central limit theorem, mentioned on page 36. They are of great value in deriving *s*-confidence limits on population parameters, based on sample data. In reliability work it is not usually necessary to derive exact s-confidence limits and therefore the approximate methods described are quite adequate.

Example 2.10

A sample of 100 values has a mean of 27.56, with a standard deviation of 1.10. Derive 95 per cent s-confidence limits for the population mean. (Assume that the sample means are s-normally distributed.)

In this case, the SD of the sample means, or standard error of the estimate, is

$$\frac{\sigma}{n^{1/2}} = \frac{1.1}{(100)^{1/2}} = 0.11$$

We can refer to the table of the s-normal c.d.f. (Appendix 1) to obtain the 95 per cent single-sided s-confidence limits. The closest tabulated value of z is 1.65. Therefore, approximately ± 1.65 SDs are enclosed within the 95 per cent single-sided s-confidence limits. Since the s-normal distribution is symmetrical, the 90 per cent *double-sided* s-confidence interval will exclude 5 per cent of values at either limit.

In the example, 1.65 SDs = 0.18. Therefore the 95 per cent s-confidence limits on the population mean are 27.56 ± 0.18, and the 90 per cent s-confidence interval is $(27.56 - 0.18)$ to $(27.56 + 0.18)$.

As a guide in s-confidence calculations, assuming a s-normal distribution:

± 1.65 SDs enclose approximately 90 per cent s-confidence limits (i.e. 5 per cent lie in each tail).
± 2.0 SDs enclose approximately 95 per cent s-confidence limits (i.e. 2.5 per cent lie in each tail).
± 2.5 SDs enclose approximately 99 per cent s-confidence limits (i.e. 0.5 per cent lie in each tail).

In many reliability situations we assume a CFR, in which case the failure data will be exponentially distributed and thus highly skewed. Therefore, the s-normal approximation for s-confidence limit estimation is inappropriate (unless n is large). We use the χ^2 distribution for estimating the confidence limits in these circumstances.

It can be shown that the lower and upper (one-sided) s-confidence limits for data which are generated by a Poisson process, such as failure data when the failure rate is constant, are given by respectively

$$\theta_1 = \frac{2T}{\chi_\alpha^2(\nu = 2n)}$$

$$\theta_u = \frac{2T}{\chi_{1-\alpha}^2(\nu = 2n)}$$

$$(2.43)$$

where θ_1 and θ_u are the lower and upper confidence limits on MTBF, T the total test time, α the confidence level and n the number of failures. Equation (2.43) is correct when the test is stopped at the nth failure, i.e. a *failure truncated test*. For *time truncated tests*, use $\nu = 2n + 2$ degrees of freedom for evaluating the lower confidence limit. Values of χ^2 for different values of ν are given in Appendix 3.

Example 2.11

Ten units were tested for a total of 1000 h and 3 failures occurred. The test was then failure-truncated. Assuming a CFR, what is the 90 per cent lower confidence limit (LCL) on MTBF?

$$\hat{\theta} = \frac{1000}{3} = 333\,h$$

From Appendix 3, for $\nu = 6$ and $\alpha = 0.9, \chi^2 = 10.6$. Therefore,

$$\theta_1 = \frac{2 \times 1000}{10.6} = 189\,h$$

Figure 2.18 may also be used for this estimation (for time-truncated tests).

s-confidence limits for discrete data

The normal and Poisson approximations to the binomial distribution may not be acceptable for evaluating s-confidence limits with sample sizes below 100. In these situations we can conveniently use the nomogram in Fig. 2.16 to derive the lower confidence limit. For example, if four failures occur in 100 tests, the 95 per cent LCL on reliability of the population is given by drawing the line from the 0.05 point on the P scale through the 100/4 point and reading the value on the p scale. In this case $p = 0.09$, i.e. the 95 per cent LCL on reliability is 0.91.

STATISTICAL HYPOTHESIS TESTING

It is often necessary to determine whether observed differences between the statistics of a sample and prior knowledge of a population, or between two sets of sample statistics, are s-significant or due merely to chance. The variation inherent in sampling makes this distinction itself subject to chance. We need, therefore, to have methods for carrying out such tests. Statistical hypothesis testing is similar to s-confidence estimation, but instead of asking the question *How s-confident are we that the population parameter value is within the given limits?* (On the assumption that the sample and the population come from the same distribution), we ask *How s-significant is the deviation of the sample?*

   In statistical hypothesis testing, we set up a *null hypothesis*, i.e. that the two sets of information are derived from the same distribution. We then derive the *s-significance* to which this inference is tenable. As in s-confidence estimation, the s-significance we can attach to the inference will depend upon the size of the sample. Many s-significance test techniques have been developed for dealing with the many types of situation which can be encountered.

Figure 2.18 Confidence limits for measurement of MTBF for time truncated tests (see Example 2.11). Key: compute confidence limits as $\hat{\theta}(1 + D/100)$, where $\hat{\theta}$ is MTBF estimate

In this section we will cover a few of the simpler methods commonly used in reliability work. However, the reader should be aware that the methods described and the more advanced techniques are readily accessible on modern calculators and as computer programs. The texts listed in the Bibliography should be used to identify appropriate tests and tables for special cases.

Tests for differences in means (z test)

A very common s-significance test is for the hypothesis that the mean of a set of data is the same as that of an assumed s-normal population, with known μ and σ. This is the z *test*. The *z-statistic* is given by

$$z = \frac{|\mu - \bar{x}|}{S_{\bar{x}}} = \frac{|\mu - \bar{x}|}{\sigma n^{-1/2}} \tag{2.44}$$

where n is the sample size, μ the population mean, $\bar{x}$ the sample mean and σ the population SD. We then derive the s-significance level from the s-normal c.d.f. table.

Example 2.12

A type of roller bearing has a time to failure which is s-normally distributed, with a mean of 6000 h and an SD of 450 h. A sample of 9, using a changed lubricant, gave a mean life of 6400 h. Has the new lubricant resulted in an s-significant change in mean life?

$$z = \frac{|6000 - 6400|}{450 \times 9^{-1/2}} = 2.67$$

From Appendix 1, $z = 2.67$ indicates a cumulative probability of 0.996. This indicates that there is only 0.004 probability of observing this change purely by chance, i.e. the change is s-significant at the 0.4 per cent level. Thus we reject the null hypothesis that the sample data are derived from the same s-normal distribution as the population, and infer that the new lubricant does provide an increased life.

s-significance is denoted by α. In engineering, an s-significance level of less than 5 per cent can usually be considered to be sufficient evidence upon which to reject a null hypothesis. An s-significance of greater than 10 per cent would not normally constitute sufficient evidence, and we might either reject the null hypothesis or perform further trials to obtain more data. The s-significance level considered sufficient will depend upon the importance of the decision to be made based on the evidence. As with s-confidence, s-significance should also be assessed in the light of engineering knowledge.

Instead of testing a sample against a population, we may need to determine whether there is an s-significant difference between the means of two samples. The SD of the distribution of the difference in the means of the samples is

$$S_{(\bar{x}_1 - \bar{x}_2)} = \frac{\sigma_1}{n_1^{1/2}} + \frac{\sigma_2}{n_2^{1/2}} \tag{2.45}$$

The SD of the distribution of the difference of the sampling means is called the *standard error of the difference*. This test assumes that the SDs are the population SDs. Then

$$z = \frac{\text{difference in sample means}}{\text{standard error of the difference}}$$

Example 2.13

In Example 2.12, if the mean value of 6000 and SD of 450 were in fact derived from a sample of 60, does the mean of 6400, with an SD of 380 from a sample of 9 represent an s-significant difference?

The difference in the means is

$$6400 - 6000 = 400$$

The standard error of the difference is

$$S_d = \frac{\sigma_1}{n_1^{1/2}} + \frac{\sigma_2}{n_2^{1/2}}$$

$$= \frac{450}{60^{1/2}} + \frac{380}{9^{1/2}} = 185$$

$$z = \frac{400}{185} = 2.16$$

$$a = 1 - \Phi(z) = 0.015(1.5 \text{per cent})$$

We can therefore say that the difference is highly *s*-significant, a similar result to that of Example 2.12.

Use of the *z* test for binomial trials

We can also use the *z* test for testing the *s*-significance of binomial data. Since in such cases we are concerned with both extremes of the distribution, we use a two-sided test, i.e. we use 2α instead of α.

Example 2.14

Two sets of tests give the results in Table 2.2. We need to know if the differences in test results are *s*-significant.

The null hypothesis that the tests are without difference is examined by combining the test results:

$$P = \frac{\text{total failed}}{\text{total tested}} = \frac{30}{527} = 0.057$$

The standard error of the difference in proportions is

$$S_d = \left[pq \left(\frac{1}{n_1} + \frac{1}{n_2} \right) \right]^{1/2}$$

$$= \left[0.057 \times 0.943 \left(\frac{1}{217} + \frac{1}{310} \right) \right]^{1/2}$$

$$= 0.02$$

Table 2.2 Results for tests in Example 2.14

Test	Number tested, n	Number failed
1	217	16
2	310	14

The proportion failed in test 1 is $16/217 = 0.074$. The proportion failed in test 2 is $14/310 = 0.045$. The difference in proportions is $0.074 - 0.045 = 0.029$. Therefore $z = 0.029/0.020 = 1.45$, giving

$$\alpha = 1 - \Phi(z) = 7.35 \text{ per cent}$$

$$2\alpha = 14.7 \text{ per cent}$$

With such a result, we would be unable to reject the null hypothesis and would therefore infer that the difference between the tests is not very s-significant.

χ^2 test for significance

The χ^2 test for the s-significance of differences is used when we can make no assumptions about the underlying distributions. The value of the χ^2 statistic is calculated by summing the terms

$$\frac{(x_i - E_i)^2}{E_i}$$

where x_i and E_i are the ith observed and expected values, respectively. This value is compared with the χ^2 value appropriate to the required s-significance level.

Example 2.15

Using the data of Example 2.14, the χ^2 test is set up as follows:

Test	Failure		Success		Totals
1	16	12.35	201	204.65	217
2	14	17.65	296	292.35	310
Totals	30		497		527

The first number in each column is the observed value and the second number is the expected value based upon the totals of the observations (e.g. expected failures in test $1 = 30/527 \times 217 = 12.35$).

$$\chi^2 = \frac{(16 - 12.35)^2}{12.35} + \frac{(201 - 204.65)^2}{204.65} + \frac{(14 - 17.65)^2}{17.65} + \frac{(296 - 292.35)^2}{292.35} = 1.94$$

The number of DF is one less than the number of different possibilities which could exist. In this case there is only one DF, since there are two possibilities – pass and fail. The value of χ^2 of 1.94 for 1 DF (from Appendix 3) occurs with a cumulative probability of between 80 and 90 per cent. Therefore α (two-sided) is between 10 and 20 per cent. The difference between the observed data sets is therefore not s-significant. This inference is the same as that derived in Example 2.14.

Tests for differences in variances

Variance ratio test (F test)

The s-significance tests for differences in means described above have been based on the assumption in the null hypothesis that the samples came from the same s-normal distribution, and therefore should have a common mean. We can also perform s-significance tests on the differences of variances. The *variance ratio, F*, is defined as

$$F = \frac{\text{greater estimate of population variance}}{\text{lesser estimate of population variance}}$$

Values of the F distribution are tabulated in Appendix 4 against the number of degrees of freedom in the two variance estimates (for a sample size n, DF $= n - 1$). The use of the F test is illustrated by Example 2.16.

Example 2.16

Life test data on two items give the results in Table 2.3.

Table 2.3 Life test data on two items

	Sample size, n	Sample standard deviation, σ	Sample variance, σ^2
Item 1	20	37	1369
Item 2	10	31	961

$$F = \frac{1369}{961} = 1.42$$

Entering the tables of F values at 19 DF for the greater variance estimate and 9 DF for the lesser variance estimate, we see that at the 5 per cent level our value for F is less than the tabulated value. Therefore the difference in the variances is not s-significant at the 5 per cent level.

NON-PARAMETRIC INFERENTIAL METHODS

Methods have been developed for measuring and comparing statistical variables when no assumption is made as to the form of the underlying distributions. These are called non-parametric (or distribution-free) statistical methods. They are only slightly less powerful than parametric methods in terms of the accuracy of the inferences derived for assumed s-normal distributions. However, they are more powerful when the distributions are not s-normal. They are also simple to use. Therefore they can be very useful in reliability work provided that the data being analysed are independently and identically distributed (IID). The implications of data not independently and identically distributed are covered on page 73 and in the next chapter.

Comparison of median values

The sign test

If a null hypothesis states that the median values of two samples are the same, then about half the values of each sample should lie on either side of the median. Therefore about half the values of $(x_i - \bar{x})$ should be positive and half negative. If the null hypothesis is true and r is the number of differences with one sign, then r has a binomial distribution with parameters n and $p = \frac{1}{2}$. We can therefore use the binomial distribution to determine critical values of r to test whether there is an s-significant difference between the median values. Table 2.4 gives critical values for r for the sign test where r is the number of less frequent signs. If the value of r is equal to or less than the tabulated value the null hypothesis is rejected.

Table 2.4 Critical values of r for the sign test

	Significance level per cent		
n	10	5	1
8	1	0	0
10	1	1	0
12	2	2	1
14	3	2	1
16	4	3	2
18	5	4	3
20	5	5	3
25	7	7	5
30	10	9	7
35	12	11	9
40	14	13	11
45	16	15	13
50	18	17	15
55	20	19	17
60	23	21	19
75	29	28	25
100	41	39	36

Example 2.17

Ten items are tested to failure, with lives

    98, 125, 141, 72, 119, 88, 64, 187, 92, 114

Do these results indicate an s-significant change from the previous median life of 125?

 The sign test result is

    $-0 + - - - - + - -$

i.e. $r = 2$, $n = 9$ (since one difference $= 0$, we discard this item).

Table 2.4 shows that r is greater than the critical value for $n = 9$ at the 10 per cent s-significance level, and therefore the difference in median values is not significant at this level.

The weighted sign test

We can use the sign test to determine the likely magnitude of differences between samples when differences in medians are significant. The amount by which the samples are believed to differ are added to (or subtracted from) the values of one of the samples, and the sign test is then performed as described above. The test then indicates whether the two samples differ s-significantly by the weighted value.

Tests for variance

Non-parametric tests for analysis of variance are given in Chapter 5.

Reliability estimates

Non-parametric methods for estimating reliability values are given in Chapter 12.

GOODNESS OF FIT

In analysing statistical data we need to determine how well the data fit an assumed distribution. The goodness of fit can be tested statistically, to provide a level of s-significance that the null hypothesis (i.e. that the data do fit the assumed distribution) is rejected. Goodness-of-fit testing is an extension of s-significance testing in which the sample c.d.f. is compared with the assumed true c.d.f.

A number of methods are available to test how closely a set of data fits an assumed distribution. As with s-significance testing, the power of these tests in rejecting incorrect hypotheses varies with the number and type of data available, and with the assumption being tested.

The χ^2 goodness-of-fit test

A commonly used and versatile test is the χ^2 goodness-of-fit test, since it is equally applicable to any assumed distribution, provided that a reasonably large number of data points is available. For accuracy, it is desirable to have at least three data classes, or *cells*, with at least five data points in each cell.

The justification for the χ^2 goodness-of-fit test is the assumption that, if a sample is divided into n cells (i.e. we have ν degrees of freedom where $\nu = n - 1$), then the values within each cell would be normally distributed about the expected value, if the assumed distribution is correct, i.e. if x_i and E_i are the observed and expected values for cell i:

$$\sum_{i}^{n} \frac{(x_i - E_i)^2}{E_i} = \chi^2 \quad \text{(with } n - 1 \text{ degrees of freedom)}$$

High values of χ^2 cast doubt on the null hypothesis. The null hypothesis is usually rejected when the value of χ^2 falls outside the 90th percentile. If χ^2 is below this value, there is insufficient information to reject the hypothesis that the data come from the supposed distribution. If we obtain a very low χ^2 (e.g. less than the 10th percentile), it suggests that the data correspond more closely to the supposed distribution than natural sampling variability would allow (i.e. perhaps the data have been 'doctored' in some way).

The application can be described by use of an example.

Example 2.18

Failure data of transistors are given in Table 2.5. What is the likelihood that failures occur at a constant average rate of 12 failures/1000 hours?

Table 2.5   Data from an overstress life test of transistors

Cell (h)	Number in cell	Cell (h)	Number in cell
0–999	18	3000–3999	12
1000–1999	14	4000–4999	6
2000–2999	10		

$$\chi^2 = \frac{(18 - 12)^2}{12} + \frac{(14 - 12)^2}{12} + \frac{(10 - 12)^2}{12} + \frac{(12 - 12)^2}{12} + \frac{(6 - 12)^2}{12} = 6.67$$

Referring to Appendix 3 for values of χ^2 with $(n - 1) = 4$ degrees of freedom, 6.67 lies between the 80th and 90th percentiles of the χ^2 distribution. Therefore the null hypothesis that the data are derived from a constant hazard rate process cannot be rejected at the 90 per cent level.

If an assumed distribution gave expected values of 20, 15, 12, 10, 9 (i.e. a decreasing hazard rate), then

$$\chi^2 = \frac{(18 - 20)^2}{20} + \frac{(14 - 15)^2}{15} + \frac{(10 - 12)^2}{12} + \frac{(12 - 10)^2}{10} + \frac{(8 - 9)^2}{9} = 1.11$$

$\chi^2 = 1.11$ lies close to the 10th percentile. Therefore we cannot reject the null hypothesis of the decreasing hazard rate distribution at the 90 per cent level.

Note that the E_i values should always be at least 5. Cells should be amalgamated if necessary to achieve this, with the degrees of freedom reduced accordingly. Also, if we have estimated the parameters of the distribution we are fitting to, the degrees of freedom should be reduced by the number of parameters estimated.

The Kolmogorov–Smirnov test

Another goodness-of-fit test commonly used in reliability work is the Kolmogorov–Smirnov (K–S) test. It is rather simpler to use than the χ^2 test and can give better results with small numbers of data points. It is also convenient to use in conjunction with probability plots (see Chapter 3), since it is based upon cumulative ranked data, i.e. the sample c.d.f. The procedure is:

1. Tabulate the ranked failure data. Calculate the values of $|x_i - E_i|$ where x_i is the ith cumulative rank value and E_i the expected cumulative rank value for the assumed distribution.
2. Determine the highest single value.
3. Compare this value with the appropriate K–S value (Appendix 5).

Example 2.19

Table 2.6 shows failure data with the ranked values of x_i. We wish to test the null hypothesis that the data do not fit a normal distribution with parameters which give the tabulated cumulative values of E_i. Therefore, in the E_i column we list the expected value of proportion failed at each failure time.

The largest value of $|x_i - E_i|$ is 0.08 (shown by *). The Kolmogorov–Smirnov table (Appendix 5) shows that, for $n = 12$, the critical value of $|x_i - E_i|$ is 0.338 at the 10 per cent s-significance level. Therefore the null hypothesis is not rejected at this level, and we can accept the data as coming from the hypothesized normal distribution.

Example 2.19 shows quite a large difference between the critical K–S value and the largest value of $|x_i - E_i|$. When the parameters of the assumed c.d.f. are being estimated from the sample data, as in this example, the critical K–S values are too large and give lower s-significance levels than are appropriate in the circumstances. In order to correct for this, the critical values should be multiplied by the factors:

Table 2.6 Failure data with ranked values of x_i

Event	Time to failure (h)	x_i	E_i	$|x_i - E_i|$
1	12.2	0.056	0.035	0.021
2	13.1	0.136	0.115	0.021
3	14.0	0.217	0.29	0.073
4	14.1	0.298	0.32	0.022
5	14.6	0.379	0.44	0.061
6	14.7	0.459	0.46	0.001
7	14.7	0.54	0.46	0.08*
8	15.1	0.621	0.58	0.041
9	15.7	0.702	0.73	0.028
10	15.8	0.783	0.75	0.033
11	16.3	0.864	0.85	0.014
12	16.9	0.94	0.95	0.006

0.70 $(\beta > 3.0)$

0.75 $(3.0 > \beta > 1.5)$

0.80 $(\beta < 1.5)$

where β is the Weibull shape parameter. Therefore, in Example 2.19, since the Weibull β value appropriate to the normal distribution is > 3, the corrected K–S critical value is $0.338 \times 0.70 = 0.237$.

Least squares test

The least squares goodness-of-fit test is used to measure the *linear correlation* of data with the equation of the straight line which best fits the plotted data. The line which most nearly fits a set of plotted data is called the *regression line*, and the goodness of fit as derived by the least squares method is called the *correlation coefficient*.

The correlation coefficient is

$$r = \frac{S_{xy}}{S_x S_y}$$

where

$$\left. \begin{aligned} S_z^2 &= \frac{1}{n}\sum_{i=1}^{n}(x_i - \bar{x})^2 \\ S_y^2 &= \frac{1}{n}\sum_{i=1}^{n}(y_i - \bar{y})^2 \end{aligned} \right\} \quad \text{(i.e. the sample variances)} \tag{2.46}$$

$$S_{xy} = \frac{1}{n}\sum_{i=1}^{n}(x_i - \bar{x})(y_i - \bar{y}) \quad \text{(the covariance)} \tag{2.47}$$

x_i, y_i are the coordinates of the plotted data. The regression line of the plotted data is given by

$$y - \bar{y} = \frac{S_{xy}}{S_x^2}(x - \bar{x}) \tag{2.48}$$

If r is positive, it indicates that the data are positively correlated, i.e. the regression line slope is positive, and vice versa; if $|r| = 1$ we have perfect correlation, with all plotted points lying on the regression line; if $r = 0$ the variates are not linearly correlated.

r^2 is often used instead of r to indicate correlation, since it provides a more sensitive indication, particularly with probability plots (see Chapter 3). It is called the *sample coefficient of determination*.

Linear regression analysis can be used for data that are not linearly correlated if the axes are transformed to linearize the equation. Therefore, the method can be used for evaluating the goodness of fit of data plotted on probability papers (Chapter 3).

SERIES OF EVENTS (POINT PROCESSES)

Situations in which discrete events occur randomly in a continuum (e.g. time) cannot be truly represented by a single continuous distribution function. Failures occurring in repairable systems, aircraft accidents and vehicle traffic flow past a point are examples of series of discrete events. These situations are called *stochastic point processes*. They can be analysed using the statistics of *event series*.

The Poisson distribution function (Eqn 2.41) describes the situation in which events occur randomly and at a constant average rate. This situation is described by a *homogeneous Poisson process* (HPP). An HPP is a *stationary* point process, since the distribution of the number of events in an interval of fixed length does not vary, regardless of when (where) the interval is sampled.

The Poisson distribution function is (from Eqn 2.41)

$$f(x) = \frac{(\lambda x)^n}{n!} \exp(-\lambda x) \quad \text{(for } n = 0, 1, 2, \ldots) \tag{2.49}$$

where λ is the mean rate of occurrence, so that λx is the expected number of events in $(0, x)$. Figure 2.8 can be used for estimating the probability of a given number of events in a given interval.

In a *non-homogeneous* Poisson process (NHPP) the point process is non-stationary, so that the distribution of the number of events in an interval of fixed length changes as x increases. Typically, the discrete events (e.g. failures) might occur at an increasing or decreasing rate.

Note that an essential condition of any homogeneous Poisson process is that the probabilities of events occurring in any period are independent of what has occurred in preceding periods. An HPP describes a sequence of independently and identically exponentially distributed (IIED) random variables. A NHPP describes a sequence of random variables which is neither independently nor identically distributed.

Trend analysis (time series analysis)

When analysing data from a stochastic point process it is important to determine whether the process has a trend, i.e. to know whether a failure rate is increasing, decreasing or constant. We can test for trends by analysing the *arrival values* of the event series. The arrival values $x_1, x_2, \ldots, x_n$ are the values of the independent variables (e.g. time) from $x = 0$ at which each event occurs. The *interarrival values* $X_1, X_2, \ldots, X_n$ are the intervals between successive events $1, 2, \ldots, n$, from $x = 0$. Figure 2.19 shows the distinction between arrival and interarrival values.

If x_0 is the period of observation, then the test statistic for trend is

$$U = \frac{\Sigma x_i / n - x_0 / 2}{x_0 \sqrt{1/(12n)}} \tag{2.50}$$

This is called the *centroid test* or the *Laplace test*. It compares the centroid of the observed arrival values with the mid-point of the period of observation. If $U = 0$

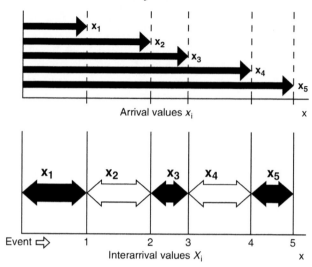

Figure 2.19 Arrival and interarrival values

there is no trend, i.e. the process is stationary. If $U < 0$ the trend is decreasing, i.e. the interarrival values are tending to become larger. Conversely, when $U > 0$ the trend is increasing, i.e. interarrival values are tending to become progressively smaller.

If the period of observation ends at an event, use $(n - 1)$ instead of n and exclude the time to the last event from the summation Σx_i.

We can test the null hypothesis that there is no trend in the chronologically ordered data by testing the value of U against the values of the standard normal variate, z. For example, using Appendix 1, if $U = 1.65$, for $z = 1.65$, $\Phi(z) = 0.95$. Therefore we can reject the null hypothesis at the 5 per cent s-significance level.

The centroid test is theoretically adequate if $n \geq 4$, when the observation interval ends with an event, and if $n \geq 3$, when the interval is terminated at a predetermined time.

The method is also called *time series analysis* (TSA).

Example 2.20

Arrival values (x_i) and interarrival values (X_i) between 12 successive failures of a component are as follows (observation ends at the last failure):

x_i	X_i	x_i	X_i
175	175	618	102
196	21	641	23
304	108	679	38
415	111	726	47
504	89	740	14
516	12	791	51

$\Sigma x_i = 5514$ (excluding 791)

$n - 1 = 11$

$$\frac{\Sigma x_i}{n - 1} = 501.3$$

$$\frac{x_0}{2} = 395.5$$

$$U = \frac{501.3 - 395.5}{791\sqrt{1/(12 \times 11)}} = 1.54 \text{ Referring to Appendix 1, for } z = 1.54, \Phi(z) = 0.94$$

Therefore we can reject the null hypothesis that there is no trend at the 6 per cent s-significance level. The interarrival times are becoming shorter, i.e. the failure rate is increasing.

The existence of a trend in the data, as in Example 2.20, indicates that the inter-arrival values *are not independently and identically distributed* (IID). This is a very important point to consider in the analysis of failure data, as will be explained in Chapter 12.

Superimposed processes

If a number of separate stochastic point process combine to form an overall process, e.g. failure processes of individual components (or sockets) in a system, these are called *superimposed processes*. If the individual random variables are IID exponential then the overall process variable is also IID exponential and the process is HPP.

If the individual variables are IID non-exponential, the overall process will tend to an HPP. Such a process is called a *renewal process*. Figure 2.20 shows these processes.

COMPUTER SOFTWARE FOR STATISTICS

Computer software is available which can be used to carry out the analytical techniques described in this chapter, and in later chapters which describe particular applications. Reference 24 is an excellent program for basic statistics training and application. See also the books homepage (Preface to the Fourth Edition, p. xxi).

PRACTICAL CONCLUSIONS

Whilst the mathematical methods described in this chapter can be useful for providing insights and for forecasting, it is important that their limitations are appreciated. They are mathematical models, and they do not necessarily reflect reality in the way that deterministic, physics-based formulae do. The important points that must be borne in mind when applying statistical methods to engineering are:

Figure 2.20   Rate of occurrence for superimposed processes

—Real variation is seldom s-normal.

—The most important variation is usually that in the tails, where there is inevitably less (or no) data, the data are more uncertain, and where conventional statistical models can be most misleading.

—Variation can change over time, so that the patterns measured at one time might not represent the true situation at another. We will cover this aspect in more detail in later chapters.

—There might be interaction effects between variables, causing combined effects that are more significant than those of individual variations. This aspect will be covered in the later chapters.

—Variation in engineering is usually made or influenced by people. People do not behave in accordance with any credible mathematical models.

—Most engineering education in statistics covers only the mathematics, and few statisticians understand the practical aspects of the engineering problems they help to solve. This leads to inappropriate analyses and conclusions, and to a distrust of statistical methods among engineers.

We must strike an appropriate balance between using deterministic and statistical methods. For example, if we conduct a test in which an item is released from a height, and if it drops the test is a success, with results as follows:

items tested (0 failures):   0 1   10   20

then we could infer that the 80 per cent s-confidence that the reliability is at least 0.9 would be:

$$0 \quad 0.90 \quad 0.98 \quad 0.99$$

This assumes that the data are entirely statistical, i.e. we have no prior knowledge of the physics or engineering. On the other hand, if we are confident that we have such knowledge (in this case, that the force of gravity will always act on the released items), then we will have 100 per cent confidence in 100 per cent reliability, even without performing any tests. In such deterministic cases statistical tests and interpretations are inappropriate.

However, in many engineering situations can range from deterministic to statistical. For example, there might be cases when the release mechanism fails to open properly. The causes and effects of variations are often uncertain (particularly when interactions exist), so we must make the best use of our knowledge and use the best methods to explore the uncertainties.

Statistical tests can, by themselves, generate misleading results. We have discussed this in the context of extrapolations beyond the range of measured data. Another example might be a series of tests that indicate that an item operated at high stress is more reliable than when operated at low stress. The results might be due to the items on the high-stress tests being manufactured using a better process, or the high stress may actually improve the reliability (e.g. a higher temperature might improve the performance of a seal), or the results might be due to chance and unrepresentative of future tests. The cause of the observed reliability difference must be ascertained and understood in engineering terms. Sometimes this can be difficult.

Ultimately, all of our understanding should be based upon real knowledge (scientific, human, etc.). The statistical methods can provide clues to help us to gain such knowledge. The quotation from Kendall and Stuart on the inside of the front cover should be the motto for all statistics applications.

BIBLIOGRAPHY

Helpful introductory sources

1. M. J. Moroney, *Facts from Figures*. Penguin (1965).
2. R. Langley, *Practical Statistics Simply Explained*, 2nd Edn. Pan (1979).
3. W. A. Shewhart, *The Economic Control of Manufactured Products*. Van Nostround (1931).
4. W. W. Hines and D. C. Montgomery, *Probability and Statistics in Engineering and Management Science* (2nd Edn). Wiley (1980).
5. L. Mann, *Applied Engineering Statistics*. Barnes & Noble (1970).
6. C. Chatfield, *Statistics for Technology*, 2nd Edn. Chapman & Hall (1978).
7. T. P. Ryan, *Statistical Methods for Quality Improvement*. J. Wiley (1989).
8. D. Wheeler, *Understanding Variation: The key to Managing Chaos*. SPC Press (1993).
9. D. C. Montgomery, G. C. Runger and M. Hubele, *Engineering Statistics* (2nd Edition). J. Wiley (2000).
10. D. R. Cox and P. A. W. Lewis, *The Statistical Analysis of Series of Events*. Chapman & Hall (1966).
11. W. J. Conover, *Practical Non-parametric Statistics*. Wiley (1971).

More advanced works

12. A. J. Duncan, *Quality Control and Industrial Statistics*, 5th Edn. Irwin, Homewood, Illinois (1986).
13. A. H. Bowker and G. J. Lieberman, *Engineering Statistics*, 2nd Edn. Prentice-Hall (1972).
14. G. J. Hahn and S. S. Shapiro, *Statistical Models in Engineering*. Wiley (1967).
15. N. L. Johnson and S. Kotz, *Distributions in Statistics*, 4 Vols. Wiley (1970).
16. M. L. Shooman, *Probabilistic Reliability—An Engineering Approach*. McGraw-Hill (1968).

17. E. S. Pearson and H. O. Hartley, *Biometrika Tables for Statisticians*. Biometrika Trust, London.
18. R. E. Barlow and F. Proschan, *Mathematical Theory of Reliability*. Wiley (1965).
19. R. E. Barlow and F. Proschan, *Statistical Theory of Reliability and Life Testing*. Holt, Rinehart & Winston (1975).
20. N. Mann, R. E. Schafer and N. D. Singpurwalla, *Methods for Statistical Analysis of Reliability and Life Data*. Wiley (1974).
21. M. Hollander and D. A. Wolfe, *Non-parametric Statistical Methods*. Wiley (1973).
22. H. Goldberg, *Extending the Limits of Reliability Theory*. Wiley (1981).
23. H. F. Martz and R. A. Waller, *Bayesian Reliability Analysis*. Wiley (1982).
24. W. Nelson, *Applied Life Data Analysis*. Wiley (1982).
25. MINITAB (General purpose statistics training and application software). Minitab Inc. 3081 Enterprise Drive, PA 16801, USA. (*The MINITAB Handbook* (Duxbury Press, Boston) is an excellent introduction to basic statistics).
26. F. W. Breyfogle III, *Statistical Methods for Testing, Development and Manufacture*. J. Wiley (1992).
27. G. J. Hahn and W. Q. Meeker, *Statistical Intervals, a Guide for Practitioners*. J. Wiley (1991).
28. S. E. Seitano, *Engineering Uncertainty and Risk Analysis*. Hydroscience Inc (2001).

QUESTIONS

1. In the test firing of a missile, there are some events that are known to cause the missile to fail to reach its target. These events are listed below; together with their approximate probabilities of occurrence during a flight:

Event	Probability
(A_1) Cloud reflection	0.0001
(A_2) Precipitation	0.005
(A_3) Target evasion	0.002
(A_4) Electronic countermeasures	0.04

The probabilities of failure if these events occur are:

$$P(F/A_1) = 0.3; \ P(F/A_2) = 0.01; \ P(F/A_3) = 0.005; \ P(F/A_4) = 0.0002.$$

Use Bayes' theorem (Eqn 2.10) to calculate the probability of each of these events being the cause in the event of a missile failing to reach its target.

$F(0) = \dfrac{25!}{0! \, 25!} (0.02)^0 (0.98)^{25}$

2. For a device with a failure probability of 0.02 when subjected to a specific test environment, use the binomial distribution to calculate the probabilities that a test sample of 25 devices will contain (a) no failures; (b) one failure; (c) more than one failure.

3. Repeat question 2 for a failure probability of 0.2.

4. Repeat questions 2 and 3 using the Poisson approximation to the binomial, and comment on the answers.

5. One of your suppliers has belatedly realized that about 10 per cent of the batches of a particular component recently supplied to you have a manufacturing fault that has

reduced their reliability. There is no external or visual means of identifying these substandard components. Batch identity has, however, been maintained, so your problem is to sort batches that have this fault ('bad' batches) from the rest ('good' batches). An accelerated test has been devised such that components from good batches have a failure probability of 0.02 whereas those from bad batches have a failure probability of 0.2. A sampling plan has been devised as follows:

(1) Take a random sample of 25 items from each unknown batch, and subject them to the test.
(2) If there are 0 or 1 failed components, decide that the batch is a good one.
(3) If there are two or more failures, decide that the batch is a bad one.

There are risks in this procedure. In particular, there are (i) the risk of deciding that a good batch is bad; and (ii) the risk of deciding that a bad batch is good. Use Bayes' theorem and your answers to questions 2 and 3 to evaluate these risks.

6. (a) Explain the circumstances in which you would expect observed failure times to conform to an exponential distribution.

(b) Explain the relationship between the exponential and Poisson distributions in a reliability context.

(c) For equipment with an MTBF of 350 h calculate the probability of surviving a 200 h mission without failure.

7. A railway train is fitted with three engine/transmission units that can be assumed to exhibit constant hazard with a mean life of 200 h. In a 15 h working day, calculate the probability of a train having: (a) no failed engine/transmission units, (b) not more than one failed unit, (c) not more than two failed units.

8. (a) Later (in Chapter 12) we shall show that failures in a complex repairable system can often be assumed to result from a Poisson process. In such a system, 1053 h of testing have been accumulated, with failures at 334 h and 891 h. Calculate (i) the current estimate of the system failure rate; (ii) the current estimate of the mean time between failures (MTBF); and (iii) the lower 90 per cent confidence limit for the MTBF.

(b) For the above system, if there is a specification requirement that the MTBF shall be at least 500 h, and this must be demonstrated at 90 per cent confidence, how much more test running of the system, without further failure, is required?

9. (a) Explain, using sketches where necessary, the meanings of the following terms used in describing the reliability behaviour of components; and show clearly how they are related to each other: (i) lifetime probability density function; (ii) cumulative distribution function; (iii) reliability function; (iv) hazard function.

(b) Write down the expression for the cumulative distribution function (c.d.f.) of the two-parameter Weibull distribution. Define its parameters and produce sketches to show how changing their values influences the c.d.f. and the hazard function.

10. Ten components were tested to failure. The ordered ages at failure (hours) were:

    70.9; 87.2; 101.7; 104.2; 106.2; 111.4; 112.6; 116.7; 143.0; 150.9.

    (a) On the assumption that these times to failure are normally distributed, estimate the component reliability and the hazard function (i) at age 100 h; and (ii) at age 150 h.

    (b) Use a Kolmogorov–Smirnov test to see whether it is reasonable to assume normality.

11. A flywheel is retained on a shaft by five bolts, which are each tightened to a specified torque of 50 ± 5 Nm. A sample of 20 assemblies was checked for bolt torque. The results from the 100 bolts had a mean of 47.2 Nm and a standard deviation of 1.38 Nm.

    (a) Assuming that torques are normally distributed, estimate the proportion below 45 Nm.

    (b) For a given assembly, what is the probability of (i) there being no bolts below 45 Nm; (ii) there being at least one bolt below 45 Nm; (iii) there being fewer than two bolts above 45 Nm; (iv) all five bolts being below 45 Nm.

    (c) In the overall sample of 100 bolts, four were actually found with torques below 45 Nm. (i) Comment on the comparison between this result and your answer to (a) above. (ii) Use this result to obtain a 90 per cent two-sided confidence interval for the proportion below 45 Nm.

    (d) Explain the meaning of the confidence interval in c (ii) above as you would to an intelligent, but non-technically-minded, manager.

    (e) The lowest torque bolt in each assembly was identified. For these 20 bolts, the mean torque was 45.5 Nm and the standard deviation 0.88 Nm. Assuming an appropriate extreme-value distribution, calculate the probability that *on a given assembly* the lowest torque will be (i) below 45 Nm; (ii) below 44 Nm.

12. The following data are the times (hours) between successive failures in a machining centre:

    96; 81; 105; 34; 92; 81; 89; 138; 75; 156; 205; 111; 177.

    Calculate the trend statistic (Eqn 2.50) and test its significance.

13. Describe four ways in which the variation of an engineering parameter might be different from that based upon assuming that the *s*-normal distribution function is the correct model. Give an example in each case.

14. In most statistical applications the results that matter most relate to the behaviour of the majority of the population being studied. Why is this not the case in most engineering applications?

15. Ten items are put on test, until all have failed. The first failure occurs at 35,000 operating cycles. Regression analysis of the times to failure shows a good fit to a two-parameter Weibull distribution, and the distribution parameters are derived. The specification states that the item should have a B_{10} life of 30,000 cycles. Discuss the practical implications assuming that the test shows compliance with the specification.

16. In question 15, how might your judgement be influenced if the item is:

    a. A steel bolt subject to fatigue loading?

    b. A plastic component in a child's toy?

    c. A lighting unit?

    d. A bearing in a gearbox?

    e. A light-emitting diode?

    (You might like to try this question after studying Chapters 8 and 9).

17. Explain what is meant by 'statistical confidence'. How might statistical confidence derived from an experiment be modified in the light of engineering knowledge?

18. In Example 2.7, describe three factors that could invalidate the assumption that tyre failures occur independently of one another. In what ways might the assumption be more valid for car tyres?

3

Probability Plotting

INTRODUCTION

It is frequently useful in reliability engineering to determine which distribution best fits a set of data and to derive interval estimates of the distribution parameters. The mathematical basis for the approaches to these problems was covered in Chapter 2. Graphical estimation methods can greatly ease this task, and probability plotting papers have been developed for this purpose. These are based upon the cumulative distribution function (c.d.f.) $F(x)$ of the distribution concerned. The axes of probability plotting papers are transformed in such a way that the true c.d.f. plots as a straight line. Therefore if the plotted data can be fitted by a straight line, the data fit the appropriate distribution. Further constructions permit the distribution parameters to be estimated. Thus data can be evaluated quickly, without a detailed knowledge of the statistical mathematics being necessary. This facilitates analysis and presentation of data.

The methods described in this chapter can be used to analyse any appropriate data, such as dimensional or parameter measurements. However, their use for analysing reliability time-to-failure data will be emphasized.

Note that probability plotting methods to derive time-to-failure distribution parameters are only applicable when the data are independently and identically distributed (IID). This is usually the case for non-repairable components and systems but may *not* be the case with failure data from repairable systems. Reliability data analysis for repairable systems will be covered in Chapter 12.

RANKING OF DATA

Mean ranking (normal probability plots)

Probability graph papers are based upon plots of the variable of interest against the cumulative percentage probability. The data therefore need to be ordered and the cumulative probability calculated. For example, consider the data on times of failure of 12 items (Table 3.1). For the first failure, the cumulative proportion is 1/12 or 8.3 per cent. For the second, the cumulative proportion is 2/12 or 16.7 per cent, and so

Table 3.1 Data on times to failure of 12 items

Order no.	Time to failure (h)	Cumulative per cent (c.d.f.)	Mean rank per cent (c.d.f.)	Median rank per cent (c.d.f.)
1	12.2	8.3 (1/12)	7.7	5.6
2	13.1	16.7 (2/12)	15.4	13.6
3	14.0	25.0 (3/12)	23.1	21.7
4	14.1	33.3 (4/12)	30.8	29.8
5	14.6	41.7 (5/12)	38.5	37.9
6	14.7	50.0 (6/12)	46.2	45.9
7	14.7	58.3 (7/12)	53.8	54.0
8	15.1	66.7 (8/12)	61.5	62.1
9	15.7	75.0 (9/12)	69.2	70.2
10	15.8	83.3 (10/12)	76.9	78.3
11	16.3	91.7 (11/12)	84.6	86.4
12	16.9	100.0 (12/12)	92.3	94.4

on to 12/12 or 100 per cent for the twelfth failure. However, for probability plotting, it is better to make an adjustment to allow for the fact that each failure represents a point on a distribution. Thus considering the whole population of which the 12 items represent a sample, the times by which 8.3, 16.7, ..., 100 per cent will have failed in several samples of 12 will all be randomly distributed. However, the data in Table 3.1 show a bias, in that the first failure is shown much further from the zero cumulative percentage point than is the last from 100 per cent (in fact it coincides). To overcome this, and thus to improve the accuracy of the estimation, *mean* or *median* ranking of cumulative percentages is used for probability plotting.

The usual method for mean ranking is to use $(n + 1)$ in the denominator, instead of n, when calculating the cumulative percentage position. Thus in Table 3.1 the cumulative percentages (mean ranks) would be

$$\frac{100}{12 + 1} = 7.7 \text{ per cent}$$

$$\frac{200}{12 + 1} = 15.4 \text{ per cent}$$

.
.
.

$$\frac{1200}{12 + 1} = 92.3 \text{ per cent}$$

These data are shown plotted on normal probability paper in Figure 3.1 (circles). The plotted points show a reasonably close fit to the straight line drawn 'by eye'. Therefore, we can say that the data appear to fit the cumulative s-normal distribution represented by the line.

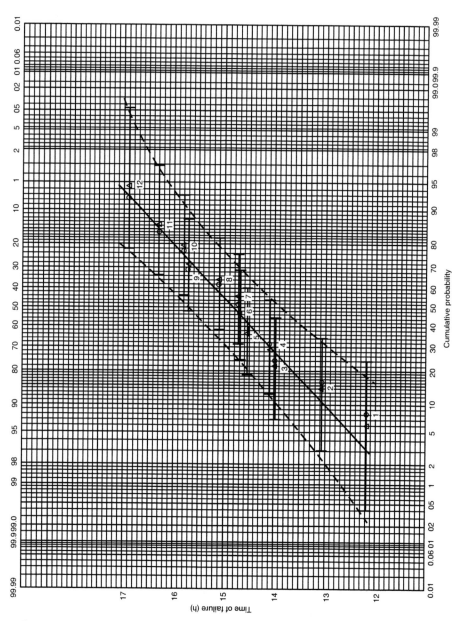

Figure 3.1 *s*-normal probability plot

Median ranking

Mean ranking is the appropriate method for a symmetrical distribution, such as the s-normal. However, for skewed distributions median ranking provides a better correction. The most common approximation used for median ranking is that due to Benard. The ith rank value is given by

$$r_i = \frac{i - 0.3}{n + 0.4}$$

where i is the ith order value and n is the sample size.

Median ranking is the method most frequently used in probability plotting, particularly if the data are known not to be normally distributed. Appendix 6 gives tables of median ranks for samples up to 50. Figure 3.1 shows the data of Table 3.1 plotted using median ranks (triangles). The effect is to alter the slope slightly, but the linearity is not affected.

PROBABILITY PLOTTING TECHNIQUES

Fitting the line

Having ranked and plotted the data, the question that often arises is *What is the best straight line fit to the data?* (assuming, of course, that there is a reasonable straight line fit). There can be a certain amount of subjectivity or even a temptation to adjust the line a little to fit a preconception. Two basic considerations apply:

1. Since the plotted data are cumulative, the points at the high cumulative proportion end of the plot are more important than the early points.
2. If the conclusions to be derived can be influenced by subjective differences in drawing the line through the data, then either there are insufficient data or there is not a good enough fit to the assumed distribution. (The same argument applies to choices of which ranking method should be used.)

Normally, a line which gives a good 'eyeball fit' to the plotted data is satisfactory, and more refined methods will give results which differ by amounts well within the 5 per cent s-significance level. However, a simple and accurate procedure to use, if rather more objectivity is desired, is to place a transparent rule on the last point and draw a line through this point such that an equal number of points lie to either side of the line.

Statistical goodness-of-fit tests can be applied, as described in Chapter 2. It is important to realize that cumulative probability plots are to a large extent self-aligning, since succeeding points can only continue upwards and to the right. Goodness-of-fit tests will nearly always indicate good correlation with any straight line drawn through such points. Therefore it is sometimes prudent to test the data against the assumed distribution using the methods described. Engineering knowledge can also be of value in determining the likely best distribution.

s-confidence limits on plotted data

s-confidence limits can be evaluated for the rank values of sample data, as shown in Chapter 2. Appendix 6 includes 5 per cent and 95 per cent median ranks, which can be used to plot the upper and lower 95 per cent s-confidence limits on the c.d.f. These are shown plotted for the data of Table 3.1 in Fig. 3.1.

Censored data

When analysing field or test failure data, we may have to consider samples in which not all items have failed. A proportion of items may have survived to the end of the test, or some may have been withdrawn prior to the end of the test. Data on such a sample are called *censored* data. Items which have survived or which have been removed are called *suspended items*. The derivation of median ranks for censored data is carried out as follows:

1. List order number (i) of failed items.
2. List increasing ordered sequence of life values (t_i) of failed items.
3. Against each failed item, list the number of items which have survived to a time between that of the previous failure and this failure (or between $t = 0$ and the first failure).
4. For each failed item, calculate the *mean order number* i_{t_i} using the formula

$$i_{t_i} = i_{t_{i-1}} + N_{t_i} \tag{3.1}$$

where

$$N_{t_i} = \frac{(n+1) - i_{t_{i-1}}}{1 + (n - \text{number of preceding items})} \tag{3.2}$$

in which n is sample size.

5. Calculate median rank for each failed item, using the formula

$$r_{t_i} = \frac{i_{t_i} - 0.3}{n + 0.4} \text{ per cent} \tag{3.3}$$

Example 3.1

Of a sample of 50 items, the running times (h) for the failed and surviving items are as shown in Table 3.2 (12 failures, 38 survivors).

From Eqns (3.1) to (3.3),

$$N_1 = \frac{50 + 1 - 0}{1 + (50 - 2)} = 1.04$$

$$i_1 = i_0 + N_1 = 0 + 1.04 = 1.04$$

Table 3.2 Running times for failed and surviving items for a sample of 50 items

Item[a]	Running time (h)	Item[a]	Running time (h)	Item[a]	Running time (h)
S1	40	S10	141	S25	165
S2	51	S11	147	S26	165
F1	54	S12	147	S27	166
F2	70	S13	150	S28	166
S3	73	F9	153	S29	166
S4	73	S14	153	S30	168
S5	80	S15	153	S31	168
F3	85	S16	154	S32	171
S6	90	S17	156	F11	173
F4	96	S18	156	S33	177
S7	102	S19	156	S34	181
F5	108	S20	158	S35	185
F6	118	S21	158	S36	188
S8	128	S22	158	F12	200
S9	128	F10	161	S37	202
F7	132	S23	162	S38	205
F8	141	S24	162		

[a]S is survivor, F is failure.

$$N_2 = \frac{50 + 1 - 1.04}{1 + (50 - 3)} = 1.04$$

$$i_2 = i_1 + N_2 = 1.04 + 1.04 = 2.08$$

$$r_1 = \frac{i_1 - 0.3}{n + 0.4} = \frac{1.04 - 0.3}{50 + 0.4} = 1.46 \text{ per cent}$$

$$r_2 = \frac{i_2 - 0.3}{n + 0.4} = \frac{2.08 - 0.3}{50 + 0.4} = 3.53 \text{ per cent}$$

These data (Table 3.3) are shown plotted on Weibull probability paper in Fig. 3.4.

'Sudden death' testing

'Sudden death' testing is a useful technique for obtaining test data quickly on a relatively large sample. The sample is divided into a number of equal groups

Table 3.3   Derivation of median ranks for Example 3.1

Failure order no.	Ordered life values, t_i(h)	No. of intervening survivors, $S(t_i)$	Mean order, i_{t_i}	Median rank, r_{t_i} per cent
1	54	2	1.04	1.46
2	70	0	2.08	3.53
3	85	3	3.19	5.74
4	96	1	4.33	7.99
5	108	1	5.50	10.31
6	118	0	6.66	12.62
7	132	2	7.89	15.07
8	141	0	8.71	16.69
9	153	4	10.12	19.47
10	161	9	12.15	23.51
11	173	10	16.47	32.08
12	200	4	25.10	49.21

(n groups, each of k items) by random selection. All items are then tested until one fails in each group. These failures are then used as the data points on the life plot.

However, the line obtained represents the distribution of times to the first failure, in a sample in which all the non-failed items are suspended. The times to failure of the failed items are ranked and plotted, giving a sudden death line. For example, Fig. 3.2 shows a plot of the times to first failure of ten samples of six each, with the rank values based on n (10). The first failure in a sample of six has a median rank of 10.91 per cent (Appendix 6). To derive the population median line, it is necessary to drop a vertical from the point where the sudden death line intersects the 50 per cent failed ordinate. A line drawn through the point where this intersects the 10.91 per cent ordinate, parallel to the sudden death line, will represent the population life distribution.

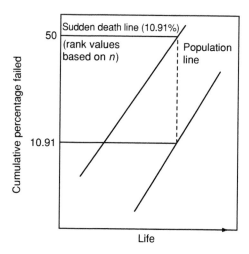

Figure 3.2   Sudden death testing

Mixed distributions

A plot of failure data may relate to one failure mode or to multiple failure modes within an item. If a straight line does not fit the failure data, particularly if an obvious change of slope is apparent, the causes of failure should be investigated to ensure that not more than one failure mode is included. For example, after a certain length of time on test, a second failure mode may become apparent, or an item may have two superimposed failure modes. In such cases each failure mode must be isolated and analysed separately, after the data on each have been appropriately censored (i.e. each A-type failure is a survival as far as B-type failures are concerned, and vice versa). However, such separation is appropriate only if the failure processes are s-independent, i.e. there is no interaction.

Screened samples

In QC and reliability work we often deal with samples which have been screened in some way. For example, machined parts will have been inspected, oversize and undersize parts removed, and electronic parts may have been screened so that no zero-life ('dead-on-arrival') parts exist. Screening can show up on probability plots, as a curvature in the tails. For example, a plot of time to failure of a fatigue specimen will normally be curved since quality control will have removed items of very low strength. In other words, there will be a positive minimum life. We will cover this aspect in more detail in the section on Weibull plotting of life data.

Estimation of parameters

We can also estimate the distribution parameters from the plot. In the example of the normal probability plot given above (Fig. 3.1) the mean life can be seen to be 14.8 h. The standard deviation can be derived by measuring the life interval covered by 34 per cent either side of the 50 per cent point on the probability scale. In this case it is 1.55 h. The 95 per cent s-confidence limits on the mean are 14 h and 15.7 h. The estimation of the parameters of other distributions will be covered in later sections of this chapter, dealing with the Weibull and extreme value distributions.

LOGNORMAL PROBABILITY PLOTS

Lognormal probability paper is used in exactly the same way as normal probability paper, except that the ordinate scale is constructed logarithmically.

WEIBULL PROBABILITY PLOTS

In reliability engineering Weibull probability analysis is the most widely used of the techniques described, due to the flexibility of the Weibull distribution in describing a number of failure patterns.

Weibull transformation

The axes of Weibull probability paper are derived by performing the transformation

$$R(t) = 1 - F(t) = \exp\left[-\left(\frac{t}{\eta}\right)^{\beta}\right]$$

assuming γ (failure-free time) is 0 (i.e. we use the two-parameter Weibull distribution). Therefore

$$\frac{1}{1 - F(t)} = \exp\left(\frac{t}{\eta}\right)^{\beta}$$

taking double logarithms,

$$\ln\,\ln\frac{1}{1 - F(t)} = \beta\ln\,t - \beta\,\ln\,\eta$$

This is a straight line of the form $y = ax + b$.

Thus the Weibull paper is constructed by having a log log reciprocal ordinate (y) scale, representing cumulative probability of failure (or failure percentage), and the abscissa (x) scale is a log scale representing the life value. The slope of a straight line plotted on this paper will be β, the shape parameter.

Examples of Weibull probability paper are shown in Figures 3.3 and 3.4, as produced by Technical and Engineering Aids to Management—TEAM (USA) and Chartwell (UK). On the TEAM paper the slope β is measured by constructing a line parallel to the line of plotted data, through the estimation point, and reading the value of β at the intersection with the β scale (1.6 in the example shown). The value of β is estimated on the Chartwell paper on the β scale by the intersection of the perpendicular from the estimation point to the line of the plotted data. η is given by the life value at the intersection with the η estimator line (63.2 per cent cumulative failure). The Chartwell paper also includes a scale for estimating the mean life μ. This is derived by reading the survival probability from the P scale, where the β estimator line intersects, and then reading the life value relative to the probability on the cumulative percentage failure scale.

Another Weibull parameter often used is the B life. This is the life by which a given percentage of the population will have failed. For example, the B_{10} life is the life at which 10 per cent of the population will have failed. It is estimated by reading the appropriate life value against the intersection of the line of the plotted data with the 10 per cent cumulative failure line. For example, the B_{10} life in Fig. 3.4 is 110 h. ($\hat{B}_{10}$ indicates that this is an estimate from the sample data.)

Probability papers are available covering various decade ranges. For example, the two papers shown in Figs 3.3 and 3.4 cover 3 and 2 decades, respectively. If the correct decade range is not available, the data may still be plotted by allowing the plot to reappear on the left-hand side of the paper or by attaching more decade ranges to the right-hand side.

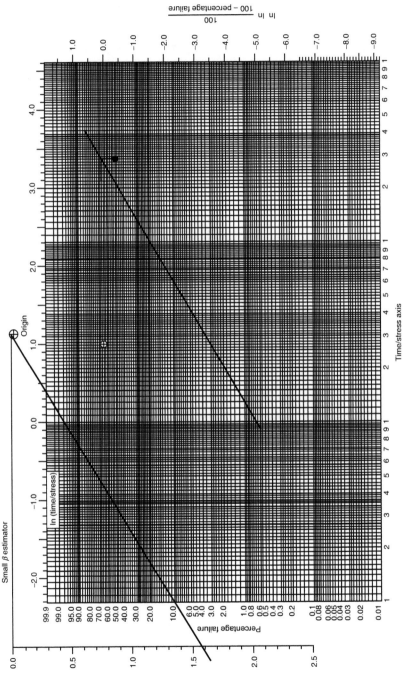

Figure 3.3 TEAM Weibull probability paper

TEAM Weibull probability(0.01–99.9) x 3-cycle logarithmic

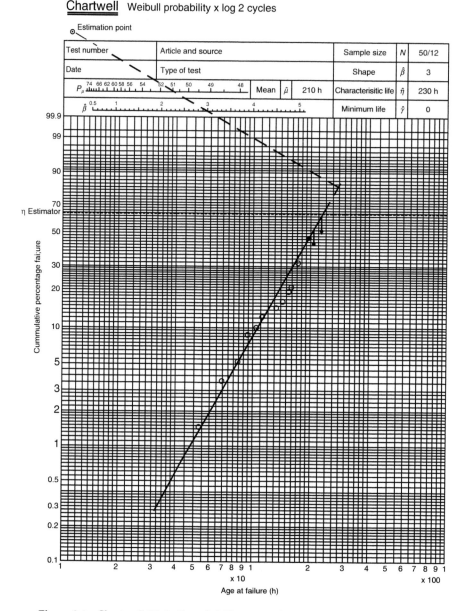

Figure 3.4   Chartwell Weibull probability paper (censored data—Example 3.1)

Failure-free life

The examples given so far all relate to cases where γ is 0, i.e. there is no failure-free time. Of course this is not necessarily the case in reliability work. If an item has a finite (positive) failure-free time under test, e.g. a fatigue test specimen, the failure data will plot as a curve, seen convex from above, since the transformation to achieve the Weibull scales assumes that the data fit a two-parameter distribution.

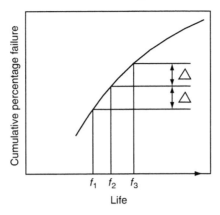

Figure 3.5 Life plot: failure-free time

The effect of a finite life is to shift the age-at-failure scale to the left. It is possible to have an apparent negative value for γ, for example, if the items under test had accumulated unrecorded operating time before the start of the test. In this case the curve will appear concave from above.

In order to transform such data into a straight line for analysis of the Weibull parameters it is necessary to transform the three-parameter distribution into a two-parameter one. To achieve this, equally spaced horizontals are drawn as shown in Fig. 3.5 with the middle line intersecting the mid-point of the plotted data. The corresponding values of t_1, t_2 and t_3 are then read off the life scale. Then

$$\hat{\gamma} = t_2 - \frac{(t_3 - t_2)(t_2 - t_1)}{(t_3 - t_2) - (t_2 - t_1)}$$

The data are then replotted, with the value of $\hat{\gamma}$ so derived subtracted from each life value. The life parameters estimated from the plot must then have the value of $\hat{\gamma}$ added to give the true life values.

Discretion must be used in interpreting data that do not plot as a straight line, since the cause of the non-linearity may be due to the existence of mixed distributions or because the data do not fit the Weibull distribution. It is quite likely to be due simply to the randomness in the sample. The failure mechanisms must be studied, and engineering judgement used, to ensure that the correct interpretations are made. It is a common error to assume that, because a straight line provides a reasonably good fit to the data, that there is no failure-free life. However, in most cases wearout failure modes do exhibit a finite failure-free life. Therefore a value for γ can sometimes be estimated from knowledge of the product and its application. Alternatively, the time to first failure is often a satisfactory estimate of γ. In these cases the procedure described above is not necessary. Generally, data on several failure modes in a system are likely to fit a two-parameter distribution ($\gamma = 0$), but single wearout failure modes are more likely to have positive values of γ.

Table 3.4 Failure data on 15 electrical relays tested to failure

Order no.	Median rank per cent	Cycles to failure ($\times 10^5$)	Cycles to failure, adjusted ($\times 10^5$)
1	4.5	6.2	2.1
2	10.9	9.0	4.9
3	17.4	10.2	6.1
4	23.9	12.1	8.0
5	30.5	12.6	8.5
6	37.0	14.4	10.3
7	43.5	14.7	10.6
8	50.0	16.1	12.0
9	56.5	18.6	14.5
10	63.0	20.5	16.4
11	69.5	20.6	16.5
12	76.1	23.0	18.9
13	82.6	26.7	22.6
14	89.1	27.6	23.5
15	95.5	34.4	30.3

Example 3.2

Failure data on 15 electrical relays tested to failure, where the criterion for failure is a predetermined level of contact resistance, are shown in Table 3.4.

The cycles to failure are plotted in Fig. 3.6. The best fit appears to be the curve shown indicating a finite minimum life. This conclusion is accepted as being valid from knowledge of the relay contact behaviour.

The geometrical mid-point on the percentage failure scale gives $t_2 = 13 \times 10^5$. t_1 and t_3 are 6.2×10^5 and 34.4×10^5, respectively.

$$\hat{\gamma} = t_2 - \frac{(t_3 - t_2)(t_2 - t_1)}{(t_3 - t_2) - (t_2 - t_1)}$$

$$= \left[13 - \frac{(34.4 - 13)(13 - 6.7)}{(34.4 - 13) - (13 - 6.7)} \right] \times 10^5 = 4.07 \times 10^5$$

Subtracting 4.1 from each of the values in the third column gives the results in the fourth column. When these are plotted in Figure 3.6 the adjusted data can be fitted by the straight line shown. From this we can read the other Weibull distribution parameters: $\hat{\beta} = 1.7$, $\hat{\eta} = 16 \times 10^5$ cycles.

s-confidence limits on the shape parameter

It is important to derive s-confidence limits on the shape parameter β, since decisions may be based upon this value. Figure 3.7 gives factors F_β against sample size for different s-confidence levels (99, 95, 90 per cent) on β. The upper and lower s-confidence limits are then

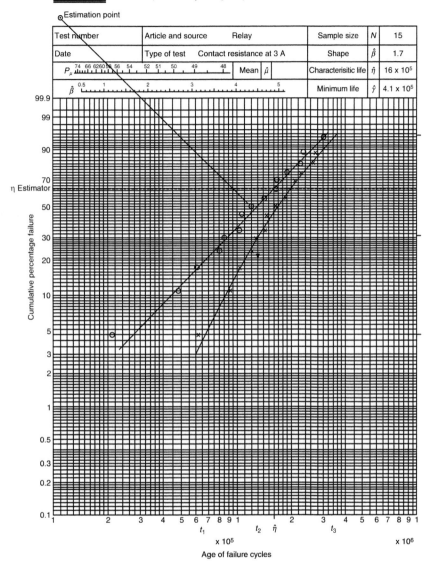

Figure 3.6 Relay life plot (see Example 3.2)

$$\beta_u = \hat{\beta} F_\beta$$

$$\beta_l = \hat{\beta} \frac{1}{F_\beta}$$

Example 3.3

Derive the upper and lower confidence limits if $n = 10$, $\beta = 1.6$ for $C = 90$ per cent (double-sided).

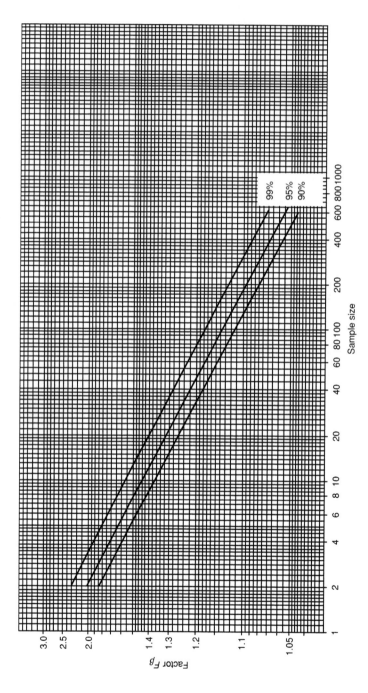

Figure 3.7 Confidence limits for shape parameter β for different confidence values

From Fig. 3.7, $F_\beta = 1.37$; therefore,

$$\beta_u = 1.6 \times 1.37 = 2.19$$

$$\beta_1 = \frac{1.6}{1.37} = 1.17$$

i.e. we have a 90 per cent s-confidence that $2.19 \geq \beta \geq 1.17$.

EXTREME VALUE PROBABILITY PLOTTING

Extreme value statistics are described in Chapter 2. TEAM extreme value probability paper is shown in Fig. 3.8. It uses a linear scale for the random variable and a cumulative probability scale. Two auxiliary scales give the *reduced variate, y,* and the *return period*. The return period is

$$\frac{1}{(1 - p)}$$

where p is the cumulative probability that a future observed value will not exceed the previous value. Thus the return period represents the period over which there is a 50 per cent probability of the event occurring. Both scales are read by drawing the vertical through the point of intersection of the value of the variable with the line of the plotted data.

Example 3.4

The breaking strength of a long wire was tested, using a sample of 15 equal lengths. Since the strength of a wire can be considered to depend upon the existence of imperfections, the extreme value distribution of minimum values might be an appropriate fit. The results are shown in Table 3.5 and plotted in Fig. 3.8.

 Since this would be a distribution of minimum value it will be left-skewed, and the data are therefore arranged in descending order of magnitude. Plotting the data the other way around would generate a convex curve, viewed from above. A plot of an extreme value distribution of maximum values would be made with the data in ascending order. In this case the probability scale represents the cumulative probability that the breaking strength will be greater than the value indicated, so that there is a 95 per cent probability that the strength of a wire *of this length* will be greater than 46 N. If the wire is longer it will be likely to be weaker, since the probability of its containing extreme value imperfections will be higher.

 The mode $\hat{\mu}$ is estimated directly from the plot. In this case $\hat{\mu} = 69.5$ N.$\hat{\sigma}$, the measure of variability, can be derived from the expression (see page 44):

$$\text{Mean} = \hat{\mu} - 0.577\hat{\sigma}$$

$$\hat{\sigma} = \frac{\hat{\mu} - \text{mean}}{0.577}$$

$$= \frac{69.5 - 65}{0.577} = 7.8 \, \text{N}$$

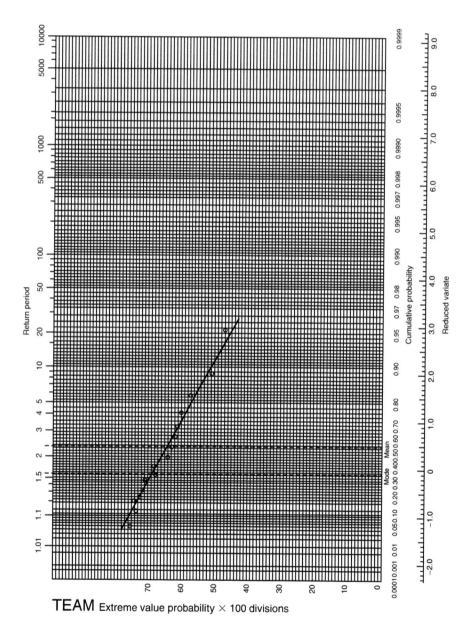

Figure 3.8 TEAM extreme value probability paper (see also Example 3.4)

Table 3.5 Breaking strengths of 15 samples of wire of equal length.

Rank order	Cumulative probability per cent (median ranks, c.d.f)	Breaking strength (N)
1	4.5	76
2	10.9	75
3	17.4	75
4	23.9	72.5
5	30.4	72
6	36.9	69
7	43.4	69
8	50.0	65
9	56.5	64
10	63.0	63
11	69.5	62
12	76.0	61
13	82.5	58
14	89.0	52
15	95.4	48

The value of $1/\sigma$ is sometimes referred to as the *Gumbel slope*. The return period $1/(1 - p)$ represents the average value of x (e.g. number of times, or time) between recurrences of a greater (or lesser) value than represented by the return period. In Example 3.4 there is a 50–50 chance that a length of wire of breaking strength less than 45 N will occur in a batch of 26 lengths. The return period is used in forecasting the likelihood of extreme events such as floods and high winds, but it is not often referred to in reliability work.

HAZARD PLOTTING

Instead of plotting the cumulative proportion of items failed, we can plot the cumulative hazard function, using hazard papers. This technique has particular advantages when dealing with censored data, as will be shown.

The theory underlying hazard plotting is as follows. From Chapter 2, we remember that the hazard rate $h(x)$ is given by

$$h(x) = \frac{f(x)}{1 - F(x)} \tag{2.21}$$

$$F(x) = \int_0^x f(x)dx \tag{2.19}$$

The cumulative hazard function

$$H(x) = \int_0^x h(x)dx$$

$$= \int_0^x \frac{f(x)}{1 - F(x)} dx \qquad (2.22)$$

$$= -\ln\left[1 - F(x)\right]$$

This relationship allows the derivation of cumulative hazard plotting paper. For example, for the Weibull distribution,

$$h(x) = \frac{\beta}{\eta^\beta} x^{\beta-1}$$

Therefore,

$$H(x) = \left(\frac{x}{\eta}\right)^\beta$$

If H is the cumulative hazard value,

$$\log x = \frac{1}{\beta}\log H + \log \eta$$

Thus, Weibull hazard paper is a log–log paper. The nominal slope of the plotted data would be $1/\beta$. When $H = 1$, $x = \eta$. Figure 3.9 is an example of Weibull hazard paper. Hazard papers for normal, lognormal, exponential and type I extreme value distributions are also available.

The hazard plotting procedure is:

1. Tabulate the times to failure in order.

2. For each failure, calculate the hazard interval

$$\Delta H_i = \frac{1}{\text{number of items remaining after previous failures/censoring}}$$

3. For each failure, calculate the cumulative hazard function

$$H = \Delta H_1 + \Delta H_2 + \cdots + \Delta H_n$$

4. Plot the cumulative hazard against life value on the chosen hazard paper.

As with Weibull probability paper, Weibull hazard paper is based upon the two-parameter Weibull distribution, and a non-zero value of failure-free life will result in a curved cumulative hazard line. A value for $\hat{\gamma}$ can be derived as described on page 84, except that equal length divisions are measured off on the abscissa instead of on the ordinate if the geometrical construction method is being used.

Hazard plotting should be used in preference to cumulative probability plotting when dealing with censored data, or when the data include multiple failure modes and we wish to analyse the overall failure distribution, as well as individual failure modes. (This is, of course, a type of censoring.) One tabulation may then be used, rather than a separate tabulation for each failure mode (see Example 3.5). However, since statistical software with plotting capabilities can automatically deal with censoring, probability plotting using such software is no more difficult with censored data.

Example 3.5

Table 3.6 gives distance to failure on a vehicle shock absorber, taken from fleet records. The data are tabulated and treated as described above. F1 is the failure mode considered. Note that, compared with the method of dealing with censored data described earlier (page 84), the calculation of plotting positions is much easier. The data are shown plotted in Fig. 3.9.

In this example, for the failure mode F1, $\hat{\beta} = 2.6$, $\hat{\eta} = 29\,000$ km (the life equivalent to 100 per cent cumulative hazard). (This is the 63.2 per cent c.d.f. on the probability scale.)

Table 3.6 Distance to failure on a vehicle shock absorber

No.	Distance (km)	F1 Δ Hazard per cent	F1 Cumulative hazard per cent
1 (38)	6 700 (F1)	2.63 (1/38)	2.63
2 (37)	6 950		
3 (36)	7 820		
4 (35)	8 790		
5 (34)	9 120 (F1)	2.94 (1/34)	5.57
6 (33)	9 660		
7 (32)	9 820		
8 (31)	11 310		
9 (30)	11 690		
10 (29)	11 850		
11 (28)	11 880		
12 (27)	12 140		
13 (26)	12 200 (F1)	3.85	9.42
14 (25)	12 870		
15 (24)	13 150 (F1)	4.17	13.59
16 (23)	13 330		
17 (22)	13 470		
18 (21)	14 040		
19 (20)	14 300 (F1)	5.00	18.59
20 (19)	17 520 (F1)	5.26	23.85
21 (18)	17 540		
22 (17)	17 890		

Probability Plotting

Table 3.6 (*continued*)

No.	Distance (km)	F1 Δ Hazard per cent	F1 Cumulative hazard per cent
23 (16)	18 450		
24 (15)	18 960		
25 (14)	18 980		
26 (13)	19 410		
27 (12)	20 100 (F1)	8.33	32.18
28 (11)	20 100		
29 (10)	20 150		
30 (9)	20 320		
31 (8)	20 900 (F1)	12.50	44.68
32 (7)	22 700 (F1)	14.29	58.96
33 (6)	23 490		
34 (5)	26 510 (F1)	20.0	78.96
35 (4)	27 410		
36 (3)	27 490 (F1)	33.3	112.29
37 (2)	27 890		
38 (1)	28 100		

CHOOSING THE DISTRIBUTION AND ASSESSING THE RESULTS

The types of failure pattern and modes of failure which can be described by the continuous statistical distributions covered in Chapter 2 and in this chapter have been mentioned in the relevant sections. The distribution which is likely to provide the best fit to a set of data is not always readily apparent, and it can be useful to follow guidelines which should lead to the most revealing presentation.

The steps to be taken are:

1. Consider the situation. What do we expect the failure pattern to be? For example, do we expect an increasing or decreasing hazard rate, or a failure-free life?
2. Plot the ranked data on the appropriate chart (probability or hazard).
3. Draw the line of best fit after estimating the failure-free life if appropriate.
4. If there is a good fit to a reasonably large sample (≥ 10 data points), estimate the parameters and determine the 5 per cent and 95 per cent confidence limits.
5. If the estimated value of β is near 3.5, try normal paper. If β is between 2.5 and 3.5, try lognormal paper, but only if the sample size is large (> 20).
6. If a good Weibull fit cannot be obtained, and the conditions make a fit to an extreme value type I or type II distribution likely, try extreme value paper. If extreme value is unlikely, look for multiple failure modes.

Statistical goodness-of-fit tests should be applied to test the fit to the assumed underlying distributions, when finer judgements are to be made and when there is a large database. The refinement inherent in such tests is not of much value for sample sizes less than about 20.

Figure 3.9 TEAM Weibull hazard paper (see also Example 3.5)

If a constant hazard rate ($\beta = 1$) is apparent, this can be an indication that multiple failure modes exist or that the time-to-failure data are suspect. This is often the case with systems in which different parts have different ages and individual part operating times are not available. A constant hazard rate can also indicate that failures are due to external events, such as faulty use or maintenance.

Whilst wearout failure modes are characterized by $\beta > 1$, situations can arise in which wearout failures occur after a finite failure-free time, and a β value of ≤ 1 is obtained. This can happen when only a proportion of the sample is defective, the defect leading to failures after a finite time. The Weibull parameters of the wearout failure mode can be derived if it is possible to identify the defective items and analyse their life and failure data separately from that of the non-defective items.

It is very important that the physical nature of failures be investigated as part of any evaluation of failure data, if the results are to be used as a basis for engineering action, as opposed to logistic action. For example, if an assembly has several failure modes, but the failure data are being used to determine requirements for spares and repairs, then we do not need to investigate each mode and the parameters applicable to the mixed distributions are sufficient. If, however, we wish to understand the failure modes in order to make improvements, we must investigate all failures and analyse the distributions separately, since two or more failure modes may have markedly different parameters, yet generate overall failure data which are fitted by a distribution which is different to that of any of the underlying ones.

It is also important to ensure that the time axis chosen is relevant to the problem, or misleading results can be generated. For example, if a number of items are tested, and the running times recorded, the failure data could show different trends depending upon whether all times to failure are taken as cumulative times from when the test on the first item is started, or if individual times to failure are analysed. If the items start their tests at different elapsed or calendar times the results can also be misleading if not carefully handled. For example, a trend might be caused by exposure to changing weather conditions, in which case an analysis based solely on running time could conceal this information. The methods of exploratory data analysis, described in Chapter 12, can be applied when appropriate.

Finally, it is essential that practical engineering criteria are applied to the analysis and to interpreting the results. For example, the earliest time(s) to failure are clearly more important than later times; in most cases, so early failures should never be considered as "outliers" unless there is adequate engineering justification for doing so.

PROBABILITY PLOTTING FOR BINOMIAL DATA

Items such as alarm systems, standby power supply systems and missile components can often be most conveniently considered as either functioning or not when needed, and the reliability data will then be binomially distributed. In these cases we are not concerned with time (or cycles, etc.) to failure, but only with the probability of failure at the instant considered. Examples in reliability using the binomial distribution were covered in Chapter 2.

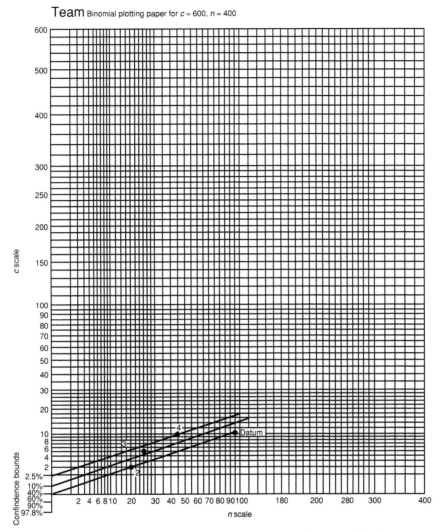

Figure 3.10　TEAM binomial plotting paper (see also Example 3.6)

Binomial probability plotting paper can be used for quickly comparing sets of data or for evaluating data against a standard. Figure 3.10 shows an example of binomial probability paper. It is constructed by having square root scales for the ordinate and the abscissa, so that points lying on a straight line through the origin have the same standard deviation.

To represent a binomial proportion on the paper, the *binomial split* is calculated and plotted, and this is connected to the origin. The binomial split is $p/(n-p)$, [or $c/(n-c)$ in the notation used on the TEAM paper], where $p(c)$ is the quantity with one attribute (e.g. defectives) and n is the sample size. For example, a proportion of 10 per cent would be plotted at 10 on the c scale and 90 on the n scale (not 100). This line then represents all samples with this proportion. s-confidence limits can be drawn parallel to the datum line through the appropriate points on the s-confidence scale.

Table 3.7   Results of tests on four samples

Trial	Sample size, n	Failures, c	$n - c$
1	32	5	27
2	20	2	18
3	20	5	15
4	55	10	45

Binomial papers with different scale ranges are available. Since the scales are square root, they may both be multiplied by multiples of 100 to provide factors of 10 in the variables.

Example 3.6

An item is required to be 90 per cent reliable. Four samples are tested and give results as shown in Table 3.7. Comment on the results.

Since not more than 10 per cent must be defective, the datum point is derived from the binomial split $10/(100 - 10) = 10/90$. The results are then plotted as shown in Fig. 3.10.

We can now use the s-confidence limit scale to interpret these results, by drawing lines parallel to the requirement line through the s-confidence value required. In this case we can see that trial 2 gives the required proportion. However, the result of trial 1 indicates with higher than 90 per cent s-confidence that it is not compliant with the requirement and trials 3 and 4 are non-compliant at the 97.5 per cent confidence level.

Binomial probability paper with an extended n scale can be used as a convenient method of monitoring test data for items with a low proportion failing, such as electronic components, as shown in Example 3.7.

Example 3.7

Two types of custom hybrid integrated circuit are used in an instrument. It is required that not more than 0.02 per cent of either type fail during production testing. Test data are shown in Table 3.8. 1 is added to the cumulative number of failures, because when plotting binomial data with a small value of p, this gives a better estimate of the population proportion $\hat{p}$. The failure data are shown plotted in Fig. 3.11. The plot shows that, at the 97.5 per cent confidence level, the type 1 circuit is outside the specification. The trend shows a statistically significant problem in lots 16 to 18. The type 2 circuit also has a higher failure ratio than specified, at the 90 per cent confidence level.

Binomial probability paper as produced by Chartwell is shown in Fig. 3.12. Sample data are plotted in the same way as on the TEAM paper. The circular scale gives the percentage defective, as $(1 - \sin^2 \alpha)$, where α is the value on the degrees scale. This value is read on the vertical scale as shown. For example, Fig. 3.12 shows a plotted binomial split of 10:150. This intersects at 75.7°, giving the percentage defective as $(1 - \sin^2 75.7°) = 6.1$ per cent. The arcsine transform is used to determine s-confidence limits, in conjunction with the standard error scale at the top of the chart. For

Table 3.8   Test data on two types of custom hybrid integrated circuit (see Example 3.7)

Lot no.	No. tested	Cum no. tested	No. failed type 1, c_1	Cum. $c_1 + 1$	No. failed type 2, c_2	Cum. $c_2 + 1$
1	1184	1184	1	2	0	1
2	1152	2336	0	2	1	2
3	1328	3664	1	3	0	2
4	1287	4951	0	3	0	2
5	1463	6414	0	3	1	3
6	1510	7924	1	4	0	3
7	1291	9215	1	5	2	5
8	1384	10 599	0	5	0	5
9	1570	12 169	0	5	0	5
10	1648	13 817	0	5	0	5
11	1586	15 403	2	7	1	6
12	1353	16 756	1	8	0	6
13	1422	18 178	0	8	0	6
14	1338	19 516	0	8	2	8
15	1546	21 062	1	9	0	8
16	1284	22 346	2	11	0	8
17	1172	23 518	2	13	0	8
18	1390	24 908	3	16	0	8
19	1521	26 429	0	16	1	9
20	1441	27 870	0	16	0	9

example, the approximate 2.5 per cent (single-sided) s-confidence limits are derived by constructing a circle about the binomial split point with a radius of two standard errors. The arcsine transform is then used in the same way to determine the s-confidence limits, in this case 10 per cent and 3 per cent. s-confidence limit lines can be drawn as on the TEAM paper by constructing lines parallel to the process line, tangential to the circle.

Bradley binomial paper

Bradley binomial plotting paper allows binomial data to be analysed and s-confidence limits to be derived graphically. Figure 3.13 shows TEAM Bradley binomial plotting paper. The data are plotted by drawing a straight line from the sample size scale to the origin, and the value of the observed probability is marked on this line. Figure 3.13 shows the plotted data for a sample of 55, of which 10 are defective, i.e. 0.18. The s-confidence limits for the population are derived by drawing a horizontal line to intersect the appropriate s-confidence level (or standard deviation) scale. In this case, the 90 per cent s-confidence limits for the defective proportion of the population would be 0.18 ± 0.086.

Bradley binomial paper allows a correction to be made to take account of finite population (lot) sizes, using the sample-to-lot ratio scale at the top of the graph. If in this case the lot size is 165, $n/N = \frac{1}{3} = 0.33$. The correction factor for the s-confidence limits is derived by connecting this point on the n/N scale to the origin and reading the value on the correction factor scale (0.82). The corrected 90 per cent s-confidence limits for the proportion defective for this lot size are therefore $0.18 \pm (0.086 \times 0.82)$.

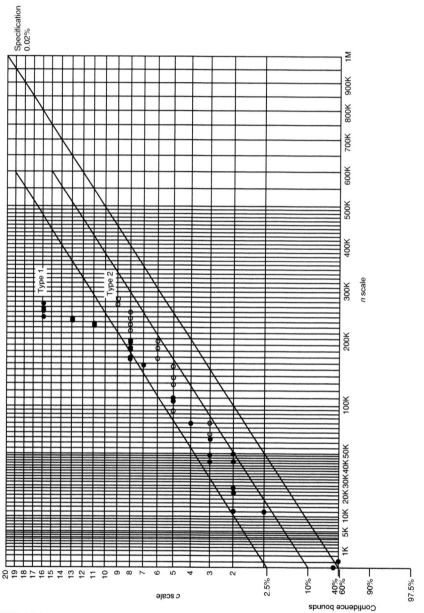

Figure 3.11   Failure plot—hybrid circuits (see Example 3.8)

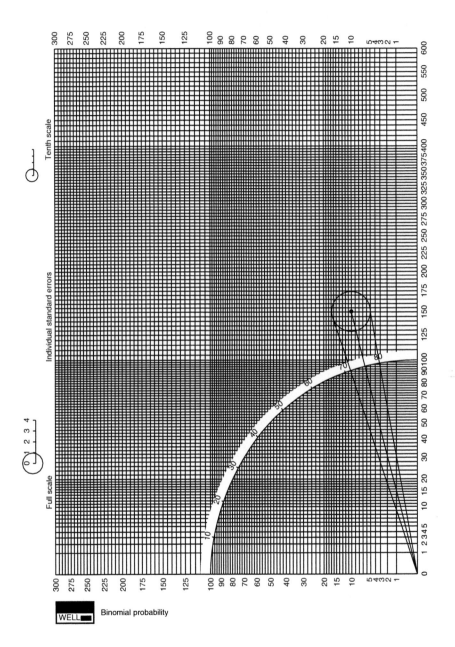

Figure 3.12 Chartwell binomial probability paper

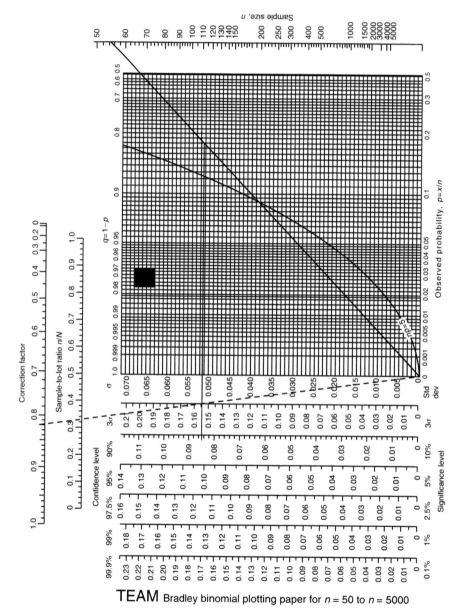

Figure 3.13   TEAM Bradley binomial paper

The *np* line on the Bradley paper indicates the boundary for points to the left of which the accuracy of *s*-confidence limit estimation is reduced.

If necessary, the appropriate Bradley plotting paper should be selected in relation to the sample size and the observed probability, to ensure that the plotted data fall to the right of the *np* line.

CONCLUSIONS

Probability plotting methods can be very useful for analysing reliability data, in the particular circumstances of items which can only fail once. This distinction is important, since the method, and the underlying statistical theory, assumes that the individual times to failure are independently and identically distributed (IID). That is, that failure of one item cannot affect the likelihood of, or time to, failure of any other item in the population, and that the distribution of times to failure is the same for all the failures considered. If these conditions do not hold, then the method can give misleading results. Techniques for analysing reliability data, when these conditions do not apply, are described in Chapter 12.

Probability plotting can also be used for analysing other IID data, such as sample measurements in quality control.

The software for statistical analysis, described in Chapter 2, includes probability plotting capabilities. Specialist software for probability plotting is also available: see the book's homepage (Preface P. xxi).

The comments on the practical interpretation of statistical data made in Chapter 2 (page 73) apply equally to the use of plotting techniques.

BIBLIOGRAPHY

1. J. R. King, *Probability Charts for Decision Making*. Industrial Press, New York (1971).
2. Notes on plotting techniques: TEAM, *Easy Analysis Methods*. Technical and Engineering Aids to Management, Tamworth, New Hampshire.
3. K. A. Dey, *Practical Statistical Analysis for the Reliability Engineer*. Reliability Analysis Center Report SOAR 2 (Rome Air Development Center). Available from National Technical Information Service (NTIS), Springfield, Virginia.
4. W. Nelson, *Applied Life Data Analysis*. Wiley (1982).

QUESTIONS

1. (a) Explain briefly (and in non-mathematical terms) why, in Weibull probability plotting, the *i*th ordered failure in a sample of *n* is plotted at the 'median rank' value rather than simply at $i n$.

   (b) Planned replacement is to be applied to a roller bearing in a critical application: the bearing is to be replaced at its B_{10} life. Ten bearings were put on a test rig and subjected to realistic operating and environmental conditions. The first seven failures occured at 370, 830, 950, 1380, 1550 and 1570 hours of operation, after which the test was discontinued. Estimate the B_{10} life from (i) the data alone; (ii) using a normal plot. Comment on any discrepancy.

2. Twenty switches were put on a rig test. The first 15 failures occurred at the following numbers of cycles of operation: 420, 890, 1090, 1120, 1400, 1810, 1815, 2150, 2500, 2510, 3030, 3290, 3330, 3710 and 4250. Plot the data on Weibull paper and give estimates of (i) the shape parameter β; (ii) the mean life μ; (iii) the B_{10} life; (iv) the upper and lower 90 per cent confidence limits on β; (v) the upper and lower 90 per cent confidence limits on the probability of failure at 2500 operations. Finally, use the Kolmogorov-Smirnov test to assess the goodness-of-fit of your data.

3. A manufacturer has progressively purchased five nominally identical special-purpose machines. There have been several instances of failure of a critical component common to each of these machines. The failure log shows the following:

Machine A—component failures at cumulative machine ages 451, 753, 968, 1255, 1429, 1997 and 2388 hours; the machine is currently at 3000 h.

Machine B—component failures at cumulative machine ages 234, 305, 509, 910, 1716 and 2110 hours; currently at 2500 h.

Machine C—component failures at cumulative machine ages 782 and 1806 hours currently at 2000 h.

Machine D—one failure at 780 hours; currently at 1500 h.

Machine E—There have been no failures; currently at 1000 h.

Use hazard plotting to estimate the Weibull shape and scale parameters.

(Hints): (1) Remember the censorings between the last recorded failures. (2) Ordinary double-log graph paper can be used if the TEAM paper is unavailable. To estimate β, simply measure the slope of the plot with a ruler (distance up divided by distance along). To estimate η, use the fact that cumulative hazard $= 1.0$ when $t = \eta$. Another approach could be to use ordinary Weibull plotting paper, using the relationship

$F(t) = 1 - \exp(-H(t))$

where $F(t)$ and $H(t)$ are respectively the c.d.f. (proportion failed) and the cumulative hazard at age (t)

4. A pump used in large quantities in a sewage works is causing problems owing to sudden and complete failures. There are two dominant failure modes, impeller failure (I) and motor failure(M). These modes are thought to be independent. Records were kept for 12 of these pumps, as follows:

Pump no	Age at failure (h)	Failure mode
1	1180	M
2	6320	M
3	1030	I
4	120	M
5	2800	I
6	970	I
7	2150	I
8	700	M
9	640	I
10	1600	I
11	520	M
12	1090	I

Estimate Weibull parameters for each mode of failure.

(Hint: consider each mode of failure separately, treating the other mode as censorings)

5. A type of pump used in reactors at a chemical processing plant operates under severe conditions and experiences frequent failures. A particular site uses five reactors which, on delivery, were fitted with new pumps.; There is one pump per reactor: When a pump fails, it is returned to the manufacturer in exchange for a reconditioned unit. The replacement pumps are claimed to be 'good as new'. The reactors have been operating concurrently for 2750 h since the plant was commissioned, with the following pump failure history:

Reactor 1—at 932, 1374 and 1997 h.
Reactor 2—at 1566, 2122 and 2456 h.
Reactor 3—at 1781 h.
Reactor 4—at 1309, 1652, 2337 and 2595 h.
Reactor 5—at 1270 and 1928 h.

(a) Calculate the Laplace statistic (Eqn 2.50) describing the behaviour of the total population of pumps as a point process.
(b) Estimate Weibull parameters for both new and reconditioned pumps.
(c) In the light of your answers to (a) and (b), comment on the claim made by the pump manufacturer.

6. A vehicle manufacturer has decided to increase the warranty period on its products from 12 to 36 months. In an effort to predict the implications of this move, it has obtained failure data from some selected fleets of vehicles over prolonged periods of operation. The data below relate to one particular fleet of 20 vehicles, showing the calendar months in which a particular component failed in each vehicle. (There is one of these components per vehicle.) If there are no entries under 'Component failure dates', that vehicle had zero failures. The data are up-to-date to October 1995, at which data all the vehicles were still in use. The component currently gives about 3 per cent failures under warranty.

Vehicle	Start date	Component failure dates
1	May 93	—
2	Jun 93	Nov 93, Jul 94
3	Jun 93	—
4	Aug 93	Feb 95
5	Oct 93	Jan 95
6	Oct 93	Oct 94
7	Oct 93	Feb 95
8	Oct 93	Sep 94, Mar 95
9	Nov 93	—
10	Nov 93	Dec 94
11	Dec 93	Jan 95, Jul 95
12	Jan 94	—
13	Jan 94	—
14	Feb 94	—
15	Feb 94	—

16	Jul 94	—
17	Jul 94	Feb 95
18	Aug 94	—
19	Dec 94	Aug 95
20	Feb 95	—

Use any suitable method to estimate the scale and shape parameters of a fitted Weibull distribution, and comment on the implications for the proposed increase in warranty period.

7. The data below refer to failures of a troublesome component installed in five similar photocopiers in a large office.

Machine no	Cumulative copies at which failures occurred	Current cumulative copies
1	13 600, 49 000	64 300
2	16 000, 23 800, 40 400	60 000
3	18 700, 28 900	46 700
4	22 200	40 600
5	6 500	39 000

(a) Estimate the parameters of a Weibull distribution describing the data.

(b) Calculate a Laplace trend statistic (Eqn 2.50) and, in the light of its value, discuss whether your answer to part (a) is meaningful.

8. The data below relate to failures of terminations in a sample of 20 semiconductor devices. Each failure results from breaking of either the wire (W) or the bond (B), whichever is the weaker. The specification requirement is that fewer than 1 per cent of terminations shall have strengths of less than 500 mg.

Failure load (mg)	B or W	Failure load (mg)	B or W
550	B	1250	B
750	W	1350	W
950	B	1450	B
950	W	1450	B
1150	W	1450	W
1150	B	1550	B
1150	B	1550	W
1150	W	1550	W
1150	W	1850	W
1250	B	2050	B

Estimate Weibull parameters for (i) termination strength; (ii) wire strength; (iii) bond strength. Comment on the results.

9. The triggering device for an automobile airbag restraint is required to have a reliability of 95 per cent when tested on an impact test rig. This reliability is to be established at 90 per cent confidence. The test is destructive. For a series of tests on a specific design, failures occurred on the 11th, 43rd, 49th, 75th, 102nd and 144th samples. All other samples passed. The 160th sample has just been tested.

(a) Plot the results on suitable binomial plotting paper (Chartwell 5703, or suitable TEAM paper, as described in the text) and estimate the current reliability and its 90 per cent lower confidence limit. Hence decide if the test requirements have been met. If not, estimate the amount of further testing required if there is no improvement to the device.

(b) Check your answer using Fig. 2.14.

(c) Consider whether doing the further testing indicated in (a) is sensible.

4

Load–strength Interference

INTRODUCTION

In Chapter 1 we set out the premise that a common cause of failure results from the situation when the applied load exceeds the strength. Load and strength are considered in the widest sense. 'Load' might refer to a mechanical stress, a voltage or internally generated stresses such as temperature. 'Strength' might refer to any resisting physical property, such as hardness, strength, melting point or adhesion. Examples are:

1. A bearing fails when the internally generated loads (due perhaps to roughness, loss of lubricity, etc.) exceed the local strength, causing fracture, overheating or seizure.
2. A transistor gate in an integrated circuit fails when the voltage applied causes a local current density, and hence temperature rise, above the melting point of the conductor or semiconductor material.
3. A hydraulic valve fails when the seal cannot withstand the applied pressure without leaking excessively.
4. A shaft fractures when torque exceeds strength.

Therefore, if we design so that strength exceeds load, we should not have failures. This is the normal approach to design, in which the designer considers the likely extreme values of load and strength, and ensures that an adequate safety factor is provided.

Additional factors of safety may be applied, e.g. as defined in pressure vessel design codes or electronic component derating rules. This approach is usually effective. Nevertheless, some failures do occur which can be represented by the load–strength model. By our definition, either the load was then too high or the strength too low. Since load and strength were considered in the design, what went wrong?

DISTRIBUTED LOAD AND STRENGTH

For most products neither load nor strength are fixed, but are distributed statistically. This is shown in Fig. 4.1(a). Each distribution has a mean value, denoted

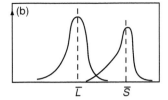

Figure 4.1 Distributed load and strength: (a) non-overlapping distributions, (b) overlapping distributions

by $\bar{L}$ or $\bar{S}$, and a standard deviation, denoted by σ_L or σ_S. If an event occurs in which the two distributions overlap, i.e. an item at the extreme weak end of the strength distribution is subjected to a load at the extreme high end of the load distribution, such that the 'tails' of the distributions overlap, failure will occur. This situation is shown in Fig. 4.1(b).

For distributed load and strength, we define two factors, the *safety margin* (SM),

$$\text{SM} = \frac{\bar{S} - \bar{L}}{(\sigma_S^2 + \sigma_L^2)^{1/2}} \tag{4.1}$$

and the *loading roughness* (LR),

$$\text{LR} = \frac{\sigma_L}{(\sigma_S^2 + \sigma_L^2)^{1/2}} \tag{4.2}$$

The safety margin is the relative separation of the mean values of load and strength and the loading roughness is the standard deviation of the load; both are relative to the combined standard deviation of the load and strength distributions.

The safety margin and loading roughness allow us, in theory, to analyse the way in which load and strength distributions interfere, and so generate a probability of failure. By contrast, a traditional deterministic safety factor, based upon mean or maximum/minimum values, does not allow a reliability estimate to be made. On the other hand, good data on load and strength properties are very often not available. Other practical difficulties arise in applying the theory, and engineers must always be alert to the fact that people, materials and the environment will not necessarily be constrained to the statistical models being used. The rest of this chapter will describe the theoretical basis of load-strength interference analysis. The theory must be applied with care and with full awareness of the practical limitations. These are discussed later.

Some examples of different safety margin/loading roughness situations are shown in Fig. 4.2. Figure 4.2(a) shows a highly reliable situation: narrow distributions of load and strength, low loading roughness and a large safety margin. If we can control the spread of strength and load, and provide such a high safety margin, the design should be intrinsically failure-free. (Note that we are considering situations where the mean strength remains constant, i.e. there is no strength degradation with time. We will cover strength degradation later.) This is the concept applied in most designs, particularly of critical components such as civil engineering

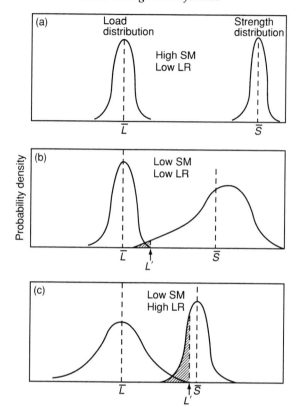

Figure 4.2   Effect of safety margin and loading roughness. Load L' causes failure of a proportion of items indicated by the shaded area

structures and pressure vessels. We apply a safety margin which experience shows to be adequate; we control quality, dimensions, etc., to limit the strength variations, and the load variation is either naturally or artificially constrained.

Figure 4.2(b) shows a situation where loading roughness is low, but due to a large standard deviation of the strength distribution the safety margin is low. Extreme load events will cause failure of weak items. However, only a small proportion of items will fail when subjected to extreme loads. This is typical of a situation where quality control methods cannot conveniently reduce the standard deviation of the strength distribution (e.g. in electronic device manufacture, where 100 per cent visual and mechanical inspection is not always feasible). In this case deliberate overstress can be applied to cause weak items to fail, thus leaving a population with a strength distribution which is truncated to the left (Fig. 4.3). The overlap is thus eliminated and the reliability of the surviving population is increased. This is the justification for high stress burn-in of electronic devices, proof-testing of pressure vessels, etc. Note that the overstress test only destroys weak items. It must not cause weakening (strength degradation) of good items. A decreasing hazard or failure rate (DFR) is characteristic of this situation, since as weak items fail ('infant mortality'), the population strength increases and the failure rate decreases.

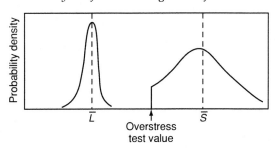

Figure 4.3 Truncation of strength distribution by screening

Figure 4.2(c) shows a low safety margin and high loading roughness due to a wide spread of the load distribution. This is a difficult situation from the reliability point of view, since an extreme stress event will cause a large proportion of the population to fail. Therefore, it is not economical to improve population reliability by screening out items likely to fail at these stresses. The options left are to increase the safety margin by increasing the mean strength, which might be expensive, or to devise means to curtail the load distribution. This is achieved in practice by devices such as current limiters in electronic circuits or pressure relief valves and dampers in pneumatic and hydraulic systems.

ANALYSIS OF LOAD–STRENGTH INTERFERENCE

The reliability of a part, for a discrete load application, is the probability that the strength exceeds the load:

$$R = P(S > L)$$

$$= \int_0^\infty f_L(L) \left[\int_L^\infty f_S(S) \, dS \right] dL$$

$$= \int_0^\infty f_S(S) \left[\int_0^S f_L(L) \, dL \right] dS$$

where $f_S(S)$ is the p.d.f. of strength and $f_L(L)$ is the p.d.f. of load.

Also, if we define y as $S - L$, y is a random variable such that

$$R = P(y > 0)$$

$$= \int_0^\infty \int_0^\infty f_S(y + L) \, f(L) \, dL \, dy \qquad (4.3)$$

s-normally distributed strength and load

If we consider *s*-normally distributed strength and load, so that the c.d.f.s are

$$F_L(L) = \Phi \left(\frac{L - \bar{L}}{\sigma_L} \right)$$

$$F_S(S) = \Phi\left(\frac{S - \bar{S}}{\sigma_S}\right)$$

if $y = S - L$, then $\bar{y} = \bar{S} - \bar{L}$ and $\sigma_y = (\sigma_S^2 + \sigma_L^2)^{1/2}$. So

$$R = P(y > 0)$$

(4.4)

$$= \Phi\left(\frac{\bar{S} - \bar{L}}{\sigma_y}\right)$$

Therefore, the reliability can be determined by finding the value of the standard cumulative s-normal variate from the s-normal distribution tables. The reliability can be expressed as

$$R = \Phi\left[\frac{\bar{S} - \bar{L}}{(\sigma_S^2 + \sigma_L^2)^{1/2}}\right]$$

(4.5)

$$= \Phi(SM) \text{ (from Eqn 4.1)}$$

Example 4.1

A component has a strength which is s-normally distributed, with a mean value of 5000 N and a standard deviation of 400 N. The load it has to withstand is also s-normally distributed, with a mean value of 3500 N and a standard deviation of 400 N. What is the reliability per load application?

The safety margin is

$$\frac{5000 - 3500}{(400^2 + 400^2)^{1/2}} = 2.65$$

From Appendix 1,

$$\Phi(2.65) = 0.996$$

Other distributions of load and strength

The integrals for other distributions of load and strength can be derived in a similar way. For example, we may need to evaluate the reliability of an item whose strength is Weibull-distributed, when subjected to loads that are extreme-value-distributed. These integrals are somewhat complex and the reader is referred to References 1 and 2 in the Bibliography for details and for tables of the evaluated integrals.

EFFECT OF SAFETY MARGIN AND LOADING ROUGHNESS ON RELIABILITY (MULTIPLE LOAD APPLICATIONS)

For multiple load applications:

$$R = \int_0^\infty f_S(S) \left[\int_0^S f_L(L)\, dL \right]^n dS$$

where n is the number of load applications.

Reliability now becomes a function of safety margin and loading roughness, and not just of safety margin. This complex integral cannot be reduced to a formula as Eqn (4.5), but can be evaluated using computerized numerical methods.

Figure 4.4 shows the effects of different values of safety margin and loading roughness on the failure probability per load application for large values of n, when both load and strength are s-normally distributed. The dotted line shows the single load application case (from Eqn 4.5). Note that the single load case is less reliable per load application than is the multiple load case.

Since the load applications are s-independent, reliability over n load applications is given by

$$R = (1 - p)^n \quad \text{(from Eqn 2.2)}$$

where p is the probability of failure per load application.

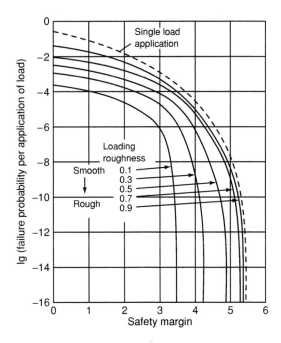

Figure 4.4 Failure probability–safety margin curves when both load and strength are s-normally distributed (for large n and $n = 1$)

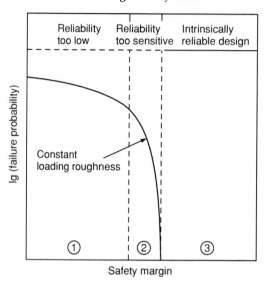

Figure 4.5   Characteristic regions of a typical failure probability–safety margin curve

For small values of p the binomial approximation allows us to simplify this to

$$R \approx 1 - np \tag{4.6}$$

The reliability for multiple load applications can then be derived, if we know the number of applications, having used Fig. 4.4 to derive the value of p. Once the safety margin exceeds a value of 3 to 5, depending upon the value of loading roughness, the failure probability becomes infinitesimal. The item can then be said to be *intrinsically reliable*. There is an intermediate region in which failure probability is very sensitive to changes in loading roughness or safety margin, whilst at low safety margins the failure probability is high. Figure 4.5 shows these characteristic regions. Similar curves can be derived for other distributions of load and strength. Figures 4.6 and 4.7 shows the failure probability—safety margin curves for smooth and rough loading situations for Weibull-distributed load and strength. These show that if the distributions are skewed so that there is considerable interference, high safety margins are necessary for high reliability. For example, Fig. 4.6 shows that, even for a low loading roughness of 0.3, a safety margin of at least 5.5 is required to ensure intrinsic reliability, when we have a right-skewed load distribution and a left-skewed strength distribution. If the loading roughness is high (Fig. 4.7), the safety margin required is 8. These curves illustrate the sensitivity of reliability to safety margin, loading roughness and the load and strength distributions.

Approximate equivalent safety margin and loading roughness values for Weibull-distributed load and strength situations can be derived by using the following approximations for the standard deviation and the mean of the 'equivalent' normal distribution (i.e. a normal distribution, the cumulative values of which coincide with the cumulative asymmetric Weibull distribution at two points in the tails: 0.9 and

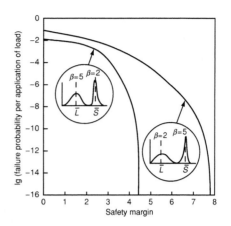

Figure 4.6 Failure probability–safety margin curves for asymmetric distributions (loading roughness = 0.3)

Figure 4.7 Failure probability–safety margin curves for asymmetric distributions (loading roughness = 0.9)

0.99 in the load distribution and 0.01 and 0.10 in the strength distribution), related to the characteristic life of the Weibull distributions. The approximations are

$$\sigma' = 0.3 \, \eta \tag{4.7}$$

$$\mu' = 0.9 \, \eta \tag{4.8}$$

(See Reference 3 in the Bibliography.)

When there is a finite minimum strength γ_S, then the mean strength

$$\bar{S} = \gamma_S + \mu'$$

Two examples of this approach to design are given to illustrate the application to electronic and mechanical engineering.

Example 4.2 (electronic)

A design of a power amplifier uses a single transistor in the output. It is required to provide an intrinsically reliable design, but in order to reduce the number of component types in the system the choice of transistor types is limited. The amplifier must operate reliably at 50 °C.

An analysis of the load demand on the amplifier gives the results in Fig. 4.8. The mean ranking of the load test data is given in Table 4.1. A type 2N2904 transistor is selected. For this device the maximum rated power dissipation is 0.6 W at 25 °C.

The load test data are shown plotted on normal probability paper (Fig. 4.9). A very good fit to the normal distribution is apparent, within the range of the data. The parameters for the load distribution are

Figure 4.8 Load data (sampled at 10 s intervals)

Table 4.1 Mean ranking of load test data

Power (W)	Cumulative percentage time (c.d.f.)
0.1	5
0.2	25
0.3	80
0.4	98.5
0.5	99.95

$$\bar{L} = 0.25 \, \text{W}$$

$$\sigma_L = 0.065 \, \text{W}$$

However, in this case we must consider the combined effects of power dissipation and elevated temperature. The temperature derating guidelines for the 2N2904 transistor advise 3.43 mW K^{-1} linear derating. Since we require the amplifier to operate at 50 °C, the equivalent combined load distribution is s-normal, with the same SD, but with a mean which is (25×3.43) mW $= 0.086$ W higher (line 1a in Fig. 4.9). The mean load $\bar{L}$ is now 0.33 W, with an unchanged standard deviation of 0.065 W.

To derive the strength distribution, 100 transistors were tested at 25 °C ambient, for 10 s at each power level (step stress), giving failure data as shown in Table 4.2. These data plot as a straight line (line 2 in Fig. 4.9), indicating a normal distribution with mean power at failure (strength) of 0.97 W and SD of 0.14 W. The data are plotted against the reciprocal (top) probability scale, as we are concerned with the tail of the distribution which intersects the load distribution.

Combining the load–strength data gives

$$\text{LR} = \frac{\sigma_L}{(\sigma_S^2 + \sigma_L^2)^{1/2}} = 0.42$$

$$\text{SM} = \frac{\bar{S} - \bar{L}}{(\sigma_S^2 + \sigma_L^2)^{1/2}} = 4.15$$

Therefore

$$R = \Phi(\text{SM})$$

$$= 0.999\,983$$

Table 4.2 Failure data for 100 transistors

Power (W)	Number failed	Cumulative percentage failure (c.d.f.) (mean ranking)
0.1	0	0
0.2	0	0
0.3	0	0
0.4	0	0
0.5	0	0
0.6	0	0
0.7	2	2
0.8	8	10
0.9	17	25
1.0	35	59
1.2	30	89
1.3	10	99

This is the reliability per application of load for a single load application. Figure 4.4 shows that for multiple load applications (large n), the failure probability per load application (p) is about 10^{-11} (zone 2 of Fig. 4.5). Over 10^6 load applications the reliability would be about 0.999 99 (from Eqn 4.6).

On the data given, therefore, this would be an unreliable design. (The analysis has been performed on the basis of best estimates. It could have been based upon a lower s-confidence limit for the strength distribution—line 2a in Fig. 4.9.)

In practice, in a case of this type the temperature and load derating guidelines described in Chapter 9 would normally be used. In fact, to use a transistor at very nearly its maximum temperature and load rating (in this case the measured highest load is 0.5 W at 25 °C, equivalent to nearly 0.6 W at 50 °C) is not good design practice, and derating factors of 0.5 to 0.8 are typical for transistor applications. The example illustrates the importance of adequate derating for a typical electronic component. The approach to this problem can also be criticized on the grounds that:

1. The failure (strength) data are sparse at the 'weak' end of the distribution. It is likely that batch-to-batch differences would be more important than the test data shown, and screening could be applied to eliminate weak devices from the population.
2. The extrapolation of the load distribution beyond the 0.5 W recorded peak level is dangerous, and this extrapolation would need to be tempered by engineering judgement and knowledge of the application.

Example 4.3 (mechanical)

A connecting rod must transmit a tension load which trials show to be Weibull distributed, with the following parameters: $\beta_L = 2.0$, characteristic load, $\eta_L = 4000\,\text{N}$, minimum load, $\gamma_L = 0$. Tests on the material to be used show a Weibull strength distribution as follows: $\beta_S = 5.0$, characteristic strength, $\eta_S = 6.8 \times 10^8$ N m^{-2}, minimum strength, $\gamma_S = 1.9 \times 10^8$ N m^{-2}. A large number of components are to be manufactured, so it is not feasible to test each one. What cross-section would

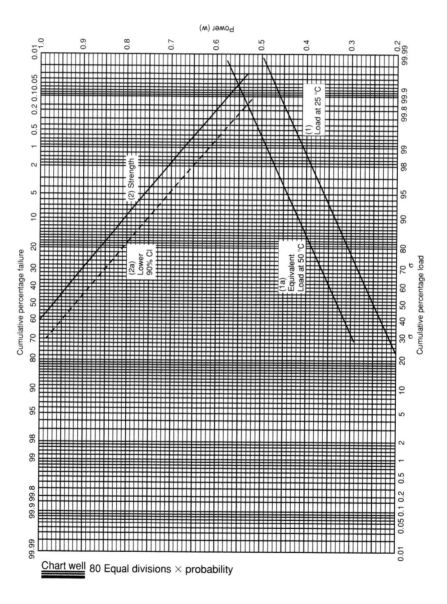

Figure 4.9 Transistor load–strength plots (see Example 4.2)

ensure intrinsic reliability, assuming that manufacturing quality control will be able to ensure that the material strength is maintained and that the connecting rods are required to operate only for a period well within the fatigue life?

The loading roughness can be derived by using Eqns (4.7) and (4.8). That is:

$$\sigma' = 0.3\,\eta$$
$$\mu' = 0.9\,\eta$$

Mean strength $= \gamma_S + \mu_S$

Therefore,

$$\sigma_S = 0.3 \times 6.8 \times 10^8 = 2.04 \times 10^8\,\mathrm{N\,m^{-2}}$$
$$\mu_S = 0.9 \times 6.8 \times 10^8 = 6.12 \times 10^8\,\mathrm{N\,m^{-2}}$$

Mean material strength $= (1.9 + 6.12) \times 10^8 = 8.02 \times 10^8\,\mathrm{N\,m^{-2}}$

$$\sigma_L = 0.3 \times 4000 = 1200\,\mathrm{N}$$
$$\mu_L = \bar{L} = 0.9 \times 4000 = 3600\,\mathrm{N}$$

The mean value of the strength of the rod

$$\bar{S} = \frac{\pi d^2}{4} \times \text{mean material strength}$$

where d is the diameter. The standard deviation of the strength distribution is

$$\frac{\pi d^2}{4}\,\sigma_S$$

Therefore, the loading roughness (LR) is given by

$$\frac{\sigma_L}{(\sigma_S^2 + \sigma_L^2)^{1/2}} = \frac{1200}{\{[2.04 \times (\pi d^2/4) \times 10^8]^2 + 1200^2\}^{1/2}}$$

and the safety margin (SM) by

$$\frac{\bar{S} - \bar{L}}{(\sigma_S^2 + \sigma_L^2)^{1/2}} = \frac{[8.02 \times (\pi d^2/4) \times 10^8] - 3600}{\{[2.04 \times (\pi d^2/4) \times 10^8]^2 + 1200^2\}^{1/2}}$$

Figures 4.6 and 4.7 show that for $\beta_L = 2.0$ and $\beta_S = 5.0$, we need a safety margin of at least 5, depending upon the loading roughness, for intrinsic reliability. We can derive a suitable strength value, and hence the correct diameter, by iteration.

Table 4.3 Results of substituting values of d in the equations for SM and LR

$d(\times 10^{-3}\,\text{m})$	5	6	7	8
SM	2.90	3.24	3.43	3.56
LR	0.29	0.20	0.15	0.12

First, select a mean strength, $\bar{S} = 6 \times \bar{L}$, say. Then

$$\bar{S} = 6 \times 3600 = 21\,600\,\text{N}$$

Therefore,

$$21\,600 = \frac{\pi d^2}{4} \times 8.02 \times 10^8$$

Thus, $d = 5.35 \times 10^{-3}\,\text{m}$

Substituting values for d in the equations for SM and LR, we can produce Table 4.3.

Since these results are derived by considering the equivalent normal distributions of load and strength, we can use Fig. 4.4 to ascertain the appropriate safety margin/loading roughness combination for intrinsic reliability. An LR of 0.3 requires an SM > 4.5 for intrinsic reliability, and therefore the 5 mm diameter would be too small. On the other hand, the 7 mm diameter gives an SM of 3.43 at an LR of 0.15, which is just about within the intrinsically reliable zone. Therefore a diameter of at least 7 mm should be chosen.

The load and strength distributions are shown plotted in Fig. 4.10, with the strength calculated for a 7 mm diameter. The plot shows the margin available between the cumulative 99.9 per cent load probability (10 000 N) and the 0.1 per cent cumulative strength probability (15 000 N). Also, the cumulative probability of the load exceeding the minimum strength of 7300 N is about 96.5 per cent which is the reliability per load application for minimum strength items.

PRACTICAL ASPECTS

The examples illustrate some of the limitations of the statistical engineering approach to design. The main difficulty is that, in attempting to take account of variability, we are introducing assumptions that might not be tenable, e.g. by extrapolating the load and strength data to the very low probability tails of the assumed population distributions. The comments in Chapter 2 (page 84) apply strongly, as do those on the interpretation of probability plots (Chapter 3 page 100). We must therefore use engineering knowledge to support the analysis, and use the statistical approach to cater for engineering uncertainty, or when we have good statistical data. For example, in many mechanical engineering applications good data exist or can be obtained on load distributions, such as wind loads on structures, gust loads on aircraft or the loads on automotive suspension components. We will call such loading situations 'predictable'.

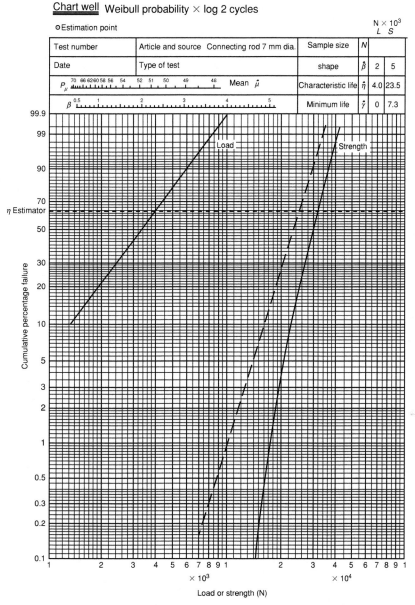

Figure 4.10   Connecting rod load–strength plots

On the other hand, some loading situations are much more uncertain, particularly when they can vary markedly between applications. Electronic circuits subject to transient overload due to the use of faulty procedures or because of the failure of a protective system, or a motor bearing used in a hand power drill, represent cases in which the high extremes of the load distribution can be very uncertain. The distribution may be multimodal, with high loads showing peaks, for instance when there is resonance. We will call this loading situation 'unpredictable'.

Obviously it will not always be easy to make a definite classification; for example, we can make an unpredictable load distribution predictable if we can collect sufficient data. The methods described above are meaningful if applied in predictable loading situations. (Strength distributions are more often predictable, unless there is progressive strength reduction, which we will cover later.) However, if the loading is very unpredictable the probability estimates will be very uncertain. When loading is unpredictable we must revert to traditional methods. This does not mean that we cannot achieve high reliability in this way. However, evolving a reliable design is likely to be more expensive, since it is necessary either to deliberately overdesign or to improve the design in the light of experience. The traditional safety factors derived as a result of this experience ensure that a new design will be reliable, provided that the new application does not represent too far an extrapolation.

Instead of considering the distributions of load and strength, we can use discrete maximum/minimum values in appropriate cases. For example, we can use a simple lowest strength value if this can be assured by quality control. In the case of Example 4.3, we could have decided that in practice the strength would not be less than 20 000 N, which would have given a safety factor of 2 above the 99.9 per cent load c.d.f. In other cases we might also assume that for practical purposes the load is curtailed, as in situations where the load is applied by a system with an upper limit of power, such as a hydraulic ram or a human operator. If the load and strength distributions are both curtailed, the traditional safety factor approach is adequate, provided that other constraints such as weight or cost do not make a higher risk design necessary.

The statistical engineering approach can lead to overdesign if it is applied without regard to real curtailment of the distributions. Conversely, traditional deterministic safety factor approaches can result in overdesign when weight or cost reduction must take priority.

In many cases, other design requirements (such as for stiffness) provide intrinsic reliability. The techniques described above should therefore be used when it is necessary to assess the risk of failure in marginal or critical applications.

In this chapter we have taken no account of the possibility of strength reduction with time or with cyclic loading. The methods described above are only relevant when we can ignore strength reduction, for instance if the item is to be operated well within the safe fatigue life or if no weakening is expected to occur. Reliability and life analysis in the presence of strength degradation is covered in Chapter 8.

Finally, it is important that reliability estimates that are made using these methods are treated only as very rough, order-of-magnitude figures.

BIBLIOGRAPHY

1. K. C. Kapur and L. R. Lamberson, *Reliability in Engineering Design*. Wiley (1977).
2. C. Lipson and N. J. Sheth, *Statistical Design and Analysis of Engineering Experiments*. McGraw-Hill Kogakusha (1973).
3. A. D. S. Carter, *Mechanical Reliability*, 2nd Edn. Macmillan (1986).

QUESTIONS

1. Describe the nature of the load and strength distributions in four practical engineering situations (use sketches to show the shapes and locations of the distributions). Comment on each situation in relation to the predictability of failures and reliability, and in relation to the methods that can be used to reduce the probabilities of failure.

2. (a) Give the formulae for safety margin and loading roughness in situations where the load applied to an item and the strength of the item are assumed to be *s*-normally distributed.

   (b) Sketch the relationship between failure probability and safety margin for different values of loading roughness, indicating approximate values for the parameters.

3. (a) If loads are applied randomly to randomly selected items, when both the loads and strengths are normally distributed, what is the expression for the reliability per load application?

   (b) Describe and comment on the factors that influence the accuracy of reliability predictions made using this approach.

4. Describe two examples each from mechanical and electronic engineering by which extreme load and strength values are curtailed, in practical engineering design and manufacture.

5. If the tests described in Example 4.2 were repeated, and

   a. One transistor failed at 0.5 W, how would you re-interpret the results?
   b. The first 10 failures occurred at 0.8 W, how would you re-interpret the results?
   c. One transistor failed at 0.5 W, how would you re-interpret the results?
   d. The first 10 failures occurred at 0.8 W, how would you re-interpret the results?

5

Statistical Experiments

INTRODUCTION

Earlier chapters have shown how statistically-distributed parameters and their effects can be analysed, when there is one variable and its effect, or two variables, e.g. load and strength. This chapter deals with the problem of assessing the combined effects of multiple variables on a measurable output or other characteristic of a product, by means of experiments. This is not a problem that is important in all designs, particularly when there are fairly large margins between capability and required performance, or for designs involving negligible risk or uncertainty, or when only one or a few items are to be manufactured. However, when designs have to be optimized in relation to variations in parameter values, processes, and environmental conditions, particularly if these variations can have combined effects, we should use methods that can evaluate the effects of the simultaneous variations. For example, it might be necessary to maximize the power output from a generator, and minimize the variation of its output, in relationship to rotational speed, several dimensions, coil geometry, and load conditions. All of these could have single or combined effects which cannot all be easily or accurately computed using theoretical calculations. Likewise, most electronic circuit designs involve the use of many components, all with distributed parameter values. These can have combined effects on performance or test yield that are difficult to evaluate owing to the number of variables and interactions involved.

Statistical methods of experimentation have been developed which enable the effects of variation to be evaluated in these types of situation. They are applicable whenever the effects cannot be easily theoretically evaluated, particularly when there is a large component of random variation or interactions between variables. For multivariable problems, the methods are much more economical than traditional experiments, in which the effect of one variable is evaluated at a time. The traditional approach also does not enable interactions to be analysed, when these are not known empirically. The rest of this chapter describes the statistical experimental methods, and how they can be adapted and applied to optimization and problem-solving in engineering.

STATISTICAL DESIGN OF EXPERIMENTS AND ANALYSIS OF VARIANCE

The statistical approach to design of experiments and the analysis of variance (ANOVA) technique was developed by R. A. Fisher, and is a very elegant, economical and powerful method for determining the s-significant effects and interactions in multivariable situations. Analysis of variance is used widely in such fields as market research, optimization of chemical and metallurgical processes, agriculture and medical research. It can provide the insights necessary for optimizing product designs and for preventing and solving quality and reliability problems. Whilst the methods described below might appear tedious, the computer software described in Chapter 2 includes facilities for ANOVA.

Analysis of single variables

The variance of a set of data (sample) is equal to

$$\frac{1}{n}\sum_{i=1}^{n}(x_i - \bar{x})^2$$

(Eqn. 2.15), where n is the sample size and $\bar{x}$ is the mean value. The population variance estimate is derived by dividing the sum of squares, $\Sigma(x_i - \bar{x})^2$, not by n but by $(n-1)$, where $(n-1)$ denotes the number of degrees of freedom (DF). Then $\hat{\sigma} = \Sigma(x_i - \bar{x})^2/(n-1)$ (Eqn. 2.16).

Example 5.1

To show how the variance of a group of samples can be analysed, consider a simple experiment in which 20 bearings, 5 each from 4 different suppliers, are run to failure. Table 5.1 shows the results.

We need to know if the observed variation between the samples is s-significant or is only a reflection of the variations of the populations from which the samples were drawn. Within each sample of five there are quite large variations. We must therefore analyse the difference between the 'between sample' (BS) and the 'within sample' (WS) variance and relate this to the populations.

Table 5.1 Times to failure of 20 bearings

Sample	Times to failure, [a]x_i (h)					Sample totals	Sample means, x_i
1	4	1	3	5	7	20	4
2	6	6	5	10	3	30	6
3	3	2	5	7	8	25	5
4	7	8	8	12	10	45	9

[a]$\Sigma x_i = 120$.

The next step is to calculate the sample totals and sample means, as shown in Table 5.1. The overall average value

$$\bar{x} = \frac{\Sigma x_i}{n} = \frac{120}{20} = 6$$

Since n is 20, the total number of degrees of freedom (DF) is 19. We then calculate the values of $(x_i - \bar{x})^2$. These are shown in Table 5.2.

Table 5.2   Values of $(x_i - \bar{x})^2$ for the data of Table 5.1

Sample			$(x_i - \bar{x})^2$			$\Sigma(x_i - \bar{x})^{2a}$
1	4	25	9	1	1	40
2	0	0	1	16	9	26
3	9	16	1	1	4	31
4	1	4	4	36	16	61

[a]Overall $\Sigma(x_i - \bar{x})^2 = 158$.

Having derived the total sum of squares and the DF, we must derive the WS and BS values. To derive the BS effect, we assume that each item value is equal to its sample mean (x_i'). The sample means for each item in samples 1 to 4 are 4, 6, 5 and 9, respectively. The BS sums of squares $(x_i' - \bar{x})^2$ are then, for each item in samples 1 to 4, 4, 0, 1 and 9, respectively, giving sample totals $\Sigma(x_i' - \bar{x})^2$ of 20, 0, 5 and 45 and an overall BS $\Sigma(x_i' - \bar{x})$ of 70. The BS DF is $4 - 1 = 3$.

Now we derive the equivalent values for the WS variance, by removing the BS effect. We achieve this by subtracting x_i' from each item in the original table to give a value x_i''. The result is shown in Table 5.3. Table 5.4 gives the values of the WS sums of squares, derived by squaring the values as they stand, since now $\bar{x}'' = 0$. The number of WS DF is $(5 - 1) \times 4 = 16$ (4 DF within each sample for a total of four samples). We can now tabulate the analysis of variance (Table 5.5).

If we can assume that the variables are s-normally distributed and that all the variances are equal, we can use the F-test (variance ratio test) (page 65) to test the null hypothesis that the two variance estimates (BS and WS) are estimates of the same common (population) variance. The WS variance represents the experimental error, or *residual* variance. F is the ratio of the variances, and a table showing the s-significance levels of variance ratios in relation to the number of degrees of freedom is given in Appendix 4. In this case

$$F = \frac{23.33}{5.50} = 4.24$$

Table 5.3   Values of $(x_i - x_i')$ for the data of Table 5.1

Sample			x_i''			$WS\Sigma x_i''$
1	0	-3	-1	1	3	0
2	0	0	-1	4	-3	0
3	-2	-3	0	2	3	0
4	-2	-1	-1	3	1	0

Table 5.4 Values of WS $(x_i'' - \bar{x}'')^2$ for the data of Table 5.1

Sample	WS $(x_i'' - \bar{x}'')^2$					WS$(x_i'' - \bar{x}'')^{2a}$
1	0	9	1	1	9	20
2	0	0	1	16	9	26
3	4	9	0	4	9	26
4	4	1	1	9	1	16

[a]Overall $\Sigma(x_i'' - \bar{x})^2 = 88$.

Table 5.5 Sources of variance for the data in Table 5.1

Source of variance	$\Sigma(.)$	DF	σ^2
BS	70	3	23.33
WS (residual)	88	16	5.50
Total	158	19	8.32

Table 5.6 Squares of the data values in Table 5.1

Sample	x_i^2					Totals
1	16	1	9	25	49	100
2	36	36	25	100	9	206
3	9	4	25	49	64	151
4	49	64	64	144	100	421

For 3 DF in the greater variance estimate and 16 DF in the smaller variance estimate, Appendix 4 shows that the 5 per cent *s*-significance level of *F* is 3.24. Since our value of *F* is greater than this, we conclude that the variance ratio is *s*-significant at the 5 per cent level, and the null hypothesis that there is no difference between the samples is therefore rejected at this level.

It is possible to simplify the analysis of variance procedure described above. Having set up the initial table of values, we produce Table 5.6 showing the squares of the values. We then use a *correction factor* equal to $(\Sigma x_i)^2/n$. In this case, $\Sigma x_i = 120$; therefore the correction factor CF = 720. The total sum of squares is then

$$\Sigma(x_i^2) - CF = 158$$

(as derived earlier). The between sample sum of squares is

$$\frac{\text{Sum of squares of sample totals}}{\text{Number of items in sample}} - CF = 70$$

The within sample sum of squares is the difference between the total sum of squares and the between sample sum of squares:

$$158 - 70 = 88$$

The degrees of freedom are derived as before.

If the sample sizes are not all the same, it is necessary to derive the between sample sum of squares by squaring each sample total, dividing by the numbers in the samples, adding these values and then subtracting the correction factor.

A further simplification can be applied when the values are not small numbers. If a constant value is subtracted from all the values, the variances (which form the subject of the analysis) are unaffected, but the arithmetic can be simplified. Since the first step in the analysis is to square the values, we can subtract a number which results in negative values. Subtracting a number close to the overall mean value provides the greatest simplification. Of course this trick will not always be necessary if an electronic calculator is being used, but even then it can speed up the calculation as it reduces the number of keystrokes, especially when dealing with the squares of fairly large numbers. The analysis results are similarly unaffected by division by a constant value.

Analysis of multiple variables (factorial experiments)

The method described above can be extended to analyse more than one source of variance. When there is more than one source of variance, interactions may occur between them, and the interactions may be more *s*-significant than the individual sources of variance. The following example will illustrate this for a three-factor situation.

Example 5.2

On a hydraulic system using 'O' ring seals, leaks occur apparently randomly. Three manufacturers' seals are used, and hydraulic oil pressure and temperature vary through the system. Argument rages as to whether the high temperature locations are the source of the problem, or the manufacturer, or even pressure. Life test data show that seal life may be assumed to be *s*-normally distributed at these stress levels, with variances which do not change with stress. A test rig is therefore designed to determine the effect on seal performance of oil temperature, oil pressure and type of seal. We set up the experiment in which we apply two values of pressure, 15 and 18 MN m^{-2} (denoted p_1 and p_2, in increasing order), and three values of temperature, 80, 100 and 120°C (t_1, t_2 and t_3, also in increasing order). We then select seals for application to the different test conditions, with two tests at each test combination. The results of the test are shown in Table 5.7, showing the time (h) to show a detectable leak.

To simplify Table 5.7, subtract 100 from each value. The coded data then appear as in Table 5.8.

The sums of squares for the three sources of variation are then derived as follows:

1. The correction factor:

$$CF = \frac{(\Sigma x_i)^2}{n} = \frac{(1409)^2}{36} = 55\,147$$

Table 5.7 Results of experiments on 'O' ring seals

Temperature	Type (T) 1		T2		T3	
	p_1	p_2	p_1	p_2	p_1	p_2
	104	209	181	172	157	178
t_1	196	132	129	151	187	211
	136	140	162	133	141	164
t_2	97	122	108	114	174	128
	108	96	99	112	122	135
t_3	121	110	123	109	130	118

Table 5.8 The data of Table 5.7 after subtracting 100 from each datum

Temperature	Type (T) 1		T2		T3		
	p_1	p_2	p_1	p_2	p_1	p_2	
t_1	4	109	81	72	57	78	
	96	32	29	51	87	111	
t_2	36	40	62	33	41	64	$\Sigma x_i(t)$
	−3	22	8	14	74	28	
t_3	8	−4	−1	12	22	35	
	21	10	23	9	30	18	

$$\Sigma x_i(\text{T})$$

2. The between types sum of squares:

$$SS_T = \frac{\Sigma x_i^2(T)}{2 \times 3 \times 2} - CF$$

$$= \frac{371^2 + 393^2 + 645^2}{12} - 55\ 147 = 3863$$

Between three types there are two DF.
3. The between pressures sum of squares:

$$SS_p = \frac{\Sigma x_i^2(p)}{3 \times 3 \times 2} - CF$$

$$= \frac{675^2 + 734^2}{18} - 55\ 147 = 96$$

Between two pressures there is one DF.
4. The between temperatures sum of squares:

$$SS_t = \frac{\Sigma x_i^2(t)}{2 \times 3 \times 2} - CF$$

$$= \frac{807^2 + 419^2 + 183^2}{12} - 55\ 147 = 16\ 544$$

Between three temperatures there are two DF.
5. The types–temperature interaction sum of squares:

$$SS_{Tt} = \frac{\Sigma x_i^2(Tt)}{2 \times 2} - CF - SS_T - SS_t$$

$$= \frac{241^2 + 95^2 + 35^2 + 233^2 + 117^2 + 43^2 + 333^2 + 207^2 + 105^2}{4}$$

$$- 55\,147 - 3863 - 16\,544$$

$$= 176$$

The T and the t effects each have 2 DF, so the interaction has $2 \times 2 = 4$ DF.
6. The types–pressure interaction sum of squares:

$$SS_{Tp} = \frac{\Sigma x_i^2(Tp)}{6} - CF - SS_T - SS_p$$

$$= \frac{162^2 + 209^2 + 202^2 + 191^2 + 311^2 + 334^2}{2 \times 3} - 55\,147 - 3863 - 96$$

$$= 142$$

The T effect has 2 DF and the p effect has 1 DF, so the Tp interaction has $2 \times 1 = 2$ DF.
7. The temperature–pressure interaction sum of squares:

$$SS_{tp} = \frac{\Sigma x_i^2(tp)}{2 \times 3} - CF - SS_t - SS_p$$

$$= \frac{354^2 + 453^2 + 218^2 + 201^2 + 103^2 + 80^2}{6} - 55\,147 - 16\,544 - 96$$

$$= 789$$

The t effect has 2 DF and the p effect 1 DF. Therefore the tp interaction has $2 \times 1 = 2$ DF.
8. The type–temperature–pressure interaction sum of squares:

$$SS_{Ttp} = \frac{\Sigma x_i^2(Ttp)}{2} - CF - SS_T - SS_t - SS_p - SS_{Tt} - SS_{tp} - SS_{Tp}$$

$$= \frac{154\,613}{2} - 55\,147 - 3863 - 16\,544 - 96 - 176 - 789 - 142$$

$$= 550$$

There are $2 \times 1 \times 2 = 4$ DF.

9. The total sum of squares:

$$SS_{tot} = \Sigma x_i^2 - CF$$
$$= 91\ 475 - 55\ 147 = 36\ 328$$

Total DF $= 36 - 1 = 35$ DF.

10. Residual (experimental error) sum of squares is

SS_{tot} − (all other SS)

$$= 36\ 328 - 3863 - 96 - 16\ 544 - 376 - 142 - 789 - 550$$
$$= 13\ 988$$

The residual DF:
Total DF − (all other DF) $= 35 - 2 - 1 - 2 - 4 - 2 - 2 - 4 = 18$ DF

Examination of the analysis of variance table (Table 5.9) shows that all the interactions show variance estimates much less than the residual variance, and therefore they are clearly not s-significant.

Having determined that none of the interactions are s-significant, we can assume that these variations are also due to the residual or experimental variance. We can therefore combine these sums of squares and degrees of freedom to provide a better estimate of the residual variance. The revised residual variance is thus:

$$\frac{176 + 142 + 789 + 550 + 13\ 988}{4 + 2 + 2 + 4 + 18} = \frac{15\ 645}{30}$$

$$= 521 \quad \text{with 30 DF}$$

We can now test the s-significance of the main factors. Clearly the effect of pressure is not s-significant. For the 'type' (T) main effect, the variance ratio $F = 1932/521$

Table 5.9 Analysis of variance table

| | Main factors | | | Interactions | | | | |
| | | | | First order | | | Second order | |
Effect	Type (T)	Pressure (p)	Temperature (t)	Tt	Tp	tp	Ttp	Residual
SS	3863	96	16 544	176	142	789	550	13 920
DF	2	1	2	4	2	2	4	18
SS/DF[a]	1932	96	8272	44	71	395	138	773
SD	44	10	91	10	8	20	9	28

[a]Variance estimate.

= 3.71 with 2 and 30 DF. The *F*-distribution table (Appendix 4) shows that this is *s*-significant at the 5 per cent level. For the temperature (*t*) main effect, $F = 8272/521 = 15.88$ with 2 and 30 DF. This is *s*-significant, even at the 1 per cent level.

   Therefore the experiment shows that the life of the seals is *s*-significantly dependent upon operating temperature and upon type. Pressure and interactions show effects which, if important, are not discernible within the experimental error; the effects are 'lost in the noise'. In other words, no type is *s*-significantly better or worse at higher pressures. Referring back to the original table of results, we can calculate the mean lives of the three types of seal, under the range of test conditions: type 1, 130 h; type 2, 132 h; and type 3, 154 h. Assuming that no other aspects such as cost predominate, we should therefore select type 3, which we should attempt to operate at low oil temperatures.

Non-normally distributed variables

In Example 5.2 above it is important to note that the method described is statistically correct only for *s*-normally distributed variables. As shown in Chapter 2, in accordance with the central limit theorem many parameters in engineering are *s*-normally distributed. However, it is prudent to test variables for *s*-normality before performing an analysis of variance. If any of the key variables are substantially non-*s*-normally distributed, non-parametric analysis methods should be used, as described later.

Two-level factorial experiments

It is possible to simplify the analysis of variance method if we adopt a two-level factorial design for the experiment. In this approach, we take only two values for each main effect, high and low, denoted by + and −. Example 5.3 below shows the results of a three-factor non-replicated experiment (such an experiment is called a 2^3 factorial design, i.e. three factors, each at two levels).

Example 5.3

From the results of Table 5.10, the first three columns of Table 5.11 represent the *design matrix* of the factorial experiment. The response value is the mean of the values at each test combination or in this case, as there is no replication, the single test value.

   The main effects can be simply calculated by averaging the difference between the response values for each high and low factor setting, using the appropriate signs in the 'factor' columns. Thus:

Main effect A $= [(15 - 13) + (13 - 12) + (14 - 11) + (14 - 12)]/4 = 2$

This is the same as: $(15 + 13 + 14 + 14 - 13 - 12 - 11 - 12)/4 = 2$
Likewise,
Main effect B $= (15 + 13 - 14 - 14 + 13 + 12 - 11 - 12)/4 = 0.5$
and,
Main effect C $= (15 - 13 + 14 - 14 + 13 - 12 + 11 - 12)/4 = 0.5$

Table 5.10 Results of a three-factor non-replicated experiment

A	B			
	−	−	+	+
		C		
	−	+	−	+
−	12	11	12	13
+	14	14	13	15

The sum of squares of each effect is then calculated using the formula

$$SS = 2^{k-2} \times (\text{effect estimate})^2$$

where k is the number of factors. Therefore,

$$SS_A = 2 \times 2^2 = 8$$
$$SS_B = 2 \times 0.5^2 = 0.5$$
$$SS_C = 2 \times 0.5^2 = 0.5$$

We can derive the interaction effects by expanding Table 5.11. Additional columns are added, one for each interaction. The signs under each interaction column are derived by algebraic multiplication of the signs of the constituent main effects. The ABC interaction signs are derived from AB × C or AC × B or BC × A.
    The AB interaction is then

$$(15 + 13 - 14 - 14 - 13 - 12 + 11 + 12)/4 = -0.5$$

and

$$SS_{AB} = 2 \times (-0.5)^2 = 0.5$$

Table 5.11 Response table and interaction of effects A, B, C

Factor			Interactions				Response
A	B	C	AB	AC	BC	ABC	
+	+	+	+	+	+	+	15
+	+	−	+	−	−	−	13
+	−	+	−	+	−	−	14
+	−	−	−	−	+	+	14
−	+	+	−	−	+	−	13
−	+	−	−	+	−	+	12
−	−	+	+	−	−	+	11
−	−	−	+	+	+	−	12

Table 5.12 Analysis of variance table

	A	B	C	AB	AC	BC	ABC	Total
				Effect				
SS	8	0.5	0.5	0.5	0.5	2	0	12
DF	1	1	1	1	1	1	1	7
SS/DF[a]	8	0.5	0.5	0.5	0.5	2	0	

[a]Variance estimate.

The other interaction sums of squares can be calculated in the same way and the analysis of variance table (Table 5.12) constructed.

Whilst the experiment in Example 5.3 indicates a large A main effect, and possibly an important BC interaction effect, it is not possible to test these statistically since no residual variance is available. To obtain a value of residual variance further replication would be necessary, or a value of experimental error might be available from other experiments. Alternatively, if the interactions, particularly high order interactions, are insignificant, they can be combined to give a residual value. Also, with so few degrees of freedom, the F test requires a very large value of the variance ratio in order to give high confidence that the effect is significant. Therefore, an unreplicated 2^3 factorial experiment may not always be sufficiently sensitive. However, the example will serve as an introduction to the next section, dealing with situations where several variables need to be considered.

Fractional factorial experiments

So far we have considered experiments in which all combinations of factors were tested, i.e. *full factorial* experiments. These can be expensive and time-consuming, since if the number of factors to be tested is f, the number of levels is L and the number of replications is r, then the number of tests to be performed is rL^f, i.e. in the hydraulic seal example $2 \times 3^2 = 18$, or in a three-level four-factor experiment with two replications $2 \times 3^4 = 162$. In a *fractional factorial* experiment we economize by eliminating some test combinations. Obviously we then lose information, but if the experiment is planned so that only those effects and interactions which are already believed to be unimportant are eliminated, we can make a compromise between total information, or experiment costs, and experimental value.

A point to remember is that higher order interactions are unlikely to have any engineering meaning or to show statistical significance, and therefore the full factorial experiment with several factors can give us information not all of which is meaningful.

We can design a fractional factorial experiment in different ways, depending on which effects and interactions we wish to analyse. Selection of the appropriate design is made starting from the full factorial design matrix. Table 5.13 gives the full design matrix for a 2^4 factorial experiment.

We select those interactions which we do not consider worth analysing. In the 2^4 experiment, for example, the ABCD interaction would not normally be considered significant. We therefore omit all the rows in the table in which the ABCD column shows – Thus we eliminate half of our test combinations, leaving a *half factorial*

Table 5.13　The full design matrix for a 2^4 factorial experiment

Test no.	1 A	2 B	3 C	4 D	5 AB	6 AC	7 AD	8 BC	9 BD	10 CD	11 ABC	12 ABD	13 ACD	14 BCD	15 ABCD
1	−	−	−	−	+	+	+	+	+	+	−	−	−	−	+
2	−	−	−	+	+	+	−	+	−	−	−	+	+	+	−
3	−	−	+	−	+	−	+	−	+	−	+	−	+	+	−
4	−	−	+	+	+	−	−	−	−	+	+	+	−	−	+
5	−	+	−	−	−	+	+	−	−	+	+	+	−	+	−
6	−	+	−	+	−	+	−	−	+	−	+	−	+	−	+
7	−	+	+	−	−	−	+	+	−	−	−	+	+	−	+
8	−	+	+	+	−	−	−	+	+	+	−	−	−	+	−
9	+	−	−	−	−	−	−	+	+	+	+	+	+	−	−
10	+	−	−	+	−	−	+	+	−	−	+	−	−	+	+
11	+	−	+	−	−	+	−	−	+	−	−	+	−	+	+
12	+	−	+	+	−	+	+	−	−	+	−	−	+	−	−
13	+	+	−	−	+	−	−	−	−	+	−	−	+	+	+
14	+	+	−	+	+	−	+	−	+	−	−	+	−	−	−
15	+	+	+	−	+	+	−	+	−	−	+	−	−	−	−
16	+	+	+	+	+	+	+	+	+	+	+	+	+	+	+

Table 5.14   Table 5.13 omitting rows where ABCD gives minus

A	B	C	D	AB	AC	AD	BC	BD	CD	ABC	ABD	ACD	BCD	ABCD
+	+	+	+	+	+	+	+	+	+	+	+	+	+	+
+	−	−	+	−	−	+	+	−	−	+	−	−	+	+
+	−	+	−	−	+	−	−	+	−	−	+	−	+	+
+	+	−	−	+	−	−	−	−	+	−	−	+	+	+
−	−	−	−	+	+	+	+	+	+	−	−	−	−	+
−	+	+	−	−	−	+	+	−	−	+	+	−	−	+
−	+	−	+	−	+	−	−	+	−	+	−	+	−	+
−	−	+	+	+	−	−	−	−	+	+	+	−	−	+

experiment. What else do we lose as a result? Table 5.14 shows what is left of the experiment.

Examination of this table shows the following pairs of identical columns

A, BCD; B, ACD; C,ABD; D,ABC; AB,CD; AC,BD: AD,BC

This means that the A main effect and the BCD interaction effect will be indistinguishable in the results. In fact in any experiment to this design we will not be able to distinguish response values for these effects; we say that they are *aliased* or *confounded*. If we can assume that the first- and second-order interactions aliased in this case are insignificant, then this will be an appropriate fractional design, reducing the number of tests from $2^4 = 16$ to $\frac{1}{2} \times 2^4 = 8$. We will still be able to analyse all the main effects, and up to three first-order interactions if we considered that they were likely to be significant. For example, if engineering knowledge tells us that the AB interaction is likely to be significant but the CD interaction not, we can attribute the variance estimate to the AB interaction.

A similar breakdown of a 2^3 experiment would show that it is not possible to produce a fractional factorial without aliasing main effects with first-order interactions. This is unlikely to be acceptable, and fractional factorial designs are normally only used when there are four or more factors to be analysed. Quarter factorial designs can be used when appropriate, following similar logic to that described above. The value of using fractional factorial designs increases rapidly when large numbers of effects must be analysed. For example, if there are seven main effects, a full factorial experiment would analyse a large number of high level interactions which would not be meaningful, and would require $2^7 = 128$ tests for no repeats. We can design a sixteenth fractional factorial layout with only eight tests, which will analyse all the main effects and most of the first-order interactions, as shown in Table 5.15. The aliases are deliberately planned, e.g. by evaluating the signs for the D column by multiplying the signs for A and B, and so on. We can select which interactions to alias by engineering judgement. It should be noted that other effects are also aliased. The full list of aliased effects can be derived by multiplying the aliased effects. For example, if D and AB are aliased, then the ABD interaction effect is also aliased, and so on. (If a squared term arises, let the squared term equal unity, e.g. $AB \times AC + A^2BC = BC$.)

Table 5.15 Sixteenth fractional factorial layout for seven main effects

Test no.	A	B	C	D	E	F	G
1	+	+	+	+	+	+	+
2	+	+	−	+	−	−	−
3	+	−	+	−	+	−	−
4	+	−	−	−	−	+	+
5	−	+	+	−	−	+	−
6	−	+	−	−	+	−	+
7	−	−	+	+	−	−	+
8	−	−	−	+	+	+	−
Alias				AB	AC	BC	ABC

Example 5.4

A jet engine fuel control system contains a number of components, seven of which are considered likely to affect the desired fuel flow ($30\,\mathrm{dm^3\,min^{-1}}$) under specified conditions. A sixteenth fractional factorial experiment is carried out with each of the seven components at high and low values within their tolerances. The results are given in Table 5.16. (The experiment is designed on the basis that components D, E, F, G and the aliased interactions are not considered likely to be significant.)

The main effects are:
A: $(6 + 4 + 4 - 2 + 2 - 0 - 2 + 4)/4 \;=\; 4$
B: $(6 + 4 - 4 + 2 - 2 + 0 - 2 + 4)/4 \;=\; 2$
C: $(6 - 4 + 4 + 2 - 2 - 0 + 2 + 4)/4 \;=\; 3$
D: $(6 + 4 - 4 + 2 + 2 - 0 + 2 - 4)/4 \;=\; 2$
E: $(6 - 4 + 4 + 2 + 2 + 0 - 2 - 4)/4 \;=\; 1$
F: $(6 - 4 - 4 - 2 - 2 - 0 - 2 - 4)/4 \;= -3$
G: $(6 - 4 - 4 - 2 + 2 + 0 + 2 + 4)/4 \;=\; 1$

$\mathrm{SS} = 2^{(7-2)} \times (\text{effect estimate})^2 = 32 \times (\text{effect estimate})^2$. Therefore

Table 5.16 Results of a sixteenth fractional factorial experiment for seven jet engine components affecting fuel flow

Test no.	Component							Response $(\mathrm{dm^3\,min^{-1}})$	Response −30
	A	B	C	D	E	F	G		
1	+	+	+	+	+	+	+	36	6
2	+	+	−	+	−	−	−	34	4
3	+	−	+	−	+	−	−	34	4
4	+	−	−	−	−	+	+	28	−2
5	−	+	+	−	−	+	−	28	−2
6	−	+	−	−	+	−	+	30	0
7	−	−	+	+	−	−	+	32	2
8	−	−	−	+	+	+	−	26	−4
Alias				AB	AC	BC	ABC		

$$SS_A = 512$$

$$SS_B = 64$$

$$SS_C = 288$$

$$SS_D = 64$$

$$SS_E = 32$$

$$SS_F = 288$$

$$SS_G = 32$$

Therefore the variance of component A is the most critical, factor in the design, with components C and F also requiring control.

NON-PARAMETRIC METHODS

Non-parametric methods may also be used for the two-way analysis of variance. Non-parametric methods can be useful when the variables cannot be assumed to be s-normally distributed since the methods imply no distribution for the data. They are also quite easy to apply. We will describe only the Kruskal–Wallis test, since other tests require fairly large numbers of observations in each group and this is often not realistic in reliability work.

In the Kruskal-Wallis test we replace the observed values by their rank order numbers. If the null hypotheses (that there is no difference in variances) is true, then the rank order numbers $1 \ldots N$ should be evenly distributed among the groups, so the mean of the rank order numbers in groups should be about the same. If T is total of rank order numbers and T_i is the total of rank order numbers in the ith group, then

$$T = \frac{N(N+1)}{2}$$

The average rank in group i is T_i/n_i (where n_i is the number of observations in group i).

The sum of squares of the rank order numbers is

$$\sum \frac{T_i^2}{n_i} - \frac{T^2}{N} = \sum \frac{T_i^2}{n_i} - \frac{N(N+1)^2}{4}$$

The Kruskal–Wallis criterion is

$$\chi^2_{\nu=(n_i-1)} = \frac{12}{N(N+1)} \times (\text{sum of squares of rank order numbers})$$

$$= \frac{12}{N(N+1)} \sum \frac{T_i^2}{n_i} - 3(N+1)$$

Table 5.17 Test data on five groups

Group	Test					
	1	2	3	4	5	6
1	4	17	21	341	35	39
2	17	26	29	34	40	42
3	25	25	33	41	41	43
4	1	5	109	22	27	34
5	14	24	31	32	33	32

Table 5.18 Rank order numbers (Example 5.5)

Group	Test						T_i
	1	2	3	4	5	6	
1	2	$6\frac{1}{2}$	8	$15\frac{1}{3}$	24	25	80.83
2	$6\frac{1}{2}$	13	$15\frac{1}{2}$	$22\frac{1}{2}$	26	29	112.33
3	$11\frac{1}{2}$	$11\frac{1}{2}$	$20\frac{1}{2}$	$27\frac{1}{2}$	$27\frac{1}{2}$	30	128.5
4	1	3	4	9	14	$22\frac{1}{2}$	53.5
5	5	10	$15\frac{1}{3}$	$18\frac{1}{2}$	$20\frac{1}{2}$	$18\frac{1}{2}$	87.83

Example 5.5

Test data on five groups are given in table 5.17 and the rank order numbers in Table 5.18.

$$N = 30$$

$$n_i = 6$$

$$\chi^2_{\nu=4} = \frac{12}{30 \times 31}\left(\frac{80.83^2 + 112.33^2 + 128.5^2 + 53.5^2 + 87.83^2}{6}\right) - (3 \times 31)$$

$$= 99.4 - 93 = 6.4$$

From Appendix 2, this is s-significant at about 20 per cent.

RANDOMIZING THE DATA

At this stage it is necessary to point out an essential aspect of any statistically designed experiment. s-significant sources of variance must be made to show their presence not only against the background of experimental error but against other sources of variation which might exist but which might not be tested for in the experiment. For instance, in Example 5.2, a source of variation might be the order in which seals are tested, or the batch from which seals are drawn. To eliminate the effects of extraneous factors and to ensure that only the effects being analysed will affect the results, it is important that the experiment is randomized. Thus the items selected for test and the sequence of tests must be selected at random, using random number tables or another suitable randomizing process. In Example 5.2, test samples should have been drawn at random from several batches of seals of each type and the test sequences should also have been randomized. It is also very important to eliminate human bias from the experiment, by hiding the identity of the items under test, if practicable.

If the items under test undergo a sequence of processes, e.g. heat treatment, followed by machining, then plating, the items should undergo each process in random order, i.e. separately randomized for each process.

Because of the importance of randomizing the data, it is nearly always necessary to design and plan an experiment to provide the data for analysis of variance. Data collected from a process as it normally occurs is unlikely to be valid for this purpose. Careful planning is important, so that once the experiment starts unforeseen circumstances do not cause disruption of the plan or introduce unwanted sources of variation or bias. A dummy run can be useful to confirm that the experiment can be run as planned.

ENGINEERING INTERPRETATION OF RESULTS

A statistical experiment can always, by its nature, produce results which conflict with the physical or chemical basis of the situation. The probability of a variance estimate being s-significant in relation to the experimental error is determined in the analysis, but we must always be on the lookout for the occurrence of chance results which do not fit our knowledge of the processes being studied. For example, in the hydraulic seal experiment we could study further the temperature–pressure interaction. Since the variance estimate for this interaction was higher than for the other interactions we might be tempted to suspect some interaction. In another experiment the seals might be selected in such a way that this variance estimate showed s-significance.

Figure 5.1 shows the interaction graphically, in two forms. If there were no interaction, i.e. the two effects acted quite independently upon seal life, the lines

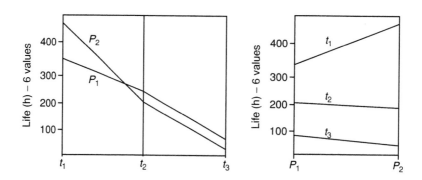

Figure 5.1 Temperature–pressure interactions

would be parallel between $t_1 t_2$, $t_2 t_3$ or $p_1 p_2$. In this case an interaction appears to exist between 80(t_1) and 100°C(t_2), but not between 100 and 120°C(t_3). This we can dismiss as being unlikely, taking into account the nature of the seals and the range of pressures and temperatures applied. That is not to say that such an interaction would always be dismissed, only that we can legitimately use our engineering knowledge to help interpret the results of a statistical experiment. The right balance must always be struck between the statistical and engineering interpretations. If a result appears highly s-significant, such as the temperature main effect, then it is conversely highly unlikely that it is a perverse result. If the engineering interpretation clashes with the statistical result and the decision to be made based on the result is important, then it is wise to repeat the experiment, varying the plan to emphasize the effects in question. In the hydraulic seal experiment, for example, we might perform another experiment, using type 3 seals only, but at three values of pressure as well as three values of temperature, and making four replications of each test instead of only two.

THE TAGUCHI METHOD

Genichi Taguchi has developed a framework for statistical design of experiments adapted to the particular requirements of engineering design. Taguchi suggested that the design process consists of three phases: *system design, parameter design*, and *tolerance design*. In the system design phase the basic concept is decided, using theoretical knowledge and experience to calculate the basic parameter values to provide the performance required. Parameter design involves refining the values so that the performance is optimized in relation to factors and variation which are not under the effective control of the designer, so that the design is 'robust' in relation to these. Tolerance design is the final stage, in which the effects of random variation of manufacturing processes and environments are evaluated, to determine whether the design of the product and the production processes can be further optimized, particularly in relation to cost of the product and the production processes. Note that the design process is considered to explicitly include the design of the production methods and their control. Parameter and tolerance design are based on statistical design of experiments.

Taguchi separates variables into two types. *Control factors* are those variables which can be practically and economically controlled, such as a controllable dimensional or electrical parameter. *Noise factors* are the variables which are difficult or expensive to control in practice, though they can be controlled in an experiment, e.g. ambient temperature, or parameter variation within a tolerance range. The objective is then to determine the combination of control factor settings (design and process variables) which will make the product have the maximum 'robustness' to the expected variation in the noise factors. The measure of robustness is the *signal-to-noise* ratio, which is analogous to the term as used in control engineering.

Figure 5.2 illustrates the approach. This shows the response of an output parameter to a variable. This could be the operating characteristic of a transistor or of a hydraulic valve, for example. If the desired output parameter value is A, setting the input parameter at A', with the tolerance shown, will result in an output centred on

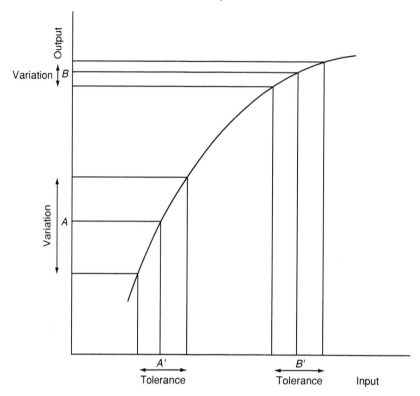

Figure 5.2   Taguchi method (1)

A, with variation as shown. However, the design would be much better, i.e. more robust to variation of the input parameter, if this were centred at *B'*, since the output would be much less variable with the same variation of the input parameter. The fact that the output value is now too high can be adjusted by adding another component to the system, with a linear or other less sensitive form of operating characteristic. This is a simple case for illustration, involving only one variable and its effect. For a multi-dimensional picture, with relationships which are not known empirically, the statistical experimental approach must be used.

Figure 5.3 illustrates the concept when multiple variations, control and noise factors affect the output of interest. This shows the design as a control system, whose performance must be optimized in relation to the effect of all variations and interactions.

The experimental framework is as described earlier, using fractional factorial designs. Taguchi argued that, in most engineering situations, interactions do not have significant effects, so that much reduced, and therefore more economical, fractional factorial designs can be applied. When necessary, subsidiary or confirmatory experiments can be run to ensure that this assumption is correct. Taguchi developed a range of such design matrices, or *orthogonal arrays*, from which the appropriate one for a particular experiment can be selected. For example, the 'L8' array is a sixteenth fractional factorial design for seven variables, each at two levels,

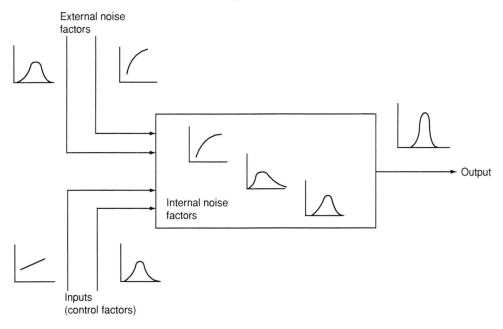

Figure 5.3 Taguchi method (2)

as shown in Table 5.14. (The 'L' refers to the Latin square derivation.) Further orthogonal arrays are given in References 9 to 11 and 16 to 18.

The arrays can be combined, to give an inner and an outer array, as shown in Table 5.19. The inner array contains the control factors, and the outer array the noise factors. The signal-to-noise ratio is calculated for the combination of control factors being considered, using the outer array, the formula depending on whether the desired output parameter must be maximized, minimized or centralized. The expressions are as follows:

$$\text{Maximum output, S/N ratio} = -10 \log\left[\frac{\Sigma(1/x^2)}{n}\right]$$

$$\text{Minimum output, S/N ratio} = -\log\left[\frac{\Sigma x^2}{n}\right]$$

$$\text{Centralized output, S/N ratio} = -10 \log\left[\hat{\sigma}^2\right]$$

where x is the mean response for the range of control factor settings, and $\hat{\sigma}$ is the estimate of the standard deviation. The ANOVA is performed as described earlier, using the S/N ratio calculated for each row of the inner array. The ANOVA can, of course, also be performed on the raw response data.

Example 5.6 shows a Taguchi implementation of the problem of Example 5.4.

Example 5.6

Table 5.19 shows the results of a Taguchi experiment on the fuel control system, with only the variation in components A, B and C being considered to be significant. These are then selected as control factors. The effects of two noise factors, X and Y, are to be investigated. The design must be robust in terms of the central value of the output parameter, fuel flow, i.e. minimal variation about the nominal value.

Figure 5.4 shows graphically the effects of varying the control factors on the mean response and signal-to-noise ratio. As in Example 5.4, variation of C has the largest effect on the mean response, with A and B also having effects. However, variation of B and C have negligible effects on the signal-to-noise ratio, but the low value of A provides a much more robust design than the higher value.

Table 5.19 Results of Taguchi experiment on fuel system components (Example 5.6)

			OUTER ARRAY (2 × 2)						
INNER ARRAY (L4: 3 × 2)			X	+	+	−	−		
			Y	+	−	+	−		
A	B	C	RESPONSE (−30)				mean	$\hat{\sigma}$	$\frac{S/N}{(-10\log\sigma^2)}$
1	+	+	+	8	6	4	4	5.5	1.91	−5.63
2	+	−	−	0	0	−2	−4	−1.5	1.91	−5.63
3	−	+	−	0	−2	0	−2	−1.0	1.15	−1.12
4	−	−	+	4	2	4	2	3.0	1.15	−1.12

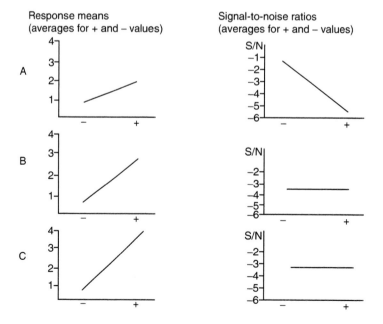

Figure 5.4 Results of Taguchi experiment (Example 5.6)

This is a rather simple experimental design, to illustrate the principles. Typical experiments might utilize rather larger arrays for both the control and noise factors.

EVOLUTIONARY OPERATION

Evolutionary operation (EVOP) is a method used to optimize a process or system. It depends upon the vital few variables having been identified by methods such as those described above, and allows the optimum values to be determined in relation to the response required.

Consider an item in which two s-significant variables X and Y have been isolated. The results of a two-factor two-level experiment can be depicted as shown in Fig. 5.5. These results indicate that the response value might be increased by further increasing the values of X and Y. We therefore plan a further experiment, using the previous high values of X and Y as the new low values. Continuing in this way, we might generate a series of responses as shown in Fig. 5.6. The experiments are continued until no further improvement is achieved. In this case the combination of X and Y values which give the response of 23 appear to be close to the optimum.

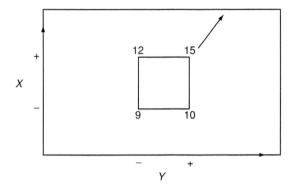

Figure 5.5 Evolutionary operation (1)

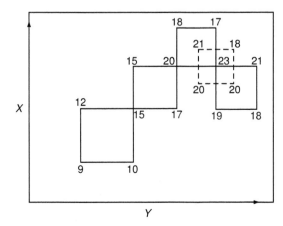

Figure 5.6 Evolutionary operation (2)

The method can be extended to include further variables. For three variables a three-dimensional chart can be drawn to show the results, but computer analysis is required for more than three variables. Further refinement can be achieved by running a further experiment, as indicated by the dotted square in Fig. 5.6. If none of the corners indicate an improvement over the previous optimum, then we have achieved the objective, unless even higher optimization is required.

In performing this type of optimization it must be remembered that we are looking for *s*-significant changes in the response. The amount of change which is *s*-significant will have become apparent from the first analysis of variance. A further point to bear in mind is that in some cases the *response surface*, considered as a contour of the response values projected perpendicular to the plane of the variables in the two-variable case, may have more than one maximum value. Evolutionary operation is also called *response surface analysis*.

CONCLUSIONS

Statistical experimental methods of design optimization and problem-solving in engineering design and production can be very effective and economic. They can provide higher levels of optimization and better understanding of the effects of variables than is possible with purely deterministic approaches, when the effects are difficult to calculate or are caused by interactions. However, as with any statistical method, they do not by themselves explain why a result occurs. A scientific or engineering explanation must always be developed, so that the effects can be understood and controlled.

It is essential that careful plans are made to ensure that the experiments will provide the answers required. This is particularly important for statistical experiments, owing to the fact that several trials are involved in each experiment, and this can lead to high costs. Therefore a balance must be struck between the cost of the experiment and the value to be obtained, and care must be taken to select the experiment and parameter ranges that will give the most information. The 'brainstorming' approach is associated with the Taguchi method, but should be used in the planning of any engineering experiment. In this approach, all involved in the design of the product and its production processes meet and suggest which are the likely important control and noise factors, and plan the experimental framework. The team must consider all sources of variation, deterministic, functional and random, and their likely ranges, so that the most appropriate and cost-effective experiment is planned. A person who is skilled and experienced in the design and analysis of statistical experiments must be a team member, and may be the leader. It is important to create an atmosphere of trust and teamwork, and the whole team must agree with the plan once it is evolved. Note the similarity with the Quality Function Deployment method described in Chapter 7. The philosophical and psychological basis is the same, and QFD should highlight the features for which statistical experiments should be performed.

Statistical experiments are equally effective for problem solving. In particular, the brainstorming approach very often leads to the identification and solution of problems even before experiments are conducted, especially when the variation is functional, as described earlier.

The results of statistical experiments should be used as the basis for setting the relevant process controls for production. This aspect is covered in Chapter 13. In particular, the Taguchi method is compatible with modern concepts of statistical process control in production, as described in Chapter 13, as it points the way to minimizing the variation of responses, rather than just optimizing the mean value. The explicit treatment of control and noise factors is an effective way of achieving this, and is a realistic approach for most engineering applications.

The Taguchi approach has been criticized by some of the statistics community and by others (see, for example, References 12 and 15), for not being statistically rigorous and for under-emphasizing the effects of interactions. Whilst there is some justification for these criticisms, it is important to appreciate that Taguchi has developed an operational method which deliberately economizes on the number of trials to be performed, in order to reduce experiment costs. The planning must take account of the extent to which theoretical and other knowledge, for example experience, can be used to generate a more cost-effective experiment. For example, theory and experience can often indicate when interactions are unlikely or insignificant. Also, full randomization of treatments might be omitted in an experiment involving different processing treatments, to save time. Taguchi recommends that confirmatory experiments should be conducted, to ensure that the assumptions made in the plan are valid.

It is arguable that Taguchi's greatest contribution has been to foster a much wider awareness of the power of statistical experiments for product and process design optimization and problem solving. The other major benefit has been the emphasis of the need for an integrated approach to the design of the product and of the production processes.

Statistical experiments can be conducted using computer-aided design software, when the software includes the necessary facilities, such as Monte Carlo simulation and statistical analysis routines. Of course there would be limitations in relation to the extent to which the software truly simulates the system and its responses to variation, but on the other hand experiments will be much less expensive, and quicker, than using hardware. Therefore initial optimization can often usefully be performed by simulation, with hardware experiments being run to confirm and refine the results.

The methods described in this chapter are appropriate for analysing cause-and-effect relationships which are linear, or which can be considered to be approximately linear over the likely range of variation. They cannot, of course, be used to analyse non-linear or discontinuous functions, such as resonances or changes of state.

The correct approach to the use of statistical experiments in engineering design and development, and for problem solving, is to use the power of all the methods available, as well as the skills and experience of the people involved. Teamwork and training are essential, and this in turn implies good management.

BIBLIOGRAPHY

1. M. J. Moroney, *Facts from Figures*. Penguin, Harmondsworth (1965).
2. G. E. Box, W. G. Hunter and J. S. Hunter, *Statistics for Experimenters*. Wiley (1978).
3. C. Lipson and N. J. Sheth, *Statistical Design and Analysis of Engineering Experiments*. McGraw-Hill (1973).
4. W. G. Cochran and G. M. Cox, *Experimental Designs*, 2nd edn. Wiley (1957).
5. R. L. Mason, R. F. Gunst and J. L. Hess, *Statistical Design and Analysis of Experiments*. Wiley (1989).
6. D. C. Montgomery, *Design and Analysis of Experiments*, 5th edn. Wiley 2000.
7. F. W. Breyfogle III, *Statistical Methods for Testing, Development and Manufacturing*. Wiley-Interscience (1992).
8. M. Hollander and D. A. Wolfe, *Non-parametric Statistical Methods*. Wiley (1963).
9. G. Taguchi, *Introduction to Quality Engineering*. Unipub/Asian Productivity Association (1986).
10. G. Taguchi, *Systems of Experimental Design*. Unipub/Asian Productivity Association (1978).
11. P. J. Ross, *Taguchi Techniques for Quality Engineering*. McGraw-Hill (1988).
12. N. Logothetis and H. P. Wynn, *Quality through Design*. Oxford University Press (1990).
13. K. R. Bhote, *World Class Quality*. American Management Association (1988).
14. D. M. Grove and T. P. Davis, *Engineering Quality and Design of Experiments*. Longman Higher Education (1992).
15. T. P. Ryan, *Statistical Methods for Quality Improvement* 2nd edn. Wiley-Interscience (2001).
16. L. W. Condra, *Reliability Improvement with Design of Experiments*. Marcel Dekker (1993).
17. S. H. Park, *Robust Design and Analysis for Quality Engineering*. Chapman & Hall (1996).
18. R. K. Roy, *Design of Experiments Using the Taguchi Approach*. Wiley-Interscience (2001).

QUESTIONS

1. A manufacturer has undertaken experiments to improve the hot-starting reliability of an engine. There were two control factors, mixture setting (M) and ignition timing (T), which were each set at three levels. The trials were numbered as below (which is a full factorial, equivalent to a Taguchi L9 orthogonal array with only two of its four columns allocated):

		Mixture 1	2	3
	1	Expt 1	Expt 2	Expt 3
Timing	2	Expt 4	Expt 5	Expt 6
	3	Expt 7	Expt 8	Expt 9

The response variable was the number of starter-motor shaft revolutions required to start the engine. Twenty results were obtained at each setting of the control factors, with averages as follows:

Expt	1	2	3	4	5	6	7	8	9
Av. revs	12.15	17.15	20.85	16.25	19.25	16.45	22.10	20.05	14.15

(a) What initial conclusions can be drawn about the control factors and their inter-action (without performing a significance test)?

(b) There are two identifiable noise factors (throttle position and clutch pedal pos-ition), both of which can be set at two levels. How would the noise factors have been incorporated into the experimental design?

(c) Do you think a Taguchi signal-to-noise ratio would be a better measure than the average revs? Which one would you use?

(d) What additional information do you need to perform an analysis of variance?

2. (a) It has been suggested that reliability testing should not be applied to proving a single design parameter but should instead be based on simple factorial experimen-tation on key factors which might improve reliability. What do you think of this idea?

(b) The following factors have been suggested as possible influences on the reliability of an electromechanical assembly:

A–electrical terminations (wrapped or soldered)
B–type of switching circuit (relay or solid-state)
C–component supplier (supplier 1 or supplier 2)
D–cooling (convection or fan).

It is suspected that D might interact with B and C.
   An accelerated testing procedure has been developed whereby assemblies are subjected to repeated environmental and operational cycles until they fail. As this testing is expensive, a maximum of 10 prototypes can be made and tested. Design a suitable experiment, and identify all aliased interactions.

3. One measure of the reliability of a portable communications receiver is its ability to give adequate reception under varying signal strengths, which are strongly influenced by climatic conditions and other environmental factors outside the user's control. It was decided that, within the design of a receiver, there were seven factors which could possibly influence the quality of reception.

   An experiment was designed using a 2^{7-3} layout. In the design shown in Table 5.13, the six factors (denoted A–G), each at two levels, were allocated respectively to columns 1, 2, 3, 4, 12, 14 and 15. The remaining columns were left free for evaluation of interactions. It was felt safe to assume that three- and four-factor interactions would not occur.
   In the experiment the 16 prototype receivers were installed in an area of known poor reception, and each was evaluated for performance on four separate occasions (between which the environmental conditions varied in a typical manner). The results are shown below (the response being an index of reception quality):

Test no.	Experimental results (reception index)			
1	6.66	5.90	6.72	4.81
2	7.76	5.77	8.36	8.62
3	5.59	6.34	7.35	8.50
4	6.36	5.37	6.17	6.46

5	7.00	6.76	5.47	5.92
6	7.52	4.71	6.69	8.14
7	7.25	5.08	5.66	5.04
8	6.18	6.47	7.55	5.92
9	7.21	5.37	7.34	4.48
10	6.95	6.96	8.36	6.87
11	7.08	5.74	6.72	6.70
12	5.34	7.56	7.22	6.89
13	8.09	8.27	5.69	5.96
14	7.72	5.62	5.77	6.79
15	6.43	6.59	6.08	5.37
16	5.52	5.82	5.82	7.29

Calculate the maximum output ('biggest is best') signal-to-noise ratios. From these, calculate the effects and sums of squares for signal-to-noise for all the factors and interactions, and carry out an analysis of variance using these values to test the significance of the various factors and interactions. What would you recommend as the final design?

6

Reliability Prediction and Modelling

INTRODUCTION

An accurate prediction of the reliability of a new product, before it is manufactured or marketed, is obviously highly desirable. Depending upon the product and its market, advance knowledge of reliability would allow accurate forecasts to be made of support costs, spares requirements, warranty costs, marketability, etc. It is arguable, on the other hand, that an accurate prediction of reliability implies such knowledge of the causes of failure that they could be eliminated. In fact, a reliability prediction can rarely be made with high accuracy or confidence. Nevertheless, it can often provide an adequate basis for forecasting of dependent factors such as life cycle costs. Reliability prediction can also be valuable as part of the study and design processes, for comparing options and for highlighting critical reliability features of designs.

If a type of fighter aircraft has on average 2.5 flying hours between failures, and a new aircraft of similar complexity and performance is being planned, 2.5 flying hours between failures might be a reasonable basis for forecasting. If the existing aircraft is twin-engined, with electronic systems based mainly on an earlier technology, whereas the new aircraft is to have one engine and more modern electronics, judgements have to be made on how these changes might affect the failure rate. Instead of comparing only the complete aircraft, we must compare components. What difference will one engine make compared with two? (In this case, trade-off decisions will have been made. For example, two engines might be safer, but a single engine might lead to lower support costs.) What will be the difference in failure rate between the old electronic systems and the new ones? The comparisons begin at the level of the overall system and as the system becomes more closely defined extend to more detailed levels.

Eventually, in principle, it is necessary to consider the reliability contributions of individual parts.

However, the lower the level of analysis, the greater is the uncertainty inherent predicting reliability. The great majority of modern engineering components are sufficiently reliable that, for practical purposes, they generate no inherent quantifiable failure rate contribution. Also, many system failures are not caused by failures of parts, and not all part failures cause system failures.

Since reliability is affected strongly by human-related factors such as training and motivation of design and test engineers, quality of production, and maintenance skills, these factors must also be taken into account. In many cases they can be much more significant than past data. Therefore reliability 'databases' must always be treated with caution as a basis for predicting the reliability of new systems. There is no intrinsic limit to the reliability that can be achieved, but the database approach to prediction can imply that there is.

FUNDAMENTAL LIMITATIONS OF RELIABILITY PREDICTION

Predictions in physics and engineering

In engineering and science we use mathematical models for prediction. For example, the power consumption of a new electronic system can be predicted using Ohm's law and the model power = current × e.m.f. Likewise, we can predict future planetary positions using Newton's laws and our knowledge of the present positions, velocities and masses. These laws are valid within the appropriate domain (e.g. Ohm's law does not hold at temperatures near absolute zero; Newton's laws are not valid at the subatomic level). However, for practical, everyday purposes such deterministic laws serve our purposes well, and we use them to make predictions, taking due account of such practical aspects as measurement errors in initial conditions.

Whilst most laws in physics, for practical predictive purposes, can be considered to be deterministic, the underlying mechanisms can be stochastic. For example, the pressure exerted by a gas in an enclosure is a function of the random motions of very large numbers of molecules. The statistical central limiting theorem, applied to such a vast number of separate random events and interactions, enables us to use the average effect of the molecular kinetic energy to predict the value we call pressure. Thus Boyle's law is really empirical, as are other 'deterministic' physical laws such as Ohm's law. It is only at the level of individual or very few actions and interactions, such as in nuclear physics experiments, that physicists find it necessary to take account of uncertainty due to the stochastic nature of the underlying processes. However, for practical purposes we ignore the infinitesimal variations, particularly as they are often not even measurable, in the same way as we accept the Newtonian view.

Of course some physical systems are comprised of very few components, actions and interactions, with no underlying stochastic mechanism. Thus laws relating to planetary motion or the momentum exchange between two billiard balls can be considered deterministic because of the small numbers involved. If, however, we attempt to use these deterministic laws to predict the behaviour of a planetary system comprised of 100 planets, or the instantaneous positions of a large number of billiard balls at some time after the first impact, the computational problems become significant and begin to degrade the credibility of the predictions. Also, the effects of small errors and variations progressively accumulate, quickly leading to increased uncertainty and to divergent behaviour.

Physical laws are, therefore, useful predictors of system behaviour either when small numbers of actions and interactions are involved or when very large numbers are involved. For systems involving moderately large numbers the predictive power of empirical and deterministic laws diminishes. This effect is observable in fields such as aerodynamics, meteorology and the study of explosions, for example. Whilst the individual physical and chemical processes are understood and predictable, the large numbers of simultaneous interacting processes swamp our capability to compute predicted values using these simple, known relationships. Very fast computing is applied, and new empirical relationships derived, in our attempts to predict the behaviour of these systems, with results that often fall short of our requirements. Prediction power is increased if we can simplify the problem to a few variables or if we complicate it to a very large number. Thus, we can apply the laws of heat transfer to predict the local atmospheric effect of a rise in sea temperature, and we can use past data to predict the average rainfall on a continent. However, short-term local rainfall cannot be predicted accurately.

Prediction power is greater in time-invariant or cyclical systems, e.g. Ohm's law in electrical circuits, or our understanding of tidal motion as a result of gravity. The systems mentioned in the last paragraph (weather, aerodynamics, explosives) are time-variant, and this leads to divergent behaviour. Thus predictions of the instantaneous state become progressively less credible as time proceeds (unless a steady state is ultimately achieved, e.g. with an explosion).

It follows from the arguments presented that one or a few measurements of a physical quantity can usually provide us with sufficient information on which to make a prediction of the future state of a physical time-invariant or cyclical system. In other words, very few data are required in order to make confident predictions. We measure the flow of electricity in wire and we assume that the current will remain at that value under the same general conditions. When the quantity is time-variant more measurements might need to be taken before we can predict confidently from the data. It is only when the system is moderately complex that one or a few measurements do not enable us to predict the future state.

For a mathematical model to be accepted as a basis for scientific prediction, it must be based upon a theory which explains the relationship. It is also necessary for the model to be based upon unambiguous definitions of the parameters used. Finally, scientists, and therefore engineers, expect the predictions made using the models to be always repeatable. If a model used in science is found not to predict correctly an outcome under certain circumstances this is taken as evidence that the model, and the underlying theory, needs to be revised, and a new theory is postulated.

Predictions in reliability

The concept of deriving mathematical models which could be used to predict reliability, in the same way as models are developed and used in other scientific and engineering fields, is intuitively appealing, and has attracted much attention. For example, failure rate models have been derived for electronic components, based upon parameters such as operating temperature and other stresses. These are described in Chapter 9. Similar models have been derived for non-electronic components, and even for computer software. Sometimes these models are as simple as a single fixed value for failure rate or reliability, or a fixed value with simple modifying factors. However, some of the models derived are quite complex, taking account of many factors considered likely to affect reliability.

Like other predictive models in science and engineering, these have been based upon considerations of what might affect the parameter of interest, in this case reliability, or, in other words, cause failure. Thus there have been attempts to create theories. However, this approach is of severely limited validity for predicting reliability.

A model such as Ohm's law is credible because there is no question as to whether or not an electric current flows when an e.m.f. is applied across a conductor. However, whilst an engineering component might have properties such as conductance, mass, etc., all unambiguously defined and measurable, it is very unlikely to have an intrinsic reliability that meets such criteria. For example, a good transistor or hydraulic actuator, if correctly applied, should not fail in use, during the expected life of the system in which it is used. Whether a particular one, or a proportion, fail, and the modes of failure and distribution of times to failure are likely to be due to factors other than those explainable in purely physical terms. Some transistors might fail because of accidental overstress, some because of processing defects, or there may be no failures at all. If the hydraulic actuators are operated for a long time in a harsh environment, some might develop leaks which some operators might classify as failures. Whilst the good transistors will never degrade in a typical operating lifetime, since no progressive physical or chemical changes occur in such devices, we can expect progressive deterioration in the hydraulic actuator due to wear.

Failure, or the absence of failure, is therefore heavily dependent upon human actions and perceptions. This is never true of laws of nature. This represents a fundamental limitation of the concept of reliability prediction using mathematical models.

The onset of failure is nearly always a discontinuous function, subject to the predictive difficulties described for physical models of the behaviour of systems which contain moderately large numbers of factors and interactions, and whose progression to a failed state is time-variant. The failure mechanisms of even relatively simple components such as electronic parts or gear-wheels exhibit this level of complexity and uncertainty, defying the validity of mathematical models based on analysis of past data. If, however, the 'part' being considered is itself a complex assembly, such as a jet engine or an electronic control system, then past data can be a more credible guide to future reliability.

We saw in Chapter 4 how reliability can vary by orders of magnitude with small changes in load and strength distributions, and the large amount of uncertainty inherent in estimating reliability from the load-strength model. These real uncertainties must be borne in mind when synthesizing the reliability of a system by considering the likely failure rates of its parts.

Another severe limitation arises from the fact that reliability models are usually based upon statistical analysis of past data. Much more data is required to derive a statistical relationship than to confirm a deterministic (theory-based) one, and even then there will be uncertainty because the sample can seldom be taken to be wholly representative of the whole population. For example, the true value of a life parameter is never known, only its distribution about an expected value, so we cannot say when failure will occur. Sometimes we can say that the likelihood increases, e.g. in fatigue testing or if we detect wear in a bearing, but we can very rarely predict the time of failure. A statistically-derived relationship can never by itself be proof of a causal connection or even establish a theory. It must be supported by theory based upon an understanding of the cause-and-effect relationship.

It is never sensible to make a prediction based on past data unless we are sure that the underlying conditions which can affect future behaviour will not change. In most fields in which statisticians work, such as actuarial, agricultural, or social studies, it is usually safe to make such an assumption. However, since engineering is very much concerned with deliberate change, in design, processes, and applications, predictions of reliability based solely on past data ignore the fact that changes might be made with the objective of improving reliability. In most modern engineering applications there is continuous effort to improve quality and reliability, so the use of past data to predict the future can be very misleading and unduly pessimistic.

Of course there are situations in which we can assume that change will not be significant, or in which we can extrapolate taking account of the likely effectiveness of planned changes. For example, in a system containing many parts which are subject to progressive deterioration, e.g. an office lighting system containing many fluorescent lighting units, we can predict the frequency and pattern of failures fairly accurately, but these are special cases.

A reliability prediction for a system containing many parts is likely to be more accurate than for small system. However, it is common for the erroneous implication to be drawn that, because the individual parts failure data can be combined to provide a reasonably accurate prediction for the reliability of the system, the parts failure data can be used with as much confidence for spare parts provisioning or decisions related to the safety of individual parts. It is important to remember that variances in reliability at the part level can be orders of magnitude greater than the variance at the system level.

Therefore, any reliability prediction based upon the use of mathematical models or standard data such as in References 2 and 3, or in reliability growth models (Chapter 12) must be treated with some scepticism. See also the comments on reliability analysis methods for repairable systems on pages 344 and 345.

A designer can design an electronic system to a power budget. The power dissipation of the new system can be predicted with credible precision, so that all systems made to the design will dissipate, within reasonable tolerance, the predicted power. However, a designer cannot design for an MTBF, unless he places as much faith in the reliability mathematical model as he does in, say, Ohm's law. The MTBF of each unit cannot be measured as can its power consumption, and there is no reason in logic to believe that they will all show the same MTBFs or patterns of failure over a period of operation, or even that the overall average MTBF or failure pattern will conform to that predicted.

Unfortunately the mathematical modelling approach to reliability prediction (including extrapolation from test data) has been given undue and insufficiently critical attention in the literature and in reliability standards. It is not uncommon to see MTBF predictions such as 5761 hours. No power dissipation prediction would attempt such precision; 6 kW is an equivalent figure more likely to be stated. The naive presentation of reliability predictions has done much to undermine the credibility of reliability engineering, particularly since so often systems achieve reliability levels far higher or far higher than the predictions.

RELIABILITY DATABASES

There are several published databases which give reliability information on engineering components and sub-systems. The most common approach is to state failure rates, expressed as failures per million hours. For non-repairable items this may be interpreted as the failure rate contribution to the system. These data are then used to synthesize system failure rate, by summation and by taking account of system configuration, as described later. Summation of part failure rates is called 'parts count' reliability prediction.

Probably the best known source of failure rate data for electronic components is US MIL-HDBK-217 (Reference 2), which is described in Chapter 9. Other data sources for electronic components have been produced and published by organizations such as telecommunications companies. Databases have also been produced for non-electronic components, e.g. Reference 3. These purport to show the failure rates of generic parts types such as hydraulic valves, electric motors, gaskets and springs. The variety of types and applications of such parts is so diverse that such data cannot be relied upon as being generally applicable. There is no general unit of operating time for a gasket or a spring, as there might be for a transistor. Also, for most non-electronic parts, the assumption of a constant hazard rate can be grossly misleading.

These databases are subject to all of the criticisms made above about the limitations of reliability prediction techniques at the part level. There is considerable controversy about their validity, and they should not be used. In fact they tend to be used mainly in connection with systems for which the customer specifies the reliability methods to be applied, e.g. in defence contracts. They are not used by most design organizations involved in more commercial engineering, such as computing, automotive, and domestic electronic systems. Some purchasing organizations, for example NASA, the US. Army and the UK Ministry of Defence specifically discourage or prohibit their use.

Reliability data can be useful in specific prediction applications, such as aircraft, petrochemical plant, photocopying machines or automobiles, when the data are derived from the area of application. However, such data should not be transferred from one application area to another without careful assessment. Even within the application area they should be used with care, since even then conditions can vary widely. An electric motor used to perform the same function, under the same loading conditions as previously in a new design of photocopier, might be expected to show the same failure pattern. However, if the motor is to be used for a different function, with different operating cycles, or even if it is bought from a different supplier, the old failure data will not be appropriate.

THE PRACTICAL APPROACH

Having identified the fundamental limitations of reliability prediction models and data, we are still left with the problem that it is often necessary to predict the likely reliability of a new system. We also know that it is possible to make reasonably credible reliability predictions for systems under certain circumstances. These are:

1. The system is similar to systems developed, built and used previously, so that we can apply our experience of what happened before.
2. The new system does not involve significant technological risk (this follows from 1).
3. The system will be manufactured in large quantities, or is very complex (i.e. contains many parts, or the parts are complex) or will be used over a long time, or a combination of these conditions applies, i.e. there is an asymptotic property.
4. There is a strong commitment to the achievement of the reliability predicted, as an overriding priority.

Thus, we can make credible reliability predictions for a new TV receiver or automobile engine. No great changes from past practice are involved, technological risks are low, they will be built in large quantities and they are quite complex, and in present markets the system must compete with established, reliable products. Likewise, we can make a credible reliability prediction for a complete new aircraft or an oil production platform, since the 'parts' (engines, pumps, sub-systems) are themselves complex, and their reliabilities are probably quite accurately known. However, for a new, high technology, weapon system these conditions may not apply, though nowadays the commitment to reliability is generally stronger than in the past.

Note that reliability predictions (in the sense of a reasonable expectation) for the TV receiver, the automobile engine, the aircraft and the production platform could be made without recourse to statistical or empirical mathematical models. Rather, they could be based upon knowledge of past performance and present conditions, and on management targets and priorities. The reliability prediction does not ensure that the reliability values will be achieved; it is not a demonstration in the way that a mass or power consumption prediction, being based on physical laws, would be. Rather, it should be used as the basis for setting the objective—one which is likely to be attained only if there is a human commitment to it. This is well illustrated by the notable reliability achievements of the US space programme and the Japanese consumer products industry.

Reliability prediction for new, high technology products must, therefore, be based upon identification of objectives and assessment of risks, in that order. This must be an iterative procedure, since objectives and risks must be balanced; the reliability engineer plays an important part in this process, since he must assess whether objectives are realistic in relation to the risks. This assessment can be aided by the educated use of appropriate models and data, which help to quantify the risks. Once the risks are assessed and the objective quantified, development must be continuously monitored in relation to the reduction of risks through analysis and tests, and to the measured reliability during tests. This is necessary to provide assurance that the objective will be met, if need be by additional management action such as provision of extra resources to solve particular problems.

Reliability predictions for systems should therefore be made 'top down', not synthesized from the parts level. Exceptions can be made in the case of large systems which are themselves made up of complex sub-systems, whose reliabilities are known.

The purpose to which the prediction will be applied should also influence the methods used and the estimates derived. For example, if the prediction will be used to determine spare item stocks or repair costs, an optimistic figure might be acceptable. However, if it is to be used as part of a safety analysis, a pessimistic figure would be more appropriate. It is good practice to indicate the likely expected ranges of uncertainty, with reasons.

In situations in which the methods to be used are imposed (e.g. MIL-HDBK-217), the reliability prediction report should state that the results are derived accordingly, but are not supported in any other way.

The predictions should always take account of objectives and related management aspects, such as commitment and risk. If management does not "drive" the reliability effort, the prediction can become a meaningless exercise. As overriding considerations, it must be remembered that there is no theoretical limit to the reliability that can be attained, and that the achievement of very high reliability does not necessarily entail higher costs.

SYSTEMS RELIABILITY MODELS

The basic series reliability model

Consider a system composed of two independent components, each exhibiting a constant hazard rate. If the failure of either component will result in failure of the system, the system can be represented by a *reliability block diagram* (RBD) (Fig. 6.1). (A reliability block diagram does not necessarily represent the system's operational logic or functional partitioning.)

If λ_1 and λ_2 are the hazard rates of the two components, the system hazard rate will be $\lambda_1 + \lambda_2$. Because the hazard rates are constant, the component reliabilities R_1 and R_2, over a time of operation t, are $\exp(-\lambda_1 t)$ and $\exp(-\lambda_2 t)$. The reliability of the system is the combined probability of no failure of either component, i.e. $R_1 R_2 = \exp[-(\lambda_1 + \lambda_2)t]$. In general, for a series of n, s-independent components:

$$R = \prod_{i=1}^{n} R_i \tag{6.1}$$

Figure 6.1 Series system

where R_i is the reliability of the ith component. This is known as the product rule or series rule (see Eqn 2.2):

$$\lambda = \sum_{i=1}^{n} \lambda_i \quad \text{and} \quad R = \exp(-\lambda t) \tag{6.2}$$

This is the simplest basic model on which *parts count* reliability prediction is based.

The failure logic model of the overall system will be more complex if there are redundant subsystems or components. Also, if system failure can be caused by events other than component failures, such as interface problems, the model should specifically include these, for example as extra blocks.

Active redundancy

The reliability block diagram for the simplest redundant system is shown in Fig. 6.2. In this system, composed of two s-independent parts with reliabilities R_1 and R_2, satisfactory operation occurs if *either* one or *both* parts function. Therefore, the reliability of the system, R, is equal to the probability of part 1 *or* part 2 surviving.

From Eqn (2.6), the probability

$$(R_1 + R_2) = R_1 + R_2 - R_1 R_2$$

This is often written

$$1 - (1 - R_1)(1 - R_2)$$

For the constant hazard rate case,

$$R = \exp(-\lambda_1 t) + \exp(-\lambda_2 t) - \exp[-(\lambda_1 + \lambda_2)t] \tag{6.3}$$

The general expression for active parallel redundancy is

$$R = 1 - \prod_{i=1}^{n} (1 - R_i) \tag{6.4}$$

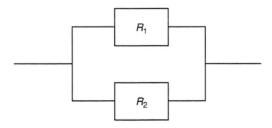

Figure 6.2 Dual redundant system

where R_i is the reliability of the ith unit and n the number of units in parallel. If in the two-unit active redundant system $\lambda_1 = \lambda_2 = 0.1$ failures per 1000 h, the system reliability over 1000 h is 0.9909. This is a significant increase over the reliability of a simple non-redundant unit, which is 0.9048. Such a large reliability gain often justifies the extra expense of designing redundancy into systems. The gain usually exceeds the range of prediction uncertainty. The example quoted is for a *non-maintained* system, i.e. the system is not repaired when one equipment fails. In practice, most active redundant systems include an indication of failure of one equipment, which can then be repaired. A maintained active redundant system is, of course, theoretically more reliable than a non-maintained one. Examples of non-maintained active redundancy can be found in spacecraft systems (e.g. dual thrust motors for orbital station-keeping) and maintained active redundancy is a feature of systems such as power generating systems and railway signals.

m-out-of-n redundancy

In some active parallel redundant configurations, m out of the n units may be required to be working for the system to function. This is called m-out-of-n (or m/n) parallel redundancy. The reliability of an m/n system, with n, s-independent components in which all the unit reliabilities are equal, is the binomial reliability function (Eqn 2.37):

$$R_{SYS} = 1 - \sum_{i=0}^{m-1} \binom{n}{i} R^i (1 - R)^{n-i} \tag{6.5}$$

or, for the constant hazard rate case:

$$R_{SYS} = 1 - \frac{1}{(\lambda t + 1)^n} \sum_{i=0}^{m-1} \binom{n}{i} (\lambda t)^{n-i} \tag{6.6}$$

Standby redundancy

Standby redundancy is achieved when one unit does not operate continuously but is only switched on when the primary unit fails. A standby electrical generating system is an example. The block diagram in Fig. 6.3 shows another. The standby unit and the sensing and switching system may be considered to have a 'one-shot' reliability R_s of starting and maintaining system function until the primary equipment is repaired, or R_s may be time-dependent. The switch and the redundant unit may have dormant hazard rates, particularly if they are not maintained or monitored.

Taking the case where the system is non-maintained, the units have equal constant operating hazard rates λ, there are no dormant failures and $R_s = 1$, then

$$R_{SYS} = \exp(-\lambda t) + \lambda t \exp(-\lambda t) \tag{6.7}$$

The general reliability formula for n equal units in a standby redundant configuration (perfect switching) is

Figure 6.3 Reliability block diagram for a missile system

$$R_{SYS} = \sum_{i=0}^{n-1} \frac{(\lambda t)^i}{i!} \exp(-\lambda t) \tag{6.8}$$

If in a standby redundant system $\lambda_1 = \lambda_2 = 0.1$ failure per 1000 h, then the system reliability is 0.9953. This is higher than for the active redundant system [$R(1000) = 0.9909$], since the standby system is at risk for a shorter time. If we take into account less than perfect reliability for the sensing and switching system and possibly a dormant hazard rate for the standby equipment, then standby system reliability would be reduced.

Further redundancy considerations

The redundant systems described represent the tip of an iceberg as far as the variety and complexity of system reliability models are concerned. For systems where very high safety or reliability is required, more complex redundancy is frequently applied. Some examples of these are:

1. In aircraft, dual or triple active redundant hydraulic power systems are used, with a further emergency (standby) back-up system in case of a failure of all the primary circuits.
2. Aircraft electronic flying controls typically utilize triple voting active redundancy. A sensing system automatically switches off one system if it transmits signals which do not match those transmitted by the other two, and there is a manual back-up system. The reliability evaluation must include the reliability of all three primary systems, the sensing system and the manual system.
3. Fire detection and suppression systems consist of detectors, which may be in parallel active redundant configuration, and a suppression system which is triggered by the detectors.

We must be careful to ensure that single-point failures which can partly eliminate the effect of redundancy are considered in assessing redundant systems. For example, if redundant electronic circuits are included within one integrated circuit package, a single failure such as a leaking hermetic seal could cause both circuits to fail. Such s-dependent failures are sometimes referred to as *common mode* (or *common cause*) failures, particularly in relation to systems. As far as is practicable, they must be identified and included in the analysis. Common mode failures are discussed in more detail later.

AVAILABILITY OF REPAIRABLE SYSTEMS

Availability is defined as the probability that an item will be available when required, or as the proportion of total time that the item is available for use. Therefore the availability of a repairable item is a function of its failure rate, λ, and of its repair or replacement rate. The proportion of total time that the item is available is the *steady-state availability*. For a simple unit, with a constant failure rate λ and a constant mean repair rate μ, where $\mu = (\text{MTTR})^{-1}$, the steady-state availability is equal to

$$A = \frac{\mu}{\lambda + \mu} = \frac{\text{MTBF}}{\text{MTBF} + \text{MTTR}} \tag{6.9}$$

The instantaneous availability, or probability that the item will be available at time t is equal to

$$A = \frac{\mu}{\lambda + \mu} + \frac{\lambda}{\lambda + \mu} \exp[-(\lambda + \mu)t] \tag{6.10}$$

which approaches the steady-state availability as t becomes large.

It is often more revealing, particularly when comparing design options, to consider system *unavailability*:

Steady-state unavailability $= 1 - A$

$$= \frac{\lambda}{\lambda + \mu} \tag{6.11}$$

and

$$\text{Instantaneous unavailability} = \frac{\lambda}{\lambda + \mu} - \frac{\lambda}{\lambda + \mu} \exp[-(\lambda + \mu)t] \tag{6.12}$$

If scheduled maintenance is necessary and involves taking the system out of action, this must be included in the availability formula. The availability of spare units for repair by replacement is often a further consideration, dependent upon the previous spares usage and the repair rate of replacement units.

Availability is an important consideration in relatively complex systems, such as power stations, satellites, chemical plant and radar stations. In such systems, high reliability by itself is not sufficient to ensure that the system will be available when needed. It is also necessary to ensure that it can be repaired quickly and that essential scheduled maintenance tasks can be performed quickly, if possible without shutting down the system. Therefore maintainability is an important aspect of design for maximum availability, and trade-offs are often necessary between reliability and maintainability features. For example, built-in test equipment (BITE) is incorporated into many electronic systems. This added complexity can degrade reliability and can also result in spurious failure indications. However, BITE can greatly reduce maintenance times, by providing an instantaneous indication of fault location, and therefore availability can be increased. (This is not the only reason for the use of BITE. It can also reduce the need for external test equipment and for training requirements for trouble-shooting, etc.)

Availability is also affected by redundancy. If standby systems can be repaired or overhauled while the primary system provides the required service, overall availability can be greatly increased.

Table 6.1 shows the reliability and steady-state availability functions for some system configurations. It shows clearly the large gains in reliability and steady-state availability which can be provided by redundancy. However, these are relatively simple situations, particularly as a constant failure rate is assumed. Also, for the standby redundant case, it is assumed that:

1. The reliability of the changeover system is unity.
2. No common-cause failures occur.
3. Failures are detected and repaired as soon as they occur.

Of course, these conditions do not necessarily apply, particularly in the case of standby equipment, which must be tested at intervals to determine whether it is serviceable. The availability then depends upon the test interval. Monitoring systems are sometimes employed, e.g. built-in test equipment (BITE) for electronic equipment, but this does not necessarily have a 100 per cent chance of detecting all failures. In real-life situations it is necessary to consider these aspects, and the analysis can become very complex. Methods for dealing with more complex systems are given at the end of this chapter, and maintenance and maintainability are covered in more detail in Chapter 14.

Example 6.1

A shipboard missile system is composed of two warning radars, a control system, a launch and guidance system, and the missiles. The radars are arranged so that either can give warning if the other fails, in a standby redundant configuration. Four missiles are available for firing and the system is considered to be reliable if three out of four missiles can be fired and guided. Figure 6.3 shows the system in reliability block diagram form, with the MTBFs of the subsystems. The reliability of each missile is 0.9. Assuming that: (1) the launch and guidance system is constantly activated, (2) the missile flight time is negligible and (3) all elements are s-independent, evaluate: (a) the reliability of the system over 24 h, (b) the steady-state availability of the system, excluding the missiles, if the mean repair time for all units is 2 h and the changeover switch reliability is 0.95.

Reliability Prediction and Modelling

Table 6.1 Reliability and availability for some systems configurations. (R. H. Myers, K. L. Wong, and H. M. Gordy, *Reliability Engineering for Electronic Systems*, Copyright © 1964 John Wiley & Sons, Inc. Reprinted by permission of John Wiley & Sons, Inc.)

Reliability configuration	Reliability (no repair) for CFR λ	General system reliability for n blocks
[single block R]	$\exp(-\lambda t)$	
[blocks R_1, R_2 in series]	$\exp[-(\lambda_1+\lambda_2)t]$	$\displaystyle\prod_{i=1}^{n} R_i$
[blocks R_1, R_2 in parallel] Active	$\exp(-\lambda_1 t)+\exp(-\lambda_2 t)$ $-\exp[-(\lambda_1+\lambda_2)t]$	$\displaystyle 1-\prod_{i=1}^{n}(1-R_i)$
Standby	$\dfrac{\lambda_2\exp(-\lambda_1 t)-\lambda_1\exp(-\lambda_2 t)^{\text{b}}}{\lambda_2-\lambda_1}$	$\displaystyle \exp(-\lambda t)\sum_{i=0}^{n-1}\frac{(\lambda t)^i}{i!}^{\text{a}}$
[three blocks R in parallel] Active 1/3	$3\exp(-\lambda t)-3\exp(-2\lambda t)$ $+\exp(-3\lambda t)$	As above (active)
Active 2/3	$3\exp(-2\lambda t)-2\exp(-3\lambda t)$	$\displaystyle 1-\sum_{i=0}^{m-1}\binom{n}{i}R^i(1-R)^{n-i}{}^{\text{c}}$
Standby 1/3	$\exp(-\lambda t)+\lambda t\exp(-\lambda t)^{\text{d}}$ $+\tfrac{1}{2}\lambda^2 t^2\exp(-\lambda t)$	$\displaystyle \exp(-\lambda t)\sum_{i=0}^{n-1}\frac{(\lambda t)^i}{i!}$

The reliabilities of the units over a 24 h period are:

Primary radar	0.9762	(failure rate $\lambda_P = 0.001$)
Standby radar	0.9762	(failure rate $\lambda_S = 0.001$)
Launch and guidance	0.9685	(failure rate $\lambda_{LG} = 0.0013$)

(a) The overall radar reliability is given by (from Table 6.1)

$$R_{\text{radar}} = \exp(-\lambda t)+\lambda t\exp(-\lambda t)$$

$$= 0.9762 + (0.001 \times 24 \times 0.9762) = 0.9996$$

The probability of the primary radar failing is $(1-R_P)$. The probability of this radar failing *and* the switch failing is the product of the two failure probabilities:

$$(1-0.9762)(1-0.95) = 0.0012$$

Therefore the switch reliability effect can be considered equivalent to a series unit with reliability

Table 6.1 *(continued)*

R for $\lambda_1 = \lambda_2 = 0.01$, $t = 100$	Steady-state availability, A, repair rate, μ, CFR, λ	General steady-state availability, A, for n blocks	A for $\lambda = 0.01$, $\mu = 0.2$
0.37	$\dfrac{\mu}{\lambda + \mu}$	–	0.95
0.14	$\dfrac{\mu_1 \mu_2}{\mu_1 \mu_2 + \mu_1 \lambda_2 + \mu_2 \lambda_1 + \lambda_1 \lambda_2}$	$\displaystyle\prod_{i=1}^{n} \dfrac{\mu_i}{\lambda_i + \mu_i}$	0.907
0.60	$\dfrac{\mu^2 + 2\mu\lambda^a}{\mu^2 + 2\mu\lambda + 2\lambda^2}$	$1 - \displaystyle\prod_{i=1}^{n} \dfrac{\lambda_i^a}{\lambda_i + \mu_i}$	0.996
0.74	$\dfrac{\mu^2 + \mu\lambda}{\mu^2 + \mu\lambda + \lambda^2}$	–	0.998
0.75	$\dfrac{\mu^3 + 3\mu^2\lambda + 6\mu\lambda^2}{\mu^3 + 3\mu^2\lambda + 6\mu\lambda^2 + 6\lambda^3}$	As above (active)	0.9999
0.31	$\dfrac{\mu^3 + 3\mu^2\lambda}{\mu^3 + 3\mu^2\lambda + 6\mu\lambda^2 + 6\lambda^3}$	$1 - \dfrac{1}{(\lambda + \mu)^n} \displaystyle\sum_{i=0}^{m-1} \binom{n}{i} \mu^i \lambda^{n-i}$	0.987
0.92	$\dfrac{\mu^3 + \mu^2\lambda + \mu\lambda^2}{\mu^3 + \mu^2\lambda + \mu\lambda^2 + \lambda^3}$	–	0.9999

[a] $\lambda_1 = \lambda_2 = \lambda$. Assumes series repair, i.e. single repair team.
[b] When $\lambda_1 = \lambda_2$ the reliability formula becomes indeterminate. When $\lambda_1 = \lambda_2$ use $R(t) = \exp(-\lambda t) + \lambda t \exp(-\lambda t)$. If $\lambda_1 \approx \lambda_2$ use $\lambda = (\lambda_1 + \lambda_2)/2$. Assumes perfect switching.
[c] For *m*-out-of-*n* redundancy.
[d] Assumes perfect switching.

$$R_{SW} = (1 - 0.0012) = 0.9988$$

The system reliability, up to the point of missile launch, is therefore

$$R_S = R_{radar} \times R_{SW} \times R_{LG}$$

$$= 0.9996 \times 0.9988 \times 0.9685 = 0.9670$$

The reliability of any three out of four missiles is given by the cumulative binomial distribution (Eqn 6.5):

$$R_M = 1 - \left[\binom{4}{0} 0.9^0 \times 0.1^4 + \binom{4}{1} 0.9^1 \times 0.1^3 + \binom{4}{2} 0.9^2 \times 0.1^2 \right]$$

$$= 0.9477$$

The total system reliability is therefore

$$R'_S = R_S \times R_M$$

$$= 0.9760 \times 0.9477 = 0.9250$$

(b) The availability of the redundant radar configuration is (see Table 6.1)

$$A_{radar} = \frac{\mu^2 + \mu\lambda}{\mu^2 + \mu\lambda + 2\lambda^2}$$

$$= \frac{0.5^2 + (0.5 \times 0.001)}{(0.5)^2 + (0.5 \times 0.001) + 2(0.001)^2}$$

$$= 0.999\,997 \qquad (\text{unavailability} = 3 \times 10^6)$$

The availability of the launch and guidance system

$$A_{LG} = \frac{\mu}{\mu + \lambda}$$

$$= \frac{0.5}{0.5 + 0.0013} = 0.9974 \qquad (\text{unavailability} = 2.6 \times 10^{-3})$$

The system availability is therefore

$$A_{radar} \times A_{LG} = 0.9974 \qquad (\text{unavailability} = 2.6 \times 10^{-3})$$

The previous example can be used to illustrate how such an analysis can be used for performing sensitivity studies to compare system design options. For example, a 20 per cent reduction in the MTBF of the launch and guidance system would have a far greater impact on system reliability than would a similar reduction in the MTBF of the two radars.

In maintained systems which utilize redundancy for reliability or safety reasons, separate analyses should be performed to assess system reliability in terms of the required output and failure rate in terms of maintenance arisings. In the latter case all elements can be considered as being in series, since all failures, whether of primary or standby elements, lead to repair action.

MODULAR DESIGN

Availability and the cost of maintaining a system can also be influenced by the way in which the design is partitioned. 'Modular' design is used in many complex products, such as electronic systems and aeroengines, to ensure that a failure can be corrected by a relatively easy replacement of the defective module, rather than by replacement of the complete unit.

Table 6.2 MTBR and replacement costs for the four modules

	MTBR (h)	Replacement costs ($)
Module 1	2500	3000
Module 2	4000	2000
Module 3	4000	2500
Module 4	10 000	1000

Table 6.3 Cost per year of replacing the modules

	Replacements per year	Cost per year ($)
Module 1	12	36 000
Module 2	7.5	15 000
Module 3	7.5	18 750
Module 4	3	3000
Total	30	$72 750

Example 6.2

An aircraft gas-turbine engine has a mean time between replacements (MTBR) – scheduled and unscheduled—of 1000 flight hours. With a total annual flying rate of 30 000 h and an average cost of replacement of $10 000, the annual repair bill amounted to $300 000. The manufacturer redesigned the engine so that it could be separated into four modules, with MTBR and replacement costs as shown in Table 6.2. What would be the new annual cost?

With the same total number of replacements, the annual repair cost is greatly reduced, from $300 000 to $72 750 (see Table 6.3).

Note that Example 6.2 does not take into account the different spares holding that would be required for the modular design (i.e. the operator would keep spare modules, instead of spare engines, thus making a further saving). In fact other factors would complicate such an analysis in practice. For example the different scheduled overhaul periods of the modules compared with the whole engine, the effect of wearout failure modes giving non-constant replacement rates with time since overhaul, etc. Monte Carlo simulation is often used for planning and decision-making in this sort of situation (see page 190).

BLOCK DIAGRAM ANALYSIS

The failure logic of a system can be shown as a reliability block diagram (RBD), which shows the logical connections between components of the system. The RBD is not necessarily the same as a block schematic diagram of the system's functional layout. We have already shown examples of RBDs for simple series and parallel systems. For systems involving complex interactions construction of the RBD can be

Figure 6.4 Block diagram decomposition

quite difficult, and a different RBD will be necessary for different definitions of what constitutes a system failure.

Block diagram analysis consists of reducing the overall RBD to a simple system which can then be analysed using the formulae for series and parallel arrangements. It is necessary to assume s-independence of block reliabilities.

The technique is also called block diagram *decomposition*. It is illustrated in Example 6.3.

Example 6.3

The system shown in Fig. 6.4 can be reduced as follows (assuming s-independent reliabilities):

$$R_s = R_1 \times R_2 \times R_B \times R_{10} \times R_C \qquad \text{(from Eqn 6.1)}$$

$$R_B = 1 - [1 - (R_3 \times R_4 \times R_5)]\,[1 - (R_6 \times R_7 \times R_8)]\,(1 - R_9) \qquad \text{(from Eqn 6.4)}$$

$$R_C = 1 - \frac{3 \times 2}{3 \times 2} R_{11}^0 (1 - R_{11})^3 + \frac{3 \times 2}{2} R_{11}(1 - R_{11})^2 \qquad \text{(from Eqn 6.5)}$$

$$= 1 - (1 - R_{11})^3 + 3R_{11}(1 - R_{11})^2$$

Cut and tie sets

Complex RBDs can be analysed using *cut set* or *tie set* methods. A cut set is produced by drawing a line through blocks in the system to show the minimum number of failed blocks which would lead to system failure. Tie sets (or *path sets*) are produced by drawing lines through blocks which, if all were working, would allow the system to work. Figure 6.5 illustrates the way that cut and tie sets are produced. In this system there are three cut sets and two tie sets.

Approximate bounds on system reliability as derived from cut sets and tie sets, respectively, are given by

$$R_s > 1 - \sum_{j}^{N} \prod_{i}^{n_j} (1 - R_i) \qquad (6.13)$$

(a)

(b)

Figure 6.5 (a) Cut sets and (b) tie sets

$$R_s < \sum_{j}^{T} \prod_{i}^{n_j} R_i \qquad (6.14)$$

where N is the number of cut sets, T is the number of tie sets and n_j is the number of blocks in the jth cut set or tie set.

Example 6.4

Determine the reliability bounds of the system in Fig. 6.5 for

$$R_1 = R_2 = R_3 = R_4 = 0.9.$$

Cut set:

$$R_s > 1 - [(1 - R_1)(1 - R_3) + (1 - R_2)(1 - R_3) + (1 - R_4)]$$
$$> 1 - [3 - R_1 - R_2 - 2R_3 - R_4 + R_1R_3 + R_2R_3]$$
$$> 1 - 0.12 = 0.88$$

Tie set:

$$R_s < R_1R_2R_4 + R_3R_4$$
$$< 1.54 \text{ (i.e. } < 1.0)$$

For comparison, the exact reliability is

$$R_s = [1 - (1 - R_1R_2)(1 - R_3)]R_4$$
$$= R_3R_4 + R_1R_2R_4 - R_1R_2R_3R_4$$
$$= 0.883$$

The cut and tie set approaches are not used for systems as simple as in Example 6.4, since the decomposition approach is easy and gives an exact result. However, since the derivation of exact reliability using the decomposition approach can become an intractable problem for complex systems, the cut and tie set approach has its uses in such applications. The approximations converge to the exact system reliability as the system complexity increases, and the convergence is more rapid when the block reliabilities are high. Tie sets are not usually identified or evaluated in system analysis, however.

Cut and tie set methods are suitable for computer application. Their use is appropriate for the analysis of large systems in which various configurations are possible, such as aircraft controls, power generation, or control and instrumentation systems for large plant installations. The technique is subject to the constraint (as is the decomposition method) that all block reliabilities must be s- independent.

Common mode failures

A common mode (or common cause) failure is one which can lead to the failure of all paths in a redundant configuration. Identification and evaluation of common mode failures is very important, since they might have a higher probability of occurrence than the failure probability of the redundant system when only individual path failures are considered. In the design of redundant systems it is very important to identify and eliminate sources of common mode failures, or to reduce their probability of occurrence to levels an order or more below that of other failure modes.

For example, consider a system in which each path has a reliability $R = 0.99$ and a common mode failure which has a probability of non-occurrence $R_{CM} = 0.98$. The system can be designed either with a single unit or in a dual redundant configuration (Fig. 6.6 (a) and (b)). Ignoring the common mode failure, the reliability of the dual redundant system would be 0.9999. However, the common mode failure practically eliminates the advantage of the redundant configuration.

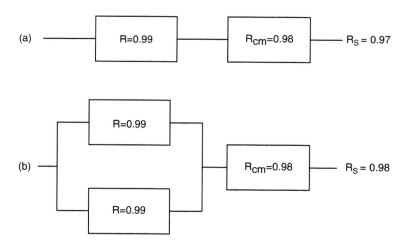

Figure 6.6 Effect of common mode failure

Examples of sources of common mode failures are:

1. Changeover systems to activate standby redundant units.
2. Sensor systems to detect failure of a path.
3. Indicator systems to alert personnel to failure of a path.
4. Power or fuel supplies which are common to different paths.
5. Maintenance actions which are common to different paths, e.g. an aircraft engine oil check after which a maintenance technician omits to replace the oil seal on *all* engines. (This has actually happened twice, very nearly causing a major disaster each time).
6. Operating actions which are common to different paths, so that the same human error will lead to loss of both.
7. Software which is common to all paths, or software timing problems between parallel processors.
8. 'Next weakest link' failures. Failure of one item puts an increased load on the next item in series, or on a redundant unit, which fails as a result.

Common mode failures can be very difficult to foresee, and great care must be taken when analysing safety aspects of systems to ensure that possible sources are identified.

Enabling events

An enabling event is one which, whilst not necessarily a failure or a direct cause of failure, will cause a higher level failure event when accompanied by a failure. Like common mode failures, enabling events can be important and difficult to anticipate, but they must be considered. Examples of enabling events are:

1. Warning systems disabled for maintenance, or because they create spurious warnings.
2. Controls incorrectly set.
3. Operating or maintenance personnel following procedures incorrectly, or not following procedures.
4. Standby elements being out of action due to maintenance.

Practical aspects

It is essential that practical engineering considerations are applied to system reliability analyses. The reliability block diagram implies that the 'blocks' are either 'failed' or operating, and that the logic is correct. Examples of situations in which practical and logical errors can occur are:

1. Two diodes (or two check valves) connected in series. If either fails open circuit (stuck closed), there will be no current (fluid) flow, so they will be in series from a reliability point of view. On the other hand, if either fails short circuit (or stuck open), the other will provide the required system function (current/fluid will flow in one direction), so they will be in parallel from a reliability point of view.

2. Two thermally activated switches (thermostats) wired in parallel to provide over-temperature protection for a system by starting a cooling fan, with backup in case of failure of one of the switches. This configuration might be modelled as two switches in parallel. However, practical engineering considerations that could make this simple model misleading or invalid are:

   —If one fails to close, the other will perform the protection function. However, there is no way of knowing that one has failed (without additional circuits or checks).
   —If one fails permanently closed, the fan would run continuously.
   —Since the two switches will operate at slightly different temperatures, one will probably do all the switching, so the duty will not be shared equally. If the one that switches first fails open, the contacts on the other might have degraded due to inactivity, so that it fails to switch.
   —If they do both switch about the same number of times, they might both deteriorate (contact wear) at the same rate, so that when one fails the other might fail soon after.

3. Data on the in-flight failure probabilities of aircraft engines are used to determine whether new types of commercial aircraft meet safety criteria (e.g. 'extended twin overwater operations' (ETOPS)). Engines might fail so that they do not provide power, but also in ways that cause consequential damage and possible loss of the aircraft. If the analyses consider only loss of power, they would be biased against twin-engine aircraft.

4. Common mode failures are often difficult to predict, but can dominate the real reliability or safety of systems. Maintenance work or other human actions are prime contributors. For example:

   —A technician fails to replace seals on aircraft engine oil sampling plugs, causing all of the engines to fail (this has actually happened twice!).
   —The Chernobyl nuclear reactor accident was caused by the operators conducting an unauthorized test.

5. Unexpected combinations of events can occur. For example:

   —The Concorde crash was caused by debris on the runway causing tyre failure, and fragments of the tyre then pierced the fuel tank.
   —The explosion of the Boeing 747 (TWA Flight 101) over Long Island Sound was caused by a combination of electrical cable insulation failure, buildup of conductive contamination on the fuel gauging probe, and pressure reduction in the empty fuel tank at altitude.

6. System failures can be caused by events other than failure of components or subsystems, such as electromagnetic interference (Chapter 9), operator error, etc.

These examples illustrate the need for reliability and safety analyses to be performed by engineers with practical knowledge and experience of the system design, manufacture, operation and maintenance.

FAULT TREE ANALYSIS (FTA)

Fault tree analysis (FTA) is a reliability/safety design analysis technique which starts from consideration of system failure effects, referred to as 'top events'. The analysis proceeds by determining how these can be caused by individual or combined lower level failures or events.

Standard symbols are used in constructing an FTA to describe events and logical connections. These are shown in Fig. 6.7. Figure 6.8 shows a simple BDA for an aircraft internal combustion engine. There are two ignition systems in an active parallel redundant configuration. The FTA (Figure 6.9) shows that the top event

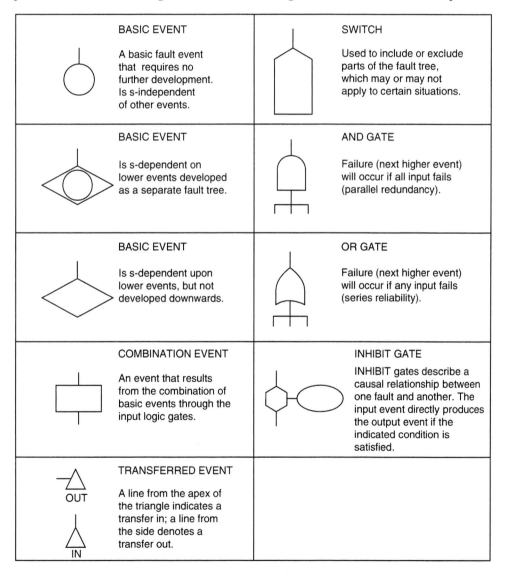

BASIC EVENT A basic fault event that requires no further development. Is s-independent of other events.	**SWITCH** Used to include or exclude parts of the fault tree, which may or may not apply to certain situations.
BASIC EVENT Is s-dependent on lower events developed as a separate fault tree.	**AND GATE** Failure (next higher event) will occur if all input fails (parallel redundancy).
BASIC EVENT Is s-dependent upon lower events, but not developed downwards.	**OR GATE** Failure (next higher event) will occur if any input fails (series reliability).
COMBINATION EVENT An event that results from the combination of basic events through the input logic gates.	**INHIBIT GATE** INHIBIT gates describe a causal relationship between one fault and another. The input event directly produces the output event if the indicated condition is satisfied.
TRANSFERRED EVENT OUT A line from the apex of the triangle indicates a transfer in; a line from the side denotes a transfer out. IN	

Figure 6.7 Standard symbols used in fault tree analysis

Figure 6.8 Reliability block diagram of engine

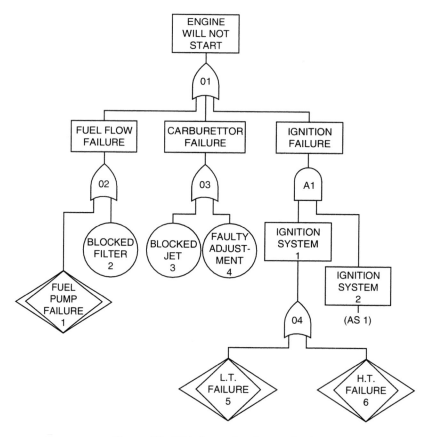

Figure 6.9 FTA for engine (incomplete)

failure to start can be caused by either fuel flow failure, carburettor failure or ignition failure (three-input OR gate). At a lower level total ignition failure is caused by failure of ignition systems 1 and 2 (two-input AND gate).

In addition to showing the logical connections between failure events in relation to defined top events, FTA can be used to quantify the top event probabilities, in the same way as in block diagram analysis. Failure probabilities derived from the reliability prediction values can be assigned to the failure events, and cut set and tie set methods can be applied to evaluate system failure probability.

Note that a different FTA will have to be constructed for each defined top event which can be caused by different failure modes or different logical connections between failure events. In the engine example, if the top event is 'unsafe for flight' then it would be necessary for both ignition systems to be available before take-off, and gate A1 would have to be changed to an OR gate.

The FTA shown is very simple; a representative FTA for a system such as this, showing all component failure modes, or for a large system, such as a flight control system or a chemical process plant, can be very complex and impracticable to draw out and evaluate manually. Computer programs are used for generating and evaluating FTAs. These perform cutset analysis and create the fault-tree graphics. The use of computer programs for FTA provides the same advantages of effectiveness, economy and ease of iterative analysis as described for FMECA. The practical aspects of system reliability modelling described earlier apply equally to FTA.

PETRI NETS

A Petri net is a general-purpose graphical and mathematical tool for describing relations existing between conditions and events. The basic symbols of Petri nets include:

○: *Place*, drawn as a circle, denotes event

──: *Immediate transition*, drawn as a thin bar, denotes event transfer with no delay time

▬: *Timed transition*, drawn as a thick bar, denotes event transfer with a period of delay time

⌇: *Arc*, drawn as an arrow, between places and transitions

●: *Token*, drawn as a dot, contained in places, denotes the data

♀: *Inhibitor arc*, drawn as a line with a circle end, between places and transitions

The transition is said to *fire* if input places satisfy an enabled condition. Transition firing will remove one token from each of its input places and put one token into all of its output places. Basic structures of logic relations for Petri nets are listed in Fig. 6.10, where there are two types of input places for the transition; namely, specified type and conditional type. The former one has single output arc whereas the latter has multiples. Tokens in the specified type place have only one outgoing destination, i.e. if the input place(s) holds a token then the transition fires and gives the output place(s) a token. However, tokens in the conditional type place have more

Logic relation	TRANSFER	AND	OR	TRANSFER AND	TRANSFER OR	INHIBITION
Description	If P then Q	If P AND Q then R	If P OR Q then R	If P then Q AND R	If P then Q OR R	If P AND Q' then R
Boolean function	Q=P	R=P*Q	R=P+Q	Q=R=P	Q+R=P	R=P*Q'
Petri nets						

Figure 6.10   Basic structures of logic relations for Petri nets (Courtesy S. Yang)

than one outgoing paths that may lead the system to different situations. For the 'TRANSFER OR' Petri nets in Fig. 6.10, whether Q or R takes over a token from P depends on conditions, such as probability, extra action, or self-condition of the place.

There are three types of transitions that are classified based on time. Transitions with no time delay due to transition are called immediate transitions, while those needing a certain constant period of time for transition are called timed transitions. The third type is called a stochastic transition: it is used for modelling a process with random time. Owing to the variety of logical relations that can be represented with Petri nets, it is a powerful tool for modelling systems. Petri nets can be used not only for simulation, reliability analysis, and failure monitoring, but also for dynamic behaviour observation. This greatly helps fault tracing and failure state analysis. Moreover, the use of Petri nets can improve the dialogue between analysts and designers of a system.

Transformation between fault trees and Petri nets

Figure 6.11 is a fault tree example in which events A, B, C, D, and E are basic causes of event 0. The logic relations between the events are described as well. The correlations between the fault tree and Petri net are shown in Fig. 6.12. Figure 6.13 is the Petri net transformation of Fig. 6.11.

Minimum cut sets

To identify the minimum cut sets in a Petri net a matrix method is used, as follows:

1. Put down the numbers of the input places in a row if the output place is connected by multi-arcs from transitions. This accounts for OR-models.
2. If the output place is connected by one arc from a transition then the numbers of the input places should be put down in a column. This accounts for AND-models.
3. The common entry located in rows is the entry shared by each row.
4. Starting from the top event down to the basic events until all places are replaced by basic events, the matrix is thus formed, called the *basic event matrix*. The column vectors of the matrix constitute cut sets.

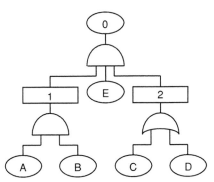

Figure 6.11 A fault tree (Courtesy S. Yang)

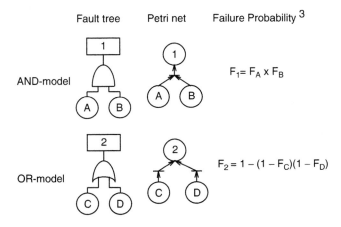

Figure 6.12 Correlations between fault tree and Petri net (Courtesy S. Yang)

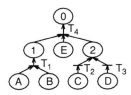

Figure 6.13 The Petri net transformation of Fig. 6.11 (Courtesy S. Yang)

5. Remove the supersets from the basic event matrix and the remaining column vectors become minimum cut sets.

For example, the basic event matrix for Fig. 6.13 is shown in Fig. 6.14
Minimum cut sets can be derived in an opposite, bottom-up, direction, i.e. from basic places to the top place. Transitions with $T = 0$ are called *immediate transitions*. If a Petri net with immediate transitions, i.e. the token transfer between places do not take time, then it can be absorbed to a simplified form called the *equivalent Petri net*.

1	E	2	2
A		C	D
B			

Minimum cut sets: ↑ ↑ ↑ ↑

Figure 6.14 Minimum cut sets of Fig. 6.13

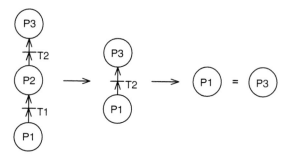

Figure 6.15 The absorption principle of equivalent Petri nets (Courtesy S. Yang)

Figure 6.15 shows the principle of absorption. After absorption, all the remaining places are basic events. The equivalent Petri net exactly constitutes the minimum cut sets, i.e. the input of each transition represents a minimum cut set.

Marking transfer

A *marking* of a Petri net is defined as the total number of tokens at each place, denoted by a column vector M. Thus vector $M_k = (n_1, n_2, \ldots, n_m)^T$ represents that token numbers of places $P_1, P_2, \ldots, P_m$ at state k are $n_1, n_2, \ldots, n_m$, respectively. Consequently, Petri nets can be expressed in state space form which gives the next state M_{k+1} from its previous state M_k:

$$M_{k+1} = \mathbf{A}M_k + \mathbf{B}U_k, \quad k = 1, 2, \ldots \tag{6.15}$$

where M_k is the marking at state k, a $m \times 1$ column vector, U_k is an input vector at state k, and $\mathbf{A}$ and $\mathbf{B}$ are matrix coefficients.

Since vector M_k represents the marking in a Petri net at state k, the failure state of a system may vary with time. Hence, the markings of a Petri net depend on time dynamically. The dynamic behaviour of system failure is defined as the system failure state with time varied, and is determined by the movement of tokens in the Petri net model. Thus, the dynamic behaviour of a system failure can be investigated by Petri nets whereas it cannot be done by fault trees. The method can be extended to include failure detection and correction logic, and repair and delay times. However, to date, there is no proprietary software that is available for creating and analysing Petri nets, so the technique is not easy to apply. Reference 16 provides an introduction, and some applications have been described, for example References 17 and 18.

STATE-SPACE ANALYSIS (MARKOV ANALYSIS)

A system or component can be in one of two states (e.g. failed, non-failed), and we can define the probabilities associated with these states on a discrete or continuous basis, the probability of being in one or other at a future time can be valuated using *state-space* (or *state-time*) analysis. In reliability and availability analysis, failure probability and the probability of being returned to an available state, failure rate and repair rate, are the variables of interest.

The best-known state-space analysis technique is Markov analysis. The Markov method can be applied under the following major constraints:

1. The probabilities of changing from one state to another must remain constant, i.e. the process must be homogenous. Thus the method can only be used when a constant hazard or failure rate assumption can be justified.
2. Future states of the system are independent of all past states except the immediately preceding one. This is an important constraint in the analysis of repairable systems, since it implies that repair returns the system to an 'as new' condition.

Nevertheless, Markov analysis can be usefully applied to system reliability, safety and availability studies, particularly to maintained systems for which BDA is not directly applicable, provided that the constraints described above are not too severe. The method is used for analysing complex systems such as power generation and communications. Computer programs can conveniently be written for Markov analysis.

The Markov method can be illustrated by considering a single component, which can be in one of two states: failed (F) and available (A). The probability of transition from A to F is $P_{A \to F}$ and from F to A is $P_{F \to A}$. Figure 6.16 shows the situation diagrammatically. This is called a *state transition* or a *state-space diagram*. All states, all transition probabilities and probabilities of remaining in the existing state (= 1− transition probability) are shown. This is a *discrete* Markov chain, since we can use it to describe the situation from increment to increment of time. Example 6.5 illustrates this.

Example 6.5

The component in Fig. 6.16 has transitional probabilities in equal time intervals as follows:

$$P_{A \to F} = 0.1$$

$$P_{F \to A} = 0.6$$

What is the probability of being available after four time intervals, assuming that the system is initially available?

This problem can be solved by using a *tree diagram* (Fig. 6.17).

The availability of the system is shown plotted by time interval in Fig. 6.18. Note how availability approaches a steady state after a number of time intervals. This is a necessary conclusion of the underlying assumptions of constant failure and repair rates and of *s*-independence of events.

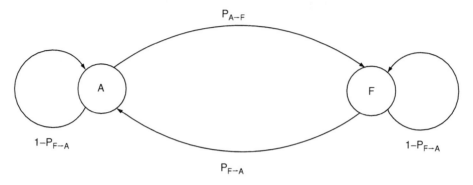

Figure 6.16 Two-state Markov state transition diagram

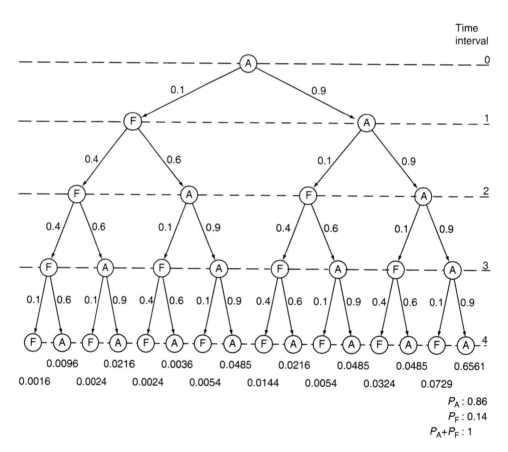

Figure 6.17 Tree diagram for Example 6.5

Whilst the transient states will be dependent upon the initial conditions (available or failed), the steady state condition is independent of the initial condition. However, the rate at which the steady state is approached is dependent upon the initial condition and on the transition probabilities.

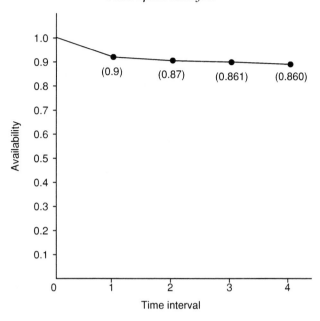

Figure 6.18 Transient availability of repaired system

Complex systems

The tree diagram approach used above obviously becomes quickly intractable if the system is much more complex than the one-component system described, and analysed over just a few increments. For more complex systems, matrix methods can be used, particularly as these can be readily solved by computer programs. For example, for a single repairable component the probability of being available at the end of any time interval can be derived using the *stochastic transitional probability matrix*:

$$P = \begin{vmatrix} P_{A \to A} & P_{A \to F} \\ P_{F \to A} & P_{F \to F} \end{vmatrix} \tag{6.16}$$

The stochastic transitional probability matrix for Example 6.5 is

$$P = \begin{vmatrix} 0.9 & 0.1 \\ 0.6 & 0.4 \end{vmatrix}$$

The probability of being available after the first time increment is given by the first term in the first row (0.9), and the probability of being unavailable by the second term in the first row (0.1). To derive availability after the second time increment, we square the matrix:

$$P^2 = \begin{vmatrix} 0.9 & 0.1 \\ 0.6 & 0.4 \end{vmatrix}^2 = \begin{vmatrix} 0.87 & 0.13 \\ 0.78 & 0.22 \end{vmatrix}$$

The probability of being available at the end of the second increment is given by the first term in the top row of the matrix (0.87). The unavailability $= 1 - 0.87 = 0.13$ (the second term in the top row).

For the third time increment, we evaluate the third power of the probability matrix, and so on.

Note that the bottom row of the probability matrix raised to the power 1, 2, 3, etc., gives the probability of being available (first term) or failed (second term) if the system started from the failed state. The reader is invited to repeat the tree diagram (Fig. 6.17), starting from the failed state, to corroborate this. Note also that the rows always summate to 1; i.e. the total probability of all states. (Revision notes on simple matrix algebra are given in Appendix 7.)

If a system has more than two states (multicomponent or redundant systems), then the stochastic transitional probability matrix will have more than 2×2 elements. For example, for a two-component system, the states could be:

State	Component 1	Component 2	
1	A	A	
2	A	$\bar{A}$	
3	$\bar{A}$	A	A: available
4	$\bar{A}$	$\bar{A}$	$\bar{A}$: unavailable

The probabilities of moving from any one state to any other can be shown on a 4×4 matrix. If the transition probabilities are the same as for the previous example, then:

$$P_{1\to1} = 0.9 \times 0.9 = 0.81$$

$$P_{1\to2} = 0.9 \times 0.1 = 0.09$$

$$P_{1\to3} = 0.1 \times 0.9 = 0.09$$

$$P_{1\to4} = 0.1 \times 0.1 = 0.01$$

$$P_{2\to1} = 0.9 \times 0.6 = 0.54$$

$$P_{2\to2} = 0.9 \times 0.4 = 0.36$$

$$P_{2\to3} = 0.1 \times 0.6 = 0.06$$

$$P_{2\to4} = 0.1 \times 0.4 = 0.04$$

$$P_{3\to1} = 0.6 \times 0.9 = 0.54$$

$$P_{3\to2} = 0.6 \times 0.1 = 0.06$$

$$P_{3\to3} = 0.4 \times 0.9 = 0.36$$

$$P_{3\to4} = 0.4 \times 0.1 = 0.04$$

$$P_{4\to1} = 0.6 \times 0.6 = 0.36$$

$P_{4\rightarrow 2} = 0.6 \times 0.4 = 0.24$

$P_{4\rightarrow 3} = 0.4 \times 0.6 = 0.24$

$P_{4\rightarrow 4} = 0.4 \times 0.4 = 0.16$

and the probability matrix is

$$
P = \begin{vmatrix}
P_{1\rightarrow 1} & P_{1\rightarrow 2} & P_{1\rightarrow 3} & P_{1\rightarrow 4} \\
P_{2\rightarrow 1} & P_{2\rightarrow 2} & \cdots & \\
P_{3\rightarrow 1} & \vdots & \vdots & \vdots \\
P_{4\rightarrow 1} & \cdots & \cdots & P_{4\rightarrow 4}
\end{vmatrix}
$$

$$
= \begin{vmatrix}
0.81 & 0.09 & 0.09 & 0.01 \\
0.54 & 0.36 & 0.06 & 0.04 \\
0.54 & 0.06 & 0.36 & 0.04 \\
0.36 & 0.24 & 0.24 & 0.16
\end{vmatrix}
$$

The first two terms in the first row give the probability of being available and unavailable after the first time increment, given that the system was available at the start. The availability after 2, 3, etc., intervals can be derived from $P^2, P^3, \ldots$, as above.

It is easy to see how, even for quite simple systems, the matrix algebra quickly diverges in complexity. However, computer programs can easily handle the evaluation of large matrices, so this type of analysis is feasible in the appropriate circumstances.

Continuous Markov processes

So far we have considered discrete Markov processes. We can also use the Markov method to evaluate the availability of systems in which the failure rate and the repair rate (λ, μ) are assumed to be constant in a time continuum. The state transition diagram for a single repairable item is shown in Fig. 6.19.

In the steady state, the stochastic transitional probability matrix is:

$$
P = \begin{vmatrix}
1 - \lambda & \lambda \\
\mu & 1 - \mu
\end{vmatrix}
\tag{6.17}
$$

The instaneous availability, before the steady state has been reached, can be derived using Eqn (6.11).

The methods described in the previous section can be applied for evaluating more complex, continuous Markov chains. Markov analysis can also be used for availability analysis, taking account of the holdings and repair rate of spares. The reader should refer to Reference 11 in the Bibliography for details of the Markov method as applied to more complex systems.

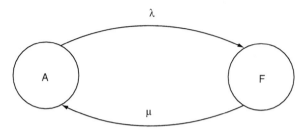

Figure 6.19   State–space diagram for a single-component repairable system

Limitations, advantages and applications of Markov analysis

The Markov analysis method suffers one major disadvantage. As mentioned earlier, it is necessary to assume constant probabilities or rates for all occurrences (failures and repairs). It is also necessary to assume that events are s-independent. These assumptions are hardly ever valid in real life, as explained in Chapter 2 and earlier in this chapter. The extent to which they might affect the situation should be carefully considered when evaluating the results of a Markov analysis of a system.

Markov analysis requires knowledge of matrix operations. This can result in difficulties in communicating the methods and the results to people other than reliability specialists. The severe simplifying assumptions can also affect the credibility of the results.

Markov analysis is fast when run on computers and is therefore economical once the inputs have been prepared. The method is used for analysing systems such as power distribution networks and logistic systems.

MONTE CARLO SIMULATION

In a Monte Carlo simulation, a logical model of the system being analysed is repeatedly evaluated, each run using different values of the distributed parameters. The selection of parameter values is made randomly, but with probabilities governed by the relevant distribution functions.

Monte Carlo simulation can be used for system reliability and availability modelling, using suitable computer programs. Since Monte Carlo simulation involves no complex mathematical analysis, it is an attractive alternative approach. It is a relatively easy way to model complex systems, and the input algorithms are easy to understand. There are no constraints regarding the nature of input assumptions on parameters such as failure and repair rates, so non-constant values can be used. It is also easy to model aspects such as queueing rules for repairs, repair priorities and 'cannibalization' (the use of serviceable spare parts from unserviceable systems).

One problem with Monte Carlo analysis is its expensive use of computer time. Since every event (failure, repair, movement, etc.) must be sampled for every unit of time, using the input distributions, a simulation of a moderately large system over a reasonable period of time can require hours of computer run-time. Also, since the simulation of probabilistic events generates variable results, in effect simulating the variability of real life, it is usually necessary to perform a number of runs in order to obtain estimates of means and variances of the output parameters of interest, such as availability, number of repairs arising and repair facility utilization. On the other hand, the effects of variations can be assessed.

A flow diagram showing the failure and repair logic for a typical system is shown in Fig. 6.20, and the results of simulation runs are shown in Fig. 6.21. References 19 and 20 describe the method in detail.

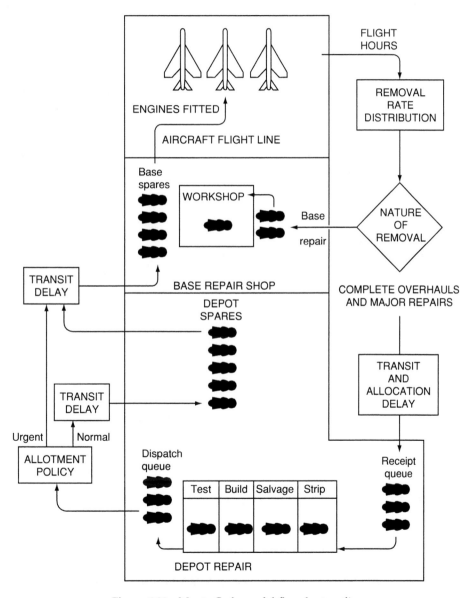

Figure 6.20 Monte Carlo model flowchart outline

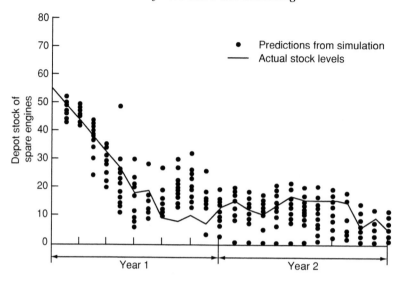

Figure 6.21   Depot stock levels

RELIABILITY APPORTIONMENT

Sometimes it is necessary to break an overall system reliability requirement down to individual subsystem reliabilities. This is common in large systems, particularly when different design teams or subcontractors are involved. The main contractor or system design team leader requires early assurance that subsystems will have reliabilities which will match the system requirement, and therefore the appropriate values have to be included in the subsystem specifications.

The starting point for reliability apportionment is a reliability block diagram for the system drawn to show the appropriate system structure. The system requirement is then broken down in proportions which take account of the complexity, risk and existing experience related to each block. It is important to take account of the uncertainty inherent in such an early prediction, and therefore the block reliabilities need not aggregate to the system requirement, but to some higher reliability. The apportionment and specifications derived from it should take account of different operating conditions of subsystems. For example, a radar system might have a subsystem which operates for only half of the total system operating time, and therefore this should be shown on the RBD, and the failure rate apportioned to it should be related clearly either to the operating time of the system or of the subsystem.

STANDARD METHODS FOR RELIABILITY PREDICTION AND MODELLING

The most commonly used standard reference for reliability prediction was US MIL-STD-756 (Reference 1 in the Bibliography).

IEEE Standard 1413 (Reference 4) has been created, to establish a framework around which reliability prediction should be performed for electronic systems, though it could be applied to any technology. Prediction results obtained from an IEEE 1413 compliant reliability prediction are accompanied by responses to a set of questions identified in IEEE 1413. In particular, an IEEE 1413 compliant prediction provides documentation of:

—the prediction results,
—the intended use of prediction results,
—the method(s) used,
—inputs required for selected method(s),
—the extent to which each input is known,
—the source of known input data,
—assumptions for unknown input data,
—the prediction result,
—figures of merit,
—conclusions and recommendations,
—confidence in prediction,
—sources of uncertainty,
—limitations and
—repeatability.

A further reference is NASA-CR-1129 (Reference 5 in the Bibliography), which describes the approaches to reliability prediction for NASA space programmes. Here the stress is laid on reliability prediction as a method of comparing designs, and there is less emphasis on quantitative methods. A British Standard, BS 5760 (Reference 6), also describes reliability prediction methods and relates them to the reliability programme. UK Defence Standard 00–41 emphasizes the top-down approach to reliability prediction. See also Chapter 9 (page 270).

CONCLUSIONS

System reliability (and safety and availability) prediction and modelling can be a frustrating exercise, since even quite simple systems can lead to complex reliability logic when redundancy, repair times, testing and monitoring are taken into account. On the other hand, the parameters used, particularly reliability values, are usually very uncertain, and can be highly variable between similar systems. This chapter has described some approaches, but it must be realized that the results can be very sensitive to parameter changes. For example, a single common-cause failure, overlooked in the analysis, might have a probability of occurrence which would completely invalidate the reliability calculations for a high reliability redundant system. In real life, availability is often determined more by spares holdings, administrative times (transport, documentation, delays, etc.) than by 'predictable' factors such as mean repair time. Therefore predictions and models of system reliability and availability should be used as a form of design review, to provide a disciplined framework for considering factors which will affect reliability and availability, and the sensitivity to changes in assumptions. Critical aspects can then be highlighted for further attention, and alternative system approaches can be compared.

Prediction and modelling are concepts which have generated much attention, literature and controversy in the reliability field. Considerable effort has been expended on the development and updating of databases and models, and a large proportion of the journal articles and conference contributions is devoted to the topics. Most of this work is, however, of only abstruse interest, since reliability is not a parameter which is inherently predictable, on the basis of the laws of nature or of statistical extrapolation. The mathematical techniques described in this chapter are useful only in so far as the values inserted into the formulae are known within reasonably close limits. The use of complicated formulae to analyse the effects of parameters which are highly uncertain is inefficient and potentially misleading. It is essential that the extent of uncertainty is always considered, and that reliability specialists are discouraged from taking prediction and modelling work beyond the point of usefulness.

Finally, the predictions must be based on the committment by the project management, and on a realistic appreciation of the technologies and risks.

BIBLIOGRAPHY

1. US MIL-STD-756: *Reliability Prediction*. National Technical Information Service, Springfield, Virginia (Now discontinued).
2. US MIL-HBK-217: *Reliability Prediction for Electronic Systems*. Available from the National Technical Information Service, Springfield, Virginia.
3. *Non-electronic Parts Reliability Data (NPRD)*: USAF Rome Air Development Center. Available from the National Technical Information Service, Springfield, Virginia.
4. IEEE Standard 1413 *Standard Methodology for Reliability Prediction and Assessment of Electronic Systems and Equipment*.
5. NASA-CR-1129: *Reliability Prediction*. Available from the National Technical Information Service, Springfield, Virginia.
6. British Standard, BS 5760: *Reliability of Systems, Equipments and Components*. British Standards Institution, London.
7. K. C. Kapur and L. R. Lamberson, *Reliability in Engineering Design*. Wiley (1977).
8. C. Singh and R. Billington, *System Reliability/Modelling and Evaluation*. Hutchinson (1977).
9. A. K. S. Jardine, *Maintenance, Replacement and Reliability*. Pitman (1973).
10. E. J. Henley and H. Kumanoto, *Reliability Engineering and Risk Assessment*. Prentice-Hall (1981).
11. R. Billington and R. N. Allen, *Reliability Evaluation of Engineering Systems*. Plenum Press (1982).
12. A. Pages and M. Gondran, *Systems Reliability Evaluation and Prediction in Engineering*. North Oxford Academic (1986).
13. A. Villemeur, *Reliability, Availability, Maintainability and Safety Analysis*. J. Wiley (1992).
14. J. D. Andrews and T. R. Moss, *Reliability and Risk Assessment*. Longman (J. Wiley in USA) (1993).
15. A. Høyland and M. Rausand, *System Reliability Theory, Models and Statistical Methods*. J. Wiley (1994).
16. V. Alexander and Yakovlev, *Hardware Design and Petri Nets*. Kluwer Academic Publishers (2000).
17. S. K. Yang and T. S. Liu, *Failure Analysis for an Airbag Inflator by Petri Nets*. Quality and Reliability Engineering International, **13**, pp. 139–151, 1997.

18. S. K. Yang and T. S. Liu, *A Petri Net Approach to Early Failure Detection and Isolation for Preventive Maintenance*. Quality and Reliability Engineering International, **14**, pp. 319–330, 1998.
19. F. F. Martin, *Computer Modelling and Simulation*. Wiley (1968).
20. J. Banks (Editor), *Handbook of Simulation*. Wiley-Interscience (1998).

QUESTIONS

1. Assume that you are responsible for the reliability aspects of a system containing both electronic and mechanical elements. The customer for the system requires that a numerical reliability prediction be provided.

   (a) Describe what is meant by a 'reliability prediction' in this context.

   (b) Identify some sources of data that can be used to assist in quantifying the prediction, and discuss the dangers that have to be guarded against in the use of such data.

   (c) Do you expect the prediction to overestimate or underestimate the reliability that could be achieved by the system? Give your reasons.

2. Two resistors are connected in parallel in an electrical circuit. They can fail either open-circuit (resistance $= \infty$) or short-circuit (resistance $= 0$). Draw reliability block diagrams for the pair of resistors (i) for the circuit failing open-circuit; (ii) for the circuit failing-short circuit.

3. The system sketched below is used to regulate the downstream pressure of gas in a chemical plant. There are two regulators whose function is to keep the downstream pressure at a constant value. The upstream pressure fluctuates, but is always much higher than the required downstream pressure. The regulators function independently, but both sense the downstream pressure at the same point (X).

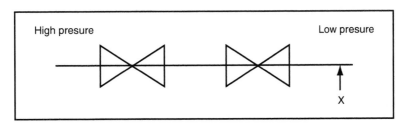

   A regulator can fail in one of two ways: 'open', in which case it becomes 'straight through' so that the flow of gas is unrestricted, and reliability in this mode is R_0; or 'closed', in which it totally blocks the flow of gas, and reliability in this mode is R_c.

   (a) On the assumption that the two types of failure are independent, produce expressions for system reliability (i) for the system failing due to total loss of flow; (ii) for the system failing due to overpressure downstream.

   (b) The times to failure of the regulators are described by the following distributions:
   Closed—exponential with mean life 2 years.
   Open—Weibull with $\beta = 1.8$ and $\eta = 1.6$ years.

What are the probabilities of the system giving one year of failure-free operation in each mode?

(c) What is the probability of obtaining one year of failure-free operation irrespective of the mode of failure?

4. It has been suggested that the reliability of the system in question 3 can be improved by adopting a twin-stream system as in the diagram below.

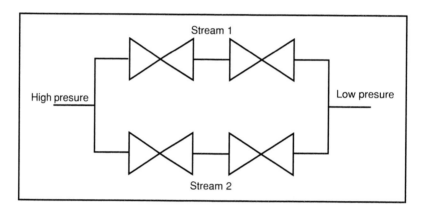

Calculate the two system reliabilities for (loss of flow and overpressure) for this configuration.

5. Draw block diagrams for both modes of failure of the 'twin stream' in question 4. Draw the cut sets and tie sets in each case.

6. Calculate the reliability over a 150 h mission for the system whose reliability block diagram is shown below.

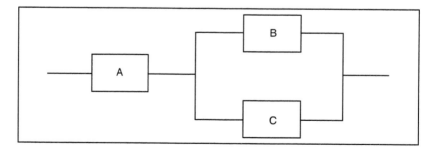

The probability distributions of lives to failure of the elements are:
A—constant hazard, mean life = 837 h
B—Weibull, with shape parameter = 2.0, characteristic life 315 h
C—normal, with mean = 420 h and standard deviation 133 h.

7. An element in a manufacturing system has a mean time between failures of 25 h. When it fails it takes, on average, 2 h to restore it to an operating condition. It has been

suggested that the problems caused by the unreliability of this element could be avoided by installing a second identical element as a standby.

(a) Calculate the probability of completing an 8 h shift without a system failure both for the single-element and for the redundant-element system.

(b) If a repair team is made available such that repairs can be started on a failed element immediately after failure, calculate the long-term system availability (assuming continuous operation) for both the single-element and the redundant-element system. State all assumptions made and comment on whether you consider them reasonable.

(c) If the additional element costs £25 000 and downtime costs £100 per hour, what is the payback period for the additional element? (Ignore discounting of cash flows.)

8. In the context of fault-tree analysis, explain the meaning of each of the following:

    an AND gate   "top" event
    an OR gate   a basic event
    a priority AND gate   an undeveloped event.

In each case, sketch the conventional symbol used and give a practical example.

9. A control system consists of an electrical power supply, a standby battery supply which is activated by a sensor and switch if the main supply fails, a hydraulic power pack, a controller, and two actuators acting in parallel (i.e. control exists if either or both actuators are functioning).

    a. Draw the system reliability block diagram.
    b. Draw the fault tree appropriate to the top event "total loss of actuator control".

10. In question 9, if the reliabilities of the separate components are as follows:

Main electrical supply: 0.99
Standby battery supply: 0.995
Sensor and switch: 0.995
Hydraulic supply: 0.95
Controller: 0.98
Actuator (each): 0.99

What is the system reliability for the top event "total loss of actuator control"?

11. This is a very simple example of simulation to explore the sort of principles involved in much more complex (and realistic) situations as in Fig. 6. It describes the common 'queuing' problem, but by using simulation we are able to circumvent the convenient, but often implausible assumptions that are usually incorporated in conventional 'queuing theory'. In particular, we do not need to assume exponentially distributed arrival and service times.

A radio taxi service operates between 9am and 7pm. Calls arrive at random at a rate of three per hour. The time to reach a customer is a Weibull distribution with $\beta = 2$, $\eta = 0.25$ h. The journey time is normally distributed with mean 30 minutes and standard deviation 6 minutes. A customer will only be offered a service if there is a taxi available, i.e. not currently on a journey.

You are invited to do one 'manual' simulation of a day's operation with a fleet of two taxis, and evaluate the reliability (i.e. proportion of customers picked up). The calculations are simple but tedious, so computer methods are used when we want many thousands of simulations of much more complex situations (as in Fig. 5.10). You will need to generate random sample values for:

t_1–the time of a customer call since the previous one
t_2–the time to reach the customer
t_3–the journey time with the customer.

These will be developed from random numbers between 0 and 1. Such numbers can be generated in various ways; e.g.

–using the ran # function on a calculator
–rolling a 20-sided die
–using tables of random numbers
–using the random number generator built into common application software (e.g. most spreadsheets).

Simulating t_1:

The reliability function (probability that the time to reach a customer exceeds $t = t_1$) is from the exponential distribution, i.e. $= \exp(-\lambda t)$, from which $t = 1/\lambda \ln[1/R(t)]$. Simply generate a random number between 0 and 1 for $R(t)$ and calculate the value of t using $\lambda = 1/3$. For example, for a random number of 0.439, the random time is $3\ln(1/0.439) = 0.823\,\text{h}$.

Simulating t_2:

This time will be simulated from a Weibull distribution, where

$$R(t) = \exp[-(t/\eta)^\beta]$$

from which

$$t = \eta\{\ln[1/R(t)]\}^{1/\beta}$$

We have $\beta = 2$ and $\eta = 0.25$; so, for example, with a random number of 0.772, the simulated value of t_2 is

$$0.25\{\ln[1/0.772]\}^{0.5} = 0.127\,\text{h}$$

Simulating t_3:

This comes from a normal distribution. There is no closed form for the reliability function of this distribution, so some alternative approach is necessary. The easiest is to use tables of standardized random normal deviates (as included in most statistics books) or an implementation in a computer spreadsheet. Suppose we obtained a value of -0.194. For our particular normal distribution the mean is 0.5 h and the standard deviation 0.1 h, so the simulated random time is

$$0.5 + (-0.194 \times 0.1) = 0.481\,\text{h}.$$

Using these ideas, 'walk through' a random sample of one day of operation of two taxis. See how many requests were made, and how many were delivered to their destinations.

12. A device used in a ground radar system has age to failure that is described approximately by a Weibull distribution with mean life 83 h, shape parameter 1.5, and location parameter zero. When it fails it takes on average 3.5 h to repair.

    (a) Calculate the reliability over a 25 h period, and the 'steady state' availability of the device.

    (b) Calculate the reliability over 25 h, and the 'steady state' availability of a subsystem that consists of two of these devices in active parallel redundancy.

    (c) Identify all assumptions made in these calculations, and discuss their validity.

    (d) Explain the meaning of the 'steady state availability' in (a) and (b) above, and consider whether it gives the most suitable measure of availability in this example.

13. Describe the practical limitations of using methods such as BDA, FTA, Markov chains, etc. for assessing the reliability, in qualitative and in quantitative terms, of the following (Hint: in each case, first define the failure you are considering):

    a. A parachute (consider the canopy, the lines, and the deployment mechanism).

    b. A microprocessor (consider the power input, the data input and output connections, and every transistor and capacitor).

    c. An electric motor.

    d. A mechanical assembly that consists of static parts bolted together.

    e. A train.

7

Reliability in Design

INTRODUCTION

The reliability of a product is strongly influenced by decisions made during the design process. Deficiencies in design affect all items produced and are progressively more expensive to correct as development proceeds. It is often not practicable or economic to change a design once production has started. It is therefore essential that design disciplines are used which minimize the possibility of failure and which allow design deficiencies to be detected and corrected as early as possible. In Chapter 4 the basic requirements for failure-free design were laid down, i.e. adequate safety margins, protection against extreme load events and protection against strength degradation. The design must also take account of all other factors that can affect reliability, such as production methods, use and maintenance, and failures not caused by load.

The design process must therefore be organized to ensure that failure-free design principles are used and that any deviations from the principles are detected and corrected. Failure-free design is the only acceptable principle for any reliability-conscious project team. Anything less will be reflected in the acceptance of failures throughout the development and production cycle, and a low rate of improvement. The designer must produce designs which will not fail if manufactured and used as specified. In order to be able to do this test data may be needed to reduce uncertainties. Any subsequent failures can then be firmly classified as design deficiencies which escaped the review or test system, or as being due to manufacturing failures or overload. Failure-free design therefore involves prevention, check and cure.

This chapter (and Chapters 8 and 9 on mechanical and electronic reliability) does not cover the full range of knowledge necessary for good design. The designer must be aware of the materials, processes, components, production methods, design rules and guidelines, costs, and much else in order to create a good design. These chapters cover only fairly general topics, which should be known by reliability and design engineers working on most types of project.

COMPUTER-AIDED ENGINEERING

Computer-aided engineering (CAE) methods are available to assist with a wide variety of design tasks. Their power, ease of use, and increasing availability due to reducing costs of computing equipment and software are resulting in increasing applications. CAE also makes possible the creation of designs which would otherwise be very difficult or uneconomic, for example complex electronic circuits. (CAE for electronic design is usually referred to as electronic design automation (EDA).) CAE can also provide enormous improvements in engineering productivity. Properly used, it can lead to the creation of more reliable designs.

Proprietary versions of the SPICE analog simulation program can be used to design electronic circuits, and to test their operation under different operating conditions. Component performance details are held in the database. The designer can, in principle, design the circuit, then 'build' it and 'test' it, all on the computer screen. The effects of parameter changes or failure modes can be quickly evaluated, and dynamic as well as static operating conditions can be tested. More advanced software exists for digital circuit design and evaluation.

In the mechanical engineering field, software is available for stress analysis, which performs finite element analysis calculations for mechanical and thermal stress calculations, and for analysis of vibration and load responses. Drafting software is used for generating manufacturing drawings and machine tool instructions, and this can also be used to optimize the design of mechanisms.

Specialist CAE software is also available for design and analysis of systems and products incorporating other technologies, such as hydraulics, magnetics, and microwave electronics. Multi-technology capability is now also being provided, so that mixed technology designs can be modelled and analysed.

CAE provides the capability for rapid assessment of different design options, and for analysing the effects of tolerances, variation, and failure modes. Therefore, if used in a systematic, disciplined way, with adequate documentation of the options studied and assessments performed, designs can be optimized for costs, producibility and reliability. CAE should be considered as a powerful aid to more cost-effective and correct design, not merely a means of speeding up the design process.

However, there are important limitations inherent in most CAE tools. The software models can never be totally accurate representations of all aspects of the design and of its operating environment. For example, electronic circuit simulation programs generally ignore the effects of electromagnetic interference between components, and drafting systems will ignore distortion due to stress or temperature. Therefore it is essential that engineers using CAE are aware of the limitations, and how these could affect their designs. The effective application of modern CAE places greater responsibility upon designers to be aware of the practical aspects and limitations of the relevant technologies. Otherwise they can be easily misled into placing undue faith in the accuracy and completeness of the software models, resulting in incorrect or unreliable designs.

CAE methods are described in Reference 1; this book. The homepage for his book (Preface, p. xxi) lists available software.

ENVIRONMENTS

The environments in which the product will be expected to be stored, operated and maintained must be carefully assessed, as well as the expected severity and durations. The assessment must include all aspects that could affect the product's operation, safety and reliability. Physical factors include temperature, vibration, shock, humidity, pressure, etc. Extreme values and, where relevant, rates of change must be considered. Other environmental conditions, such as corrosive atmospheres, electrical interference, power supply variation, etc., must also be taken into account. Where appropriate, combined environmental conditions, such as temperature/corrosive atmosphere and vibration/contamination, should be assessed. An aspect of the environment often neglected is the treatment of the product by people, in storage, handling, operation and maintenance.

Environmental aspects should be reviewed systematically, and the review should be properly documented.

The protective measures to be taken must be identified, as appropriate to storage, transport, handling, operation and maintenance. Protective measures include packaging, provision of warning labels and instructions, protective treatment of surfaces, and design features. Detailed design aspects are covered in Chapters 8 and 9.

Resistance to environmental conditions must be confirmed by test when hardware is available. Test aspects are covered in Chapters 11 and 12.

Load protection

Protection against extreme loads is not always possible, but should be considered whenever practicable. In many cases the maximum load can be pre-determined, and no special protection is necessary. However, in many other loading situations extreme external loads can occur and can be protected against. Standard products are available to provide protection against, for example, overpressure in hydraulic or pneumatic systems, impact loads or electrical overload. When overload protection is provided, the reliability analysis is performed on the basis of the maximum load which can be anticipated, bearing in mind the tolerances of the protection system. In appropriate cases, loads which can occur when the protection system fails must also be considered.

However, in most practical cases it will be sufficient to design to withstand a predetermined load and to accept the fact that loads above this will cause failure. The probability of such loads occurring must be determined for a full reliability analysis to be performed. It may not always be practicable to determine the distribution of such extreme events, but data may be available either from failure records of similar items, or from test or other records.

Where credible data are not available, the worst design load case must be estimated. The important point is that the worst design case is estimated and specified. A common cause of failure is the use of safety factors related to average load conditions, without adequate consideration having been given to the extreme conditions which can occur during use of the product.

Protection against strength degradation

Strength degradation, in its many forms, can be one of the most difficult aspects to take into account in design reliability analysis. Strength degradation due to fatigue in metals is fairly well understood and documented, and therefore reliability analysis involving metal fatigue, including the effects of stress raisers such as notches, corners, holes and surface finish, can be performed satisfactorily, and parts can be designed to operate below the fatigue limit, or for a defined safe life.

However, other weakening mechanisms are often more complex. Combined stresses may accelerate damage or reduce the fatigue limit. Corrosion and wear are dependent upon environments and lubrication, the effects of which are therefore often difficult to forecast. If complete protection is not possible, the designer must specify maintenance procedures for inspection, lubrication or scheduled replacement.

Reliability analysis of designs with complex weakening processes is often impracticable. Tests should then be designed to provide the required data by generating failures under known loading conditions.

Chapter 8 covers these aspects in more detail.

DESIGN ANALYSIS METHODS

Despite discipline, training and care, it is inevitable that occasional oversights or errors will occur in new designs. Design analysis methods have been developed to highlight critical aspects and to focus attention on possible shortfalls.

Design analyses are sometimes considered tedious and expensive. In most cases the analysis will show that nearly all aspects of the design are satisfactory, and much more effort will have been expended in showing this than in highlighting a few deficiencies. However, the discovery of a very few deficiencies at an appropriately early stage can save far more than the costs that might be incurred by having to modify the design at a later stage, or by having to live with the consequences of the defect. Therefore, well-managed design analyses are extremely cost-effective. The tedium and expense can be greatly reduced by good planning and preparation and by the use of computerized methods. In this section, we will describe the main design analysis techniques available. Their place in the overall design review process and the way they should be managed are also covered. The main reliability design analysis techniques described are:

1. Quality function deployment.
2. Reliability prediction (covered in Chapter 6).
3. Load–strength analysis.
4. Failure modes, effects and criticality analysis.
5. Fault tree analysis (described in Chapter 6).
6. Hazard and operability study.
7. Parts materials and process review.
8. Others, including human aspects manufacturing, maintenance, etc.

QUALITY FUNCTION DEPLOYMENT (QFD)

Quality Function Deployment (QFD) is a horrible expression for a simple technique to identify all of the factors which might affect the ability of a design or product to satisfy the customer, and the methods and responsibilities necessary to ensure control. QFD goes beyond reliability, as it covers aspects such as customer preferences for feel, appearance, etc., but it is a useful and systematic way to highlight design and process activities and controls necessary to ensure reliability.

QFD begins by a team consisting of the key marketing, design, production, reliability and quality staff working their way through the project plan or specification, and identifying the features that will require to be controlled, the control methods applicable, and the responsible people. Constraints and risks are also identified, as well as resources necessary. At this stage no analysis or detailed planning is performed, but the methods likely to be applied are identified. These methods are described later in this chapter and in others.

QFD makes use of charts which enable the requirements to be listed, and controls, responsibilities, constraints, etc., to be tabulated, as they relate to design, analysis, test, production, etc. An example is shown in Fig. 7.1.

This shows requirements rated on an importance scale (1–5), and the design features that can affect them. Each feature is in turn rated against its contribution to each requirement, and a total rating of each feature is derived by multiplying each rating by the importance value, and adding these values. Thus the bearing selection, housing construction, and mounting design come out as the most critical design features.

The 'benchmark' column is used to rate each requirement, as perceived by potential customers, against those of competitive products. Only one benchmark is shown here; more can be added for other competitors. Benchmarking is a useful method for putting requirements into a sound marketing perspective.

The correlation matrices indicate the extent to which requirements and features interact: plus sign(s) indicate positive correlation, and minus negative correlation. For example, magnet material and stator winding design might interact strongly. The minus signs in the requirements matrix indicate conflicting requirements.

The options available are shown. In some cases further modelling or experiments are required, and this part of the chart can be used to indicate the variables that need to be included in such work.

The shape of the QFD chart has led to its being called the 'house of quality'. Of course quality here is used in the widest sense, to include all aspects of the product that will affect its reputation and cost. Figure 7.1 is a top level chart: lower level charts are used to analyse more detailed aspects, for example, more detailed design and component characteristics, and production processes and tolerances, always against the same set of requirements. Thus every aspect of design and production, including analysis, test, production process control, final inspection, packaging, maintenance, etc., is systematically evaluated and planned for, always in relation to the most important product requirements. Requirements and features that are not important are shown up as such, and this can be a very important contribution to cost reduction and reliability improvement.

Like other conceptual analysis methods, there is no single standard approach, and users should be encouraged to develop formats and methods appropriate to their products and problems.

Reference 2 describes the method in detail.

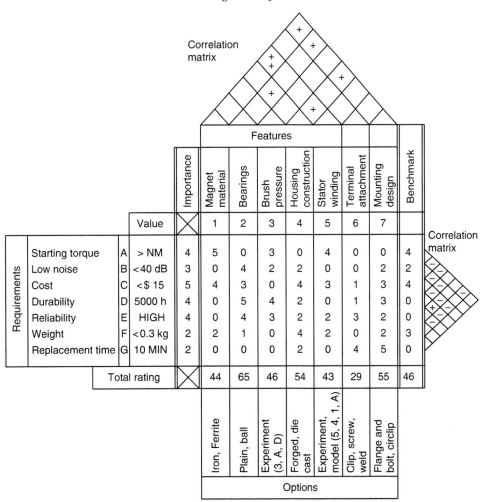

Figure 7.1 Quality function deployment for electric motor design

LOAD–STRENGTH ANALYSIS (LSA)

Load–strength analysis (LSA) is a procedure to ensure that all load and strengh aspects have been considered in deriving the design, and if necessary in planning of tests. The load–strength analysis (LSA) should include the following:

—Determine the most likely worst case values and patterns of variation of load and strength.
—Evaluate the safety margin for intrinsic reliability.
—Determine protection methods (load limit, derating, screening, other quality control methods).
—Identify and analyse strength degradation modes.

—Test to failure to corroborate, analyse results.
—Correct or control (redesign, safe life, quality control, maintenance, etc.).

Table 7.1 is an example of a hypothetical load–strength analysis for a mechanical and electrical assembly. The example shows approaches that can be used for different aspects of the analysis. Event probabilities can be expressed as full distributions, or as the likelihood of a particular limiting case being exceeded. The former is more appropriate when the load(s) can cause degradation, or if a more detailed reliability assessment is required. Both examples show typical, though rather simple, cases where the effects of combined loads might have been overlooked but for the analysis. For example, the solenoid might be supplied with a manufacturer's rating of 28 V operating, ±2 V, and a maximum ambient temperature of 45 °C. A room temperature test of the solenoid might have confirmed its ability to function with a 32 V supply without overheating. However, the combined environment of +45°C and 32 V supply, albeit an infrequent occurrence, could lead to failure.

The mechanical example is less easy to analyse and testing is likely to be the best way of providing assurance, if the assembly is critical enough to warrant it. Where the load–strength analysis indicates possible problems, further analysis should be undertaken, e.g. use of probabilistic methods as described in Chapter 4, and CAE methods. Tests should be planned to confirm all design decisions.

FAILURE MODES, EFFECTS AND CRITICALITY ANALYSIS (FMECA)

Failure modes effects and criticality analysis (FMECA) (or failure modes and effects analysis (FMEA)), is probably the most widely used and most effective design reliability analysis method. The principle of FMECA is to consider each mode of failure of every component of a system and to ascertain the effects on system operation of each failure mode in turn. Failure effects may be considered at more than one level, e.g. at subsystem and at overall system level. Failure modes are classified in relation to the severity of their effects.

An FMECA may be based on a hardware or a functional approach. In the hardware approach actual hardware failure modes are considered (e.g. resistor open circuit, bearing seizure). The functional approach is used when hardware items cannot be uniquely identified or in early design stages when hardware is not fully defined. In this approach function failures are considered (e.g. no feedback, memory lost). Note that a functional failure mode can become a hardware failure effect in a hardware-approach FMECA. An FMECA can also be performed using a combination of hardware and functional approaches.

Figures 7.2 and 7.3 show typical worksheets taken from US MIL-HBBK-1629 (Reference 3). Method 101 is a non-quantitative method, which serves to highlight failure modes whose effects would be considered important in relation to severity, detectability, maintainability or safety.

Method 102 (criticality analysis) includes consideration of failure rate or probability, failure mode ratio and a quantitative assessment of criticality, in order to provide a quantitative criticality rating for the component or function. The *failure mode criticality number* is

Table 7.1 Load–strength analysis example

Item (Matl, function)	Worst case load/combined load	Frequency/probability of occurrence	Data source	Combined effect	Strength	Remarks
Rivet (×4) (aluminium, fixing bracket to plastic frame)	1. 50 N total, axial 2. 40 N, lateral impact 3. Temperature 0–35°C	Continuous See load distribution annex 1	— Operating data	— Combine with 1. *Degradation* NIL	Plastic frame is weak link NIL effect	Life test to confirm Thickness of plastic frame at bracket attachment may be critical feature
Solenoid coil	1. 32 V (at 27°C) 2. 45°C ambient	$1/10^4$ h $1/10^2$ h	Data on power supply variation	72°C	Insulation limited to 70°C	Overvoltage protection or improved cooling needed

Reliability in Design

System _____
Indenture level _____
Reference drawing _____
Mission _____

Date _____
Sheet _____ of _____
Compiled by _____
Approved by _____

Identification number	Item/functional identification (nomenclature)	Function	Failure modes and causes	Mission phase/ operational mode	Failure effects			Failure detection method	Compensating provisions	Severity class	Remarks
					Local effects	Next higher level	End effects				

Figure 7.2 MIL-STD-1629 worksheet for method 101

System _____

Indenture level _____

Reference drawing _____

Mission _____

Date _____

Sheet _____ of _____

Compiled by _____

Approved by _____

Identification number	Item/functional identification (nomenclature)	Function	Failure modes and causes	Mission phase/ operational mode	Severity class	Failure probability / Failure rate data source	Failure effect probability	Failure mode ratio (a)	Failure rate (λ_p)	Operating time (t) (β)	Failure mode crit $C_m = \beta a \lambda_p{}'$	Item crit $C_r = \Sigma (C_m)$	Remarks

Figure 7.3   MIL-STD-1629 worksheet for method 102

$$C_m = \beta\alpha\lambda_p t \hspace{8cm} (7.1)$$

where β = conditional probability of loss of function or mission
          α = failure mode ratio (for an item, $\Sigma_\alpha = 1$)
          λ_p = part failure or hazard rate
          t = operating or at-risk time of item

$\lambda_p t$ can be replaced by failure probability, $1 - \exp(-\alpha\lambda_p t)$.

The *item criticality number* is the sum of the failure mode criticality numbers for the item.

Variations on the MIL-HDBK-1629 methods are also used, for example Reference 4.

Steps in performing an FMECA

An effective FMECA can be performed only by an engineer or team of engineers having thorough knowledge of the system's design and application. The first step therefore is to obtain all the information available on the design. This includes specifications, drawings, computer-aided engineering (CAE) data, stress analysis, test results, etc., to the extent they are available at the time. For a criticality analysis, the reliability prediction information must also be available or it might be generated simultaneously.

A system functional block diagram and reliability block diagram (Chapter 6) should be prepared, if not already available, as these form the basis for preparing the FMECA and for understanding the completed analysis.

If the system operates in more than one phase in which different functional relationships or item operating modes exist, these must be considered in the analysis. The effects of redundancy must also be considered by evaluating the effects of failure modes assuming that the redundant subsystem is or is not available.

An FMECA can be performed from different viewpoints, such as safety, mission success, availability, repair cost, failure mode or effect detectability, etc. It is necessary to decide, and to state, the viewpoint or viewpoints being considered in the analysis. For example, a safety-related FMECA might give a low criticality number to an item whose reliability seriously affects availability, but which is not safety critical.

The FMECA is then prepared, using the appropriate worksheet, and working to the item or subassembly level considered appropriate, bearing in mind the design data available and the objectives of the analysis. For a new design, particularly when the effects of failures are serious (high warranty costs, reliability reputation, safety, etc.) the analysis should take account of all failure modes of all components. However, it might be appropriate to consider functional failure modes of subassemblies when these are based upon existing designs, e.g. modular power supplies in electronic systems, particularly if the design details are not known.

The FMECA should be started as soon as initial design information is available. It should be performed iteratively as the design evolves, so that the analysis can be used to influence the design and to provide documentation of the eventually completed design. Design options should be separately analysed, so that reliability implications can be considered in deciding on which option to choose. Test results should be used to update the analysis.

FMECA is not a trivial task, and can involve many hours or weeks of work. It can also be difficult to trace the effects of low-level failures correctly through complex systems. If the system has been designed or modelled using CAE (or electronic design automation: EDA) software, this can be used to assist in the analysis, thus aiding the task of working out the effects of component-level failures on the operation of complex systems. Even with aids such as these, FMECA can be an inappropriate method for some designs, such as digital electronic systems in which low-level failures (e.g. of transistors within integrated circuits) are very, but uniformly, unlikely, and the effects are dynamic in the sense that they could differ widely depending upon the state of the system. FMECA is not appropriate for software designs.

FMECA is widely used in many industries, particularly in those for which failures can have serious consequences, such as military, aerospace, automotive, medical equipment, etc. Some industries have established standardized approaches (the US Military Standard is MIL-Handbook-1629, and the US automotive companies have also produced a guidance document (References 3 and 4). However, these present rather rigid approaches, which furthermore are not appropriate for systems involving modern digital electronics, so they should be used only to the extent required by contracts and as valid for the technology.

Uses for FMECA

FMECAs can be used very effectively for several purposes, in addition to the prime one of identifying safety or reliability critical failure modes and effects. These include:

1. Identifying features to be included in the test programme (Chapter 11).
2. Preparation of diagnostic routines such as flowcharts or fault-finding tables. The FMECA provides a convenient listing of the failure modes which produce particular failure effects or symptoms, and their relative likelihoods of occurrence.
3. Preparation of preventive maintenance requirements. The effects and likelihood of failures can be considered in relation to the need for scheduled inspection servicing or replacement. For example, if a failure mode has an insignificant effect on safety or operating success, the item could be replaced only on failure rather than at shceduled intervals, to reduce the probability of failure. See Chapter 14.
4. Design of built-in test (BIT), failure indications and redundancy. The failure detectability, including BIT, viewpoint is an important one in FMECA of system which include these features.
5. For analysis of testability, particularly for electronic subassemblies and systems, to ensure that hardware can be economically tested and failures diagnosed, using automatic or manual test equipment.
6. For development of software for automatic test and BIT.
7. For retention as formal records of the safety and reliability analysis, to be used as evidence if required in reports to customers or in product safety litigation.

8. An FMECA can be performed specifically to consider the possibility of production-induced failures, e.g. wrong diode orientation. Such a *production* FMECA can be very useful in test planning and in design for ease of production.

It is important to coordinate these activities, so that the most effective use can be made of the FMECAs in all of them, and to ensure that FMECAs are available at the right time and to the right people.

Computerized FMECA

Computer programs have been developed for performance of FMECA. Using a computer program instead of FMECA worksheets allows FMECAs to be produced more quickly and accurately, and greatly increases the ease of editing and updating to take account of design changes, design options, different viewpoints, and different input assumptions. Like any other computer-aided design technique, computerized FMECA frees engineers to concentrate on engineering, rather than on tedious compilation, so that for the same total effort designs can be more thoroughly investigated, or less effort can be expended for the same depth of analysis. Also, by eliminating the more tedious and time-consuming aspects of the work, engineers' motivation and effectiveness is increased.

Computerized FMECA enables more perceptive analysis to be performed. Failure effects can be ranked in criticality order, at different system levels, in different phases of system operation and from different viewpoints. Report preparation can be partly automated and sensitivity analyses quickly performed. Figure 7.4 shows part of a computerized FMECA, performed using the PREDICTOR® (TM Management Sciences Inc.) FMECA program, on a hypothetical anti-aircraft missile control system.

Other FMECA software is also available. Computerized FMECA is also included in the software for integrated logistic support (ILS), as described in Chapter 14.

It is also possible, and effective, to use a spreadsheet to create an FMECA. This method has the advantage that the format and type of analysis can be designed to suit the particular design and methods of analysis. Modern integrated software also allows a wide range of graphic presentations to be created, databases to be used for reliability information, and for the analysis to be incorporated into wordprocessors for report preparation.

Computer-aided engineering (CAE) software, when used to create and analyse designs, can be used to perform FMECAs. Use of CAE permits very detailed, objective analyses to be made, covering transient and dynamic conditions, which can be very difficult to analyse manually. For example, EDA software can be used to determine the effects on circuit operation of component failure modes, by running the simulation with parameter values set at the failure conditions. Mechanical design CAE software can be similarly used in some cases, for example finite element analysis (FEA) software can be used to analyse the effects of mechanical or thermal stress. The SABER (TM Avanti Corp.) program enables FMECAs of mixed-technology designs to be created, and for the FMECA to be automated using a database of component failure modes.

Mission failure effects (MFE)

MFE	Effect	Loss probability
1	No effect	0
2	True no control	1
3	False no control	1
4	Intermittent control	0.4
5	Unstable control	0.7
6	Control flutter	0.5

SFE no.	System effect	MFE no.	System effect failure probability	Percent of all groups failure probability	Mission loss prob.	System criticality	Percent contribution to all groups criticality
2	No drive to control fins	2	0.3713E-05	19.69	1.0	0.3713E-05	24.75
7	T no output to solenoid	2	0.2495E-05	13.23	1.0	0.2495E-05	16.63
5	Drive to one extremity of travel	2	0.1500E-05	7.95	1.0	0.1500E-06	9.99
8	F no output to solenoid	3	0.1500E-05	7.95	1.0	0.1500E-05	9.99
10	Unstable output to solenoid	5	0.1976E-05	10.48	0.7	0.1383E-05	9.22
11	Low output to solenoid	4	0.2686E-05	14.24	0.4	0.1074E-05	7.16
13	No fin movement	2	0.1000E-05	5.30	1.0	0.1000E-05	6.66
3	Intermittent drive	5	0.1383E-05	7.33	0.7	0.9661E-06	6.45
12	Jitter on output to solenoid	6	0.9000E-06	4.77	0.5	0.4500E-06	2.99
4	Slow drive	4	0.1080E-05	5.72	0.4	0.4320E-06	2.87
9	Incorrect sensing of input demand	2	0.3952E-06	2.09	1.0	0.3952E-06	2.63
6	Drive ceases prematurely	4	0.2240E-06	1.18	0.4	0.8962E-07	0.59

Figure 7.4 Part of output listing from computerized FMECA (PREDICTOR®, TM Management Sciences Inc., Albuquerque, NM, USA)

RELIABILITY PREDICTIONS FOR FMECA

Since FMECAs are performed primarily to identify critical failure modes and to evaluate design options, failure rate or reliability values which could be considered as realistic worst cases should be used. Standard methods sometimes stipulate the reliability prediction methods to be used, e.g. MIL-HDBK-217 for electronics. However, it is very important to appreciate the large amount of uncertainty inherent in reliability prediction, particularly at the level of individual failure events (see Chapter 6). Therefore, worst-case or pessimistic reliability values should always be used as input assumptions for failure modes which are identified as critical, or which might be critical if the pessimistic assumption proved to be realistic. Alternatively, and preferably unless credible quantitative data are available, a value scale such as 0–1 should be used, with prearranged assignment (e.g. 1 = will definitely occur, 0.5 = will occur occasionally, 0.1 = will rarely occur, 0 = will never occur). Generally, the more critical the failure mode the more pessimistic should be the worst-case reliability assumptions.

HAZARD AND OPERABILITY STUDY (HAZOPS)

Hazard and operability study (HAZOPS) is a technique for the systematic determination of the potential hazards that could be generated by a system, and of the methods that should be applied to remove or minimize them. It is used in the development of systems such as petrochemical plant, railway systems, etc., and usually is part of the mandatory safety approval process. Table 7.2 shows an example of the format used.

For the failure/deviation column, a set of 'guidewords' is sometimes applied to help in the identification of things that could possibly go wrong. The usual guidewords are:

—no/not   —part of
—more   —reverse
—less   —other than
—as well as

Table 7.2   HAZOPS on motion system (partial)

Component/ Function	Failure/ Deviation	Possible Cause/s	Consequences/ Event No	Safeguards	Action
Electrical power	No power	1. Main power fail 2. Connector	System failure (1)	Provide standby power?	System design
Hydraulic supply	Main AND standby fail	1. Main AND pressure sensor fail.	System failure (2)	Checks on maintenance	Maintenance schedule
		2. Main AND changeover valve fail	System failure (2)		
PWM circuit	Permanent 'on'	See FMEA	System failure (3)	*To be determined*	Analysis, test
Solenoid valve	Stuck open	Corrosion	System failure (3)	*To be determined*	Test

HAZOPS should cover the whole range of potential failure causes, including natural hazards, human failures, etc.

PARTS, MATERIALS AND PROCESSES (PMP) REVIEW

All new parts, materials and processes called up in the design should be identified. 'New' in this context means new to the particular design and production organization. The designer is likely to assume that a part or material will perform as specified in the brochure and that processes can be controlled to comply with the design. The reliability and quality assurance (QA) staff must ensure that this faith is well-founded. New parts, materials and processes must therefore be assessed or tested before being applied, so that adequate training for production people can be planned, quality control safeguards set up and alternative sources located. New parts, materials and processes must be formally approved for production and added to the approved lists.

Materials and processes must be assessed in relation to reliability. The main reliability considerations include:

1. *Cyclical loading.* Whenever loading is cyclical, including frequent impact loads, fatigue must be considered.
2. *External environment.* The environmental conditions of storage and operation must be considered in relation to factors such as corrosion and extreme temperature effects.
3. *Wear.* The wear properties of materials must be considered for all moving parts in contact.

There is such a wide variation of material properties, even among categories such as steels, aluminium alloys, plastics and rubbers, that it is not practicable to generalize about how these should be considered in relation to reliability. Material selection will be based upon several factors; the design review procedure should ensure that the reliability implications receive the attention appropriate to the application. Chapter 8 covers mechanical design for reliability in more detail.

NON-MATERIAL FAILURE MODES

Most reliability engineering is concerned with material failure, such as caused by load–strength interference and strength degradation. However, there is a large class of failure modes which are not related to this type of material failure, but which can have consequences which are just as serious. Examples of these are:

1. Fasteners which secure essential panels and which can be insecurely fastened due to wear or left unfastened without being detected.
2. Wear in seals, causing low pressure leaks in hydraulic systems.
3. Resistance increase of electrical contacts due to arcing and accretion of oxides.
4. Failure of protective surfaces, such as paints, metal plating or anodized surfaces.

5. Distortion of pins, or intermittent contact, on multipin electrical connectors.
6. Drift in electronic component parameter values.
7. Electromagnetic interference (EMI) and timing problems in electronic systems.
8. Other personnel-induced failures such as faulty maintenance, handling or storage, e.g. omitting to charge electrolytic capacitors kept in long-term storage, which can result in reduced charge capacity in use.
9. Interface problems between sub-systems, due to tolerance mismatch.

All of these modes can lead to perceived failures. Failure reporting systems always include a proportion of such failures. However, there is usually more scope for subjective interpretation and for variability due to factors such as skill levels, personal attitudes and maintenance procedures, especially for complex equipment.

Non-material failures can be harder to assess at the design stage, and often do not show up during a test programme. Design reliability assessments should address these types of failure, even though it may be impracticable to attempt to predict the frequency of occurrence in some cases, particularly for personnel-induced failures.

HUMAN RELIABILITY

The term 'human reliability' is used to cover the situations in which people, as operators or maintainers, can affect the correct or safe operation of systems. In these circumstances people are fallible, and can cause component or system failure in many ways.

Human reliability must be considered in any design in which human fallibility might affect reliability or safety. Design analyses such as FMECA and FTA should include specific consideration of human factors, such as the possibility of incorrect operation or maintenance, ability to detect and respond to failure conditions, and ergonomic or other factors that might influence them.

Attempts have been made to quantify various human error probabilities, but such data should be treated with caution, as human performance is too variable to be credibly forecastable from past records. Human error probability is usually very dependent on training, supervision, and motivational factors, so these must be considered in the analysis. Of course in many cases the design organization has little or no control over these factors, but the analyses can be used to highlight the need for specific training, independent checks, or operator and maintainer instructions and warnings.

DESIGN ANALYSIS FOR PROCESSES

The processes that will be used to manufacture and maintain the product must be understood and optimized. It is essential that all of the processes are capable of being performed correctly and efficiently, whether by people or by machines. Therefore the designers must know the methods that will be used and their capabilities and limitations, and must design both the product and the processes accordingly. The test programme must include tests of all of the processes that have been shown by the analyses to be critical or important.

The applicable analysis methods that can be used are described below.

Variation

As discussed in Chapter 2, all manufacturing processes are subject to variation, as are parameter values and dimensions of parts and subassemblies. Production people and processes inevitably vary in their performance in terms of accuracy and correctness. The design must take all of these into account, and must minimize the possibility of failure due to production-related causes.

The use of FMECA to help to identify such causes has been mentioned above. Techniques for evaluating the effects of variation in production processes are described later. These methods are sometimes referred to as *process* design, to distinguish them from those aspects of the design which address the product specification and its environment. Product and process design, using an integrated approach, including the test and analysis techniques described later, are sometimes referred to as *off-line quality control*. Process design leads to the setting of the correct controls on the production processes, for monitoring as part of *on-line quality control*, described in Chapter 13.

There are two possible approaches to designing for parameter variation and tolerances, the 'worst case' approach, and statistical methods. The traditional approach is to consider the worst case. For example, if a shaft must fit into a bore, the shaft and bore diameters and tolerances might be specified as: shaft, 20 ± 0.1 mm; bore, 20.2 ± 0.1 mm, in order to ensure that all shafts fit all bores. If the tolerance limits are based upon machining processes which produce parts with *s*-normally distributed diameters, of which 2.5 per cent are oversize and 2.5 per cent are undersize (2σ limits), the probability of a shaft and bore having an interference would be

$$0.025 \times 0.025 = 0.000625$$

On the other hand, most combinations will result in a fairly loose fit. Figure 7.5 shows the situation graphically. Twenty-five per cent of combinations will have fits greater than 0.2 mm.

If, however, statistical tolerancing were used, we could design for much closer nominal diameters and still have an acceptably low probability of interference. If the shaft nominal diameter was set at 20.1 mm (dotted line on Fig. 7.5), the interference probability P_I can be calculated as:

$$P_\mathrm{I} = 1 - \Phi\left[\frac{D_1 - D_2}{(\sigma_1^2 + \sigma_2^2)^{1/2}}\right]$$

In this case, if 2σ is 0.1 mm then σ is 0.05 mm. Therefore,

$$P_\mathrm{I} = 1 - \Phi\left[\frac{20.2 - 20.1}{(0.05^2 + 0.05^2)^{1/2}}\right]$$

$$= 0.08$$

Figure 7.5   Shaft–bore interference

This is a very simple example to illustrate the principle. Reference 7 in the Chapter 2 Bibliography covers statistical tolerancing in more detail. For systems such as electronic circuits, where tolerances of many components must be considered, statistical analysis of parameter tolerances or drift can provide more economic designs, since the probability of several parts being near their specification limits is much lower than for one or two parts. Statistical tolerancing can also result in lower production costs, since part and subassembly test specifications do not need to be as tight and thus there will be fewer test rejects. However, statistical tolerancing must be based on the correct models, and it is not always safe to assume that variables are s-normally distributed. For example, many electronic parts are sorted, with parts whose values lie close to the nominal being sold as 'precision' parts. Thus the distribution of values of a lot from which such parts have been removed would be bimodal, with no parts being within the sorted range of the nominal value (see Chapter 2 (page 49) and Fig. 9.10).

Methods for analysing multiple simultaneous variations were described in Chapter 5.

Process FMECA

A *process FMECA* (PFMECA) is performed as described above for FMECA, but instead of asking 'how can the component or function fail?', we ask 'how can the process fail?'. Each process task is considered in turn, with the objective of identifying potential problems so that improved methods or controls can be set up.

'Poka Yoke'

Poka Yoke is the Japanese expression for 'mistake proofing'. It is a design approach that considers the ways in which processes might be incorrectly performed, and then making it difficult or impossible to do so. Examples are templates to ensure that directional components cannot be connected the wrong way round.

Testability analysis

Electronic circuits and systems must be tested after assembly to ensure that they are correct, and to indicate the sources of failure. Testability and testability analysis are described in Chapter 9.

Test yield analysis

By using the Monte Carlo capabilities of design software and knowledge of test or measurement criteria, yield prediction analyses can be performed on designs that are described in appropriate models.

Maintainability analysis

Maintenance tasks that might be necessary, such as lubrication, cleaning, replenishment, calibration, failure diagnosis and repair, must all be analysed to ensure that they can be performed correctly by the people likely to be involved. Aspects that should be covered include physical accessibility, time to perform the tasks, skill levels, training requirements, special tools and equipment, and the need for special facilities. Maintenance and maintainability are covered in Chapter 14.

CRITICAL ITEMS LIST

The critical items list is a summary of the items shown by the other analyses to be likely either to have an appreciable effect on the product's reliability or to involve uncertainty. Its purpose is to highlight these items and summarize the action being taken to reduce the risks. The initial list will be based upon the design analyses, but updates will take account of test results, design changes and service data as the project develops. The critical items list is a top document for management reporting and action as it is based upon the 'management by exception' principle and summarizes the important reliability problems. Therefore, it should not usually include more than ten items and these should be ranked in order of criticality, so that management attention can be focused upon the few most important problems. It could be supported by a Pareto chart (Chapter 12) to show the relative importance, when there are sufficient data. The critical items list should provide only identification of the problem and a very brief description and status report, with references to other relevant reports.

SUMMARY

All of the methods described above are expensive in engineering time, and this can be a constraint on their effective application. They all involve the detailed consideration of many aspects, such as product characteristics in QFD, system simulation, LSA, failure modes in FMECA, sneak analysis (SA) (described in Chapter 9) and FTA, and process analysis. However, in addition to improving reliability, they can help to reduce overall project time and cost if they are applied effectively, particularly since the prime objective is always to identify and prevent or correct problems early. If product timescales are very tight, which is often the case, it is important that early decisions are made on which methods will be applied and how the results will be used. It might be appropriate to limit the scope of the analyses. For example, the QFD might be limited to a small number of critical design requirements, including variations if appropriate, and the LSA and FMECA to a few identified critical components rather than applied to all.

For most engineering projects a high-level QFD, design analysis using appropriate simulation software and FMECA should be the minimal approach. The top-down FMECA approach mentioned above can be particularly effective.

An important result of the development of modern engineering design and analysis software is the possibility of reducing or even eliminating the need for development testing. This can be a very worthwhile objective, since projects are nearly always under pressure to reduce development costs and time. However, no design analysis software can deal with the whole range of possible operating stresses, environments, variations and degradation mechanisms that can cause failures. All design analysis and simulation methods involve assumptions and simplifications that can, to varying degrees, generate erroneous or misleading results. The methods imply that we, and the software used to model the components and the system, know and fully understand all of the cause and effect relationships and interactions that affect performance, reliability, etc. Placing faith in the results of analysis can present risks which depend upon the design being analysed, the methods and software used, and the skill and knowledge of the engineers involved.

Whilst it is always possible in theory to analyse the effects of variation by performing analyses with parameter values set at, say, tolerance limits or over tolerance ranges, and most CAE software includes facilities for tolerance analysis, it is often difficult and time-consuming to perform such analyses effectively. In particular, analysis implies that distribution parameters and interaction effects are known. As explained in Chapter 2, these aspects are often very uncertain.

Ultimately, only the actual hardware embodies the whole truth of the design, particularly aspects which might have been neglected or misrepresented in the analysis. *The need for testing is a direct consequence of the uncertainty gaps that arise as a result of the limitations inherent in all design analysis.* Since the analytical methods described indicate the aspects of designs that are critical or might present risks, they provide an excellent basis for optimizing the test programme. Therefore the results of the analyses should be used to help to plan and prioritize the tests, and the engineers involved should be part of the test team. Test methods are described in Chapter 11, and we will discuss the management aspects of integrating the design and test activities in Chapter 15.

Information on the most popular software for design reliability and safety analysis, with Internet contact information is provided on the book's homepage (page xxi).

MANAGEMENT OF DESIGN REVIEW

The review techniques described must be made part of a disciplined design sequence, or they will merely generate work and not advance the objective of more reliable design. To be effective, they must be performed by the people who understand the design. This does not necessarily mean the designer, for two reasons. First, the analyses are an audit of his work and therefore an independent assessment is generally more likely to highlight aspects requiring further work than would be the case if the designer were reviewing his own work. Second, the analyses are not original work in the same sense as is the design. The designer is paid to be creative and time spent on reassessing this effort is non-productive in this sense. The designer may, however, be the best qualified to perform much of the analysis, since he knows the problem, assessed the options, carried out all the design calculations and created the solution. On the other hand, the creative talent may not be the best at patiently performing the rather tedious review methods.

The best solution to this situation is for the person performing the reviews to work closely with the designer and to act as his 'Devil's advocate' during the creative process. In this way, the designer and the reviewer work as a team, and problem areas are highlighted as early as possible. The organization of reliability engineering staff to provide this service is covered in Chapter 15. The reviewer should ideally be a reliability engineer who can be respected by the designer as a competent member of a team whose joint objective is the excellence of the design. Since the reliability engineer is unlikely to spend as much time on one design as the designer, one reliability engineer can usually cover the work of several designers. The ratio obviously depends upon the reliability effort considered necessary on the project and on the design disciplines involved.

By working as a team, the design and reliability staff can resolve many problems before the formal analysis reports are produced, and agreement can be reached on recommendations, such as the tests to be performed. Since the reliability engineer should plan and supervise the tests, the link is maintained. Also, the team approach makes it possible for designs to be adequately reviewed and analysed before drawings are signed off, beyond which stage it is always more difficult and more expensive to incorporate changes.

Unfortunately, this team approach is frequently not applied, and design and reliability staff work separately in preparing analyses and criticizing one another's work at a distance, either by memo or over the conference table. Design review techniques then lose credibility, as do reliability staff. The main victim is the design itself, since the protagonists usually prosper within their separate organizations. If the organization is such that it is not possible to set up a team approach to design reliability, it is usually better not to try to apply the design review methods at all, since the effect will almost certainly be too little, too late.

To be of continuing value, the design analyses must be updated continually as design and development proceed. Each formal review must be based upon analyses of the design as it stands and supported by test data, parts assessments, etc. The analyses should be scheduled as part of the design programme, with design reviews scheduled at suitable intervals. The reviews should be planned well in advance, and the designer must be fully aware of the procedure. All people attending must also be briefed in advance, so they do not waste review time by trying to understand basic features. To this end, all attendees must be provided with a copy of all formal analysis reports (reliability prediction, load–strength analysis, PMP review, maintainability analysis, critical items list, FMECA, FTA) and a description of the item, with appropriate design data such as drawings. The designer should give a short presentation of the design and clear up any general queries. Each analysis report should then form a separate agenda item, with the queries and recommendations as the subjects for discussion and decision. If experience has generated a checklist appropriate to the design, this could also be run through, but see the comments below.

With this procedure, nearly all aspects requiring further study or decision will have been discussed before, during the continuous, informal process of the team approach to preparing the analyses. The formal review then becomes a decision-making forum, and it is not bogged down with discussion of trivial points. This contrasts markedly with the type of design review meeting which is based largely upon the use of checklists, with little preparatory work. Such reviews become a stolid march through the checklist, many of whose questions might be irrelevant to the design. They can become a substitute for thinking.

Three golden rules for the use of checklists should be:

1. Use them in the design office, not during the formal design review meetings.
2. Ensure that they are relevant and up to date.
3. Avoid vague questions such as 'Has maintenance been considered?', or even 'Are the grease points accessible?' 'What access is provided for lubrication?' would be a better question, since it calls for detailed response, not a simple affirmative.

The design review team should consist of staff from sales, production, QA, and specialists in key design areas. The people on the spot are the designer and his reliability engineer team member (who may belong to the QA department). The chairman should be the project manager or another person who can make decisions affecting the design, e.g. the chief designer. Sometimes design reviews are chaired by the procuring agency, or it may require the option of attending. A design review which is advisory and has no authority is unlikely to be effective, and therefore all those attending must be concerned with the project (apart from specialists called in as advisers).

Formal design review meetings should be scheduled to take place when sufficient information is available to make the meeting worthwhile and in time to influence future work with the minimum of interference with project schedules and budgets. Three reviews are typical, based upon initial designs, completion of development testing and production standard drawings. Each review authorizes transition to the next phase, with such provisos deemed to be necessary, e.g. design changes, additional tests. The design reviews should be major milestones in a project's evolution. They are not concerned solely with reliability, of course, but reliability engineers have considerably influenced the ways that modern design reviews are conducted, and design reviews are key events in reliability programmes.

CONFIGURATION CONTROL

Configuration control is the process whereby the exact design standard of a system is known. Configuration control applies to hardware and to software. Effective configuration control ensures that, for example, the specifications and sources of components, and the issue numbers of drawings, can be readily identified for a particular system. Configuration control is very important in the development and production of systems, and it is mandatory for projects such as in aerospace (civil and military) and defence. Formal control should start after the first design review.

Configuration control is important to reliability, since it allows failures to be traced back to the appropriate design standard. For example, failures might occur in a component machined to a particular tolerance; the configuration control system should enable this cause to be identified.

BIBLIOGRAPHY

1. P. D. T. O'Connor, *Test Engineering*. J. Wiley (2001).
2. Y. Akao, *Quality Function Deployment*. Productivity Press (1990).
3. US MIL HDDK 1629: *Failure Mode and Effects Analysis*. National Technical Information Service, Springfield, Virginia. (Now discontinued).
4. SAE J-1739, *Potential Failure Mode and Effects Analysis*. Society of Automotive Engineers, USA.
5. British Standard, BS 5760: *Reliability of Systems, Equipment and Components*. British Standards Institution, London.
6. J. D. Andrews and T. R. Moss, *Reliability and Risk Assessment*. Longman (J. Wiley in USA) (1993).
7. A. Pages and M. Gondran, *Systems Reliability Evaluation and Prediction in Engineering*. North Oxford Academic (1986).
8. D. Clausing, *Total Quality Development: a Step by Step Guide to World Class Concurrent Engineering*. ASME Press (1994).

QUESTIONS

1. Produce a failure mode and effect analysis (FMEA) for five components in *one* of the following systems: (i) a domestic washing machine; (ii) the braking system of a motorcar; (iii) a simple camera; (iv) a portable transistor radio; or (v) any other system with which you are familiar (giving a brief explanation of its function). Your answer should be properly laid out as if it formed part of a complete FMEA on the system. Explain the additional considerations that would be included to convert your FMEA into a FMECA.

2. Describe the main uses to which a completed FMECA can be applied.

3. Describe three methods that can be used to analyse and improve the processes that will be used for the manufacture of a new product design.

4. Explain briefly, using diagrams if appropriate, the following methods:
    a. Quality function deployment.
    b. Process FMEA.
    c. Hazard and operability study.
    d. Poka yoke.
    e. Critical items list.

5. Comment on the values you would apply to the likelihood of failures if you are performing a FMECA for each of the following (consider the availability of data, its credibility and the purpose of the analysis):

   a. Operation of a car seat positioning system.
   b. Operation of a train braking system.
   c. Test coverage of an electronic circuit being used in a mobile telephone.
   d. Operation of the fuse of a hydrogen bomb.

6. Give four questions that would be appropriate for the reliability aspects of a design review of either a high speed mechanism involving bearings, shafts, and gears, and high mechanical stress, or for design of a DC electrical power supply unit which uses a standard AC power input.

7. Discuss the ways by which the design review process should be managed in order to provide the most effective assurance that new product designs are reliable, produceable and maintainable. Comment on the organisational and procedural aspects, as well as the actual conduct of the review.

8. Discuss the factors you would consider in producing a Reliability Critical Items List for

   a) a modern microcomputer controlled washing machine.
   b) a fighter aircraft electronic box.

8

Reliability of Mechanical Components and Systems

INTRODUCTION

Mechanical components can fail if they break as a result of applied mechanical stresses. Such failures occur primarily due to two causes:

1. Overstress leading to fracture. Stresses may be tension, compression or shear. Bending stresses cause tensile and compressive forces, but fracture usually occurs in tension.
2. Degradation of strength, so that working stresses cause fracture after a period of time.

For example, a pressure vessel will burst if the pressure exceeds its design burst strength, or if a crack or other defect has developed to weaken it sufficiently.

Mechanical components and systems can also fail for many other reasons, such as (though this list is by no means exhaustive):

— Backlash in controls, linkages and gears, due to wear, excessive tolerances, or incorrect assembly or maintenance.
— Incorrect adjustments on valves, metering devices, etc.
— Seizing of moving parts in contact, such as bearings or slides, due to contamination, corrosion, or surface damage.
— Leaking of seals, due to wear or damage.
— Loose fasteners, due to incorrect tightening, wear, or incorrect locking.
— Excessive vibration or noise, due to wear, out-of-balance rotating components, or resonance.

Designers must be aware of these and other potential causes of failure, and must design to prevent or minimize their occurrence. Appreciation of 'Murphy's Law' ('if a thing can go wrong, it will') is essential, particularly in relation to systems which are maintained and which include other than simple operator involvement. This chapter will describe overload and strength deterioration, and relevant aspects of component and material selection and manufacturing processes.

225

MECHANICAL STRESS, STRENGTH AND FRACTURE

Mechanical stress can be either *tensile, compressive* or *shear*. Tensile stress is caused when the material is pulled, so that the stress attempts to overcome the internal forces holding the material together. Typical material behaviour in tension is shown in Fig. 8.1. This shows that, as stress increases, the material stretches proportionally to the stress (the *elastic region*), then begins to stretch more rapidly (the *plastic region*), and finally fractures. In the elastic region the material will return to the original unstressed length if the stress is removed. The amount of deformation is called the *strain*. In the plastic region the material will retain some or all of the deformation if the stress is removed. Fracture occurs when sufficient energy has been applied to overcome the internal forces.

Stress is the load per unit cross-sectional area, conventionally expressed as σ, and is measured in kg/m^2, lbs/in^2 (psi), or pascals (Pa) (N/m^2). The strain (ε) is the ratio of the change in length to the original length. The relationship between stress and strain is described by *Hooke's Law*:

$$\sigma = E\varepsilon \qquad\qquad (8.1)$$

where E is *Young's Modulus*, or the *modulus of elasticity* for the material. A high value of E indicates that the material is *stiff*. A low value means that the material is soft or *ductile*.

The strength of a material in tension is measured by its *Yield Strength* (the stress at which irreversible plastic deformation begins) or *Ultimate Tensile Strength* (UTS), the stress at which fracture occurs. Note that the UTS might be lower than the yield strength.

When a specimen is subjected to tensile stress narrowing or 'necking' occurs, so that the cross-sectional area is reduced. This causes the 'true' stress level to increase compared with the 'engineering' stress calculated on the basis of the original

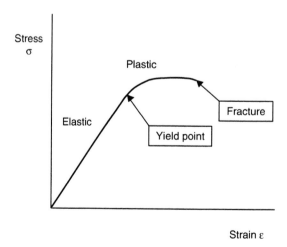

Figure 8.1 Material behaviour in tensile stress

unstressed cross-section. However, engineering design practice generally limits stress to not more than about 0.2 per cent strain, so the engineering stress–strain relationships are mostly used.

The elastic/plastic/fracture behaviour of a material is determined by its atomic or molecular structure. Atoms in solids are bound together by the interatomic or intermolecular attractive forces. E is proportional to the interatomic spacing, and it is reduced if temperature is increased. Elastic deformation extends the interatomic distances. Plastic deformation occurs when the energy applied is sufficient to cause the atomic planes, for example in a crystal, to slip along the lattice structure and take up new stable conditions. Fracture occurs when the stress exceeds the UTS.

Material surfaces contain energy, in the same way as the surface of a liquid possesses surface tension, due to the fact that the interatomic attractive forces between atoms at the surface can act in only two dimensions and so do not cancel as they do within the bulk material. In solids this energy is much higher than in liquids. When fracture occurs, two new surfaces are created. This extra energy is imparted by the applied stress which causes the fracture. Knowing the surface energy of a material enables us to determine the theoretical strength. This far exceeds what we actually measure, by factors of 1000 to over 10 000. The reason for the difference is that some plastic deformation occurs at stresses much lower than the theoretical elastic limit, as actual materials contain defects that create stress concentrations, for example dislocations within crystal planes of crystalline materials (metals, metal alloys, silicon, carbon, etc.), and between molecular boundaries in amorphous materials like plastics. Very pure single crystals, such as carbon fibres, can be produced with strengths that approach the theoretical values. The practical strength of a material can be determined only by tests to failure, though theoretical knowledge of aspects such as crystal structure, uniformity, etc., enable materials scientists to make approximate forecasts of strength.

Another important material property is *toughness*. Toughness is the opposite of *brittleness*. It is the resistance to fracture, measured as the energy input per unit volume required to cause fracture. This is a combination of strength and ductility, which is represented by the area under the stress–strain curve. Figure 8.2 shows this schematically (and very generally) for different material types.

The different patterns of behaviour represent the properties of ductility, brittleness and toughness. A ductile, weak material like pure copper will exhibit considerable strain for a given stress, and will fracture at low stress. A tough material like kevlar or titanium will have little strain and a high UTS. A brittle material like cast iron, glass or ceramic will show very little strain, but lower resistance to rapid stress application such as impact loads. Material properties, especially of metals, vary widely as a result of processes such as heat treatment and machining. In practice materials are applied so that the maximum stress is always well below the yield strength, by a factor of at least 2.

A crack will grow if the energy at the crack tip is sufficient to overcome the interatomic forces, and thus open it further. *Griffith's Law* expresses this:

$$\sigma = \sqrt{(2E\gamma/\pi a)} \tag{8.2}$$

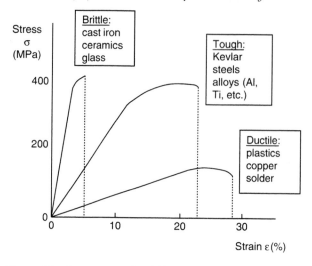

Figure 8.2   Stress–strain for different materials (generalized)

where:

σ = maximum stress at crack tip (note: not the average stress in the material)
E = modulus of elasticity
γ = surface energy
a = half crack length

The maximum stress at the tip of a crack (or at any other defect or *stress raiser*) is proportional to the applied total stress, the size of the crack or defect, and the sharpness of the tip around which the stress is applied. The ratio of maximum stress to applied stress is the *stress concentration factor*. Whilst the total or average stress can usually be determined quite accurately, using methods such as finite element analysis, the maximum stress around a stress raiser, and thus the strength, is often much less certain. Stress concentrations can be reduced by designing to provide adequate radii of curvature on corners of stressed components, ensuring that material surfaces are smooth, and, in the case of cracks in sheet material, by drilling a hole at the tip of the crack to increase the radius.

Compressive strength is much more difficult to analyse and predict. It depends upon the mode of failure (usually buckling for most engineering materials and components such as steel or aluminium alloy vehicle panels, struts, and electrical connector pins) and the shape of the component. Compressive fracture can also occur, particularly in brittle materials.

Structures that have bending loads applied are subjected to both tensile and compressive stress. The upper part of a loaded cantilevered beam will be in tension, whilst the lower part will be in compression.

Stress can also be applied in *shear*. A common practical example is solder joints that connect surface-mounted electronic components (integrated circuit packages) to circuit boards: during operation the temperature rise in the component causes thermal expansion relative to the circuit board, thus applying shear stresses to the solder joints.

The discussion above has presented a very brief overview of the topic. In most cases of applied stress the material behaviour is more complex, since combined effects occur. For example, a component in tensile stress will be caused to be compressed in the directions perpendicular to the tensile stress, so there will be a compressive stress also. Bending loads cause varying tensile and compressive stress from top to bottom of the beam, and therefore shear stress within the beam. Fracture in compression might be caused by shear stresses generated in the material. Finite element analysis (FEA), using modern software, can be used to analyse complex loading situations. However, it is nearly always necessary to test structural components to determine their true strength, especially if the designs are not simple.

FATIGUE

Fatigue damage within engineering materials is caused when a repeated mechanical stress is applied, the stress being above a limiting value called the fatigue limit. Fatigue damage is cumulative, so that repeated or fatigue limit above the fatigue limit will eventually result in failure. For example, a spring subjected to cyclic extension beyond the fatigue limit will ultimately fail in tension.

Fatigue is a very important aspect of reliability of structures subject to repetitive stress, e.g. from repeated load application, wave loading, and vibration, since the critical stress can be less than a quarter of the static fracture strength, and fracture can occur after 10^7 to 10^8 cycles when the applied stress is less than half the static strength.

The fatigue damage mechanism is the formation of microcracks resulting from the energy imparted to crystal boundaries by the cyclic stresses. The cracks then continue to extend along these lines of weakness, which act as stress concentrators. Like static fracture mechanics, quantification and prediction is largely empirical and based on experiment, but the degree of uncertainty is much higher.

Initiation and growth rate of the cracks varies depending upon the material properties and on surface and internal conditions. The material property that imparts resistance to fatigue damage is the toughness. As described above, the stresses around the tip of a crack or other defect (such as a machining scratch on a component or a void or inclusion in a casting or forging) are much higher than those in the bulk of the material, so concentrating the energy at these locations. We can demonstrate this easily by repeatedly bending a straightened paperclip through 180°. Being of ductile material, the clip will not fracture on the first bending. However, the alternating tensile and compressive stresses will generate cumulative fatigue damage, leading to fracture after typically about 20 cycles. If we now repeat the experiment, but now test paperclips which have been lightly cut with a sharp modelling knife, they will fracture in typically five cycles or fewer.

Figure 8.3 shows the general, empirical relationship between stress and cycles to fracture. This is called the *S–N* curve. Below the fatigue limit the life is indefinite, but higher stress levels induce cumulative damage, leading ultimately to failure. The S–N curve indicates the cycles to failure at any cyclic stress value between the ultimate stress and the fatigue limit S'. The curve indicates the mean value of cyclic load for a given number of cycles to failure (or vice versa). The population cycles to failure would in fact be distributed.

Figure 8.3   *S–N* curve

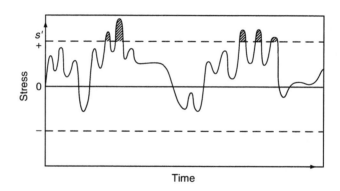

Figure 8.4   Random overload

The basic *S–N* curve shows the simplest situation, in which a uniform cyclic load is applied. In the more general practical case, with randomly distributed stresses as shown in Fig. 8.4, the population distribution of cycles to failure will have an additional variability and we will not know how much damage has been inflicted.

The fatigue life of an item subject to varying stress can be estimated using Palmgren-Miner's Law. This is expressed as

$$\frac{n_1}{N_1} + \frac{n_2}{N_2} + \frac{n_3}{N_3} + \ldots + \frac{n_k}{N_k} = 1$$

$$\sum_{i=1}^{k} \frac{n_i}{N_i} = 1 \qquad\qquad (8.3)$$

where n_i is the number of cycles at a specific stress level, above the fatigue limit, and N_i is the median number of cycles to failure at that level, as shown on the *S–N* curve.

The fatigue life of an item subject to an alternating stress with a mean value of zero is

$$N_e = \sum_{i=1}^{k} n_i \qquad (8.4)$$

N_e is called the *equivalent life*, and when used with the *S–N* diagram gives an equivalent steadily alternating stress, at which damage will occur at the same rate as under the varying stress conditions.

Example 8.1

Load data on a part indicate that there are three values which exceed the fatigue limit stress of $4.5 \times 10^8 \text{N m}^{-2}$. These values occur during operation in the following proportions:

$5.5 \times 10^8 \text{N m}^{-2} : 3$

$6.5 \times 10^8 \text{N m}^{-2} : 2$

$7.0 \times 10^8 \text{N m}^{-2} : 1$

Evaluate the equivalent constant dynamic stress.

The *S–N* diagram for the material is shown in Fig. 8.5. The cycles to failure at each overstress level are:

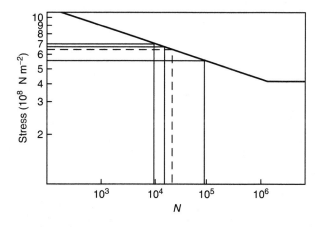

Figure 8.5 *S–N* diagram for the part in Example 8.1

$5.5 \times 10^8 \text{N m}^{-2}: 9.5 \times 10^4 \text{cycles}$

$6.5 \times 10^8 \text{N m}^{-2}: 1.5 \times 10^4 \text{cycles}$

$7.0 \times 10^8 \text{N m}^{-2}: 0.98 \times 10^4 \text{cycles}$

Therefore, from Eqn (8.3) where C is an arbitrary constant,

$$\frac{3C}{9.5 \times 10^4} + \frac{2C}{1.5 \times 10^4} + \frac{1C}{0.98 \times 10^4} = 1$$

$$C = 3746$$

From Eqn (8.4)

$$N_e = 3C + 2C + C$$

$$= 2.25 \times 10^4$$

From the *S–N* diagram, the equivalent constant dynamic stress is $6.3 \times 10^8 \text{N m}^{-2}$.

Figure 8.6 shows an *S–N* diagram for a population of items, with the strength and applied stress distributions also shown. The stress distribution tail extends beyond S', thus generating fatigue damage, and the mean of the strength distribution is therefore reduced. The strength distribution variance increases as items incur different amounts of fatigue damage. At N', the tails of the load and strength distribution interfere, and we enter the increasing hazard rate period.

Population times to failure in fatigue are typically log normal or Weibull distributed, as shown by the pdf. The variance is large, typically an order of magnitude even under controlled test conditions, and much larger under random service

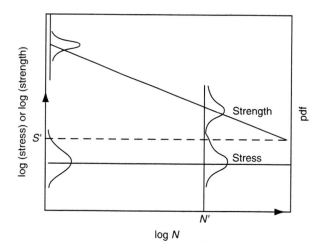

Figure 8.6 Strength deterioration with cyclic stress

environments, particularly when other factors such as temperature stress, corrosion, damage, or production variation extend the left-hand tail of the life distribution. Therefore fatigue lives are predicted conservatively, particularly for critical components and structures. However, reliability values calculated in this way are subject to considerable uncertainty. The usual practice in designing for a safe life is to estimate the equivalent cycles to failure and to assign a safe life based upon the expected variation in N_e. However, the predicted safe life should always be confirmed by carrying out life tests, using simulated or actual environments, and actual production items.

Analyses of time to failure in test and service situations can be performed using Weibull probability plots. Weibull distributions of times to failure show a positive failure-free life (γ) and slope (β) values greater than 1 (typically 2–3.5), i.e. an increasing hazard rate with failures starting after the safe life interval. This life is sometimes quoted as the 'B-life', i.e. the life interval at which B% of the items have been shown to have failed on test. (B-lives are also used to define the lives of components subject to wear, e.g. bearings).

Generally for metals the fatigue life is not affected by the rate at which stress cycling is applied. This is due to the fact that, since they are good thermal conductors, any energy converted to heat is readily conducted away so there is little or no temperature rise. However, plastics generally are more likely to be locally heated by high rates of stress reversals, and this, coupled with their lower melting points and other properties, such as the glass transition temperature, can result in reduced fatigue lives at high cycle rates.

Composite materials, such as fibre-reinforced structural components, can be designed and manufactured to have tailored mechanical properties, since the stresses are transmitted primarily through the fibres rather than through the bulk material. Failure of composite components can be due to delamination or separation of the fibres, or fracture of the whole component.

Fatigue life is affected by other factors, mainly temperature and corrosion. High temperature accelerates crack growth rates, by maintaining the critical energy levels at the crack tips. Corrosion can greatly accelerate crack propagation. Complex loading situations, for example vibration superimposed on a static load, can also reduce fatigue life.

The fracture surfaces of fatigue failures typically show characteristic 'rings' spreading out from the initial fatigue crack as it progressively grows, and a granular area where the final fracture occurs (Fig. 8.7).

The fatigue behaviour of designs can be analysed using software that combines FEA and material property data. However, the software assumes (unless otherwise instructed) that material surfaces are smooth and not damaged, and that no other effects such as corrosion are present. Software for fatigue life prediction evaluates expected (average) lifetimes and variations around these, not the possible time to the first failure. The correctness of the outputs depends on the correctness of the input descriptions such as surface conditions, the adequacy of the mesh being used, and understanding of the underlying mechanics and physics. Small errors or omissions in the mesh or other inputs can diverge and result in large errors in the predicted behaviour. References 1–5 provide good introductions to the material properties described above.

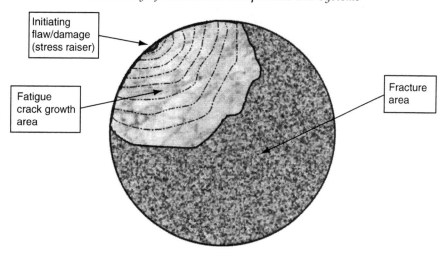

Initiating
flaw/damage
(stress raiser)

Fatigue
crack growth
area

Fracture
area

Figure 8.7 Typical fatigue failure (schematic)

Design against fatigue

Design for reliability under potential fatigue conditions means either ensuring that the distributed load does not exceed the critical load or designing for a limited 'safe life', beyond which the item is not likely to be used or will be replaced in accordance with a maintenance schedule. If we can ensure these conditions, then failure should not occur.

However, in view of the wide variation of fatigue lives and the sensitivity to stress and other environmental and material conditions, this is not easy. The following list gives the most important aspects that must be taken into account:

1. Knowledge must be obtained on the material fatigue properties, from the appropriate data sources, and, where necessary, by test. This knowledge must be related to the final state of the item, after all processes (machining, etc.) which might affect fatigue.
2. Stress distributions must be controlled, by careful attention to design of stress concentration areas such as holes, fixings, and corners and fillets. The location of resonant anti-nodes in items subject to vibration must be identified. Finite element and nodal analysis methods are used for this work. (See later section on vibration (p. 238)).
3. Design for 'fail safe', i.e. the load can be taken by other members or the effect of fatigue failure otherwise mitigated, until the failed component can be detected and repaired or replaced. This approach is common in aircraft structural design.
4. Design for ease of inspection to detect fatigue damage (cracks), and for ease of repair.
5. Use of protective techniques, such as surface treatment to relieve surface stresses (shot peening, heat treatment), increasing surface toughness (nitriding of steels, heat treatment), or provision of 'crack stoppers', fillets added to reduce the stress at crack tips.

6. Care in manufacture and maintenance to ensure that surfaces are not damaged by scratches, nicks, or impact.

Maintenance of fatigue-prone components

It is very important that critical components subject to fatigue loading can be inspected to check for crack initiation and growth. Maintenance techniques for such components include:

1. Visual inspection.
2. Non-destructive test (NDT) methods, such as dye penetrants, X-ray, and acoustic emission tests.
3. Where appropriate, monitoring of vibration spectra.
4. Scheduled replacement before the end of the fatigue life.

The scheduling and planning of these maintenance techniques must be based upon knowledge of the material properties (fatigue life, crack propagation rates, variability), the load duty cycle, the effect of failure, and test data. See Chapter 14 for a more detailed discussion of maintenance planning principles.

CREEP

Creep is the gradual increase in length of a component that is subjected to combined continuous or cyclic tensile stress and high temperature. Creep is a plastic (i.e. permanent) deformation, which occurs when the material temperature exceeds about 50 per cent of the melting point, on the absolute temperature scale. The effect is significant with components like turbine discs and blades in gas turbine engines, due to the combined very high temperatures and centrifugal forces. It has recently become a problem in electronics assemblies using surface mount components. Since solder melts at about 183 °C, system operating temperatures are generally within the creep temperature range. Therefore, permanent deformation takes place due to the shear stresses imposed by thermal cycling. The deformation in turn can result in higher shear stresses, thus accelerating the fatigue mechanism.

References 1–5 describe creep.

WEAR

Wear mechanisms

Wear is the removal of material from the surfaces of components as a result of their movement relative to other components or materials. Wear can occur by a variety of mechanisms, and more than one mechanism may operate in any particular situation. The science and methods related to understanding and controlling wear in engineering comprise the discipline of *tribology*. The main wear mechanisms are described below.

Adhesive wear occurs when smooth surfaces rub against each other. The contact load causes interactions between the high spots on the surfaces and the relative motion creates local heating and dragging between the surfaces. This results in particles being broken or scraped off the surfaces, and loose particles of wear debris are generated.

Fretting is similar to adhesive wear, but it occurs between surfaces subject to small oscillatory movements. The small movements prevent the wear debris from escaping from the wear region, so the particles are broken up to smaller sizes and might become oxidized. The repeated movements over the same parts of the surface also result in some surface fatigue, and corrosion also contributes to the mechanism.

Abrasive wear occurs when a relatively soft surface is scored by a relatively hard surface. The wear mechanism is basically a cutting action often with displacement of the soft material at the sides of grooves scored in the soft material.

Fluid erosion is caused to surfaces in contact with fluids, if the fluid impacts against the surfaces with sufficient energy. For example, high velocity fluid jets can cause this type of damage. If the fluid contains solid particles the wear is accelerated. Cavitation is the formation and violent collapse of vacuum bubbles in flowing liquids subject to rapid pressure changes. The violent collapse of the vacuum bubbles on to the material surfaces causes fluid erosion. Pumps, propellers and hydraulic components can suffer this type of damage.

Corrosive wear involves the removal of material from a surface by electrolytic action. It is important as a wear mechanism because other wear processes might remove protective films from surfaces and leave them in a chemically active condition. Corrosion can therefore be a powerful additive mechanism to other wear mechanisms.

Methods of wear reduction

The main methods of wear reduction are:

1. Minimize the potential for wear in a design by avoiding as far as practicable conditions leading to wear, such as contact of vibrating surfaces.
2. Selection of materials and surface treatments that are wear-resistant or self-lubricating.
3. Lubrication, and design of efficient lubricating systems and ease of access for lubrication when necessary.

When wear problems arise in use, an essential starting point for investigation is examination of the worn surfaces to determine which of the various wear mechanisms, or combinations of mechanisms, is involved. For example, if a plain bearing shows signs of adhesive wear at one end, the oil film thickness and likely shaft deflection or misalignment should be checked. If the problem is abrasive wear the lubricant and surfaces should be checked for contamination or wear debris.

In serious cases design changes or operational limitations might be needed to overcome wear problems. In others a change of material, surface treatment or change of lubricant might be sufficient. It is also important to ensure that lubricant filtration, when appropriate, is effective.

Maintenance of systems subject to wear

The life and reliability of components and systems subject to wear are very dependent upon good maintenance. Maintenance plans should be prepared, taking into account cleaning and lubrication requirements, atmospheric and contamination conditions, lubricant life and filtration, material properties and wear rates, and the effects of failure. In appropriate cases maintenance also involves scheduled monitoring of lubricant samples, using magnetic plugs to collect ferrous particles and spectroscopic oil analysis programmes (SOAP) to identify changes in levels of wear materials. Vibration or acoustic monitoring is also applied. These techniques are used in systems such as industrial and aero engines, gearboxes, etc.

References 6 and 7 are excellent introductions to tribology and wear.

CORROSION

Corrosion affects ferrous and some other non-ferrous engineering metals, such as aluminium and magnesium. It is a particularly severe reliability problem with ferrous products, especially in damp environments. Corrosion can be accelerated by chemical contamination, for example by salt in coastal or marine environments.

The primary corrosion mechanism is oxidation. Some metals, particularly aluminium, have oxides which form as very hard surface layers, thus providing protection for the underlying material. However, ferrous alloys do not have this property, so oxidation damage is cumulative.

Galvanic corrosion can also be a problem in some applications. This occurs when electromotive potentials are built up as a result of dissimilar metals being in contact and conditions exist for an electric current to flow. This can lead to the formation of intermetallic compounds and the acceleration of other chemical action. Also, electrolytic corrosion can occur, with similar results, in electrical and electronic systems when induced currents flow across dissimilar metal boundaries. This can occur, for example, when earthing or electrical bonding is inadequate. Electroylytic corrosion affects the most electrically active element in the circuit.

Stress corrosion is caused by a combination of tensile stress and corrosion damage. Corrosion initiates surface weaknesses, leading to crack formation. Further corrosion and weakening occurs at crack tips, where the metal is in a chemically active state and where the high temperatures generated accelerate further chemical action. Thus the combined effect can be much faster than either occurring alone.

Design methods to prevent or reduce corrosion include:

1. Selection of materials appropriate to the application and the expected environments.

2. Surface protection, such as anodizing for non-ferrous metals, plasma spraying, painting, metal plating (galvanizing, chrome plating), and lubrication.
3. Other environmental protection, such as the use of dryers or desiccators.
4. Avoidance of situations in which galvanic or electrolytic corrosion can occur.
5. Awareness and avoidance of conditions likely to generate stress corrosion.

Correct maintenance is essential to ensure the reliability of corrosion-prone components. Maintenance in these situations involves ensuring the integrity of the protective measures described above. Since corrosion damage is usually extremely variable, scheduled maintenance should be based upon experience and criticality.

Reference 8 describes corrosion in detail.

VIBRATION AND SHOCK

Components and assemblies can be subjected to vibration and shock inputs, during use, transport or maintenance. Vibration and shock can cause:

—Fracture due to fatigue, or due to mechanical overstress.
—Wear of components such as bearings, connectors, etc.
—Loosening of fasteners, such as screws, bolts, etc.
—Leaks in hydraulic and pneumatic systems, due to wear of seals or loosening of connectors.
—Acoustic noise (10–10 000 Hz).

Common vibration inputs are:
—Reciprocating or rotating machinery. The dominant vibration frequency (Hz) generated by rotating masses will be r.p.m./60.
—Wheel vibration, on road and rail vehicles.
—Aerodynamic effects on aircraft and missile structures.
—Pressure fluctuations in hydraulic and pneumatic systems.
—Acoustic noise.

Vibration of a structure may occur at a fixed frequency, at different frequencies over time, or simultaneously over a range of frequencies. Vibration over a wide range of simultaneous frequencies is called *broad band* vibration. Vibration can occur in or about different linear and rotating axes.

The important units in relation to sinusoidal vibration are:

—Frequency (Hz)
—Displacement (mm), defined as peak or peak-to-peak values.
—Velocity (m/s), defined as peak values.
—Acceleration (m/s^2 or g_n), defined as peak values.

Every structure has one or more resonant frequencies, and if the vibration input occurs at these, or at harmonics, vibration displacements will be maximized. The locations at which zero vibration displacements occur are called *nodes*, and maximum displacement amplitudes occur at the *anti-nodes*. There may be more than one resonant frequency within the expected environmental range, and these may exist along different axes. There may also be more complex resonance modes, such as torsional or combinations of mechanical, acoustic, rotational or electromechanical modes. Sometimes simultaneously occurring resonances might be important, for example two components that vibrate in different modes or at different frequencies and in so doing impact one another. Common examples are electronic circuit boards, hydraulic pipes, and vehicle panels.

The resonant frequency is proportional to the stiffness of the structure, and inversely proportional to the inertia. Therefore, to ensure that resonant frequencies are well above any input vibrations that might be applied, structures must be sufficiently stiff, especially where there are relatively heavy parts, such as large components on circuit boards.

The vibration amplitude at any frequency is reduced by damping. Damping can also change the resonant frequency. Damping is provided in hydraulic and pneumatic systems by accumulators, in suspension and steering systems by mechanical dampers, and by using anti-vibration mountings for motors, electronic boxes, etc.

The pattern of vibration as a function of other parameters, such as engine speed, can be shown on a *waterfall plot*. Figure 8.8 is an example. Waterfall plots help to indicate the sources of vibration and noise. For example, a resonance that is independent of speed shows as a vertical line, and one that is generated at a particular speed shows peaks running horizontally. The peak heights (or colours on colour map displays) indicate the amplitudes.

Shock loads can cause vibration, though the amplitude is usually attenuated due to inherent or applied damping. Shock loads are only a particular type of vibration input: relatively high intensity and frequency, for short intervals.

Reference 9 provides a comprehensive treatment of the subject. Testing methods for vibration and shock will be covered in Chapter 11.

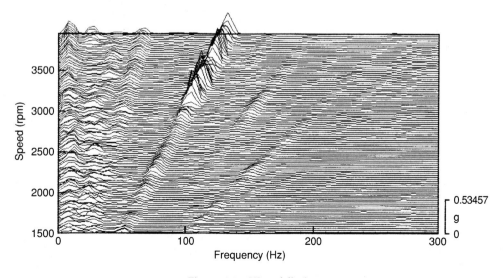

Figure 8.8 Waterfall plot

TEMPERATURE EFFECTS

Failures can be caused by materials being subjected to high or to low temperatures. The main high temperature failure modes are:

—Softening and weakening (metals, some plastics).
—Melting (metals, some plastics).
—Charring (plastics, organic materials).
—Other chemical changes.
—Reduced viscosity or loss of lubricants.
—Interaction effects, such as temperature-accelerated corrosion.

   Low-temperature effects can include embrittlement of plastics, increasing viscosity of lubricants, condensation, and freezing of condensation or coolants.
   Most temperature effects are deterministic (melting points, condensation temperatures, freezing points, viscosities). Effects such as these are not cumulative, so time and numbers of temperature cycles do not directly affect reliability. However, secondary effects might be cumulative, for example the effects of lubricant viscosity on rate of wear.
   All materials have a *thermal coefficient of expansion* (TCE). If two components with different TCEs are attached to one another, or two attached components can experience different temperatures, then mechanical stresses will be set up. An important example of this situation is the attachment of electronic components to circuit boards or other substrates, particularly surface mount integrated circuit packages that are attached to the circuit board by solder joints around the package periphery, as described in Chapter 9. When the IC is powered and operated heat is generated, so the package temperature rises. The heat is transferred through the package to the circuit board (which might include a 'heat plane', to improve heat dissipation). The thermal resistance of the heat flow path from the package to the eventual heat sink will result in the package being hotter than the board, and in the board temperature rise lagging that of the package. If power is cycled, the temperature differences will be also. This will result in cyclic shear stresses being imparted to the solder joints. The magnitude of these stresses can lead to fatigue failures, in the form of cracks running through the joint. These in turn can cause electrical failure, usually of an intermittent nature, after a sufficient number of cycles. This type of failure is particularly important in electronic systems that must withstand many on–off cycles, such as engine control systems. If the systems are also subjected to vibration, the combined effects of thermal and vibration cycling can be highly interactive.
   Chemical reactions, gaseous and liquid diffusion and some other physical processes are accelerated by increasing temperature. Arrhenius' Law expresses this phenomenon:

$$R = K \exp[-E/kT] \qquad\qquad (8.5)$$

where:
   R = process rate

K = constant
E = activation energy for the process
K = Boltzmann's constant
T = Temperature K

Typically, chemical process rates increase by a factor of 2 for every 10–20 °C rise in temperature. An important group of processes that can be thermally accelerated is corrosion, particularly rusting of iron and steel.

Humidity and condensation

Damp environments can cause or accelerate failure processes such as corrosion and mould growth. Temperature and humidity are closely related, humidity being inversely proportional to temperature until the dew point is reached, below which moisture condenses on to surfaces. Liquid water can cause further failures, including:

—Chemical corrosion, if contamination is also present.
—Electrolytic corrosion, by providing an electrolyte.
—Short circuiting of electrical systems, particularly within connectors.
—Mould growth.

Plastics are generally hygroscopic, that is, they absorb moisture, whether above or below the dew point. Therefore any components that are encapsulated in plastics, particularly electronic components and assemblies, are in principle prone to moisture ingress. This presented until fairly recently a major limitation on the application of plastic encapsulated components, since they suffered corrosion of the aluminium conductor metalization when used in high humidity environments. Their use in military and aerospace systems was banned. However, modern components such as integrated circuits have much improved protection against moisture, due to better control of the chip's surface protective layer and control of the plastic material purity and the encapsulating process, so that today there are few limitations on their application, and moisture-related failures are very rare.

MATERIALS

Selection of appropriate materials is an important aspect of design for reliability, and it is essential that designers are aware of the relevant properties in the application environments. With the very large and increasing range of materials available this knowledge is not easy to retain, and designers should obtain data and application advice from suppliers as well as from handbooks and other databases. A few examples of points to consider in selecting engineering materials for reliability are given below. The list is by no means exhaustive, and it excludes obvious considerations such as strength, hardness, flexibility, etc., as appropriate to the application.

Metal alloys

1. Fatigue resistance.
2. Corrosion environment, compatibility.
3. Surface protection methods.
4. Electrochemical (electrolytic, galvanic) corrosion if dissimilar metals in contact.

Plastics, rubbers

1. Resistance to chemical attack from materials in contact or in the local atmosphere (lubricants, pollutants, etc.).
2. Temperature stability (dimensional, physical), and strength variation at high and low temperature.
3. Sensitivity to ultraviolet radiation (sunlight).
4. Moisture absorption (all plastics are hygroscopic).

Ceramics

Brittleness, fracture toughness.

Composites, adhesives

1. Impact strength.
2. Erosion.
3. Directional strength.

Reference 11 is an excellent source of information on selection of engineering materials.

COMPONENTS

The range of mechanical components is vast, ranging from springs, seals and bearings to engines, pumps and power transmission units. Even among the most basic components there is little standardization, and new products and concepts are constantly being developed. It would not be feasible to attempt to provide guidance on the detailed reliability aspects of such a range in this chapter, but some general principles should be applied.

1. All relevant aspects of the component's application must be carefully evaluated, using the techniques described in the previous chapters. Where experience exists of application in another system, all data on past performance should be used, such as modes and causes of failure, application conditions, durability, etc. It is essential to discuss the application fully with the supplier's applications engineers, to the extent of making them effectively part of the design team, with commitment to the success of the product.
2. Use mature components in preference to new ones unless there are clear overriding reasons of cost, performance, etc. Novelty, even when the risks seem insignifi-

cant, often introduces unpleasant surprises. All new components should be placed on the critical items list (see Chapter 6).

3. Minimize the number of components and of component types. Whether a spring or a hydraulic pump, this approach not only reduces costs of the product and of assembly, but also can improve reliability. For example, where a mechanism such as a paper feed requires springs, cams and levers, careful study of the problem can often reveal ways in which one component can perform more than one function.

4. Pay attention to detail. It is very often the simple design problems which lead to unreliability, because insufficient attention was paid to them. For example, spring attachment lugs on plastic components that break as a result of the hard spring material cutting through (a metal bush could be a solution), and the location of components so that they suffer contamination from water or oil, or are difficult to fit and adjust, are common examples of failure to apply design skills and experience to the 'simple' jobs.

PROCESSES

Designers must be aware of the reliability aspects of the manufacturing processes. Machine processes create variations in dimensions, which can affect wear and fatigue properties. Processes designed to improve material properties must be considered and designed for. For example, heat treatment, metal plating, anodizing, chemical treatment, and painting require careful control if they are to be effective, and the design of the product and of its methods of assembly must ensure that these processes can be applied correctly and efficiently.

Other processes that can affect reliability include:

Fasteners

A huge range of different fastening methods and systems is available, including rivets, bolts and nuts, clamps, adhesives, etc. Fasteners can loosen under vibration or as a result of temperature cycling. Fasteners can fail due to fatigue, and fatigue cracks can start at holes for rivets and bolts.

Bolts and nuts can be combined with locking devices to prevent loosening. These include deformable plastic inserts, spring washers, crush washers, split pin retainers, adhesives, locking wires, etc. The integrity of many locking devices can be degraded if they are used more than once. Bolts and nuts used in some applications must be accurately torque loaded to ensure that the correct holding force is applied, and that the fasteners are not over-stressed on assembly.

Adhesives

Adhesives are used for many assembly operations, including aircraft and other vehicle structures, electronic component mountings on to heat sinks, locking of bolts and nuts, etc. The most commonly used industrial adhesives are epoxy plastics and cyanoacrylics. Epoxies are two-component adhesives which must be mixed shortly before use. Cyanoacrylics are contact adhesives that form an instant bond. Other adhesive compounds and systems include elastomerics (used in applications such as vibration isolation) and adhesive tapes.

All adhesives require careful preparation and cleaning of the surfaces to be bonded, and they all have limitations in relation to the kinds of materials they can bond. Adhesives also have temperature limits, and generally cannot withstand temperatures above 200 °C.

Welding and soldering

Metals can be joined by welding, and several different welding methods are used, depending on the materials and the application. Steel structures are welded with electric arcs or oxy-acetylene gas torches. Alloys such as of aluminium and magnesium, which burn in oxygen, are arc welded in inert gas (argon). Car assemblies are spot welded by robots applying pressure and high electric current to form resistance welds. Surfaces can also be welded by friction (high pressure and vibration, including ultrasonic welding of gold wire bonds on microelectronic assemblies).

Tin–lead solder is by far the most common method for connecting electronic and electrical components within systems. It also serves as a structural connection. Soldering for electronics assembly is described in Chapter 9.

Seals

Seals are used to prevent leaks in systems such as water, oil hydraulic and pneumatic components and pipe connections, around rotating shafts and reciprocating actuator rams, and to protect items in sealed containers. Special seals include those to block electromagnetic radiation from or into electronic equipment enclosures.

The effectiveness of seals is always influenced by control of assembly operations, and often also by maintenance. They are always affected by usage (wear, erosion, etc.), so they tend to degrade over time and use.

Reference 10 is a good introduction to engineering seals.

Chapter 13 covers the control of manufacturing processes. However, it is essential that the capabilities and problems of these are given as much consideration in design as aspects such as performance and cost. The manufacturing operations affect these aspects also, so a fully integrated approach, as described in Chapters 7 and 15, must be followed. Production and quality engineers must be included in the design team, and not left to devise production methods and quality standards after the design has been finalized.

BIBLIOGRAPHY

Fracture mechanics

1. J. E. Gordon, *The New Science of Strong Materials*. Penguin Books (1968).
2. R. W. Hertzberg, *Deformation and Fracture Mechanics of Engineering Materials*. J. Wiley (1976).

3. J. A. Collins, *Failure of Materials in Mechanical Design*. J. Wiley (1981).
4. N. E. Dowling, *Mechanical Behaviour of Materials*, 2nd edn. Prentice-Hall (1993)
5. A. D. S. Carter, *Mechanical Reliability and Design*. Macmillan (1997).

Wear

6. C. Lipson, *Wear Considerations in Design*. Prentice-Hall (1967).
7. J. D. Summers-Smith, *An Introductory Guide to Industrial Tribology*. Mechanical Engineering Publications (1994).

Corrosion

8. H. H. Uhlig and R. W. Revie, *Corrosion and Corrosion Control*, 3rd edn. J. Wiley (1985).

Vibration and shock

9. C. E. Harris and C. E. Crede, *Shock and Vibration Handbook*. McGraw-Hill (1976).

Materials and components

10. J. D. Summers-Smith (ed.), *Mechanical Seal Practice for Improved Performance*, 2nd edn. Mechanical Engineering Publications (1992).
11. F. A. A. Crane and J. A. Charles, *Selection and Use of Engineering Materials*, 2nd edn. Butterworths (1989).
12. W. Brostow and R. D. Corneliussen, *Failure of Plastics*. Hanser (1986).
13. *CenBASE Materials Database*. J. Wiley.

QUESTIONS

1. Sketch and annotate the general strain behaviour of materials subjected to tensile stress. Show how this differs for brittle, tough and ductile engineering materials, and give examples of each.

2. Explain why the actual mechanical strength of engineering components is very much less than the theoretical strength. How does this difference affect the predictability of strength?

3. Briefly describe the three most common causes of strength degradation of mechanical components. Give examples of each, with descriptions of methods used to prevent or reduce the chances of failure.

4. Miner's law is used to predict the expected time to failure in fatigue.

   (a) Write down the mathematical expression for Miner's law.

   (b) A component was tested in the laboratory to determine its fatigue life. The test results were as follows:

Stress level ($\times 10^8$ Nm^{-2})   6.8 8.0 10.0

Mean cycles to failure ($\times 10^5$) 12.7 4.2   0.6

The component will be used in service with these stress levels occurring in the following proportions, respectively:

Proportion of cycles   0.5 0.3 0.2

What will be the expected time to failure in service, if the stress cycle rate is 1000 per hour?

(c) Comment on the factors that would influence the accuracy of this prediction.

5. A component designed for a cyclic mechanical stress application has been analysed to determine its likely fatigue life. Comment on the approach that you would apply for ensuring that failures do not occur if the component is:

   a. A steel mounting bracket for an actuator on an earthmoving machine.
   b. An aluminium alloy mounting bracket for a flight control actuator on an aircraft.
   c. A plastic part in a copying machine.

6. Two basic approaches can be applied in the design of components and structures that can fail as a result of fatigue damage. These are the fail-safe and safe-life approaches. Describe these, discuss the factors that would determine which approach is appropriate, and give examples of their application.

7. Describe briefly three methods that can be applied to reduce the likelihood of failure of components and structures owing to fatigue.

8. Describe three types of wear processes that can lead to failure of surfaces in moving contact. Describe how one of these can be minimized by designers.

9. Corrosion can cause failure of metallic parts. Describe three corrosion processes. How can each be minimized by designers?

10. You are designing an electronic unit that will be used on an agricultural machine. What failures might be caused by the vibration environment? What steps would you take to minimize these?

11. Describe briefly the effect of temperature on (give temperature values where appropriate, and consider also the effects of temperature cycles):

    a. The strength of a solder joint used to retain a heavy electronic component.
    b. The properties of a lubricating oil.
    c. An electronic unit located outdoors.
    d. Corrosion.

9

Electronic Systems Reliability

INTRODUCTION

Reliability engineering and management grew up largely in response to the problems of the low reliability of early electronic equipment, and many of the techniques have been developed from electronics applications. The design and construction of an electronic system, more than any other branch of engineering, involves the utilization of very large numbers of components which are similar, but over which the designer and production engineer have relatively little control. For example, for a given logic function a particular integrated circuit device might be selected. Apart from choosing a functionally identical device from a second source, the designer usually has no option but to use the catalogued item. The reliability of the device used can be controlled to a large extent in the procurement and manufacturing phases but, as will be explained, mainly by quality control methods. The circuit designer generally has little control over the design reliability of the device. This trend has become steadily more pronounced from the time that complex electronic systems started to be produced. As the transistor gave way to the integrated circuit (IC) and progressively with the advent of large scale integration (LSI) and very large scale integration (VLSI), the electronic system designer's control over some of the major factors influencing reliability has decreased. However, this is changing in some respects as system designs are increasingly implemented on custom-designed or semi-custom integrated circuits. This aspect is covered in more detail later. This is not to say that the designer's role is diminished in relation to reliability. Rather, the designer of an electronic system must be, more than in most other branches of engineering, a member of a team, involving people from other technologies, production, quality control, test planning, reliability engineering and others. Without such a team approach, his or her functionally correct design could be highly unreliable. It is less likely that the designer of a functionally correct hydraulic or mechanical system could be as badly let down. It is important to understand the reasons for this difference.

For the great majority of electronic components and systems the major determinant of reliability is quality control of all of the production processes. This is also true of non-electronic components (subject in both cases to the items being used within specification). However, most non-electronic equipment can be inspected and tested to an extent sufficient to assure that they will operate reliably. Electronic components cannot be easily inspected, since they are nearly always encapsulated. In fact, apart from X-ray inspection of parts for ultra-high reliability products, internal inspection is not generally possible. Since electronic components, once encapsulated, cannot be inspected, and since the size and quantity of modern components dictates that very precise dimensions be held at very high production rates, it is inevitable that production variations will be built into any component population. Gross defects, i.e. failure to function within specification, can easily be detected by automatic or manual testing. However, it is the defects that do not immediately affect performance that are the major causes of electronic component unreliability.

Consider a typical failure mechanism in an electronic component: a weak mechanical bond of a lead-in conductor to the conducting material of the device. This may be the case in a resistor, a capacitor, a transistor or an IC. Such a device might function satisfactorily on test after manufacture and in all functional tests when built into a system. No practical inspection method will detect the flaw. However, because the bond is defective it may fail at some later time, due to mechanical stress or overheating due to a high current density at the bond. Several other failure mechanisms lead to this sort of effect, e.g. flaws in semiconductor material and defective hermetic sealing. Similar types of failure occur in non-electronic systems, but generally they do not predominate.

The typical 'electronic' failure mechanism is a wearout or stress induced failure of a defective item. In this context, 'good' components do not fail, since the application of specified loads during the anticipated life will not lead to failure. While every defective item will have a unique life characteristic, depending upon the nature of the defect and the load(s) applied, it is possible to generalize about the nature of the failure distributions of electronic components. Taking the case of the defective bond, its time to failure is likely to be affected by the voltage applied across it, the ambient temperature around the device and mechanical loading, e.g. vibration. Other failure mechanisms, say a flaw in a silicon crystal, may be accelerated mainly by temperature variations. Defects in devices can result in high localized current densities, leading to failure when the critical value for a defective device is exceeded.

Electronic components characteristically have a decreasing hazard rate, due to the fact that as weaker components fail and are rejected the proportion of defectives in the population is reduced. Similarly, repairable electronics systems usually show a decreasing failure rate trend as defective components fail and are replaced or system manufacturing defects such as solder defects show up and are repaired. Wearout is seldom a characteristic of 'good' electronic components or systems.

Of course, by no means all electronic system unreliability is due to defective components. The designer still has the task of ensuring that the load applied to components in the system will not exceed rated (or derated values), under steady-state, transient, test or operating conditions. Since electronic component reliability can be affected by temperature, design to control temperatures, particularly localized 'hot spots', is necessary. Thus the designer is still subject to the reliability disciplines covered in Chapter 7.

Electronic system failures can be caused by mechanisms other than load exceeding strength. For example, parameter drifts in components, short circuits due to solder defects or inclusions in components, high resistance relay or connector contacts, tolerance mismatches, and electromagnetic interference are examples of failures which may not be caused by load. We will consider these failure modes later, appropriate to the various components and processes which make up electronic systems.

RELIABILITY OF ELECTRONIC COMPONENTS

Electronic components can be caused to fail by most of the same mechanisms (fatigue, creep, wear, corrosion, etc.) described in the previous chapter. Fatigue is a common cause of failure of solder joints on surface mounted components and on connections to relatively heavy or unsupported components such as transformers, switches and vertically mounted capacitors. Wear affects connectors. Corrosion can attack aluminium conductors on integrated circuits, connectors and other components. Electrical and thermal stresses can also cause failures that are unique to electronics. The main electrical stresses that can cause failures of electrical and electronic components and systems are current, voltage, and power. For all of these failure modes there are strong interactions between the electrical and thermal stresses, since current flow generates heat.

It is important to appreciate the fact that the great majority of electronic component types do not have any mechanisms that will cause degradation or failure during storage or use, provided that they are:

—Properly selected and applied, in terms of performance, stress and protection.
—Not defective or damaged when assembled into the circuit.
—Not overstressed or damaged in use.

The quality of manufacture of modern electronic components is so high that the proportions that might be defective in any purchased quantity are typically of the order of less than ten per million for complex components like ICs, and even lower for simpler components. Therefore the potential reliability of well-designed, well-manufactured electronic systems is extremely high, and there are no practical limits to the reliability that can be achieved with reasonable care and expenditure. This, however, has not always been the case. In the 1950s, when military electronic systems were becoming more complex and semiconductor components were in their infancy, component failures were the major cause of systems failure.

Stress effects

Current

Electrical currents cause the temperatures of conductors to rise. If the temperature approaches the melting point, the conductor will fuse. (Of course fuses are used as protective devices to prevent other, more serious failures from occurring.) Heat in conductors is transferred to other components and to insulation materials, primarily by conduction and convection, so thermal damage can be caused to these.

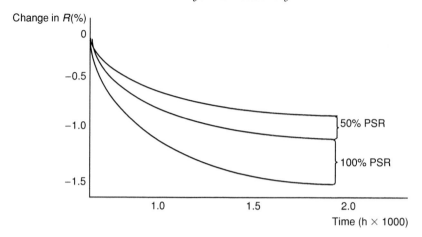

Figure 9.1 Parameter drift

High currents can also cause component parameter values, such as resistance, to drift over time. This effect is also accelerated by high operating temperatures. Figure 9.1 shows an example of this.

Electric currents also create magnetic fields. If oscillating, they can generate acoustic noise and electromechanical vibration. We will discuss electrical interference effects later.

Voltage

Voltage stress is resisted by the dielectric strength of the material between the different potentials. The most common examples are the dielectric material between capacitor plates, and the insulation (air or other insulator) between conductors. Potential differences generate currents in conductors and components, and if the current carrying capacity is insufficient the conductor or component will fail, in which case the failure mechanism is current, though the cause might be that the voltage is too high. For example, an integrated circuit might fail due to current overstress if a high electrostatic voltage is accidentally applied to it, and a 110 V appliance might fail for the same reason if connected to a 240 V supply.

High voltage levels can be induced by:

—Electrostatic discharge (ESD), caused by charge accumulation on clothing, tools, etc.
—Other electrical overstress, such as high voltage transients on power lines, unregulated power supplies, circuit faults that lead to components being overstressed, accidental connection of high voltages to low power components, etc. This is referred to as electrical overstress (EOS).

Another effect of voltage stress is *arcing*, which can occur whenever contacts are opened, for example in switches and relays. Arcing can also occur between brushes and commutators of motors and generators. Arcing generates electromagnetic noise, and also progressively damages the contact surfaces, leading eventually to failure. Arcing can also cause damage to components such as bearings, if the component provides a current path due to incorrect design or maintenance of the electrical circuit.

Arcing can be reduced by using voltage suppression components, such as capacitors across relay or switch contacts. Arcing becomes more likely, and is more difficult to suppress, if atmospheric pressure is reduced, since the dielectric constant of air is proportional to the pressure. This is why aircraft and spacecraft electrical systems operate at relatively low voltage levels, such as 28 V DC and 115 V AC.

Corona discharge can occur at sharp points at moderate to high voltage levels. This can lead to dust or other particles collecting in the area, due to ionization.

Some components can fail due to very low or zero current or voltage application. Low-power relay contacts which pass very low DC currents for long periods can stick in the closed position due to cold welding of the contact surfaces. Electrical contacts such as integrated circuit socket connectors can become open due to build-up of a thin dielectric layer caused by oxidation or contamination, which the low voltage stress is unable to break down.

Temperature

The Arrhenius formula that relates physical and chemical process rates to temperature has been used to describe the relationship between temperature and time to failure for electronic components, and is the basis of methods for predicting the reliability of electronic systems, as will be described later. However, this is an erroneous application, since, for the great majority of modern electronic components, most failure mechanisms are not activated or accelerated by temperature increase. And the materials and processes used are stable up to temperatures well in excess of those recommended for use. The reason why the relationship seemed to hold was probably because, in the early years of microcircuit technology, quality control standards were not as high, and therefore a fairly large proportion of components were observed to fail at higher temperatures. However, current data do not show such a relationship. This has major implications for thermal design, since the erroneous impression that 'the cooler the better' is widely held.

The true relationship between temperature and failure is as shown in Fig. 9.2. Most electronic components can be applied at temperatures well in excess of the figures stated in databooks. For example, databook package temperature limits for industrial grade plastic encapsulated integrated circuits and transistors are typically 85 °C, and for ceramic or metal packaged devices 125 °C. These do not, however, relate to any physical limitations, but are based more on the conventions of the industry.

High temperatures can cause damage or incorrect operation in electrical and electronic components. Wire coils in components such as inductors, solenoids, transformers and motors can become short circuited if high temperature, caused by high current or high ambient temperature, causes insulation charring.

Low temperatures can also cause components to fail, usually due to parametric changes in electrical characteristics. Typical low temperature limits for most components are −20 °C to −60 °C. However, such failures are usually reversible, and correct function is regained if the temperature rises.

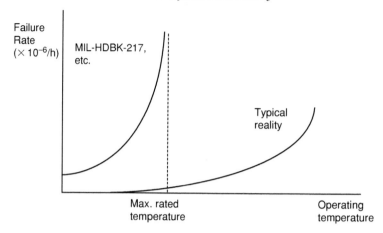

Figure 9.2 Temperature vs. reliability for electronic components

Repeated temperature changes can be more damaging than continuous operation at high temperatures. Components such as power transistors and processors can suffer fatigue damage of internal wire bonds and of attachments to heat sinks. Temperature changes also cause fatigue damage and creep deformation of solder joints on surface-mounted electronic components, as described in Chapter 8.

Power

Electrical power generates heat ($W = I^2R$). All 'active' electronic components, such as transistors, integrated circuits and amplifiers all generate heat, and therefore increased temperature. So do components like coils, voltage dropping resistors, etc. The steady-state component temperature will be the sum of the ambient temperature and the temperature generated by internal heating. The internal heat is dissipated by conduction through the device connections and mountings, and the thermal resistance between the active part of the device and the ultimate heat sink will determine the steady-state temperature. We discussed the effects of temperature on electronic device reliability earlier.

Some passive devices, such as resistors and capacitors, are also susceptible to failure due to power stress, if it causes overheating (rather than fusing due to excess current). Power stress over long periods can also cause drift of parameter values such as resistance or capacitance. Application instructions for such components generally include power stress limits.

Power stress cycling can lead to failure due to induced thermal cycling and thus fatigue, as described earlier.

COMPONENT TYPES AND FAILURE MECHANISMS

The main categories of electronic component types and their most common failure mechanisms are described in the sections below.

Integrated circuits (ICs)

Integrated circuits (ICs) are not really 'components'. They are in fact sub-systems, containing tens to millions of transistors, capacitors, and other discrete components within the IC structure. Conventionally, ICs are classified as follows: small scale integration (SSI): up to 100 logic gates; medium scale integration (MSI): up to 1000 gates; large scale integration (LSI): up to 10,000 gates; very large scale integration (VLSI): more than 10000 gates. Currently (2001) microprocessors contain over 10 million transistors. VLSI chips possess very many or even an unlimited range of potential functions.

Construction of ICs starts with the selective diffusion into the silicon wafer of areas of different charge level, either by ionic diffusion in a furnace or by implantation by a charged particle accelerator. The latter method is used for large-scale production. In the diffusion process, the areas for treatment are defined by masks. A sequence of treatments, with intermediate removal by chemical etch of selected areas, creates the structure of transistors and capacitors.

Different layers of the diffusion process are electrically isolated by layers of silicon dioxide (SiO_2). This is called *passivation*. Finally, the entire die surface, apart from connector pads for the wire bonds, is protected with a further SiO_2 or Si_3N_4 layer. This process is called *glassivation*.

The connections between transistors on the chip and to the input and output pins are made via a network of aluminium conductor tracks, by depositing a thin layer of metal (*metallization*) on to the surface, through a mask.

Finally, the assembly is packaged in either a plastic moulding or in a hermetic (ceramic or metal) package.

When ICs were first produced in the 1970s and 1980s they were mainly fairly simple analogue and digital circuits with defined functions (op-amps, adders, flip-flops, logic gates, etc.), and they were produced to closely defined generic specifications. For example, a 7408 is a 2-input AND gate, and it might have been manufactured by several suppliers. In parallel with the rapid growth in complexity and functionality in the years since, several different classes of IC have been developed. The classes that are available today include:

—'Standard' ICs. These are the components that appear in generic specifications or in manufacturers' catalogues. Examples are logic, memories, microprocessors, analogue to digital converters, signal processing devices, op-amps, etc.
—Programmable logic devices (PLDs), field programmable gate arrays (FPGAs). These are standard circuits which can be 'programmed' by selective opening of fusible links.
—Mixed signal ICs. These have both digital and analogue circuits integrated into the same chip.
—Microwave monolithic ICs (MMICs). These are used in systems such as telecommunications, and contain microwave radio circuits.
—Complex multifunction devices (*'system on a chip'*), which might contain mixed technologies, such as processing, memory, analog–digital conversion, optical conversion, etc., and a mixture of new and 'legacy' circuit designs.

Application-specific ICs

There is an increasing trend for ICs to be designed for specific applications. Standard ICs such as microprocessors and memories will always be used, but many circuits can be more economically implemented by using ICs which have been designed for the particular application. These are called *application-specific* ICs (ASICs).

In a *semi-custom* ASIC, all fabrication processes on the chip are previously completed, leaving arrays of transistors or cells to be interconnected by a conductor pattern designed for the particular application. In a full custom design, however, the chip is designed and manufactured entirely for the specific application.

Semi-custom ASICs are more economical than full custom ICs in relatively low quantities, but there is less flexibility of design and the utilization of chip area is less economical. Full custom ASICs are usually economical when large quantities are to be used, since the design and development costs are high. Both design approaches rely heavily on EDA, though the semi-custom method is easier to implement.

ASICs introduce important reliability aspects. The electronic system designer is no longer selecting 'black boxes', i.e. standard ICs, from a catalogue, but is designing the system at the component (or functional group) level. Reliability (and testability) analysis must be performed to this level, not just to input and output pins. Since design changes are very expensive, it is necessary to ensure that the circuit is reliable and testable the first time. Particular aspects that need to be considered are:

1. Satisfactory operation under the range of operating inputs and outputs. It is not usually practicable to test a LSI or VLSI design exhaustively, due to the very large number of different operating states (analogous to the problem of testing software, see Chapter 10), but the design must be tested under the widest practicable range of conditions, particularly for critical functions.
2. The effects of failures on system functions. Different failure modes will have different effects, with different levels of criticality. For example, some failure modes might have no effect until certain specific conditions are encountered, whilst others might cause total and obvious loss of all functions. FMECAs and stress analyses should therefore be performed on new IC designs, and the designs should be formally reviewed.
3. The effects of system software on the total system operation. For example, the extent to which the software is designed to compensate for specified hardware failures, by providing failure indications, selecting alternate operating paths, etc.
4. The need for and methods for providing built-in redundancy, both to increase production test yield and to improve reliability, particularly for critical applications.
5. Testability of the design. The ease with which circuits can be tested can greatly influence production costs and reliability, since untested functions present particular reliability hazards.

The EDA systems used for IC design include facilities for assessing reliability and testability. For example, failure modes can be simulated at the design stage and the effects evaluated, so stress analysis and FMECA can be integrated with the design process. Design analysis methods for electronic circuits are described in more detail later.

Microelectronics packaging

There are two main methods of packaging IC chips. In hermetic packaging the die is attached to the package base, usually by soldering, wire bonds are connected between the die connector pads and the leadout conductors, and the package is then sealed with a lid. Packages are either ceramic or metal. Plastic encapsulated ICs (PEICs or PEDs) use an epoxy or silicone encapsulant.

PEICs are cheaper than hermetic ICs, and therefore tend to be used in domestic and much commercial and industrial equipment. However, PEICs are not usually recommended for high temperature operation (above 85 °C case). They can also suffer a life dependent (wearout) failure mode due to moisture ingress, either by absorption through the encapsulation material or along the plastic/metal boundary of the leads. The moisture provides a medium for electrolytic corrosion at the interfaces of conductor tracks and wire bonds, or of the conductor tracks themselves through any gaps or holes in the glassivation layer. No plastic encapsulant is totally impervious to moisture ingress, though modern materials and process controls have greatly reduced the problem. Therefore when PEICs are used in high temperature or moisture environments or where long life is important, e.g. in military, automotive or aerospace applications, particular care should be taken to ensure their suitability.

For many years the most common package form was the dual-in-line package (DIL or DIP), with pin spacing of 0.1 inch (2.5 mm). The pins are inserted into holes in the printed circuit board (PCB) and soldered, or into a DIL socket which allows easy removal and reinsertion.

The packaging techniques used for the first 20 years or so of IC manufacture are giving way to new methods, primarily in order to enable more circuitry to be packaged in less volume. The conventional DIL package requires a PCB area many times larger than the chip surface area. The leadless chip carrier (LCC) package and the small outline IC (SOIC) are *surface-mounted devices* (SMD). These have leadouts around the periphery, which are reflow soldered to the PCB conductor tracks (or to a ceramic substrate which is in turn soldered to the PCB conductor tracks) rather than being inserted through PCB holes as with DIP. (In reflow soldering the components are placed on the PCB or substrate, and the assembly is heated in an infra-red or vapour phase oven to melt the solder.) The leadouts are on a 0.05 inch (1.25 mm) spacing or less. Use of SMDs reduces PCB surface area by about four times.

The new packaging techniques have also been developed with automation of the assembly processes in mind. The components, including the very small 'chip' packaged discrete components such as transistors, diodes, capacitors, etc., are too small, and the solder connections too fine, to be assembled manually, and automatic placement and soldering systems are used.

More recent developments are the pin grid array (PGA) and ball grid array (BGA) packages. Leadouts are taken to an array of pins on the underside of the package, on a 0.05 inch (1.25 mm) (or less) grid. As with the LCC package, connection is made to the

PCB or substrate by reflow soldering. Other packaging methods are used for special applications, such as vertical mounting for high density circuitry, direct mounting of the chip onto the PCB or substrate ('flip chip'), and chip scale packaging (CSP). Figure 9.3 shows examples of the packaging techniques described above.

The main reliability implication of the new IC packaging technologies is the fact that, as the volume per function is decreased, the power dissipation per unit volume increases. This can lead to difficult thermal management problems in order to prevent junction temperatures attaining levels above which reliability would be seriously affected. Liquid cooling of assemblies is now necessary in some applications such as in some military and high speed computing and test equipment.

Another reliability aspect of the new methods is the use of reflow soldering to the surface of the PCB or substrate. The solder connections cannot be inspected as can DIL solder connections; with a PGA or BGA they cannot be seen at all. Also, repeated thermal cycling can cause the solder joints to fail in shear since there is no mechanical compliance as with the pins on a DIL. Therefore the solder process

Figure 9.3　Packaging techniques for integrated circuits. (a) Dual in-line (DIL), (b) leadless chip carrier, (c) pin grid array (PGA), (d) microhybrid. (Courtesy National Semiconductor Corporation)

must be very carefully controlled, and subsequent burn-in and reliability tests (see later chapters) must be designed to ensure that good joints are not damaged by the test conditions.

Hybrid microelectronic packaging/multichip modules

Hybrid microelectronic packaging is a technique for mounting unencapsulated semiconductor and other devices on a ceramic substrate. Resistors are made by screen-printing with conductive in and laser-trimming to obtain the desired values. Connections from the conducting tracks to the device pads are made using fine gold or aluminium wire and ultrasonic bonding in the same way as within an encapsulated IC. The complete assembly is then encased in a hermetic package. See Fig. 9.3 (d).

Hybrid packaging provides certain advantages over conventional printed circuit board construction, for special applications. It is very rugged, since the complete circuit is encapsulated, and it allows higher density packaging than PCB mounting of components. However, it is not practicable to repair hybrid circuits, since clean room conditions and special equipment are necessary. Therefore they are suited for systems where repair is not envisaged except by replacing hybrid modules and discarding defective ones. Hybrid circuits are used in missile electronics, automotive engine controls and severe environment industrial conditions, and in many other applications where compact assembly is required, such as for high frequencies. Hybrid circuits can be custom designed and manufactured or standard catalogue hybrids can be used.

Due to their relatively large size, the number of internal bonds and the long package perimeter, hybrids tend to suffer from inclusion of contamination and conducting particles, bond failure and sealing problems more than do packaged ICs on equivalent PCB circuits. Therefore, very stringent production and quality control and required if the potential reliability of hybrid circuits is to be realized. MIL-STD-883 includes the same screening techniques for hybrid microelectronics as for discrete ICs, and the European and UK specifications (CECC, BS 9450) include similar requirements, as described later.

Multichip module (MCM) packaging is a more recent development of the micro-hybrid approach.

See References 12–14 for detailed descriptions of packaging technologies and their reliability aspects.

Microelectronic component attachment

Microelectronic components in DIP and LCC packages can either be soldered on the PCBs or plugged into IC sockets which are soldered in place. Plugging ICs into sockets provides three major advantages from the test and maintenance points of view:

1. Failed components can easily be replaced, with less danger of damaging the PCB or other components.
2. Testing and diagnosis is usually made much easier and more effective if complex devices such as microprocessors are not in place.

3. It is much easier to change components which are subject to upgrades or modifications, such as memories and ASICs.

On the other hand, there are some drawbacks which can override these advantages in certain circumstances. These are:

1. Heat transfer is degraded, so it might not be possible to derate junction temperatures adequately.
2. There might be electrical contact problems in high vibration, shock or contamination environments.
3. There is a risk of damage to the IC and the socket due to handling.

IC sockets are therefore used on many repairable systems, particularly mass-produced systems for which test costs must be kept low. However, they are not used to the same extent in high reliability, few-of-a-kind systems, such as spacecraft, or in systems such as avionics and mobile military equipment.

Microelectronic device failure modes

The main failure modes of ICs are:

—*Electrical overstress/electrostatic damage* (EOS/ESD). ICs are susceptible to damage from high voltage levels, which can be caused by transient events such as switching or electrostatic discharge from people or equipment. Most integrated circuits contain built-in EOS/ESD protection circuits, which will typically protect them against short-duration overstress conditions (typically up to 1000 V and 500 µJ).
—*Latchup* is the creation of a low resistance path between the power input and ground in an IC. CMOS ICs are prone to this failure mechanism when subjected to transient voltage overstress caused by ESD or other transient pulses from circuit operation, test equipment, etc. The effect is permanent complete failure of the device.
—*Electromigration* (EM) is a failure mechanism that is becoming increasingly important as the aluminium conductor strips (*metalization*) on ICs are made to extremely narrow dimensions (currently down to about 0.2 µm across and 0.1 µm deep). Such cross-sectional areas mean that the current density (amps/metre^2), even at the very low current and voltage levels within such circuits, can be very high. EM is the bulk movement of conductor material, at the level of individual metal crystals, due to momentum interchange with the current-carrying electrons. This can result in local narrowing of the conductor track, and thus increased local current density and eventual fusing. Also, the displaced material can form conducting whiskers, which can cause a short circuit to a neighbouring track. The EM process can be quantified using *Black's Law*:

$$T_m = A(W)J^{-N}\exp[E_A/kT] \tag{9.1}$$

where:

J   = current density (A/m^2)
N   = empirical constant, between 1 and 3
$A(W)$ = material constant, a function of line width
E_A = activation energy

EM is an important failure mode in electronic systems which must operate for long times, particularly if operating temperatures are high, such as in engine controls, spacecraft, and telecommunications systems (repeaters, switches, etc.).

—*Time-dependent dielectric breakdown* (TDDB) is a failure mode of the capacitors within ICs caused by whiskers of conductive material growing through the dielectric (silicon dioxide), and eventually short-circuiting the device. The effect is accelerated by voltage stress and by temperature.

—*Slow trapping* is the retention of electrons in the interstitial boundaries between Si and SiO_2 layers in ICs. These cause incorrect switching levels in digital logic and memory applications. Susceptibility to slow trapping is primarily dependent on device manufacturing processes.

—*Hot carriers* are electrons (or holes) that have sufficient energy to overcome the energy barrier of the Si–Si and SiO_2 boundary, and become injected into the SiO_2. It occurs in sub-micron ICs in which the electric field strengths can be sufficiently high. The effects are to increase switching times in digital devices and to degrade the characteristics of analogue devices. Hot carrier effects can be reduced by process design techniques and by circuit design, both to reduce voltage stress at critical locations.

—*Soft errors* are the incorrect switching of a memory cell caused by the passage of cosmic ray particles or alpha particles. Cosmic rays create such effects in circuits in terrestrial as well as space applications. Alpha particles are generated by trace heavy metal impurities in device packaging materials. The errors can be corrected by refreshing the memory.

—Processing problems in manufacture (diffusion, metallization, wire bonding, packaging, testing, etc.) can cause a variety of other failure mechanisms. Most will result in performance degradation (timing, data loss, etc.) or complete failure.

References 6–11 describe microelectronic component reliability physics.

It is important to appreciate that, despite the many ways by which microelectronic devices can fail, and their great complexity, modern manufacturing processes provide very high quality levels, with defective proportions being typically 0–100 per million. Also, appropriate care in system design, manufacture and use can ensure adequate protection against externally-induced failures. As a result, only a small proportion of modern electronic system failures are due to failures of microelectronic devices.

Microelectronic device specifications

In order to control the quality and reliability of microelectronic devices for military purposes, US Military Specification M-38510 was developed. This describes general controls, and separate sections ('slash sheets') give detailed specifications of particular device types. Similar international (International Electrotechnical Commission—IEC), European (CECC) and British (British Standards Institution—BS 9400) specifications have since been generated. These specifications are generally 'harmonized', so that there is little if any difference between them for a particular device type. Components produced to these specifications are referred to as 'approved' components.

Military system specifications usually require that electronic components are manufactured and tested to these standard specifications, in order to provide assurance of reliability and interchangeability. However, with the rapid growth in variety of device types, the specification systems have not kept pace, so that many of the latest device types available on the market do not have such specifications. In order to cope with this problem, an approach called *capability approval or qualified manufacturers list* (QML) provides generic approval for a device manufacturer's processes, covering all similar devices from that line. US MIL-STD-PRF 38535C describes the system for US military application. Capability approval is also appropriate for ASIC manufacture.

The general improvements in process quality have also resulted in removing the quality gap between 'approved' and commercial-grade components, so the justification for the specification systems is questionable. In fact, most manufacturers of high-reliability non-military electronic systems use commercial grade components, relying on the manufacturers' specifications and quality control.

A major difference between the different microelectronic device specifications is temperature rating. Military grade devices are rated between $-55\,°C$ and $125\,°C$, so that hermetic (ceramic or metal) packaging is necessary. Industrial grades are rated typically between $-25\,°C$ and $+85\,°C$, and commercial grade between 0 and $70\,°C$. (All operating temperatures, measured on the package.)

Microelectronic device screening

Screening is the name given to the process of finding by test which of a batch of components or assemblies is defective, without weakening or causing failure of good items. It is justified when:

1. The expected proportion defective is sufficiently high that early removal will improve yield in later tests and reliability in service.
2. The cost of screening is lower than the consequential costs of not screening.

The assumption that no weakening of good items will occur implies that the hazard rate will be decreasing.

Figure 9.4 shows the three categories of component that can be manufactured in a typical process. Most are 'good', and are produced to specification. These should not fail during the life of the equipment. Some are initially defective and fail when first tested, and are removed. They therefore do not cause equipment failures. However, a proportion might be defective, but nevertheless pass the tests. The defects will be potential causes of failure at some future time. Typical defects of this type are weak wire bond connections, silicon, oxide and conductor imperfections, impurities, inclusions, and non-hermetic packages. These components are called *freaks*.

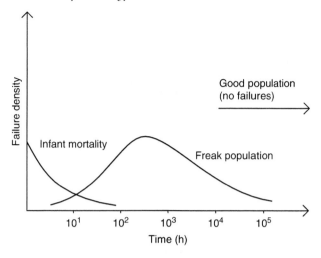

Figure 9.4 Typical failure density functions of electronic components when no component burn-in has been carried out

Screening techniques have been developed specifically for microcircuit devices. The original standard for these is US MIL-STD-883: *Test Methods and Procedures for Microelectronic Devices* (Reference 16). The other national and international standards mentioned above include very similar methods. There are three basic screen levels, as summarized in Table 9.1. The 'A' level screen (also referred to as 'S', for spacecraft application) is the most severe, and the most expensive. 'B' level screening is typically applied to microcircuits to be used in military, avionic and other severe-environment, high-integrity systems, particularly if a long operating life is required. 'C' level is a more relaxed specification, which does not include burn-in, as described below.

Burn-in is a test in which the components are subjected to high temperature operation for a long period, to stimulate failure of defective components by accelerating the stresses that will cause failure due to these defects, without damaging the good ones. In MIL-STD-883 the temperature to be used is 125°C (package temperature), for 168 hours duration. The electrical test conditions are also specified.

Component manufacturers and users have developed variations of the standard screens and burn-in methods. The main changes are in relation to the burn-in duration, since 168 hours has been shown to be longer than necessary to remove the great majority of defectives (the only justification for 168 hours is merely that it is the number of hours in a week). Also, more intensive electrical tests are sometimes applied, beyond the simple reverse-bias static tests specified. Dynamic tests, in which gates and conductors are exercised, and full functional tests with monitoring, are applied to memory devices, other VLSI devices, and ASICs, when the level of maturity of the process or design and the criticality of the application justify the additional costs.

Plastic encapsulated components cannot be burned-in at 125°C, so lower temperatures are used. Also, in place of leak tests they are tested for moisture resistance, typically for 1000 hours in an 85°C/85 per cent relative humidity (RH) chamber. This is not, however a 100 per cent screen, but a sample test to qualify the batch.

Table 9.1 Microelectronic device screening requirements[a]

		Screen level applicability		
		MIL-STD-883/BS		9400
Screen	Defects effective against	A	B	C
Pre-encapsulation visual inspection (30–200 × magnification)	Contamination, chip surface defects, wire bond positioning	100%	100%	100%
Stabilization bake	Bulk silicon defects, metallization defects (stabilizes electrical parameters)	100%	100%	100%
Temperature cycling	Package seal defects, weak bonds, cracked substrate	100%	100%	100%
Constant acceleration ($20\,000\,g$)	Chip adhesion, weak bonds, cracked substrate	100%	100%	100%
Leak tests	Package seal	100%	100%	100%
Electrical parameter tests (pre-burn-in)	Surface and metallization defects, bond failure, contamination/particles	100%	100%	100%
Burn-in test (168 h, 125°C with applied a.c. voltage stress)	Surface and metallization defects	100%	100%	–
	Weak bonds	240 h	168 h	
Electrical test (post-burn-in)	Parameter drift	100%	100%	–
X-ray	Particles, wire bond position	100%	–	–

[a]This table is not comprehensive, and the reader should refer to the appropriate standard to obtain full details of applicable tests.

A more severe test, using a non-saturating autoclave at 100°C and 100 per cent RH, is also used, as the 85°C/85 per cent RH test is not severe enough for the latest encapsulating processes.

The recent trends in microelectronic device quality have to a large extent eliminated the justification for burn-in by component users. Most component manufacturers burn-in components as part of their production processes, particularly for VLSI components and ASICs, using variations of the standard methods. Also, the new packaging technologies are not suitable for handling other than by automatic component placement machines, so user burn-in is generally inadvisable, as the handling involved can lead to damage and can degrade the solderability of the contacts.

Burn-in methods are described in Reference 17.

Other electronic components

Other electronic component types are primarily 'active' devices, such as transistors and diodes, and 'passive' devices such as resistors, capacitors, inductors, PCBs, connectors, etc. In general these discrete components are very reliable and most have no inherent degradation mechanisms (exceptions are light-emitting diodes, relays, some vacuum components, etc.). Factors that can affect reliability include thermal and electrical stress and quality control of manufacture and assembly processes. Standard specifications exist, as for microcircuits, but screening is not normally applied, apart from the manufacturers' functional tests.

Application guidelines for reliability are given in References 3 and 5 and in component manufacturers' databooks. The main component types are discussed in more detail below.

Discrete semiconductors

—Processing problems in manufacture (diffusion, surface condition, metallization, wire bonding, packaging, testing, etc.) can cause a variety of failure mechanisms. Most will result in performance degradation or complete failure.

—For power devices the uniformity and integrity of the thermal bond between the chip and the package is an important feature to ensure reliability at high power and thermal stress.

'Passive' components

—Resistors, capacitors, inductors and other components can fail due to fabrication problems, damage on assembly into circuits, ESD, and other causes. These usually cause the component to become open circuit or high resistance.

—Component parameter values can be out of tolerance initially, or parameter values can drift over time due to applied stresses, as described earlier.

—Components can be electrically 'noisy', due to intermittent internal contacts or impurities in materials.

Capacitors

—High voltage overstress generally causes the capacitor to become open circuit. High voltage/high power capacitors might even explode if they are short-circuited.

—Low voltage, low power capacitors, such as those built into integrated circuits as memory storage devices, can suffer long-term failures due to the mechanism of dendritic metal whisker growth through the dielectric.

—Capacitors that use a liquid or paste dielectric (electrolytic capacitors) degrade over time if no voltage stress is applied, and then fail short-circuit when used. They must be 're-formed' at intervals to prevent this. Capacitors kept in storage, or units such as power supplies which contain such components and which are stored or kept idle for long periods, must be appropriately maintained and checked.

—Electrolytic capacitors are damaged by reverse or alternating voltage, and so must be correctly connected and if necessary protected by diodes. Miniature tantalum capacitors are also degraded by ripple on the applied voltage, so they should not be used with unsmoothed voltage levels.

Electro-optical components

Many modern systems use optical fibres, connectors and electro-optical (EO) components for data transmission. Optical frequencies permit very high data rates. A major reliability advantage of EO systems is that they do not create EMI and they are immune to it. EO components are also used to provide over-voltage protection on data lines, by converting electrical signals to optical signals and back again to electrical. Failure modes of EO components are:

—Breakage of optical fibres.
—Misalignment of optical fibres at connections within connectors and to components. The connecting ends must be accurately cut across the length and aligned with one another.
—Degradation of light output from light-emitting diodes (LEDs).

Solder

Tin–lead solder (typically Sn63Pb37, melting point 183 °C) is by far the most common material used for attaching electronic components to circuit boards. However, because of the toxicity of lead, efforts are being made to develop lead-free soldering, using alloys of tin with silver, copper and other metals. These all have higher melting points, up to 235 °C, which in turn can affect component reliability. Solder joints can be made manually on relatively simple circuits which do not utilize components with fine pitches (less than 2.5 mm) between connections. However, for the vast majority of modern electronic systems solder connections are made automatically. The main techniques used are:

—*Through-hole mounting and manual or wave solder.* Components are mounted onto circuit boards, with their connections inserted through holes. DIP ICs are mounted this way, as are many types of discrete components. The hole spacing is typically 2.5 mm. In wave soldering, the boards are passed over a standing wave in a bath of liquid solder, so that each connection on the underside is immersed briefly in the wave. These methods are now used mainly for components and circuits which do not need to utilize the most compact packaging technologies, such as power circuits
—*Surface mount and infra-red or vapour phase solder.* The circuit boards are printed with solder paste at the component connection positions, and the SMT components are mounted on the surface by automatic placement machines. BGA solder connections are made by solder balls (typically 1 mm diameter) being accurately positioned on the solder paste on the board, then the BGA package is positioned on top. The 'loaded' boards are passed through 'reflow' ovens that melt the solder for just long enough to wet the solder so that intermetallics are formed, before the solder solidifies. The ovens are heated either with infra-red radiation, or, in the more common vapour phase or convection ovens, with gas heated to above the solder melting point. In the latter, the latent heat of condensation of the gas is transferred to the solder, and heating is very even.
—Laser soldering is also used to a limited extent, but further developments are likely.

The different types of component and solder methods are sometimes used in combination on circuit boards.

A reliable solder joint must provide good mechanical and electrical connection. This is created by the formation of intermetallic alloys at the interfaces between the solder and the surfaces being joined. The most common reasons for solder joint failure are:

—Inadequate solder wetting of the surfaces to be joined, due to surface contamination or oxidation. Components must be carefully stored and protected before placement. Components should not be stored for long periods before assembly, and fine pitch components should be handled only by placement machines.
—Insufficient heat (time or temperature) applied. The solder might be melted sufficiently to bond to one or both surfaces, but not enough to form intermetallics. Such joints will conduct, but will be mechanically weak.
—Fatigue due to thermally induced cyclic stress or vibration, as described in Chapter 8.
—Creep due to thermally induced cyclic stress , as described in Chapter 8.

All of these can lead to operating failures that show up on production test or in service. Failures can be permanent or intermittent. Modern electronic circuits can contain tens of thousands of solder connections, all of which must be correctly made. Control of the solder processes is a major factor in ensuring quality and reliability, and inspection and test of joint quality is an important feature of modern production test systems.

References 18–20 describe soldering methods and problems.

Cables and connectors

Electrical power and signals must be conducted within and between circuits. Cables and connectors are not usually perceived as high-technology or high-risk components, but they can be major contributors to unreliability of many systems if they are not carefully selected and applied.

The most common cable systems are copper wires in individual or multiconductor cables. Multiconductor cables can be round or flat (ribbon). Cable failure mechanisms include damage during manufacture, use or maintenance, and fatigue due to vibration or movement. Failures occur mainly at terminations such as connections to terminals and connectors, but also at points where damage is applied, such as by repeated bending around hinges. The failure modes are either permanent or intermittent open circuit. Cable runs should be carefully supported to restrict movement and to provide protection. Testing should address aspects such as the possibility of damage during assembly and maintenance, chafing of insulation due to vibration, fatigue due to vibration or other movement, etc.

The main types of connectors are circular multipin, for connecting round cables, and flat connectors for connecting ribbon cables and circuit boards. Individual wires are often connected by soldering or by using screw-down or push-on terminals. Low-cost connectors may not be sufficiently robust for severe environments (vibration, moisture, frequent disconnection/reconnection, etc.) or long-life applications. Connectors for important or critical applications are designed to be rugged, to protect the connector pin surfaces from moisture and contamination, and the connecting surfaces are gold plated.

In many modern electrical and electronic systems connectors contribute a high proportion, and often the majority of failures in service. The most common failure modes are permanent or intermittent open circuit due to damage or buildup of insulation on the mating surfaces due to oxidation, contamination or corrosion. Therefore it is important that they are carefully selected for the application, protected from vibration, abuse, corrosion and other stresses, and that their failure modes are taken into account in the test programme.

Data signals are also transmitted as pulses of light through optical fibre conductors and connectors. Light-emitting diodes operating in the infra-red part of the spectrum act as transmitting and receiving devices at the ends of the fibres. Special optical connectors are used to connect the ends of fibres, which must be accurately aligned and mated to ensure transmission. Optical fibres can break as a result of bending stresses.

Insulation

Insulation is as important in electrical and electronic systems as is conduction. All conductors, in cables and connectors and on circuit cards, must be insulated from one another. Windings in coils (solenoids, motors, generators, etc.) also require insulation. Insulation also provides protection against injury or death from human contact with high voltages.

Insulators can degrade and fail due to the following main causes:

—Mechanical damage, being trapped, chafed, cut, etc.
—Excessive temperature, causing charring and hence loss of dielectric strength. We are all familiar with the smell of burning shellac when an electrical appliance such as a drill or a microwave oven suffers a short circuit in a motor or transformer coil.
—Embrittlement and then fracture, caused by exposure to high temperatures, UV radiation, or chemical contamination. (Oil contamination can cause some cable insulator materials to swell and soften.)
—Rodent attack. Some insulator materials are enjoyed by mice and rats which take care of agricultural machinery during the winter.

Degradation of insulation is nearly always a long-term phenomenon (10 or more years typically).

SUMMARY OF DEVICE FAILURE MODES

For electronic devices used in a system, the most likely failure modes must be considered during the design so that their effects can be minimized. Circuit FMECAs and system reliability block diagrams should also take account of likely failure modes. Table 9.2 summarizes the main failure modes for the most common device types. The failure modes listed are not exhaustive and the device types listed are only

Table 9.2   Device failure modes

Type	Main failure modes	Typical approximate proportions (per cent)
Microcircuits		
Digital logic	Output stuck at high or low	80
	No function	20
Linear	Parameter drift	20
	No output	70
	Hard over output	10
Transistors		
	Low gain	20
	Open-circuit	30
	Short-circuit	20
	High leakage collector-base	30
Diodes		
Rectifier, general purpose	Short-circuit	10
	Open-circuit	20
	High reverse current	70
Resistors		
Film, fixed	Open-circuit	30
	Parameter change	70
Composition, fixed	Open-circuit	10
	Parameter change	90
Variables	Open-circuit	30
	Intermittent	10
	Noisy	10
	Parameter change	50
Relays		
	No transfer	20
	Intermittent	70
	Short-circuit	10
Capacitors		
Fixed	Short-circuit	60
	Open-circuit	20
	Excessive leakage	10
	Parameter change	10
Solder, connectors	Open circuit	50
	Short circuit	20
	Intermittent	30

a summary of the range. The failure mode proportions can vary considerably depending on type within the generic headings, application, rating and source. Devices used within a particular design should be individually assessed, e.g. a resistor rated very conservatively is likely to have a reduced relative chance of failing open-circuit.

Circuit design should take account of the likely failure modes whenever practicable. Capacitors in series will provide protection against failure of one causing a short-circuit and resistors in parallel will provide redundancy against one failing open. Blocking diodes are often arranged in series to protect against shorts.

Reference 6 describes reliability aspects of most types of electronic component.

CIRCUIT AND SYSTEM ASPECTS

Distortion and Jitter

Distortion is any change in the shape of a waveform from the ideal. Distortion can be caused by several factors, including mismatched input and output impedances, crossover distortion in transistors, optical devices and op-amps, transistor saturation, interference (see below), thermal effects, etc. All waveforms (power, audio, HF to microwave signals, digital signals, etc.) can be affected, and the problems grow as frequencies increase. Circuit designs should minimize distortion, but it can cause failures, or it can be the symptom of component failures or parameter variations.

Jitter is a form of distortion that results in an intermittent variation of a waveform from its ideal position, such as timing, period or phase instabilities. It can affect the operation of high-speed circuits, particularly the timing of digital pulses, and can therefore cause corruption of signals and data.

Timing

Timing is an important aspect of most digital electronic circuit design. To function correctly, the input and output voltage pulses to and from circuit elements must appear at the correct times in the logic sequences. This is relatively easy to arrange by design for simple circuits which operate at relatively low speed (clock rate or frequency). However, as speeds and complexity have increased, and continue to do so, it becomes increasingly difficult to ensure that every pulse occurs at the correct time and sequence. The integrity of the pulse waveform also becomes more difficult to assure at higher frequencies. Any digital circuit will have a speed above which it will begin to perform incorrectly. For example, in a production batch of microprocessors, all are made to the same design and using the same processes, but they will have different maximum operating speeds, due to variations in the manufacturing processes, and the final functional test is used to determine which are sold as 266 MHz, 400 MHz, 800 MHz, etc., units. At higher assembly levels, such as telecommunications or control systems, further limitations on speed of operation can be caused by inductive and capacitive effects within the circuit and by propagation delays along conductors.

Electromagnetic interference and compatibility

Electromagnetic interference (EMI) is the disturbance of correct circuit operation caused by changing electromagnetic fields or other electrical stimuli, which are then received by signal lines and so generate spurious signals. EMI is also called 'noise'? Electromagnetic compatibility (EMC) is the ability of circuits and systems to withstand these effects. EMC is sometimes referred to as electromagnetic immunity.

EMI can be generated by many sources, such as:

—High frequency radiation within the system, from switching within components, transmissions from data lines, and oscillators. Every component and conductor is a potential transmitter and a potential receiver. This is called 'cross-coupling'.
—Transient differences between voltage potentials on different parts of the circuit ground plane, due to inductive or capacitive effects.
—Electromagnetic emissions from RF components or subsystems when these are part of the system, or from other systems, such as radios, radars, engine ignition systems, electric motors, arcing across relays, lightning strikes, etc.
—Switching of inductive or capacitive loads such as motors on the same power circuit.

The operating frequencies of modern digital systems are in the radio frequency range (500 MHz to over 1 GHz), and harmonics are also generated. The design objective is to ensure that all signals travel through the circuit conductors, but at such high frequencies there are inevitably radiated emissions, which can then be received by other conductors and components. Circuit design to prevent EMI is a difficult and challenging aspect of all modern system designs. Methods are described later.

Intermittent failures

A large proportion of the failures of modern electronic systems are in fact of an intermittent nature. That is, the system performs incorrectly only under certain conditions, but not others. Such failures are most often caused by connectors that fail to connect at some times, such as under vibration or at certain temperatures, broken circuit card tracks that are intermittently open circuit, tolerance buildup effects between component parameters, etc. It is not uncommon for more than 50 per cent of reported failures of systems to be diagnosed on investigation as *'no fault found'* (NFF) or *'retest OK'* (RTOK), mainly due to the effects of intermittent failures. Worse, since the causes of the failures are mostly not detected, the faulty units are not repaired, and can cause the same system failure when reinstalled. These can therefore generate high costs of system downtime, repair work, provision of spare units, etc.

Other failure causes

There are many other causes of failure of electrical/electronic components and systems. It is impracticable to attempt to try to provide a comprehensive list, but examples include:

—Failure of vacuum devices (CRTs, light bulbs and tubes, etc.) due to seal failures.
—Failures of light bulbs due to sputtering of filament material and eventual loss of current carrying capability. This mechanism generates a quite specific life limiting feature, typically about 1000 hours.

—Failures due to non-operating environments, such as storage or standby conditions. Reference 21 covers these aspects.

ELECTRONIC SYSTEM RELIABILITY PREDICTION

The most commonly used standard database of failure rates for electronic components is that developed in the USA by the USAF Rome Air Development Center, and published as US MIL-HDBK-217 (*Reliability Prediction for Electronic Systems*) (Reference 30). This provides constant hazard rate models for all types of electronic component, taking into account factors that are considered likely to affect reliability.

MIL-HDBK-217 in fact quotes 'failure rate' models, not hazard rate models. Since the method is mainly applied to repairable systems, the models really relate to the component failure rate contribution to the system, i.e. the failure rate of the 'socket' into which that component type fits. MIL-HDBK-217 assumes independent, identically exponentially distributed (IID exponential) times to failure for all components.

The general MIL-HDBK-217 failure rate model is of the form:

$$\lambda_p = \lambda_b \pi_Q \pi_E \pi_A \ldots \tag{9.2}$$

where λ_b is the base failure rate related to temperature and π_Q, π_E, π_A, ... are factors which take account of part quality level, equipment environment, application stress, etc.

The relationship of failure rate to temperature is based upon the model for temperature dependent physical and chemical processes (diffusion, reaction), due to Arrhenius. The MIL-HDBK-217 version is:

$$\lambda_b = K \, \exp\left(\frac{-E}{kT}\right) \tag{9.3}$$

where λ_b is the process rate (component 'base' failure rate), E the activation energy (eV) for the process, k Boltzmann's constant (1.38×10^{-23}J K^{-1} or 8.63×10^{-5}eVK^{-1}), T the absolute temperature and K is a constant.

Detailed models are provided for each part type, such as microcircuits, transistors, resistors, connectors, etc. For example, the microelectronic device failure rate model is:

$$\lambda_p = \pi_Q \pi_L [C_1 \pi_T + C_2 \pi_E]/10^6 h \tag{9.4}$$

where π_Q and π_E are as described above; π_L is a learning factor, equal to unity except for new, relatively unproved devices, when $\pi_L = 10$; C_1 is a complexity factor based upon the chip gate count, or bit count for memory devices, or transistor count for linear devices; C_2 is a complexity factor based upon packaging aspects (number of pins, package type); and π_T is the temperature factor, derived from Eqn 9.2.

The quality factors are based upon specification standard and screening. Table 9.3 shows quality factors for microcircuits.

Table 9.3 Microcircuit device screening effect on quality factor

Screen level (MIL-HDBK-217)	Specification	π_Q
A	MIL-M-38510, Class S	0.25
B,B-0	As above, Class B	1
B-1	MIL-STD-883, Class B	2
—	Other commercial or unknown screening levels	10*

Note: Where screening levels for commercial components are known a scoring method can be used to reduce π_Q, by amounts that are related to the screening methods applied.

The criticisms of using such databases and models for system-level reliability prediction as described in Chapter 6 apply to the MIL-HDBK-217 method. In addition, MIL-HDBK-217 is fundamentally in error in other major respects:

1. Experience shows that only a very small proportion of failures of modern electronic systems are due to components failing owing to internal causes (typically 1 to 10 per cent).
2. Experience shows that the reliability of 'higher-grade' components, such as MIL SPEC and to high screen levels, is no longer higher than good commercial grade components purchased to manufacturers' specifications.
3. The temperature dependence of failure rate is not supported by modern experience or by considerations of physics of failure.
4. Several other parameters used in the models are of doubtful validity. For example, it is not true that failure rate increases significantly, if at all, with increasing complexity, as continual process improvements counteract the effects of complexity.
5. The models do not take account of many factors that do affect reliability, such as transient overstress, temperature cycling, variation, EMI/EMC, and control of assembly, test and maintenance.

Other databases and methods for predicting the reliability of electronic systems have been published, mostly based upon MIL-HDBK-217. The Bellcore method is a notable example, used in telecommunications systems. These are all subject to the same criticisms.

The only correct way to predict the likely reliability of electronic systems is to work top-down, as described in Chapter 6. IEEE Standard 1413 (Reference 4 in Chapter 6) also provides guidelines.

RELIABILITY IN ELECTRONIC SYSTEM DESIGN

Introduction

The designer of an electronic system must consider the following main aspects in order to create an inherently reliable design:

1. Electrical and other stresses, particularly thermal, on components, to ensure that no component can be overstressed during operation or testing.
2. Variation and tolerances of component parameter values, to ensure that circuits will function correctly within the range of likely parameter values.
3. The effects of non-stress factors, such as electrical interference, timing, and parasitic parameters. These are particularly important in high frequency and high gain circuits.
4. Ease of manufacture and maintenance, including design for test.

In addition to these primary considerations, there are other aspects of circuit and system design which can be applied to improve reliability. By reducing the number of different part types, the parts selection effort can be reduced and designs become easier to check. This also generates cost savings in production and use. Redundancy can also be designed into circuits. Whenever practicable the need for adjustments or fine tolerances should be avoided.

Not all of the means of achieving reliable electronic design are complementary. For example, redundancy and the inclusion of additional protective devices or circuits are not compatible with reducing complexity and the number of part types. The various design options relevant to reliability must be considered in relation to their effectiveness, cost and the consequences of failure.

The sections which follow outline the most important methods available to ensure high reliability. They are by no means comprehensive: circuit designers should consult their own organizations' design rules, component application notes, the Bibliography to this chapter and other relevant sources. However, what follows is intended as a guide to the points which reliability engineers and circuit designers need to consider.

Transient voltage protection

Electronic components are prone to damage by short duration high voltage transients, caused by switching of loads, capacitive or inductive effects, electrostatic discharge (ESD), incorrect testing, etc. Small semiconductor components such as ICs and low power transistors are particularly vulnerable, owing to their very low thermal inertias. MOS devices are very vulnerable to ESD, and require special protection, both externally and on-chip.

Logic devices which interface with inductive or capacitive loads, or which 'see' test connections, require transient voltage protection. This can be provided by: a *capacitor* between the voltage line to be protected and ground, to absorb high frequency transients (buffering), *diode protection*, to prevent voltages from rising beyond a fixed value (clamping), and *series resistances*, to limit current values. Figures 9.5 and 9.6 show typical arrangements for the protection of a logic device and a transistor. IC protection is also provided by transmitting logic signals via a light-emitting diode (LED) and optical transducer combination, called an opto-isolator or opto-coupler.

The transient voltage levels which can cause failure of semiconductor devices are referred to as VZAP. VZAP values depend upon transient duration. Maximum safe transient voltages are stated in manufacturers' databooks, and standard tests have been developed, for example in MIL-STD-883.

Figure 9.5 Logic device protection. Diode D1 prevents the input voltage from rising above the power supply voltage. Capacitor C1 absorbs high frequency power supply transients

Figure 9.6 Transistor protection. Resistor R1 limits the base current I_B and capacitor C1 absorbs power supply high frequency transients

Passive devices can also be damaged by transient voltages, but the energy levels required are much higher than for small semiconductor devices. Therefore passive devices do not normally need individual protection.

Very high electrostatic potentials, up to 5000 V, can be generated by triboelectrical effects on clothing, packaging material, automatic handling and assembly equipment, etc. If these are discharged into ESD sensitive components, either directly by contact with their pins, or via conductors in the system, damage or destruction is likely. Therefore it is essential that components are handled with adequate ESD precautions at all stages, and that protection is designed into circuits to safeguard components after assembly. Thereafter, care must be taken during test and maintenance, though the components will no longer be as vulnerable.

ESD can damage or destroy components even when they are unpowered, so precautions are necessary during all operations involving handling. Warning labels should be fixed to packages and equipments, and workbenches, tools and personnel must all be electrically grounded during assembly, repair and test.

Reference 32 is a good source for information on ESD.

Thermal design

It is important to control the thermal design of electronic systems, so that maximum rated operating temperatures are not exceeded under worst cases of environment and load, and so that temperature variations within the system are not severe.

The reasons are that high temperatures can accelerate some failure modes in marginally defective components, and temperature cycling between ambient and high values can cause thermal fatigue of bonds and component structures, particularly if there are high local temperature gradients.

The maximum temperature generated within a device depends on the electrical load and the local ambient temperature, as well as the thermal resistance between the active part of the device and the external environment. Temperature at the active area of a device, for example the junction of a power transistor, or averaged over the surface of an IC, can be calculated using the formula

$$T_J = T_A + \theta W \qquad\qquad\qquad\qquad (9.5)$$

where T_J is the local temperature at the active region, T_A is the ambient temperature around the component, W is the power dissipation, and θ is the thermal resistance between the active region and the ambient, measured in °C per Watt.

For devices that consume significant power levels in relation to their heat dissipation capacity, it is necessary to provide additional thermal protection. This can be achieved by mounting the device on a heat sink, typically a metal block, with fins to aid convective and radiant heat dissipation. Some devices, such as power transistors have integral heat sinks. Small fans are also used, for example to cool microprocessors. Further protection can be provided by temperature-sensitive cut-off switches or relays; power supply and conversion units often include such features.

It is sometimes necessary to consider not only the thermal path from the component's active area to the local ambient, but to design to allow heat to escape from assemblies. This is essential in densely packaged systems such as avionics, military electronics, and high-speed computers. In such systems a copper heat plane is usually incorporated into the PCB, to enable heat to flow from the components to the case of the equipment. Good thermal contact must be provided between the edge of the heat plane and the case. In turn the case can be designed to dissipate heat effectively, by the use of fins. In extreme cases liquid cooling systems are used, fluid being pumped through channels in the walls of the case or over heat sinks.

Temperature control can be greatly influenced by the layout and orientation of components and sub-assemblies such as PCBs. Hot components should be positioned downstream in the heat flow path (heat plane or air flow), and PCBs should be aligned vertically to allow convective air flow. Fans are often used to circulate air in electronic systems, to assist heat removal.

When additional thermal control measures are employed, their effects must be considered in evaluating component operating temperatures. The various thermal resistances, from the component active area to the external environment, must all be taken into account, as well as heat inputs from all heat-generating components, external sources such as solar radiation, and the effects of convection or forced cooling measures. Such a detailed thermal evaluation can best be performed with thermal modelling software, using finite element methods. Such software can be used to produce thermal maps of PCBs, taking into account each component's power load and all thermal resistances.

Thermal evaluation is important for any electronic design in which component operating temperatures might approach maximum rated values. Good detailed guidelines on thermal design for electronic systems are given in References 23 to 29.

Stress derating

Derating is the practice of limiting the stresses which may be applied to a component, to levels below the specified maxima, in order to enhance reliability. Derating values of electrical stress are expressed as ratios of applied stress to rated maximum stress. The applied stress is taken as the maximum likely to be applied during worst case operating conditions. Thermal derating is expressed as a temperature value.

Derating enhances reliability by:

1. Reducing the likelihood that marginal components will fail during the life of the system.
2. Reducing the effects of parameter variations.
3. Reducing long-term drift in parameter values.
4. Providing allowance for uncertainty in stress calculations.
5. Providing some protection against transient stresses, such as voltage spikes.

Typical derating guidelines are shown in Table 9.4, which gives electrical and thermal derating figures appropriate to normal and for critical (Hi-rel) applications such as spacecraft, or for critical functions within other systems. Such guidelines should usually be taken as advisory, since other factors such as cost or volume might be overriding. However, if stress values near to rated maxima must be used it is important that the component is carefully selected and purchased, and that stress calculations are doubled-checked.

Since thermal stress is a function of the surrounding temperature and power dissipation, combined temperature–power stress derating, as shown in Fig. 9.7, is often advised for components such as power transistors. The manufacturers' databooks should be consulted for specific guidelines, and References 2–5 also provide information.

Electromagnetic interference and Compatibility (EMI/EMC)

Circuit design to prevent EMI is a difficult and challenging aspect of all modern electronic system design.

The main design techniques are:

1. The use of filter circuits to decouple noise and transients from or to the power supply.
2. Circuits and conductors can be shielded by enclosing them in grounded, conductive boxes (Faraday shields), or, in the case of cables, grounded conductive screens. Cables can also be made less susceptible to picking up noise by using a twisted pair arrangement.

Table 9.4 Device derating guidelines

Device type	Parameter	Max. rating	
		Normal	Hi-rel
Microelectronics			
Digital	Power supply, input voltages	Derated but within	
	Output current (load,	performance spec.	
	fan-out)	0.8	0.8
	Junction temp. (hermetic): TTL	130°C	100°C
	(hermetic): CMOS	110°C	90°C
	(plastic)	100°C	70°C
	Speed	0.8	0.8
Linear	Power supply	Derated but within	
	Voltage	performance spec.	
	Input voltage	0.8	0.7
	Junction temp. (hermetic)	110°C	90°C
	(plastic)	100°C	70°C
Transistors			
Silicon (general purpose)	Collector current	0.8	0.5
	Voltage V_{cc}	0.8	0.6
	Junction temp. (hermetic)	120°C	100°C
	(plastic)	100°C	80°C
Silicon (power)	Collector current	0.8	0.6
	Voltage V_{cc}	0.8	0.6
	Voltage (reverse bias)	0.9	0.8
	Junction temp. (hermetic)	130°C	110°C
	(plastic)	110°C	90°C
Diodes			
Silicon (general purpose)	Forward current, voltages	0.8	0.5
Zener	Junction temp.	120°C	100°C
Resistors			
	Power dissipation	0.8	0.5
	Operating temp.	Rated–20°C	Rated–40°C
Capacitors			
	Voltage	0.8	0.5
	Operating temp.	Rated–20°C	Rated–40°C
Relays and switches			
Resistive or capacitive load	Current	0.8	0.5
Inductive load	Current	0.5	0.3
Motor	Current	0.3	0.2
Filament	Current	0.2	0.1

3. Circuit impedances should be balanced, e.g. between power supplies and loads, so that any noise pickup will be the same in each conductor, and will thus be self-cancelling.

4. All circuit grounds must be at the same electrical potential during circuit operation, and therefore all ground connections must provide a low impedance path back to

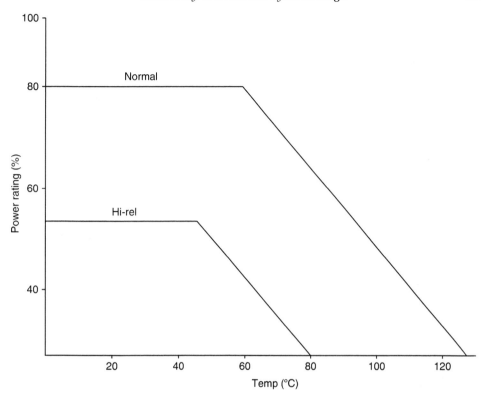

Figure 9.7 Temperature–power derating for transistors and diodes (typical)

the current source. This is particularly important in high frequency digital systems.

5. Contacts which make or break during circuit operation, e.g. microswitches and relays, must be selected to minimize EMI, and if necessary filter circuits must be designed around them.

6. Digital systems must include noise filters at the PCB power input and near to each IC. The normal approach is to use decoupling capacitors. The capacitance value must be selected in relation to the circuit frequency, so that the resonant frequency of the local *L–C* circuit (see Fig. 9.8) is well above the circuit operating frequency (to prevent resonance), but with a large enough capacitance to supply the transient current needed by the IC for its switching function.

7. Optical fibres used for data transmission are immune to electromagnetic effects.

8. In software-driven systems, coding methods can be used to provide EMI protection, see Chapter 10.

Software is available for EMI/EMC analysis.

National and International regulations exist to set standards for electromagnetic and power line emissions and associated test methods.

Most of the protection techniques against transient high voltages, described earlier, are useful in relation to EMI, and vice versa. Therefore the topics are often

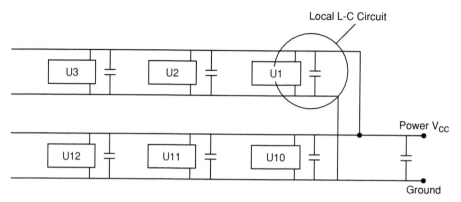

Figure 9.8   Digital circuit noise decoupling

combined in specialist texts and training. References 31 and 32 are good introductions to the whole field of EMI and EMC and Reference 33 describes ESD.

Redundancy

Redundancy techniques were covered in Chapter 6. In electronic circuit and system design it is possible to apply redundancy at any level from individual components to subsystems. Decisions on when and how to design-in redundancy depend upon the criticality of the system or function and must always be balanced against the need to minimize complexity and cost. However, it is often possible to provide worthwhile reliability improvements by using redundant circuit elements, at relatively little cost, owing to the low cost of most modern devices. The most likely component failure modes must be considered; for example, resistors in parallel will provide redundant, though possibly degraded operation if one becomes open circuit, and short circuit is an unlikely failure mode. Opposite considerations apply to capacitors.

Design simplification

Like all good engineering, electronic system designs must be kept as simple as practicable. The motto often quoted is KISS—'Keep it simple, stupid'. In electronics, design simplification is mainly a matter of minimizing the number of components to perform a required function. Reducing the number of components and their connections should improve reliability as well as reduce production costs. However, the need to provide adequate circuit protection and component derating, and where necessary redundancy, should normally take priority over component count reduction.

Minimizing the number of component types is also an important aspect of design simplification. It is inevitable that when a number of designers contribute to a system, different solutions to similar design problems will be used, resulting in a larger number of component types and values being specified than is necessary. This leads to higher production costs, and higher costs of maintenance, since more part types must be bought and stocked. It can also reduce reliability, since quality control of bought-in parts is made more difficult if an unnecessarily large number of part types must be controlled.

Design rules can be written to assist in minimizing component types, by constraining designers to preferred standard approaches. Component type reduction should also be made an objective of design review, particularly of initial designs, before prototypes are made or drawings frozen for production.

Sneak analysis

A sneak circuit is an unwanted connection in an electrical or electronic circuit, not caused by component failure, which leads to an undesirable circuit condition or which can inhibit a desired condition. Sneak circuits can be inadvertently designed into systems when interfaces are not fully specified or understood, or when designers make mistakes in the design of complex circuitry. Sneak analysis is a technique developed to identify such conditions in electrical and electronic circuits, and in operating software.

It is based on the identification within the system of 'patterns' which can lead to sneak conditions. The five basic patterns are shown in Fig. 9.9.

Any circuit can be considered as being made of combinations of these patterns. Each pattern is analysed to detect if conditions could arise, either during normal operation or due to a fault in another part of the system, that will cause a sneak. For example, in the power dome or the combination dome the power sources could be reversed if S1 and S2 are closed.

Sneak circuits are of five types:

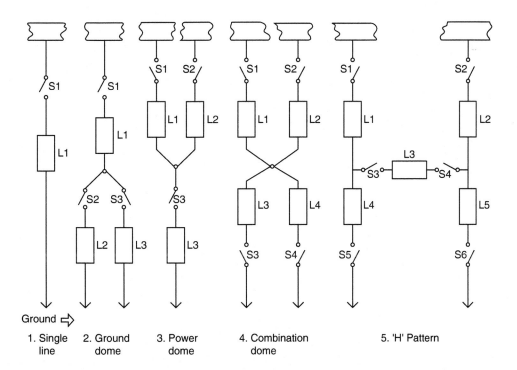

Figure 9.9 Sneak analysis basic patterns (hardware)

1. *Sneak paths.* Current flows along an unexpected route.
2. *Sneak opens.* Current does not flow along an expected route.
3. *Sneak timing.* Current flows at the incorrect time or does not flow at the correct time.
4. *Sneak indications.* False or ambiguous indications.
5. *Sneak labels.* False, ambiguous or incomplete labels on controls or indicators.

When potential sneak conditions are identified, they must be validated by test or detailed investigation, reported, and corrective action must be considered.

Sneak analysis is a tedious task when performed on relatively large systems. However, it has proved to be very beneficial in safety analysis and in assessing the integrity of controls in aircraft and industrial systems. Computer programs have been developed. Software applications are described in Chapter 10.

Sneak circuits can be avoided by careful design, or detected by adequate testing. The formal analysis technique is appropriate for critical systems, particularly when there are complex interfaces between sub-systems.

PARAMETER VARIATION AND TOLERANCES

Introduction

All electrical parameters of electronic components are subject both to initial component-to-component variation, and sometimes to long-term drift. Parameter values can also vary as a result of other factors, particularly temperature. Whether these variations are important or not in a particular design depends upon the requirements for accuracy of the parameters in that application. For example, the resistance value of a resistor in a feedback circuit of a high gain amplifier might be critical for correct operation, but would not need to be as closely controlled in a resistor used as a current limiter.

Initial variation is an inevitable consequence of the component production processes. Most controlled parameters are measured at the end of production, and the components are assigned to tolerance bands, or rejected if they fall outside the limits. For example, typical resistors are provided in tolerance ranges of 1 per cent, 5 per cent, 10 per cent and 20 per cent about the nominal resistance. Since the selection is often from the same batch, which may have had a parameter distribution as shown in Fig. 9.10, the parameter distributions of the selected tolerance ranges would be as shown, assuming only two tolerance bands had been selected. Depending upon the application, knowledge of the shape of the parameter distribution might be important.

For many component parameters, for example transistor characteristics, maximum and minimum values are stated. It is also important to note that not all parameter values are controlled in manufacture and selection. Some parameters are given only as 'typical' values. It is never a good idea to design critical circuit operation around such parameters, since they are not usually measured, and therefore are not guaranteed.

Since conductance, both of conductor and of semiconductor materials, varies with temperature, all associated parameters will also vary. Therefore resistance of resistors,

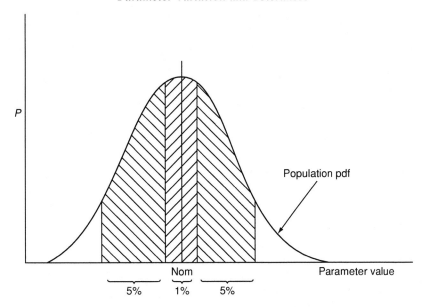

Figure 9.10 Parameter distributions after selection

and gain and switching time of transistors, are typical of temperature-dependent parameters. High temperature can also increase noise outputs. Other parameters can interact, for example the capacitance values between transistor connections is affected by bias voltage.

Parameter drift with age is also usually associated with changes in conductance, as well as in dielectric performance, so that resistors and capacitors are subject to drift, at rates which depend upon the type of materials and construction used, operating temperature and time.

Another source of circuit parameter variation arises from what are referred to as *parasitic* parameters. These are electrical parameters that are not intrinsic to the theoretical design of the component or the circuit, but which are due to construction and layout features. For example, wire wound resistors are inductive, the inductance depending upon the type of construction, PCB conductor tracks have mutual inductance and capacitance, and integrated circuit lead frames are inductive. Parasitic effects can be very important, and difficult to control, in high gain and high frequency systems.

Parameter variation can affect circuits in two ways. For circuits which are required to be manufactured in quantity, a proportion might not meet the required operating specification, and production yield will then be less than 100 per cent, adding to production cost. Variation can also cause circuits to fail to work correctly in service. Initial component-to-component variation mainly affects yield, and stress or time related variation mainly affects reliability.

Tolerance design

Every electronic circuit design (for that matter, not only electronic circuits, but any system) must be based on the nominal parameter values of all the components that will contribute to correct performance. This is called *parameter design*. It encompasses the qualities of knowledge and inventiveness necessary for the solution of design problems. However, having created the correct functional design, it is necessary to evaluate the effects of parameter variation on yield, stability and reliability. This is called *tolerance design*.

The first step in tolerance design is to determine which parameter values are likely to be most sensitive in affecting yield and performance. This can be performed initially on the basis of experience and system calculations made during the parameter design stage. However, a more systematic approach, using the techniques described below, should be used for serious design.

The next step is to determine the extent of variation of all of the important parameter values, by reference to the detailed component specifications. This step is sometimes omitted by designers using lists of 'preferred parts', which list only primary nominal values and tolerances, without giving full details of other data relevant to the application. All relevant parasitic parameters should also be evaluated at this stage.

It is possible to design-in compensating features for some variation, for example by using temperature compensation components such as thermistors or adjustable components such as variable resistors. However, these add complexity and usually degrade reliability, since they are themselves prone to drift, and adjustable components can be degraded by wear, vibration, and contamination. Wherever possible the design should aim for minimum performance variation by the careful selection of parameter values and tolerances.

Analysis methods

The analysis of tolerance effects in general design situations was covered in Chapter 5. This section introduces further methods available to analyse the effects of variation and tolerances in electronic circuits.

Worst case analysis

Worst case analysis (WCA) involves evaluating the circuit performance when the most important component parameter values are at their highest and lowest tolerance values. It is a straightforward extension of the parameter design calculations. However, it is only realistic for simple circuits. Also, purely digital circuits are relatively easy to analyse in this respect, so long as frequency and timing requirements are not too severe. However, if there are several parameters whose variation might be important, particularly if there are interactions or uncertainty, then more powerful methods of analysis should be used.

The transpose circuit

The sensitivity of the output of a circuit to parameter changes can be analysed using the *transpose circuit* method. A transpose circuit is a circuit which is topologically identical to that under investigation, except that the forward and reverse transmission coefficients of each 3-terminal device (e.g. transistor) are interchanged, the input is replaced by an open circuit, and the output by a current source of 1A. Figure 9.11 shows a circuit and a possible transpose of it. Here we are looking only at those components whose parameter variations we consider might most affect the output, in this case V_0, when the input is I_i. Note that these are instantaneous values of terms in the frequency domain (or DC values).

The sensitivity of V_0 to small changes in the conductance G of the resistor is given by

$$\frac{\partial V_0}{\partial G} = -V_G V_{G/T} \tag{9.6}$$

where V_G is the voltage across the conductance in the actual circuit and $V_{G/T}$ is the voltage across the conductance in the transpose circuit. Similar relations hold for other component parameters, if the two circuits are analysed at the same frequency. This is *Tellegen's theorem*. Using these relationships, the sensitivities to all the critical

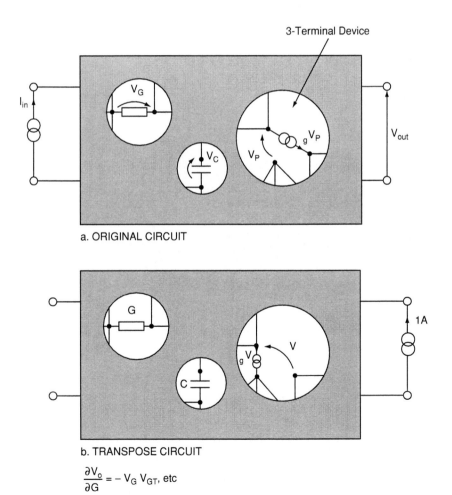

a. ORIGINAL CIRCUIT

b. TRANSPOSE CIRCUIT

$$\frac{\partial V_0}{\partial G} = -V_G V_{GT}, \text{ etc}$$

Figure 9.11   Transpose circuit

(a) Specification on the performance   (b) Some samples of a mass-produced circuit
    of a filter   violate the specification

Figure 9.12   Monte Carlo analysis of filter circuit (from Reference 36)

parameters can be evaluated, by analysing only two different circuits, using for example circuit simulation software. This is an extremely efficient technique for analysing the effects of small, single variations.

Simulation

Another method for analysing the effects of tolerances and variation is simulation, using the Monte Carlo method. The principles of Monte Carlo simulation were described in Chapter 6. Most modern circuit simulation software includes Monte Carlo techniques, which allow parameter variations to be set, either as a range or as a defined distribution shape (e.g. *s*-normal, bimodal), and the program will then randomly 'build' circuits using parameter values from the distributions, and will analyse these. Monte Carlo circuit simulation evaluates the effects of multiple simultaneous variations. Refinements allow the most critical parameters to be identified, and statistically designed experiments (see Chapter 5) can be run. Circuits can be analysed in the time and frequency domains, and there are no practical limitations, apart from the time required to run each analysis, on the number of parameters, input conditions, and simulations.

Figure 9.12 shows the results of a number of Monte Carlo simulations of a filter circuit, in relation to the specification. This shows that some parameter combinations give performance outside the specification. Carrying out a number of runs provides an estimate of production yield, and the particular parameter value combinations that caused circuits to be outside the specification can be identified.

See Reference 36 for a detailed description of the topics covered in this section.

DESIGN FOR PRODUCTION, TEST AND MAINTENANCE

Electronic circuits should be designed to be testable, using the test methods that will be applied in production. These methods depend upon the manufacturing test policy and economics, as discussed in Chapter 13. Testability is an important design feature, which can make a significant impact on production costs. Testability also affects reliability, since production defects which are not detected by the tests can lead to failure in service, and circuits which are difficult to diagnose are more likely to be inadequately or incorrectly repaired.

It is important that the circuit designer is aware of the test methods that will be used, and the requirements that they impose on the design to ensure effective and economic test coverage. Test methods for electronics are described in Reference 37.

Circuit designs must allow the ATE to initialize the operating states of components, control the circuit operation, observe and measure output states and values, and partition the circuit to reduce test program complexity. It is good practice to conduct a careful review, with the test engineers, of the testability of the design before it is finalized. Modern EDA software includes facilities for performing testability analysis. Reference 38 covers the subject of design for test in detail.

Design of electronic systems for production, test and maintainability should be included in design rules and design review. Aspects of good electronic design which contribute to these are listed below:

1. Avoid the necessity for adjustments, e.g. potentiometers, whenever possible. Adjustable components are less reliable than fixed-valued components and are more subject to drift.
2. Avoid 'select on test' situations, where components must be selected on the basis of measured parameter values. Specify components which can be used anywhere within the tolerance range of the applicable parameters and do not rely on parameters which are listed as typical rather than guaranteed. Where 'typical' tolerances must be used, ensure that the appropriate component screening is performed before assembly.
3. Ensure that adjustments are easily accessible at the appropriate assembly level.
4. Partition circuits so that subassemblies can be tested and diagnosed separately. For example, if several measured values require amplification, analogue-to-digital conversion, logic treatment and drivers for displays, it might be better to include all functions for each measured value on one PCB rather than use a PCB for each function, since fault diagnosis is made easier and repair of one channel does not affect the performance or calibration of others. However, other factors such as cost and space must also be considered.

BIBLIOGRAPHY

General

1. P. Horowitz and W. Hill, *The Art of Electronics*, 2nd edn. Cambridge University Press (1989).
2. M. Pecht (ed.), *Product Reliability, Maintainability and Supportability Handbook*. ARINC Research Corp. (1995).
3. N. B. Fuqua, *Reliability Engineering for Electronic Design*. M. Dekker (1987).
4. J. E. Arsenault and J. A. Roberts (eds), *Reliability and Maintainability of Electronic Systems*. Computer Science Press Inc., Rockville, Maryland (1980).
5. US MIL-HDBK-338: *Electronic Reliability Design Handbook*. Available from the National Technical Information Service, Springfield, Virginia.

Components

6. T. J. Bajenescu and M. I. Bazu, *Reliability of Electronic Components*. Springer-Verlag (1999).
7. M. G. Pecht, R. Radojcic and G. Rao, *Guidebook for Managing Silicon Chip Reliability*. CRC Press (1999).
8. F. Jensen, *Electronic Component Reliability*. J. Wiley (1995).
9. M. J. Howes and D. V. Morgan, *Reliability and Degradation in Semiconductor Devices and Circuits*. Wiley (1981).
10. E. A. Amerasekera and F. N. Najm, *Failure Mechanisms in Semiconductor Devices*, 2nd edn. Wiley (1997)
11. E. R. Hnatek, *Integrated Circuit Quality and Reliability*, 2nd edn. M. Dekker (1991).
12. Kraus, R. Hannemann and M. Pecht (Ed.), *Semiconductor Packaging: A Multidisciplinary Approach*. J. Wiley (1994).
13. J. H. Lau, C. P. Wong, J. L. Prince and W. Nakayama, *Electronic Packaging: Design, Materials, Process, and Reliability*. McGraw-Hill (1998).
14. F. N. Sinnadurai, *Handbook of Microelectronics Packaging and Interconnection Technologies*. Electrochemical Publications (1985).
15. British Standard, BS 9000: *Components of Assessed Quality*. British Standards Institution, London. (And equivalent US MIL, European (CECC) and International (IEC) standards.)
16. US MIL-STD-883: *Test Methods and Procedures for Microelectronic Devices*. Available from the National Technical Information Service, Springfield, Virginia.
17. F. Jensen and N. E. Peterson, *Burn-in*. Wiley (1983).
18. J. H. Lau (ed.), *Thermal Stress and Strain in Microelectronic Packaging*. Van Nostrand Reinhold (1993).
19. M. Pecht (ed.), *Soldering Processes and Equipment*. J. Wiley (1993).
20. K. Brindley and M. Judd, *Soldering in Electronics Assembly*. Newnes (1999).
21. J. Pecht and M. Pecht (ed.), *Long-Term Non-Operating Reliability of Electronic Products*. CRC Press (1995).
22. A. J. Bilotta, *Connections in Electronic Assemblies*. M. Dekker (1985).
23. R. W. Woodgate, *The Handbook of Machine Soldering: SMT and TH*, 3rd edn. Wiley-Interscience (1996).

Mechanical and thermal effects and design

24. D. S. Steinberg, *Vibration Analysis for Electronic Equipment*. J. Wiley (1988).
25. P. Lall, M. G. Pecht and E. B. Hakim, *Influence of Temperature on Microelectronics and System Reliability*. CRC Press (1997).
26. A. D. Kraus and A. Barr-Cohen, *Thermal Analysis and Control of Electronic Equipment*. McGraw-Hill (1983).
27. *Thermal Guide for Reliability Engineers*. Rome Air Development Center Report TR-82-172. Available from the National Technical Information Service, Springfield, Virginia.
28. D. J. Dean, *Thermal Design of Electronic Circuit Boards and Packages*. Electrochemical Publications (1985).
29. Patrick McCluskey, R. Grzybowski and T. Podlesak (eds.), *High Temperature Electronics*. CRC Press (1997).

Reliability prediction

30. US MIL-HDBK-217: *Reliability Prediction for Electronic Systems*. Available from the National Technical Information Service, Springfield, Virginia.

EMI/EMC/ESD

31. P. A. Chatterton and M. A. Houlden, *EMC: Electromagnetic Theory to Practical Design*. J. Wiley (1991).

32. H. W. Ott, *Noise Reduction Techniques in Electronic Systems*, 2nd edn. Wiley-Interscience (1988).

33. W. D. Greason, *Electrostatic Damage in Electronic Devices and Systems*. Research Studies Press/Wiley (1987).

34. US MIL-STD-461D, *Requirements for the Control of Electromagnetic Interference Emisions and Susceptibility*. NTIS, Springfield VI.

35. US MIL-E-6051D, *Electromagnetic Compatibility Requirements, Systems*. NTIS, Springfield, VI.

Tolerance design

36. R. Spence and R. S. Soin, *Tolerance Design in Electronic Circuits*. Addison-Wesley (1988).

Test

37. P. D. T. O'Connor, *Test Engineering*. J. Wiley (2001).

38. J. Turino and H. F. Binnendyk, *Design to Test*, 2nd edn. Logical Solutions Inc., Campbell, CA. (1991). (Also *Testability Advisor* software).

QUESTIONS

1. How does increasing temperature affect the reliability of electronic components? Illustrate your answer with three examples of specific failure mechanisms and component types.

2. Briefly describe three failure modes that can occur in modern integrated circuits. In each case, explain how they are influenced by temperature (high/low), electrical stresses, and manufacturing quality.

3. Describe the design, manufacturing and application factors that influence the following failure modes of integrated circuits: (i) electromigration; (ii) latch-up; (iii) electrostatic damage.

4. Screening is a process used to improve the quality and reliability of integrated circuits. Explain the engineering justification for IC screening, and describe briefly the tests that are typically applied in the screening process.

5. How have the recent developments in microcircuit device packaging affected reliability? How can the risks of failure be minimized?

6. What are the main factors to consider, from the reliability point of view, when selecting the type of packaging for the integrated circuits to be used in a circuit designed for mass production? Consider hermetic versus plastic packaging, and through-hole versus surface mounting.

7. Suggest ways in which the capability approval process is a more useful approach to ASIC line approval than the approved components system.

8. Use Eqn (9.4) to calculate the device *hazard rate* for a device screened to level B-1 from the following data: $\pi_L = 1$, $C_1 = 0.12$, $\pi_T = 3.7$, $\pi_V = 1$, $C_2 = 0.01$, $\pi_E = 4.2$. What is the use of the hazard rate calculated in this way?

9. Discuss the reliability of electronic components as an overall contributor to system reliability in modern systems.

10. Describe the ways in which solder connections can fail. How can circuit designers minimize solder failures?

11. Describe the ways in which electrical cables and connectors can fail. How can circuit designers minimize these failures?

12. What are 'no fault found' failures? Describe the main causes of such failures. Why are they important?

13. Why are the thermal aspects of electronic system design important for reliability? What methods can designers use to reduce the operating temperatures of electronic components?

14. A supplier has presented you with a derating policy based on deriving the derating curves from power stress versus base hazard rate curves from MIL-HDBK-217 data. State your reservations concerning this approach.

15. (a) for a small plastic transistor operating at 120 mW, estimate T_J if $\theta = 0.4\,°C\ mW^{-1}$ above 25°C, if the ambient temperature is 50°C.

    (b) If the maximum junction temperature is 150°C, estimate what power the transistor will dissipate at an ambient temperature of 60°C.

16. Using Fig. 9.7 and Table 9.4, estimate the percentage power allowed for a general-purpose silicon plastic-sealed transistor in a Hi-rel application.

17. State some of the advantages of employing thermal derating techniques in an electronic component based design.

18. What are the main sources of electromagnetic interference (EMI) that can affect electronic systems? Describe three methods that can be used to protect circuits from EMI.

19. In a normal domestic kitchen containing a fluorescent light fitting and a washing machine, list the EMI sources you may find and how as a designer you may mitigate these effects.

20. Describe three methods for analysing the effects of component parameter variations on the performance of an electronic circuit. For each, describe how the variations and their effects can be minimized by the designer.

10

Software Reliability

INTRODUCTION

Software is now part of the operating system of a very wide range of products and systems, and this trend continues to accelerate with the opportunities presented by low cost microcontroller devices. Software is relatively inexpensive to develop, costs very little to copy, weighs nothing, and does not fail in the ways that hardware does. Software also enables greater functionality to be provided than would otherwise be feasible or economic. Performing functions with software leads to less complex, cheaper, lighter and more robust systems. Therefore software is used increasingly to perform functions that otherwise would be performed by hardware, and even by humans. Recent examples are aircraft flight control systems, robotic welders, engine control systems, domestic bread-making machines, etc.

The software 'technology' used today is the same basic sequential digital logic first applied in the earliest computers. The only significant changes have been in the speed and word length capability of processors and the amount of memory available, which in turn have enabled the development of high-level computer languages and modern operating systems. Some attempts have been made to develop radically different approaches such as parallel processing and fuzzy logic, but these remain fringe applications. Therefore, the basic principles of software development, to ensure that programs are correct, safe and reliable, remain largely unchanged since they were first described in the 1970's (References 1–3).

Every copy of a computer program is identical to the original, so failures due to variability cannot occur. Also, software does not degrade, except in a few special senses,** and when it does it is easy to restore it to its original standard. Therefore, a correct program will run indefinitely without failure, and so will all copies of it. However, software can fail to perform the function intended, due to undetected errors. When a software error does exist, it exists in all copies of the program, and if it is such as to cause failure in certain circumstances, the program will always fail when those circumstances occur.

**Data or programs stored in some media can degrade. Magnetic media such as discs and tape are susceptible to electromagnetic fields, or even to being closely packed for long periods. VLSI semiconductor devices can suffer changes in the voltage state of individual memory cells due to naturally occurring alpha-particle bombardment. In each case a refresh cycle will restore the program.

Software failures can also occur as a function of the machine environment, e.g. machines can be restarted and the software 'fixed' by clearing queues, removing memory leaks, and refreshing the state of the machine. So identical copies can behave differently depending on their 'age' since robooting.

Software errors can cause system failure effects that can range from trivial to catastrophic. Therefore, software reliability and safety effort must be directed at the consequences of errors, not just at the prevention or removal of most of them.

Since most programs consist of very many individual statements and logical paths, all created by the efforts of humans, there is ample scope for errors. Therefore we must try to prevent the creation of errors, and maximize the likelihood of detecting and correcting those that are created, by imposing programming disciplines, by checking and by testing.

When software is an integral part of a hardware-software system, system failures might be caused by hardware failures or by software errors. When humans are also part of the system they can also cause failures (e.g. the Airbus crash during a low flying display, which some 'experts' immediately blamed on the new flight control software, but which the investigation concluded was caused by the pilot putting the aircraft into a situation from which the system could not prevent the crash). In some cases, it might be difficult to distinguish between hardware, software and human causes.

There are several ways by which hardware and software reliability differ. Some have already been mentioned. Table 10.1 lists the differences.

Table 10.1   Comparison of Hardware and Software Reliability Characteristics

Hardware	Software
1 Failures can be caused by deficiencies in design, production, use and maintenance.	Failures are primarily due to design faults. Repairs are made by modifying the design to make it robust against the condition that triggered the failure.
2 Failures can be due to wear or other energy-related phenomena. Sometimes warning is available before failure occurs (e.g. system can become noisy indicating degradation and impending failure).	There are no wearout phenomena. Software failures occur without warning, although very old code can exhibit an increasing failure rate as a function of errors introduced into the code while making functional code upgrades.
3 No two items are identical. Failures can be caused by variation.	There is no variation: all copies of a program are identical.
4 Repairs can be made to make equipment more reliable. This would be the case with preventive maintenance where a component is restored to an as new condition.	There is no repair. The only solution is redesign (reprogramming), which, if it removes the error and introduces no others, will result in higher reliability.
5 Reliability can depend on burn-in or wearout phenomena; that is, failure rates can be decreasing, constant or increasing with respect to time.	Reliability is not so time-dependent. Reliability improvement over time may be affected, but this is not an operational time relationship. Rather, it is a function of reliability growth of the code through detecting and correcting errors.

Table 10.1 *(contd.)*

Hardware	Software
6 Reliability may be time-related, with failures occurring as a function of operating (or storage) time, cycles, etc.	Reliability is not time related. Failures occur when a specific program step or path is executed or a specific input condition is encountered, which triggers a failure.
7 Reliability may be related to environmental factors (temperature, vibration, humidity, etc.)	The external environment does not affect reliability except insofar as it might affect program inputs. However, the program reliability is a function of the internal machine environment (queues, memory leakage, etc.)
8 Reliability can be predicted, in principle but mostly with large uncertainty, from knowledge of design, parts, usage, and environmental stress factors.	Reliability cannot be predicted from any physical bases, since it entirely depends on human factors in design. Some approaches exist based on the development process used and the extent of the code, but these are controversial.
9 Reliability can be improved by redundancy. The successful use of redundancy presumes ready detection, isolation, and switching of assets.	Reliability cannot be improved by redundancy if the parallel paths are identical, since if one path fails, the other will have the error. It is possible to provide redundancy by having diverse parallel paths with different programs written by different teams.
10 Failures can occur in components of a system in a pattern that is, to some extent, predictable from the stresses on the components and other factors. Reliability critical lists are useful to identify high risk items.	Failures are rarely predictable from analyses of separate statements. Errors are likely to exist randomly throughout the program, and any statement may be in error. Most errors lie on the boundary of the program or in its exception handling. Reliability critical lists are not appropriate.
11 Hardware interfaces are visual; one can see a Ten-pin connector.	Software interfaces are conceptual rather than visual.
12 Computer-aided design systems exist that can be used to create and analyse designs.	There are no computerized methods for software design and analysis. Software design is more of an 'art form' lacking the provability of hardware, except to a limited extent through *formal methods* (see later).
13 Hardware products use standard components as basic building blocks.	There are no standard parts in software, although there are standardised logic structures. Software reuse is being deployed, but on a limited basis.

SOFTWARE IN ENGINEERING SYSTEMS

The software that forms an integral part or sub-system of an engineering system is in some important ways different from software in other applications, such as banking, airline booking, logistics, CAE, PC operating systems and applications, etc. The differences are:

—Engineering programs are 'real time': they must operate in the system timescale, as determined by the system clock and as constrained by signal propagation and other delays (switches, actuators, etc.). A chess program or a circuit simulation program, for example, will run when executed, and it is not critical exactly how long it takes to complete the run. However, in an operational system such as a process controller or an autopilot, it is essential that the software is ready to accept inputs and completes tasks at the right times. The software must be designed so that functions are correctly timed in relation to the system clock pulses, task execution times, interrupts, etc. Timing errors are a common cause of failure in real-time systems, particularly during development. They are often difficult to detect, particularly by inspection of code. Timing errors can be caused by hardware faults or by interface problems. However, logic test instruments (logic analysers) can be used to show exactly when and under what conditions system timing errors occur, so that the causes can be pinpointed.
—Engineering programs share a wider range of interfaces with the system hardware. In addition to basic items such as processors, memory, displays and keyboards, other engineering interfaces can include measurement sensors, A/D and D/A converters, signal analysers, switches, connectors, etc.
—Engineering programs might be 'embedded' at different levels within a system: for example the main operating program might be loaded and run from disc, tape or accessible PROM devices, but other software might be embedded in components which are less accessible, such as ASICs, programmable gate arrays, signal processing ICs and flash memory devices. The BIOS chip in a PC is also an example of software embedded in this way.
—There is often scope for alternative solutions to design problems, involving decisions on which tasks should be performed by hardware (or humans) and which by software.
—Engineering software must sometimes work in electrically 'noisy' environments, so that data might be corrupted.
—Engineering programs are generally, though not always, rather smaller and simpler than most other applications

Therefore, it is very important that engineering software is developed (specified, designed, programmed, tested, managed) in close integration with the hardware and overall system work. The not uncommon practice of writing a software specification then 'outsourcing' the program development work should not be an option for important engineering software.

SOFTWARE ERRORS

Software errors ('bugs') can arise from the specification, the software system design and from the coding process.

Specification errors

Typically more than half the errors recorded during software development originate in the specification. Since software is not perceivable in a physical sense, there is little scope for common sense interpretation of ambiguities, inconsistencies or incomplete statements. Therefore, software specification must be very carefully developed and reviewed. The software specification must describe fully and accurately the requirements of the program. The program must reflect the requirements exactly. There are no safety margins in software design as in hardware design. For example, if the requirement is to measure $9\,V+/-0.5\,V$ and to indicate if the voltage is outside these tolerances, the program will do precisely that, if the specification was incorrectly formulated, e.g. if the tolerances were not stated, the out-of-tolerance voltage would be indicated at this point every time the measured voltage varied by a detectable amount from $9\,V$, whether or not the tolerances were exceeded. Depending upon the circumstances this might be an easily detectable error, or it might lead to unnecessary checks and adjustments because the out-of-tolerance indication is believed. This is a relatively simple example. Much more serious errors, such as a misunderstanding or omission of the logical requirement of the program, can be written into the specification. This type of error can be much harder to correct, involving considerable reprogramming, and can be much more serious in effect.

The Eurospace Ariane 5 spacecraft launcher failure was caused by such an error: the guidance computer and the intertial measurement unit used different bit formats for numerical data, but, even though this fact was known, no compensation was made because it had not resulted in failures on previous Ariane launchers. The new launcher's greater rocket thrust led to an overflow when the inertial unit measured velocities higher than experienced before. The NASA Mars Polar Orbiter spacecraft collided with the planet because part of the system was designed using measurements in miles while an interfacing subsystem used kilometres.

The specification must be logically complete. Consider the statement: 'Sample inputs A, B and C. If any one exceeds by $> \pm10$ units the average of the other two, feed forward the average of these two. Indicate failure of the out-of-tolerance input. If the out-of-tolerance condition does not exist, feed forward the average of the three inputs.'

This is an example of a two-out-of-three majority voting redundant system. The logic is shown in the flow diagram, Fig. 10.1. Consider the situation when the values of A, B and C are 100, 90 and 120. The values of the derived parameters, and route taken by the program, are shown in Fig. 10.1. In this case, two fault conditions exist, since both B and C exceed the average of the other two inputs. The program will indicate a B failure, as the algorithm compares B before C. The specification has not stated what should happen in the event of more than one input being out of tolerance. The program will work as shown in the algorithm, in the sense that an input will always be available, but the system may not be safe. The flowchart complies with the specification, but it probably does not reflect the real wishes of the specification writer. A software specification must cover all the possible input conditions and output requirements, and this usually requires much more detailed consideration than for a hardware specification.

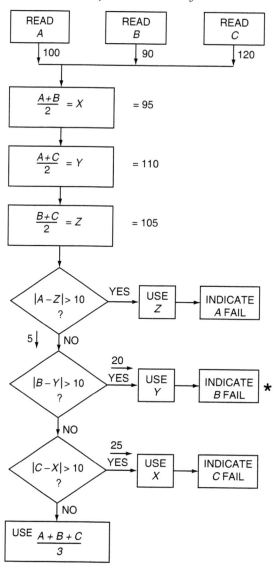

Figure 10.1 Voting redundant system

The specification must be consistent. It must not give conflicting information or use different conventions in different sections (e.g. miles and kilometres).

The specification must not include requirements that are not testable, e.g. accuracy or speed requirements that are beyond the capability of the hardware.

The specification should be more than just a description of program requirements. It should describe the structure to be used, the program test requirements and documentation needed during development and test, as well as basic requirements such as the programming language, and inputs and outputs. (Program structure, test and documentation will be covered later.) By adequately specifying these aspects, a framework for program generation will be created which minimizes the possibilities for creating errors, and which ensures that errors will be found and corrected.

Software system design

The software system design follows from the specification. The system design may be a flowchart and would define the program structure, test points, limits, etc. Errors can occur as a result of incorrect interpretation of the specification, or incomplete or incorrect logic. Errors can also occur if the software cannot handle data inputs that are incorrect but possible, such as missing or incorrect bits.

An important reliability feature of software system design is *robustness*, the term used to describe the capability of a program to withstand error conditions without serious effect, such as becoming locked in a loop or 'crashing'. The robustness of the program will depend upon the design, since it is at this stage that the paths to be taken by the program under error conditions are determined.

Software code generation

Code generation is a prime source of errors, since a typical program involves a large number of code statements. Typical errors can be:

—Typographical errrors. (sic).
—Incorrect numerical values, e.g. 0.1 for 0.01.
—Omission of symbols, e.g. parentheses.
—Inclusion of variables which are not declared, or not initialized at the start of program run.
—Inclusion of expressions which can become indeterminate, such as division by a value which can become zero.
—Accidental shared use of memory locations.

Changes to code can have dire consequences. The likelihood of injecting new faults can run as high as 50 per cent, and tends to be highest for small code changes. The injected faults tend to be more obscure and harder to detect and remove. Changes can be in conflict with the original architecture and increase code complexity.

We will briefly describe the methods that can be used to minimize the creation of errors, and to detect errors that might have been created.

PREVENTING ERRORS

Specification

The overall system specification and the software specification must be prepared in harmony. Both should allow flexibility in relation to allocation of functions and should encourage integrated engineering.

Software specifications should be more than just descriptions of requirements. They must describe the functions to be performed, in full and unambiguous detail, and the operating environment (hardware, memory allocation, timing, etc.). They should also describe explicitly all of the conditions that must NOT be allowed to occur. They should describe the program structure to be used, the program test requirements and documentation needed during development, as well as basic requirements such as the programming language, memory allocations, inputs and outputs. By adequately specifying these aspects, a framework for program generation will be created which minimizes the possibilities for creating errors, and which ensures that errors will be found and corrected.

The specifications must be carefully reviewed, to ensure that they meet all of the requirements described above, and contain no ambiguities. Specification review must be performed by the project team, including the programmers and engineers whose work will be driven by the specifications.

SOFTWARE STRUCTURE AND MODULARITY

Structure

Structured programming is an approach that constrains the programmer to using certain clear, well-defined approaches to program design, rather than allowing total freedom to design 'clever' programs which might be complex, difficult to understand or inspect, and prone to error. A major source of error in programs is the use of the GOTO statement for constructs such as loops and branches (decisions). The structured programming approach therefore discourages the use of GOTOs, requiring the use of control structures which have a single entry and a single exit. For example, the simple branch instruction in Fig. 10.2 can be programmed (using BASIC) in either an unstructured or a structured way as shown. The unstructured approach can lead to errors if the wrong line number is given (e.g. if line numbers

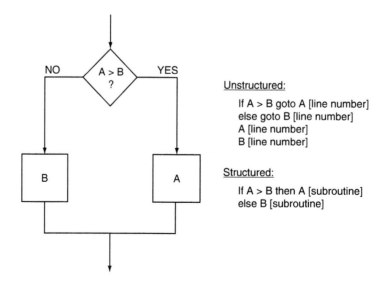

Figure 10.2 Structured versus unstructured programming

are changed as a result of program changes), and it is difficult to trace the subroutines (A, B) back to the decision point.

On the other hand, the structured approach eliminates the possibility of line number errors, and is much easier to understand and to inspect.

Structured programming leads to fewer errors, and to clearer, more easily maintained software. On the other hand, structured programs might be less efficient in terms of speed or memory requirements.

Modularity

Modular programming breaks the program requirement down into separate, smaller program requirements, or modules, each of which can be separately specified, written and tested. The overall problem is thus made easier to understand and this is a very important factor in reducing the scope for error and for easing the task of checking. The separate modules can be written and tested in a shorter time, thus reducing the chances of changes of programmer in mid-stream.

Each module specification must state how the module is to interface with other parts of the program. Thus, all the inputs and outputs must be specified. Structured programming might involve more preparatory work in determining the program structure, and in writing module specifications and test requirements. However, like good groundwork in any development programme, this effort is likely to be more than repaid later by the reduced overall time spent on program writing and debugging, and it will also result in a program which is easier to understand and to change. The capability of a program to be modified fairly easily can be compared to the maintainability of hardware, and it is often a very important feature. Program changes are necessary when logical corrections have to be made, or when the requirements change, and there are not many software development projects in which these conditions do not arise.

The optimum size of a module depends upon the function of the module and is not solely determined by the number of program elements. The size will usually be determined to some extent by where convenient interfaces can be introduced. As a rule of thumb, modules should not normally exceed 100 separate statements or lines of code in a high level language, and less in assembler code.

Requirements for structured and modular programming

Major software customers specify the need for programs to be structured and modular, to ensure reliability and maintainability. These disciplined approaches can greatly reduce software development and life cycle costs. In the United Kingdom, MASCOT (modular approach to software construction, operation and test) and, primarily in the United States, APSE (Ada programming and support environment) are methods imposed to ensure that the great difficulties and costs generated by the unstructured approaches used in the past are obviated. Reference 13 covers structured and modular programming in more detail.

Software re-use

Sometimes existing software, for example from a different or previous application, can be used, rather than having to write a new program or module. This approach can lead to savings in development costs and time, as well as reducing the possibility of creating new errors. However, be careful! Remember Ariane 5 and Mars Polar Orbiter!

PROGRAMMING STYLE

Programming style is an expression used to cover the general approach to program design and coding. Structured and modular programming are aspects of style. Other aspects are, for example, the use of "remark" statements in the listing to explain the program, 'defensive' programming in which routines are included to check for errors, and the use of simple constructs whenever practicable. Obviously, a disciplined programming style can have a great influence on software reliability and maintainability, and it is therefore important that style is covered in software design guides and design reviews, and in programmer training.

FAULT TOLERANCE

Programs can be written so that errors do not cause serious problems or complete failure of the program. We have mentioned 'robustness' in connection with program design, and this is an aspect of fault tolerance. A program should be able to find its way gracefully out of an error condition and indicate the error source. This can be achieved by programming internal tests, or checks of cycle time, with a reset and error indication if the set conditions are not met. Where safety is a factor, it is important that the program sets up safe conditions when an error occurs. For example, a process controller could be programmed to set up known safe conditions and indicate a problem, if no output is generated in two successive program cycle times or if the output value changes by more than a predetermined amount.

These software techniques can also be used to protect against hardware failures, such as failure of a sensor which provides a program input. Examples of this approach are:

—checks of cycle time for a process (e.g. time to fill a tank), and automatic shutdown if the correct time is exceeded by a set amount. This might be caused by failure of a sensor or a pump, or by a leak.
—failure of a thermostat to switch off a heating supply can be protected against by ensuring that the supply will not remain on for more than a set period, regardless of the thermostat output.

—checks for rates of change of input values. If a value changes by more than a predetermined amount take corrective action as above. For example, a pressure measurement might abruptly change to zero because of a transducer or connector failure, but such an actual pressure change might be impossible. The system should not be capable of inappropriate response to a spurious input.

—allow two or more program cycles for receipt of input data, to allow for possible data loss, interruption or corruption.

Features such as these can be provided much more easily with software than with hardware, at no extra material cost or weight, and therefore, the possibility of increasing the reliability and safety of software controlled systems should always be analysed in the specification and design stages. Their provision and optimization is much more likely when the software development is managed as part of an integrated, system approach.

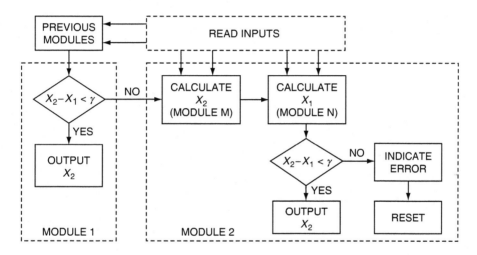

Figure 10.3 Fault tolerant algorithm

REDUNDANCY/DIVERSITY

Fault tolerance can also be provided by program redundancy. For high integrity systems separately coded programs can be arranged to run simultaneously on separate but connected controllers, or in a time-sharing mode on one controller. A voting or selection routine can be used to select the output to be used. This approach is also called *program diversity*. The effectiveness of this approach is based on the premise that two separately coded programs are very unlikely to contain the same coding errors, but of course this would not provide protection against a specification error. Redundancy can also be provided within a program by arranging that critical outputs are checked by one routine, and if the correct conditions are not present then they are checked by a different routine (Fig. 10.3).

LANGUAGES

The selection of the computer language to be used can affect the reliability of software. There are three main approaches which can be used:

1. *Machine code programming.*
2. *Assembly level programming.*
3. *High level* (or *high order) language (HLL or HOL) programming.*

Machine code programming is the creation of the microcode that the processor runs. However, programming at this level should not be used, since it confers no advantages in speed or memory, is very prone to creation of errors, is extremely difficult to check, and has no error trap capabilities.

Assembly level programs are faster to run and require less memory than HLLs. Therefore they can be attractive for real-time systems. However, assembly level programming is much more difficult and is much harder to check and to modify than HLLs. Several types of error which can be made in assembly level programming cannot be made, or are much less likely to be made, in a HLL. Therefore, assembly level programming is not favoured for relatively large programs, though it might be used for modules in order to increase speed and to reduce memory requirements. *Symbolic assemblers*, however, have some of the error-reduction features of HLLs.

Machine code and assembly programming are specific to a particular processor, since they are aimed directly at the architecture and operating system.

HLLs are processor-independent, working through a *compiler* which converts the HLL to that processor's operating system. Therefore, HLLs require more memory (the compiler itself is a large program) and they run more slowly. However, it is much easier to program in HLLs, and the programs are much easier to inspect and correct. The older HLLs (FORTRAN, BASIC) do not encourage structured programming, but the more recently developed ones (PASCAL, Ada, C, C++) do.

Since HLLs must work through a compiler, the reliability of the compiler can affect system reliability. Compilers for new HLLs and for new processors sometimes cause problems for the first few years until all errors are found and corrected. Generally speaking, though, compilers are reliable once fully developed, since they are so universally used. Modern compilers contain error detection, so that many logical, syntactical or other errors in the HLL program are displayed to the programmer, allowing them to be corrected before an attempt is made to load or run it. Automatic error correction is also possible in some cases, but this is limited to certain specific types of error.

Fuzzy logic is used to a limited extent in some modern systems. The ways in which fuzzy logic programs can fail are basically the same as for conventional logic.

Programmable logic controllers (PLCs) are often used in place of processors, for systems such as machine tools, factory automation, train door controls, etc. Programming of PLCs is much easier than for microprocessors, since only basic logic commands need to be created. PLC-based systems also avoid the need for the other requirements of processor-based systems, such as operating system software, memory, etc., so they can be simpler and more robust, and easier to test.

DATA RELIABILITY

Data reliability (or information integrity) is an important aspect of the reliability of software-based systems. When digitally coded data are transmitted, there are two sources of degradation:

1. The data might not be processed in time, so that processing errors are generated. This can arise, for example, if data arrive at a processing point (a 'server', e.g. a microprocessor or a memory address decoder) at a higher rate than the server can process.
2. The data might be corrupted in transmission or in memory by digital bits being lost or inverted, or by spurious bits being added. This can happen if there is noise in the transmission system, e.g. from electromagnetic interference or defects in memory.

System design to eliminate or reduce the incidence of failures due to processing time errors involves the use of queueing theory, applied to the expected rate and pattern of information input, the number and speed of the 'servers', and the queueing disciplines (e.g. first-in-first-out (FIFO), last-in-first-out (LIFO), etc.). Also, a form of redundancy is used, in which processed data are accepted as being valid only if they are repeated identically at least twice, say, in three cycles. This might involve some reduction in system processing or operating speed.

Data corruption due to transmission or memory defects is checked for and corrected using error detection and correction codes. The simplest and probably best known is the parity bit. An extra bit is added to each data word, so that there will always be an even (or odd) number of ones (even (or odd) parity). If an odd number of ones occurs in a word, the word will be rejected or ignored. More complex error detection codes, which provide coverage over a larger proportion of possible errors and which also correct errors, are also used. Examples of these are Hamming codes and BCH codes.

Ensuring reliable data transmission involves trade-offs in memory allocation and operating speed.

SOFTWARE CHECKING

To confirm that the specification is satisfied, the program must be checked against each item of the specification. For example, if a test specification calls for an impedance measurement of $15 \pm 1\Omega$, only a line-by-line check of the program listing is likely to discover an error that calls for a measurement tolerance of $+1\Omega$, -0Ω. Program checking can be a tedious process, but it is made much easier if the program is structured into well-specified and understandable modules, so that an independent check can be performed quickly and comprehensively. Like hardware design review procedures, the cost of program checking is usually amply repaid by savings in development time at later stages. The program should be checked in accordance with a prepared plan, which stipulates the tests required to demonstrate specification compliance.

Formal program checking, involving the design team and independent people, is called a *structured walkthrough*, or a *code review*.

SOFTWARE DESIGN ANALYSIS METHODS

Several so-called 'formal' software design methods have been developed, with the objective of setting up a disciplined framework for specification and programming that will reduce the chances of errors being created. Included among these are the Vienna Development Method (VDM), developed by IBM, the Z notation, and MASCOT (mentioned earlier).

'Formal' methods for automatically checking programs have also been developed. These check the program for consistency, structure, and errors, by assessing the source code against the specification. This is called 'static analysis'.

These methods are described briefly in Reference 6. They are still being developed, and their use is somewhat controversial, since they set out to prove logically and mathematically that the specification meets the requirements and that the program is correct. Properly used, they can lead to better specifications and program designs, and can show up errors. However, they are often expensive to apply, and cannot provide total assurance against human fallibility.

FMECA

It is not practicable to perform a failure modes, effects and criticality analysis (FMECA) (Chapter 7) on software, since software 'components' do not fail. The nearest equivalent to an FMECA is a code review, but whenever an error is detected it is corrected so the error source is eliminated. With hardware, however, we cannot eliminate the possibility of, say, a transistor failure.

In performing a FMECA of an engineering system that combines hardware and software it is necessary to consider the failure effects in the context of the operating software, since system behaviour in the event of a hardware failure might be affected by the software, as described above. This is particularly the case in systems utilizing built-in-test software, or when the software is involved in functions such as switching redundancy, displays, warnings and shut-down.

Software sneak analysis

The sneak analysis (SA) method described in Chapter 9 for evaluating circuit conditions that can lead to system failure is also applicable to software. Since a section of code does not fail but performs the programmed functions whether or not they are the intended ones, there is an analogy with an erroneous circuit design.

The program must be reduced to a set of topological patterns, as for hardware SA. Since a program of reasonable size is very difficult to reduce in this way, this step is usually computerized.

Six basic sneak patterns exist, as shown in Fig. 10.4. Note that most software sneak patterns are related to branching instructions, such as GOTO or IF THEN/ELSE

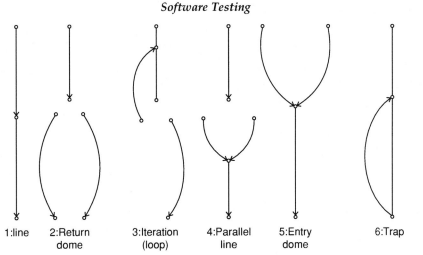

| 1 :line | 2:Return dome | 3:Iteration (loop) | 4:Parallel line | 5:Entry dome | 6:Trap |

Figure 10.4 Software sneak patterns

statements. The conditions leading to and deriving from such statements, as well as the statements themselves, are important clues in the SA.

Software sneak conditions are:

1. *Sneak output.* The wrong output is generated.
2. *Sneak inhibit.* Undesired inhibit of an input or output.
3. *Sneak timing.* The wrong output is generated because of its timing or incorrect input timing.
4. *Sneak message.* A program message incorrectly reports the state of the system.

Figure 10.1 illustrates a potential sneak message condition, since the program will not indicate that C has failed if A and B have failed. This failure is brought about by an incorrect line pattern. The program correctly identifies the correct A value and the incorrect B value, and proceeds to the output with no chance of testing C.

SOFTWARE TESTING

The objectives of software testing are to ensure that the system complies with the requirements and to detect remaining errors. Testing that a program will operate correctly over the range of system conditions is an essential part of the software and system development process. Software testing must be planned and executed in a disciplined way since, even with the most careful design effort, errors are likely to remain in any reasonably large program, due to the impracticability of finding all errors by checking, as described above. Some features, such as timing, overflow conditions and module interfacing, are not easy to check.

Few programs run perfectly the first time they are tested. The scope for error is so large, due to the difficulty that the human mind has in setting up perfectly logical structures, that it is normal for some time to be spent debugging a new program until all of the basic errors are eliminated.

There are limitations to software testing. It is not practicable to test exhaustively a reasonably complex program. The total number of possible paths through a program with n branches and loops is 2^n, analogous to the digital circuit testing problem discussed in Chapter 9. However, there is no 'ATE' for software, so all tests must be set up, run, and monitored manually. It is not normally practicable to plan a test strategy which will provide high theoretical error coverage, and the test time would be exorbitant. Therefore, the tests to be performed must be selected carefully to verify correct operation under the likely range of operating and input conditions, whilst being economical.

The software test process should be iterative, whilst code is being produced. Code should be tested as soon as it is written, to ensure that errors can be corrected quickly by the programmer who wrote it, and it is easier to devise effective tests for smaller, well-specified sections of code than for large programs. The earliest testable code is usually at the module level. The detection and correction of errors is also much less expensive early in the development programme. As errors are corrected the software must be re-tested to confirm that the redesign has been effective and has not introduced any other errors. Later, when most or all modules have been written and tested, the complete program must be tested, errors corrected, and retested. Thus, design and test proceed in steps, with test results being fed back to the programmers.

It is usual for programmers to test modules or small programs themselves. Given the specification and suitable instructions for conducting and reporting tests, they are usually in the best position to test their own work. Alternatively, or additionally, programmers might test one another's programs, so that an independent approach is taken. However, testing of larger sections or the whole program, involving the work of several programmers, should be managed by a person with system responsibility, though members of the programming team should be closely involved. This is called *integration testing*. Integration testing covers module interfaces, and should demonstrate compliance with the system specification.

The software tests must include:

—All requirements defined in the specification ('must do' and 'must not do' conditions).
—Operation at extreme conditions (timing, input parameter values and rates of change, memory utilization.)
—Ranges of possible input sequences.
—Fault tolerance (error recovery).

Since it may not be practicable to test for the complete range of input conditions it is important to test for the most critical ones and for combinations of these. Random input conditions, possibly developed from system simulation, should also be used when appropriate to provide further assurance that a wide range of inputs is covered.

Software can be tested at different levels:

—*White box* testing involves testing at the detailed structural level, for aspects such as data and control flow, memory allocation, look-ups, etc. It is performed in relation to modules or small system elements, to demonstrate correctness at these levels.

—*Verification* is the term sometimes used to cover all testing in a development or simulated environment, for example, using a host or lab computer. Verification can include module and integration testing.

—*Validation* or *black box* testing covers testing in the real environment, including running on the target computer, using the operational input and output devices, other components and connections. Validation is applicable only to integration testing, and it covers the hardware/software interface aspects, as described earlier.

These terms are not defined absolutely, and other interpretations are also applied.

Managing software testing

Software testing must be managed as an integral part of the overall system test plan. It is essential to plan the software tests with full understanding of how software can fail and in relation to the interfaces with the system hardware. In the system test plan, the software should be treated as a separate subsystem (verification) and as part of the overall system (validation).

The test specifications (for modules, integration/verification, validation) must state every test condition to be applied, and the test reports must indicate the result of each test.

Formal 100 per cent error reporting should also be started at this stage, if it is not already in operation. Obviously, all errors must be corrected, and the action taken must also be reported (changes to specification, design, code, hardware, as appropriate). The relevant test must be repeated to ensure that the correction works, and that no other errors have been created.

Formal configuration control should be started when integration testing commences, to ensure that all changes are documented and that all program copies at the current version number are identical.

For validation and other system tests, failures caused by software should be reported and actioned as part of an integrated engineering approach, since the most appropriate solutions to problems could involve hardware or software changes.

Each software test needs to be performed only once during development, unless there have been changes to the program or to related hardware. Of course there is no need to test software as part of the system production test process, since the software cannot vary from copy to copy or change over time.

There are some who argue that software testing (and checking) should be performed by people who are entirely independent of the programming team. This is called the '*cleanroom*' approach to software development. The approach is controversial, and is not consistent with the philosophy on which this book is based.

References 7 to 10 provide more information on software testing.

ERROR REPORTING

Reporting of software errors is an important part of the overall program documentation. The person who discovers an error may not be the programmer or system designer, and therefore all errors, whether discovered during checking, testing or use, need to be written up with full details of program operating conditions at the time. The corrective action report should state the source of the error (specification, design, coding) and describe the changes made. Figure 10.5 shows an example of a

```
Program

Module

Error conditions:

        Input conditions:

    Description of failure:

Effect/importance:

Execution time since last failure:          Total run time:

Date:              Time:              Signed:

Program statement(s) involved:

    Line                              Statement



Error source:

        Code:          Design:        Specification:

Correction recommended:

    Code:

    Design:

    Specification:

    Date:          Signed:          Approved:

    Correction made/tested:      Date:    Time:    Signed:

    Program master amended:      Date:    Time:    Signed:
```

Figure 10.5   Software error reporting form

software error reporting form. A software error reporting and corrective action procedure is just as important as a failure reporting system for hardware. The error reports and corrective action details should be retained with the module or program folder as part of the development record.

SOFTWARE RELIABILITY PREDICTION AND MEASUREMENT

Introduction

Efforts to quantify software reliability usually relate to predicting or measuring the probability of, or quantity of, errors existing in a program. Whilst this is a convenient starting point, there are practical difficulties. The reliability of a program depends not only upon whether or not errors exist but upon the probability that an existing error will affect the output, and the nature of the effect. Errors which are very likely to manifest themselves, e.g. those which cause a failure most times the program is run, are likely to be discovered and corrected during the development phase. An error which only causes a failure under very rare or unimportant conditions may not be a reliability problem, but the coding error that caused the total loss of a spacecraft, for example, was a disaster, despite all the previous exhaustive checking and testing.

Error generation, and the discovery and correction of errors, is a function of human capabilities and organization. Therefore, whilst theoretical models based upon program size might be postulated, the derivation of reliability values is likely to be contentious. For example, a well-structured modular program is much easier to check and test, and is less prone to error in the first place, than an unstructured program designed for the same function. A skilled and experienced programming team is less likely to generate errors than one which is less well endowed. A further difficulty in software reliability modelling is the fact that errors can originate in the specification, the design and the coding. With hardware, failure is usually a function of load, strength and time, and whether a weakness is due to the specification, the design or the production process, the physics of failure remain the same. With software, however, specification, design and coding errors are often different in nature, and the probability of their existence depends upon different factors. For example, the number of coding errors might be related to the number of code statements, but the number of specification errors might not have the same relationship. One specification error might lead to a number of separate program errors.

The following sections briefly outline some of the statistical models which have been proposed for software reliability. See Reference 11 for a detailed discussion of software reliability prediction and measurement. However, it is important to appreciate that the utility of any statistical software reliability model depends upon acceptable values for the distribution parameters being available. Unlike hardware failure statistics, there is no physical basis for parameter estimation and the data that have been analysed to date are very limited compared with the wealth of available hardware failure data. Since software reliability is so dependent upon human performance and other non-physical factors, data obtained on one program or group of programs are unlikely to be accepted as being generally applicable, in the way that data on material properties are.

The logical limitations inherent in the prediction of reliability as described in Chapter 6 apply equally to software. Indeed they are even more severe, since there are no physical or logical connections between past data and future expectation, as there are with many hardware failure modes. Therefore the methods described in this section are of mainly academic interest, and they have not been generally accepted or standardized by the software engineering community.

The Poisson model (time-related)

It is assumed that errors can exist randomly in a code structure and that their appearance is a function of the time the program is run. The number of errors occurring in time t is $N(t)$. If the following conditions exist:

1. $N(0) = 0$,
2. not more than one error can occur in the time interval $(t, t + dt)$,
3. the occurrence of an error is independent of previous errors,

then the occurrence of errors is described by the non-homogeneous Poisson distribution:

$$P[N(t) = n] = \frac{[m(t)]^n}{n!} \exp[-m(t)] \quad (n \geq 0) \tag{10.1}$$

where

$$m(t) = \int_0^t \lambda(s) \, ds$$

$m(t)$ is the mean (s-expected) number of errors occurring in the interval $(0, t)$:

$$m(t) = a[1 - \exp(-bt)]$$

where a is the total number of errors and b is a constant. The number of errors remaining after time t, assuming that each error which occurs is corrected without the introduction of others, is

$$\bar{N}(t) = a \exp(-bt) \tag{10.2}$$

The reliability function, after the most recent error occurs and is corrected at time s, is

$$R(t) = \exp[[-a\{\exp(-bs) - \exp[-b(s + t)]\}]] \tag{10.3}$$

In using a time-related model, the question arises as to what units of time should be used. The Poisson model has been tested against software error data using calendar time during which errors were detected and corrected and values for the parameters a and b derived. However, since software errors are not time-related in the way that physical (hardware) failure processes are, the use of time-related models for software errors is problematical.

The Musa model

The Musa model uses program execution time as the independent variable. A simplified version of the Musa model is

$$n = N_0 \left[1 - \exp\left(\frac{-Ct}{N_0 T_0} \right) \right] \tag{10.4}$$

where N_0 is the inherent number of errors, T_0 the MTTF at the start of testing (MTTF is mean time to failure) and C the 'testing compression factor' equal to the ratio of equivalent operating time to testing time.
 The present MTTF:

$$T = T_0 \exp\left(\frac{Ct}{N_0 T_0} \right)$$

gives

$$R(t) = \exp\left(\frac{-t}{T} \right) \tag{10.5}$$

From these relationships we can derive the number of failures which must be found and corrected, or the program execution time necessary, to improve from T_1 to T_2:

$$\Delta_n = N_0 T_0 \left(\frac{1}{T_1} - \frac{1}{T_2} \right) \tag{10.6}$$

$$\Delta_t = \left(\frac{N_0 T_0}{C} \right) \ln\left(\frac{T_2}{T_1} \right) \tag{10.7}$$

Example 10.1

A large program is believed to contain about 300 errors and the recorded MTTF at the start of testing is 1.5 h. The testing compression factor is assumed to be 4. How much testing is required to reduce the remaining number of errors to ten? What will then be the reliability over 50 h of running?

From Eqns (10.6) and (10.7),

$$(300 - 10) = 300 \times 1.5 \left(\frac{1}{1.5} - \frac{1}{T_2} \right)$$

$$t = \left(\frac{300 \times 1.5}{4}\right) \ln\left(\frac{T_2}{1.5}\right)$$

Therefore

$$T_2 = 45\,\text{h}$$

and

$$\Delta_t = 382.6\,\text{h}$$

giving

$$R_{50} = \exp\left(\frac{-50}{45}\right) = 0.33$$

The Jelinski–Moranda and Schick–Wolverton models

Two other exponential-type models which have been suggested are the Jelinski–Moranda (JM) model and the Schick–Wolverton (SW) model. In the JM and SW models, the hazard function h(t) is given respectively by:

$$h(t_i) = \phi[N_0 - n_{i-1}] \tag{10.8}$$

$$h(t_i) = \phi[N_0 - n_{i-1}]t_i \tag{10.9}$$

where t_i is the length of the ith debugging interval, i.e. the time between the $(i-1)$th and the ith errors, and ϕ is a constant.

Littlewood models

Littlewood attempts to take account of the fact that different program errors have different probabilities of causing failure. If ϕ_1, ϕ_2, ..., ϕ_N are the rates of occurrence of errors 1, 2, ..., N, the p.d.f. for the program time to failure, after the ith error has been fixed, is

$$f(t) = \lambda \exp(-\lambda t) \tag{10.10}$$

where λ is the program failure rate

$$\lambda = \phi_1 + \phi_2 + \cdots \phi_{N-i}$$

ϕ is assumed to be gamma-distributed, i.e. errors do not have constant rates of occurrence but rates which are dependent upon program usage. If the gamma distribution parameters are (α, β) (equivalent to $(a, 1/\lambda)$ in Eqn 2.28) then it can be shown, using a Bayes approach, that

$$f(t) = \frac{(N - i)\alpha(\beta + t')^{(N-i)\alpha}}{(\beta + t' + t)^{1+(N-i)\alpha}}$$ (10.11)

where t' is the time taken to detect and correct i errors. From this

$$R(t) = \left(\frac{\beta + t'}{\beta + t' + t}\right)^{(N-i)\alpha}$$ (10.12)

and

$$\lambda(t) = \frac{(N - i)\alpha}{\beta + t' + t}$$ (10.13)

At each error occurrence and correction, $\lambda(t)$ falls by an amount $\alpha/(\beta + t')$. It is assumed that all detected errors are corrected, without further errors being introduced.

Example 10.2

A large program is assumed to include a total of 300 errors, of which 250 have been detected and corrected in 20 h of execution time. Assuming the Littlewood model holds and the distribution parameters are $\alpha = 0.005$, $\beta = 4$, what is the expected reliability over a further 20 h?

From Eqn (10.12),

$$R(20) = \left(\frac{4 + 20}{4 + 20 + 20}\right)^{(300-250)0.005}$$

$$= 0.86$$

Point process analysis

Since a program can be viewed as a repairable system, with errors being detected and corrected in a time continuum, the method of point process analysis described in Chapter 2 can be applied to software reliability measurement and analysis.

HARDWARE/SOFTWARE INTERFACES

In software controlled systems, failures can occur which are difficult to diagnose to hardware or software causes, due to interactions between the two. We have already covered examples of such situations, where hardware elements provide program inputs. The software design can minimize these possibilities, as well as provide automatic diagnosis and fault indication. However, there are other types of failure which are more difficult, particularly when the hardware/software interface is less clearly defined.

Hardware meets software most closely within electronic devices such as processors and memories. A failure of a memory device, say of an individual memory cell which always indicates a logic state 1 (i.e. stuck at 1) regardless of the input, can cause failures which appear to be due to software errors. If the program is known to work under the input conditions, electronic fault-finding techniques can be used to trace the faulty device. There are times, particularly during development, when the diagnosis is not clear-cut, and the software and hardware both need to be checked. Timing errors, either due to device faults or software errors can also lead to this situation.

Memory devices of all types, whether optical or magnetic media or semiconductor memory devices, can cause system failures. Memory media and devices belong to the class of equipment sometimes called 'firmware', to indicate their interface status. Since many memory media are dynamic, i.e. the same data are handled in different locations at different times during program execution, firmware failures can lead to system failures which occur only under certain operating conditions, thus appearing to be due to software errors. Such failures can also be intermittent. Software and memory or microprocessor devices can be designed to protect the system against such failures. For example, the redundancy techniques described above could provide protection against some types of dynamic memory failure (other than a catastrophic failure of, say, a complete memory device). Redundancy can be provided to data or logic held in memory by arranging for redundant memory within the operating store or by providing independent, parallel memory devices. The program logic then has to be designed to store and access the redundant memory correctly, so the program becomes more complex.

CONCLUSIONS

The versatility and economy offered by software control can lead to an underestimation of the difficulty and cost of program generation. It is relatively easy to write a program to perform a simple defined function. To ensure that the program will operate satisfactorily under all conditions that might exist, and which will be capable of being changed or corrected easily when necessary, requires an effort greater than that required for the basic design and first-program preparation. Careful groundwork of checking the specification, planning the program structure and assessing the design against the specification is essential, or the resulting program will contain many errors and will be difficult to correct. The cost and effort of debugging a large, unstructured program containing many errors can be so high that it is cheaper to scrap the whole program and start again.

Software that is reliable from the beginning will be cheaper and quicker to develop, so the emphasis must always be to minimize the possibilities of early errors and to eliminate errors before proceeding to the next phase. The essential elements of a software development project to ensure a reliable product are:

1. Specify the requirements completely and in detail (system, software).
2. Make sure that all project staff understand the requirements.

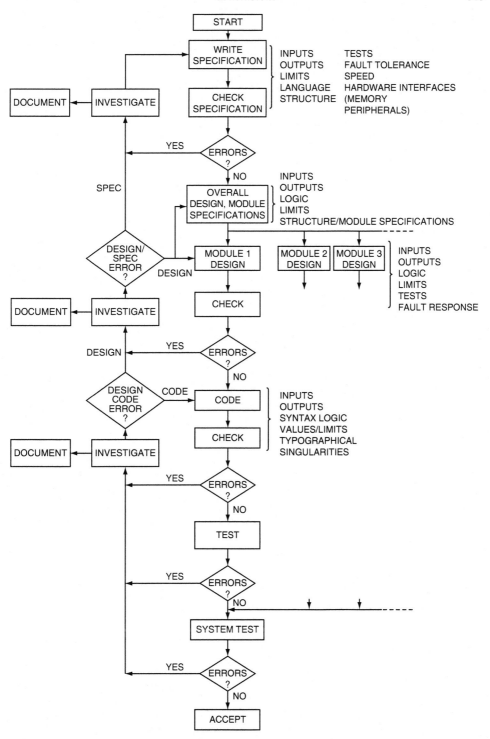

Figure 10.6 Software development for reliability

3. Check the specifications thoroughly. Keep asking 'what if...?'
4. Design a structured program and specify each module fully.
5. Check the design and the module specifications thoroughly against the system specifications.
6. Check written programs for errors, line by line.
7. Plan module and system tests to cover important input combinations, particularly at extreme values.
8. Ensure full recording of all development notes, tests, checks, errors and program changes.

Figure 10.6 shows the sequence of development activities for a software project, with the emphasis on reliability. References 4 and 5 provide excellent overviews of software reliability and safety engineering.

BIBLIOGRAPHY

1. G. J. Myers, *Software Reliability: Principles and Practice*. J. Wiley (1976).
2. B. W. Kernighan and P. J. Plauger, *The Elements of Programming Style*. McGraw-Hill (1974).
3. E. Yourdon, *Techniques of Program Structure and Design*. Prentice-Hall (1975).
4. J. D. Musa, *Software Reliability Engineering*. McGraw-Hill (1999).
5. N. G. Leveson, *Safeware – System Safety & Computers*. Addison Wesley (1995).
6. D. J. Smith and K. B. Wood, *Engineering Quality Software*, 2nd edn. Elsevier (1989).
7. G. J. Myers, *The Art of Software Testing*. J. Wiley (1979).
8. M. A. Ould and C. Unwin, *Testing in Software Development*. Cambridge University Press (1986)
9. B. Beizer, *Black-Box Testing*, J. Wiley and Sons (1995).
10. C. Kaner, H. Quoc Nguyen and J. Falk, *Testing Computer Software*, 2nd edn. J. Wiley and Sons (1999).
11. J. D. Musa, A. Iannino and K. Okumoto, *Software Reliability Prediction and Measurement*. McGraw-Hill (1987).
12. US MIL-S-52779(AD): *Software Quality Assurance Requirements*. Available from the National Technical Information Service, Springfield, Virginia.
13. UK Defence Standard 00–16: *Guide to the Achievement of Quality in Software*. HMSO.

QUESTIONS

1. Discuss the main differences between the ways in which software and hardware can fail to perform as required. Give four examples to illustrate these differences.

2. Explain the ways in which software in engineering systems can be different to other kinds of system.

3. What are the three principal stages in software development that can lead to errors in programs? Give one example of the type of software error that can be created in each stage.

4. What is structured and modular design in the context of software? Describe the main advantages and disadvantages of these approaches.

5. How can software be used to protect against hardware failures in systems that embody both? Give two examples of how software can be used to provide such protection.

6. Describe the essential points to be considered in setting up a test programme for newly developed software. Include the distinction between verification and validation.

7. Several methods have been postulated for predicting and measuring the reliability of software. What are the two main categories of software reliability model? Briefly describe one model in each category, and discuss their main assumptions in relation to predicting the reliability of a new program.

11

Reliability Testing

INTRODUCTION

Testing is an essential part of any engineering development programme. If the development risks are high the test programme becomes a major component of the overall development effort, in terms of time and other resources. For example, a new type of hydraulic pump or a new model of a video recording system will normally undergo exhaustive tests to determine that the design is reliable under the expected operating environments and for the expected operating life. Reliability testing is necessary because designs are seldom perfect and because designers cannot usually be aware of, or be able to analyse, all the likely causes of failure of their designs in service. The disciplines described in earlier chapters, when systematically applied, can contribute to a large extent to inherently reliable design. They can also result in fewer failures during testing, and thus reduce the time and cost of the test programme.

Reliability testing should be considered as part of an integrated test programme, which should include:

1. Functional testing, to confirm that the design meets the basic performance requirements.
2. Environmental testing, to ensure that the design is capable of operating under the expected range of environments.
3. Statistical tests, as described in Chapter 5, to optimize the design of the product and the production processes.
4. Reliability testing, to ensure that the product will operate without failure during its expected life.
5. Safety testing, when appropriate.

It is obviously impracticable to separate entirely the various categories of test. All testing will provide information on performance and reliability, and there will be common requirements for expertise, test equipment and other resources. The different categories of test do have certain special requirements. In particular, statutory considerations often determine safety tests, some of which may have little in common with other tests.

To provide the basis for a properly integrated development test programme, the design specification should cover all criteria to be tested (function, environment, reliability, safety). The development test programme should be drawn up to cover assurance of all these design criteria. It is important to avoid competition between people running the different categories of test, with the resulting arguments about allocation of models, facilities, and priorities. An integrated test programme reduces the chances of conflict.

The development test programme should include:

1. Model allocations (components, sub-assemblies, system)
2. Requirements for facilities such as test equipment
3. A common test and failure reporting system
4. Test schedule

One person should be put in charge of the entire programme, with the responsibility and authority for ensuring that all specification criteria will be demonstrated.

There is one conflict inherent in reliability testing as part of an integrated test programme, however. To obtain information about reliability in a cost-effective way, i.e. quickly, it is necessary to generate failures. Only then can safety margins be ascertained. On the other hand, failures interfere with functional and environmental testing. The development test programme must address this dilemma. It can be very tempting for the people running the development test programme to minimize the chance of failure occurring, in order to make the programme run smoothly and at least cost. However, weaknesses in the design (or in the way it is made) must be detected and corrected before the production phase. This can realistically be achieved only by generating failures. An ideal test programme will show up every failure mode which might occur in service.

The development test dilemma should be addressed by dividing tests into two main categories:

1. Tests in which failures are undesirable.
2. Tests which deliberately generate failures.

Statistical testing, functional testing and most environmental testing are in category 1. Most reliability testing (and some safety testing) are in category 2. The two categories of test must be run in parallel, with good communication between them and, whenever practicable, using common approaches. In particular, there must be a common reporting system for test results and failures, and for action to be taken to analyse and correct failure modes. Test and failure reporting and corrective action are covered in more detail later.

The category 2 testing should be started as soon as hardware (and software, when appropriate) is available for test. The effect of failures on schedule and cost increases progressively, the later they occur in the development programme. Therefore tests should be planned to show up failure modes as early as is practicable.

Engineering development testing methods are described in more detail in Reference 1.

PLANNING RELIABILITY TESTING

Using design analysis data

The design analyses performed during the design phase (CAE, reliability prediction, FMECA, stress analysis, parameter variation analysis, sneak circuit analysis, FTA) described in Chapter 6, as well as any earlier test results, should be used in preparing the reliability test plan. These should have highlighted the risks and uncertainties in the design, and the reliability test programme should specifically address these. For example, if the FMECA shows a particular failure mode to be highly critical, the reliability test programme should confirm that the failure is very unlikely to occur within the use environment and lifetime. Inevitably the test programme will also show up failure modes and effects not perceived during the design analyses, otherwise there would be little point in testing. Therefore, the test programme must cover the whole range of use conditions, including storage, handling, testing, repair and any other aspect which might affect reliability.

Considering variability

We have seen (Chapters 4 and 5) how variability affects the probability of failure. A major source of variability is the range of production processes involved in converting designs into hardware. Therefore the reliability test programme must cover the effects of variability on the expected and unexpected failure modes. If parameter variation analyses or statistical tests have been performed, these can be very useful in planning reliability tests to confirm the effects of variation. However, to ensure that the effects of variability are covered as far as is practicable, it is important to carry out reliability testing on several items. The number of systems to be tested must be determined by considering:

1. The extent to which the key variables can be controlled
2. The criticality of failure
3. The cost of test hardware and of testing

Only rarely will fewer than four items be adequate. For fairly simple systems (transistors, fasteners, hydraulic check valves) it might be relatively easy to control the few key variables and the criticality of failures might be relatively low. However, it is not expensive to test large quantities. For systems of moderate complexity (e.g. automobiles, TV sets, machine tools) it is much harder to control key variables, since there are so many. Every interface within the system introduces further sources of variability which can affect reliability. Therefore it is very important to test a relatively large number, 5 to 20 being a typical range. For complex systems (aero engines, aircraft, etc.) hardware and test cost tend to be the major constraints, but at least four items should be subjected to a reliability test. Reliability testing of fewer than four might be appropriate for large, expensive, complex systems which will be manufactured in very small quantities (e.g. spacecraft, ships, power stations), of which the items tested will be used operationally.

The effects of known sources of variability can sometimes be assessed by testing items in which variable parameters (e.g. dimensions, process variables) have been deliberately set at worst case values. Statistical design of experiment, (DOE) and other statistical engineering optimization techniques, as described in Chapter 5, should be used to analyse the effects of multiple sources of variation.

Durability

The reliability test programme must take account of the pattern of the main failure modes with respect to time (or cycles, distances, etc., with which the time dimension is associated).

If the failure modes have increasing hazard rates, testing must be directed towards assuring adequate reliability during the expected life. Therefore reliability tests must be of sufficient duration to demonstrate this, or they must be accelerated. Accelerated testing is covered later. Generally speaking, mechanical components and assemblies are subject to increasing hazard rates, when wear, fatigue, corrosion or other deterioration processes can cause failure. Systems subject to repair and overhaul can also become less reliable with age, due to the effects of maintenance, so the appropriate maintenance actions must be included in the test plan.

For items and systems which are not subject to an increasing hazard or failure rate, endurance testing is less important. Electronic components and systems are in this category. There is usually little to be gained by subjecting such hardware to long duration reliability testing, and it is far more effective to test several items for relatively short periods. Since such items are not prone to wearout, the tested items can usually be used for other development work without having to be overhauled.

TEST ENVIRONMENTS

The reliability test programme must cover the range of environmental conditions which the product is likely to have to endure. The main reliability-affecting environmental factors, affecting most products, are:

Temperature
Vibration
Shock
Humidity
Power input and output
Dirt
People

In addition, electronic equipment might be subjected to:

Electromagnetic effects (EMI).
Voltage transients, including electrostatic discharge (ESD).

Certain other environments can affect reliability in special cases. Examples are:

Radiation (ultraviolet, cosmic, X-rays)
Lubricant age or contamination
High altitude
Space vacuum
Industrial pollution
Electromagnetic pulse (lightning, nuclear)
Salt spray
Fungus
High intensity noise

US MIL-STD-810, UK Defence Standard 07-55 and ISO/IEC60068 (References 2, 3 and 4) provide test methods appropriate to most of these environmental conditions. However, these standards do not address reliability directly, since the objective is to show that the product will not fail or incur damage under the test conditions. Also, most of the tests do not require that the equipment be operating during the tests, and the tests are single-environment, not combined.

The environmental test programme will address the formal environmental test requirements, particularly when these are necessary in order to comply with legal or contractual requirements. The environmental aspects of the reliability test programme must take account of the environmental requirements stated in the design specification and of the planned environmental test. However, to be effective as a means of ensuring a reliable product, the environmental aspects of reliability testing must be assessed in much greater detail.

The environmental aspects of reliability testing must be determined by considering which environmental conditions, singly and in combination with others, are likely to be the most critical from the reliability point of view. In most cases, past experience and codes of practice will provide adequate guidelines. For example, US MIL-HDBK-781 (Reference 5) provides information on how to assess environmental conditions and to design the tests accordingly. Typically, a reliability test environment for an electronic system to be used in a vehicle or aircraft might be as shown in Fig. 11.1. Such testing is known as *combined environmental reliability testing* (CERT).

Test chambers are available for CERT testing, particularly for electronic systems. These include facilities for temperature cycling and for vibration input to the unit under test by locating the chamber over a floor-mounted vibrator, with a movable or flexible floor for the chamber. Electrical signals, (power, control and monitoring) can be fed through connectors in the chamber wall. Special chambers can be provided with other facilities, such as humidity and reduced pressure. Control of the chamber conditions can be programmed, and the unit under test can be controlled and monitored using external equipment, such as programmable power supplies, data loggers, etc. Figure 11.2 shows a typical CERT facility.

When past experience on standard methods is inappropriate, e.g. for a high risk product to be used in a harsh environment, the test environments must be carefully evaluated, particularly:

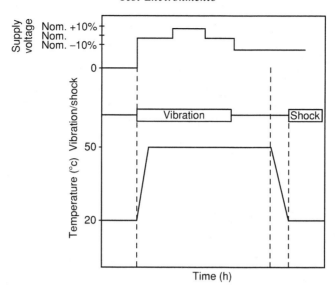

Figure 11.1 Typical CERT environmental cycles: electronic equipment in a vehicle application

Figure 11.2 CERT test facility (Courtesy Thermotron Industries Inc.)

1. Rate of change of conditions, not just maximum and minimum values. For example, a high rate of change of temperature can cause fracture or fatigue due to thermal mismatch and conductivity effects.
2. Operating and dormant conditions in relation to the outside environment. For example, moisture-assisted corrosion might cause more problems when equipment is idle than when it is operating.

3. The effects of combined environments, which might be much more severe than any one condition. Statistical experiments (Chapter 5) can be used to evaluate these effects.
4. Direction and modes of vibration and shock. This is dealt with in more detail later.
5. Particular environmental conditions applicable to the product, such as handling, storage, maintenance and particular physical conditions.

Vibration testing

Adequate vibration testing is particularly important for products which must survive vibration conditions. However, specifying and obtaining the right conditions can be difficult, and it is easy to make expensive mistakes.

The main principles of effective vibration testing are:

1. Vibration should be input to the unit under test (UUT) through more than one axis, preferably simultaneously.
2. Vibration inputs should cover the complete range of expected frequencies and intensities, so that all resonances will be excited.
3. Vibration input should be random, rather than swept frequency, so that different resonances will be excited simultaneously (see below).
4. Test fixtures to mount the UUT to the vibration tables should be designed so that they do not alter the vibration output (no fixture resonances or damping). Whenever practicable, the UUT should be mounted directly on to the vibrator platform.

The simplest vibration test is a fixed frequency 'shake', usually with a sine wave input. However, this is of little value in reliability testing. Modern vibrators can be programmed to generate any desired profile.

Swept frequency sine testing is useful for resonance searches, to enable the design to be modified if unacceptable resonances are detected.

Peak acceleration for a given frequency of sine wave vibration can be calculated using the formula:

$$A = 0.002\,f^2 D \qquad\qquad\qquad (11.1)$$

where
A = peak acceleration (g)
f = frequency (Hz)
D = peak-to-peak displacement (mm)

for example, if $f = 50\,\text{Hz}$ and $D = 2\,\text{mm}$ then $A = 10\,\text{g}$.

Alternatively, the spectrum could be a random input within a specified range and density function. Random vibration testing in which the input contains many frequencies is more effective than swept frequency for reliability testing, to show up vibration-induced failure modes, since it simultaneously excites all resonances. It is also more representative of real life.

The unit of measurement for random vibration inputs with continuous spectra is *power spectral density* (PSD). The units are g^2Hz^{-1}. Typically inputs of up to

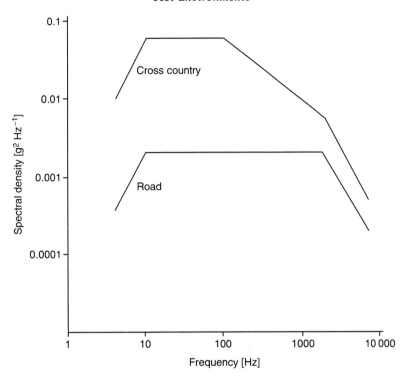

Figure 11.3   Road transport vibration levels

$0.1\,g^2Hz^{-1}$ are used for equipment which must be shown to withstand fairly severe vibration, or for screening tests on assemblies such as electronic equipment. A typical random vibration spectrum is shown in Fig. 11.3.

It is important to apply power to electrical or electronic equipment and to monitor its performance while it is being vibrated, so that intermittent failures can be detected and investigated.

Since dynamic responses are usually affected more by resonances within the product, due to the design and to production variation, than by the input spectrum from the vibrator, it is seldom cost-effective to simulate accurately the operating environment, even if it is known in detail. Since the objective in vibration reliability testing is to excite resonances simultaneously, and since almost any random spectrum will do this, test costs can be minimized by permitting large spectral tolerances, e.g. ± 6 dB.

Vibration and shock testing are described in more detail in References 1, 6 and 7.

Temperature testing

Temperature testing for reliability is a less complex subject than vibration testing. The only aspects that need to be considered are:

Extreme values (high and low)
Rate of change

As for any other environmental condition, these should be selected by considering the expected product environment and the likely effects. Temperature testing for electrical and electronic equipment is particularly important, since reliability can be affected by operating temperature (Chapter 9). Equipment should be powered and operated during temperature testing, otherwise the tests will be unrepresentative of the thermal patterns and gradients in use. It should also be monitored continuously to ensure that intermittent failures are detected.

Electromagnetic compatibility (EMC) testing

EMC testing is very important for electronic systems, since data corruption due to electromagnetic interference (EMI) or from voltage transients in power supplies can have serious consequences (see Chapter 9). The equipment must be subjected to EMI and transients to confirm that it will perform without failure under these conditions. The levels of EMI and transient waveforms must be ascertained by evaluating or measuring the operating environment, or they might be specified.

Internally induced EMI and transients must also be protected against, and tests to ensure that transmitted EMI is within limits might also be necessary.

EMI/EMC testing methods are described in References 8 and 9.

Other environments

See Reference 1.

Customer simulation testing

Functional, environmental, reliability and safety tests are all designed to demonstrate that the equipment will meet its design parameters. In general these tests are carried out by people with extensive engineering backgrounds. Such people are, with the best will in the world, often far removed from the average user of the equipment. It is therefore important, particularly in the field of consumer products (televisions, copiers, washing machines, etc.), that some reliability testing is conducted using people who are more nearly representative of typical customers or by trial customers. This is called "beta" testing. This approach is very useful in highlighting failure modes that do not show up when the equipment is used by experienced personnel or in non-representative environments.

TESTING FOR RELIABILITY AND DURABILITY: ACCELERATED TEST

In Chapters 8 and 9 we reviewed how mechanical, electrical and other stresses can lead to failures, and in Chapter 4 how variations of strength, stresses and other conditions can influence the likelihood of failure or duration (time, distance, cycles, etc.) to failure. In this section we will describe how tests should be designed and conducted to provide assurance that designs and products are reliable and durable in service.

For most engineering designs we do not know what is the *'uncertainty gap'* between the theoretical and real capabilities of the design and of the products made to it, for the whole population, over their operating lives and environments. The effects of these uncertainties can seldom be evaluated with confidence by any of the design analysis methods described in Chapter 6. How then can we plan a test programme that will reduce the uncertainty gap to an extent that we can be assured of reliability and durability, whilst taking due account of practical constraints like cost and time?

The conventional approach to this problem has been to treat reliability as a functional performance characteristic that can be measured, by testing items over a period of time whilst applying simulated or actual in-service conditions, and then calculating the reliability achieved on the test. For example, time of operation divided by the number of failures is the estimated mean time between failures (MTBF). This approach will be discussed in the next chapter. However, these methods are fundamentally inadequate for providing assurance of reliability. The main reason is that they are based on measuring the reliability achieved during the application of simulated or actual stresses that are within the specified service environments, in the expectation (or hope) that the number of failures will be below the criterion for the test. This is the wrong answer to the problem expressed above.

The correct answer is straightforward: *we must test to cause failures, not test to demonstrate successful achievement*. This concept is well accepted in a few test situations, particularly in mechanical strength testing. To derive the strength and fatigue properties of materials, samples are tested to failure. As explained in Chapter 8, we cannot accurately determine the strength of, say, an alloy or a plastic material by theoretical analysis, only by testing samples to failure. If we design a component using such a material, we can analyse the stresses using methods like FEA and we can calculate the strength using the material properties derived from the tests to failure. If the design is simple and there is an adequate margin between stress and strength, we might decide that no further testing is necessary. If, however, constraints such as weight force us to design with smaller margins, and if the component's function is critical (like supporting an aircraft engine), we might well consider it prudent to test some quantity to failure. We would then expect that failures would occur only well beyond the expected maximum stress/minimum life, to provide an adequate margin of safety to take account of the uncertainties and variations in this kind of design and application.

However, let us assume that an electronic system is being designed, and the specified maximum temperature for satisfactory operation is 40 °C. Up to what temperature should the prototype be tested? Some inexperienced (and some experienced) engineers answer that 40 °C should be the maximum test temperature, because any temperature above that would not be "representative" of specified conditions. Therefore any failures that occur above that temperature would not be considered relevant.

However, suppose that a prototype was tested at 42 °C, and failed. Should we ignore this? Might this failure occur at 35 °C on another unit built to the same drawings (effect of variability), or might it occur on this unit 6 months into the warranty period (effect of a time-dependent failure mechanism)? Might it occur at a combination of 35 °C and a small, within-specification, increase in supply voltage? Can we really be sure that the failure at 42 °C is not relevant, just because the thermal stress applied was not "representative"?

If the failure occurred at a temperature 2 °C above the specified limit it is unlikely that it would be ignored (though this does happen). Suppose, however, that failure occurred at 50 °C, or 60 °C? At what stress do we decide that the level is so high that we can ignore failures? Should we even be testing at stresses so much higher than the maximum specified values?

The answer is that these are the wrong questions. The clue is in the earlier questions about the possible cause of the failure. When failures occur on test we should ask whether they could occur in use. The question of relevance can be answered only by investigating the actual physical or chemical cause of failure. Then the questions that must be asked are:

1. Could this failure occur in use (on other items, after longer times, at other stresses, etc.)?
2. Could we prevent it from happening in use?

The stress/es that were applied are relevant only in so far as *they were the tools to provide the evidence that an opportunity exists to improve the design. We have obtained information on how to reduce the uncertainty gap.* Whether the opportunity is taken is a management issue, in which other aspects such as cost, weight, time, etc. must be considered.

The uncertainty described above is shown in Fig. 11.4. If we consider only one stress and the failures it might cause, the stress to failure distribution of production items might be as shown. As a simple example, this might be the operating temperature at which an electronic component malfunctions, or the pressure at which a seal begins to leak. As discussed in Chapter 2, the exact nature of this distribution is almost always uncertain, particularly in the tails, which are the most important areas as far as reliability and durability are concerned.

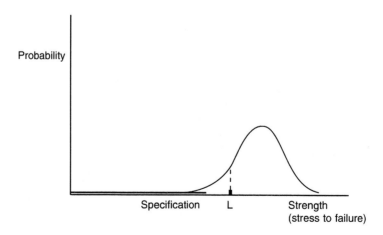

Figure 11.4   Stress, strength and test failures (1)

Suppose that the first test failure occurs at stress level *L*. At this stage we might have only a few items to test, maybe only one. We can state that the strength of this item represents a point on the distribution, but we cannot say whether it was an average strength item, a strong one or a weak one. The only way to find out the nature of the strength distribution is to test more items to failure, and to plot and analyse the results. Inevitably most of the items tested will be near to the average, because that is where most of the population will lie. Therefore, it is unlikely that any item tested will represent the weakest in the future population.

However, if we analyse the actual cause of the failure, by whatever means is appropriate, and take action to strengthen the item, then in effect we will move the strength distribution to the right. We still do not know its shape, but that is not what is important. We just want to move it out of the way. We are engineers, not theoretical scientists or statisticians, *so we can use high stresses in place of large samples*. Whilst scientific knowledge of the cause and effect relationships that affect reliability and durability is obviously necessary in order to create designs *and to determine how to improve them*, this is appropriate to determining where distributed values are centred, and sometimes the variation near the centre. In the electronics example, we might use a higher rated component, or add a heat sink. For the seal we might change the material or the dimensions, or add a second seal, or reduce the pressure. The system will therefore be made more reliable.

For many items subjected to stress tests, particularly electronic systems, two types of failure can occur: *transient (or operating) failures* and *permanent failures*. If a transient failure occurs at some stress level the correct operation will be restored if the stress is reduced. Permanent failures are those from which the operation does not recover when the stress is reduced. For stresses like temperature, power voltage level, etc., which might have low or negative limiting values, failures might occur at high and at low levels, and these will of course have different physical causes from those at the high levels. For a population of items the stresses at which the failures occur will be distributed. The general case is illustrated in Fig. 11.5.

If the failure is due to a wearout mechanism, say wear of a bearing or fatigue of a component attachment, the horizontal axis of the distribution will represent time (or

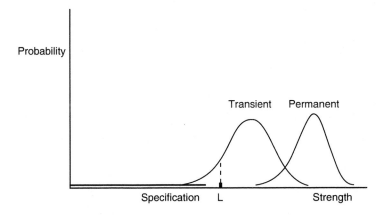

Figure 11.5 Stress, strength and test failures (2)

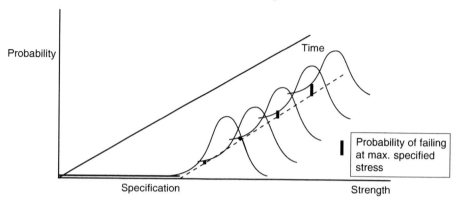

Figure 11.6 Stress, strength and test failures (3): wearout failures

cycles), for any particular stress value. In addition to the uncertainty regarding the stress and strength, we now have the further uncertainty of time. Failure on test after time t at stress level L will represent one point on an unknown three-dimensional distribution (Fig. 11.6). As Fig. 11.6 illustrates, an important feature of wearout mechanisms is that the resulting distributions of times to failure become wider as damage accumulates, thus further increasing the uncertainty. To obtain a full understanding might require more testing, with larger samples. However, the same principle applies: we are not really interested in the shapes of distributions. We just want to make the design better, if it is cost effective to do so.

In most engineering situations failures are caused by combinations of stresses and strength values, not just by one stress and one strength variable. Some might be time-dependent, others not. Using the examples above:

—The stresses applied to the electronic component could be a combination of high temperature operation, high temperature rate of change after switch-on, rate of on–off cycles, humidity when not operating, power supply voltage level, and vibration. The resisting strengths might be mechanical integrity of the internal connections, thermal conductivity of the encapsulating material, absence of defects, etc.

—For the seal, the stresses and other variables might be oil temperature, pressure, pressure fluctuations, oil conditions (viscosity, cleanliness, etc.), shaft axial and radial movement, vibration, tolerances between moving parts, tolerances on seal grooves and seal dimensions, etc.

Between any two or more of these variables there might also be interactions, as described in Chapter 5.

Therefore, even for relatively simple and common failure situations like these, there is not just one distribution that is important, but a number of possible distributions and interactions. What might cause a transistor, capacitor or seal to fail in one application might have a negligible effect in another, or a component that has worked well in previous applications might cause problems in a new but similar one. This is how life is in engineering!

This reasoning leads to the main principle of development testing for reliability. *We should increase the stresses so that we cause failures to occur, then use the information to improve reliability and durability.* Clearly there will be practical limits to the stresses applied. These limits are set by:

—The fundamental limits of the technology. For example, there is no point in testing an electronic system at temperatures above the melting point of the solder used.
—The limits of the test capability, such as the maximum temperature of the test chamber.

The logic that justifies the use of very high 'unrepresentative' stresses is based upon four aspects of engineering reality:

1. The causes of failures that will occur in the future are often very uncertain.
2. The probabilities of and durations to failures are also highly uncertain.
3. Time spent on testing is expensive, so the more quickly we can reduce the uncertainty gap the better.
4. Finding causes of failure during development and preventing recurrence is far less expensive than finding new failure causes in use.

It cannot be emphasized too strongly: testing at 'representative' stresses, in the hope that failures will not occur, is very expensive in time and money and is mostly a waste of resources. It is unfortunate that nearly all standardized approaches to stress testing (these standards are discussed later) demand the use of typical or maximum specified stresses. This approach is widely applied in industry, and it is common to observe prototypes on long-duration tests with 'simulated' stresses applied. For example, engines are run on test beds for hundreds of hours, cars are run for thousands of miles around test tracks, and electronic systems are run for thousands of hours in environmental test chambers. Tests in which the prototype does not fail are considered to be 'successes'. However, despite the long durations and high costs involved, relatively few opportunities for improvement are identified, and failures occur in service that were not observed during testing.

An important point to realise in this context is that a failure that is generated by a stress level during test *might be generated by a different stress (or stresses) in service.* For example, a fatigue failure caused by a few minutes vibration on test might be caused by months or years of temperature cycling in service. The vibration stress applied on test might be totally unrepresentative of service conditions. Once again, though, the principle applies that the "unrepresentative" test stress might stimulate a relevant failure. Furthermore, it will have done so much more rapidly than would have been the case if temperature cycling had been applied.

Accelerated stress testing can provide quantitative cause and effect information when the mechanisms of failure are already understood (e.g. material fatigue), and when the tests are planned specifically to provide such information. We can perform statistically designed experiments (Chapter 5), applying accelerated stresses to explore their effects. However, tests to provide such information require larger samples, more detailed planning and more time, and therefore would cost more than would be the case in the accelerated test approach described above. We must decide whether we need the extra information that more 'scientific' or statistical tests can provide. In many cases the information from accelerated tests, as described above, coupled with engineering knowledge, is sufficient to enable us to take appropriate action to improve designs and processes. However, sometimes we need to obtain more detailed information, especially when the cause and effect relationships are uncertain

Accelerated stress testing has been performed in some other technology and system applications for many years, but it has been unusual for the stresses to be extended considerably beyond the specified in-service values. Electronic systems have been subjected to accelerated temperature and vibration stresses, but typically to levels of the order of 20–50 percent above specifications, in development and for production units. This is called *environmental stress screening* (ESS). Other names have been used for the same approach, including STRIFE (stress + life). ESS methods have been standardized to some extent as a result of the guidelines published by the US Institute of Environmental Sciences and Technology (IEST) (Reference 10). Also, apart from the fairly limited stress combinations applied in ESS or CERT, the stresses are usually applied singly. The objective is to simulate the expected worst-case service environments, or to accelerate them by only moderate amounts. The general principle usually applied to all of these methods has been to test the item to ensure that it does not fail during the test. *This is not consistent with the approach discussed above, and in more detail below.*

Highly accelerated life testing

This approach to test is called *Highly Accelerated Life Testing* (HALT). HALT extends the principle of step-stress accelerated testing to the logical conclusions described above. The principle was developed by an American engineer, Dr Gregg Hobbs, and it is fully described in Reference 12. In HALT we make no attempts to simulate the service environment, except possibly as the starting point for the step-stress application. No limits are set to the types and levels of stresses to be applied. We apply whatever stresses might cause failures to occur as soon as practicable, whilst the equipment is continually operated and monitored. We then analyse the failures as described above, and improve the design.

By applying stresses well in excess of those that will be seen in service, failures are caused to occur very much more quickly. Typically the times or cycles to failure in HALT will be several orders of magnitude less than would be observed in service. Failures which might occur after months or years in service are stimulated in minutes by HALT. Also, the very small sample usually available for development test will show up failure modes that might occur on only a very small proportion of manufactured items. *Therefore we obtain time compression of the test programme by orders of magnitude, and much increased effectiveness. This generates proportional reductions in test programme cost and in time to market, as well as greatly improved reliability and durability.*

It is important to appreciate that reliability/durability values cannot be demonstrated or measured using HALT. An accelerated stress test can provide such information only if the cause of failure is a single predominant mechanism such as fatigue, we know exactly what single type of stress was applied, and we have a credible mathematical relationship to link the two. Such relationships exist for some failure mechanisms, as described in Chapters 8 and 9. However, since HALT applies a range of simultaneous stresses, and since the stress profiles (particularly the vibration inputs) are complex and unrecorded, such relationships cannot be derived. In HALT we are trying to stimulate failures as quickly as possible, using highly 'unrepresentative' stresses, so it is impossible and misleading to relate the results to any quantitative reliability/durability requirement such as MTBF, MTTF, etc.

The HALT approach can be applied to any kind of product or technology. For example:

—*Engines, pumps, power transmission units such as gearboxes, etc.*
  —Start tests with old lubricants or other fluids (coolants, hydraulics, etc.), rather than new.
  —Run at low fluid levels.
  —Use fluids that are heated, cooled or contaminated.
  —Use old filters.
  —Misalign shafts, bearings, etc.
  —Apply out-of-balance to rotating components.

—*Electro-mechanical assemblies such as printers, document, material or component handlers, etc.*
  —Apply high/low temperatures, vibration, humidity/damp, etc.
  —Use components with out-of-tolerance dimensions.
  —Misalign shafts, bearings, etc.
  —Use papers/documents/materials/components that exceed specifications (thickness, weight, friction, etc.).

—*Small components or assemblies such as electronic packages, mechanical latches, switches, transducers, etc.*
  —Apply high/low temperatures, vibration, humidity/damp, etc.
  —Apply high frequency vibration by fixing to suitable transducers, such as loudspeaker coils, and driving with an audio amplifier.

Test approach for accelerated test

The approach that should be applied to any accelerated test programme for reliability/durability should be:

1. Try to determine, as far as practicable, what failures might occur in service. This should have been performed during design analysis and review, particularly during the quality function deployment (QFD) and failure modes, effects and criticality analysis (FMECA) (Chapter 6).
2. List the application and environmental stresses that might cause failures. Use Chapters 8 and 9 for guidance.
3. Plan how the stresses that might stimulate foreseeable and unforeseen failures can most effectively be applied in test. Set up the item (or items) to be tested in the test chamber or other facility so that it can be operated and monitored.

4. Apply a single stress, at or near to the design maximum, and increase the level stepwise until the first failure is detected. This approach is called *step-stress accelerated testing*.
5. Determine the cause and take action to strengthen the design so that it will survive higher stresses. This action might be a permanent improvement, or a temporary measure to enable testing to be continued.
6. Continue increasing the stress/es to discover further failure causes (or the same cause at a higher stress), and take action as above.
7. Continue until all of the transient and permanent failure modes for the applied stress are discovered and, as far as technologically and economically practicable, designed out. Repeat for other single stresses.
8. Decide when to stop (fundamental technology limit, limit of stress that can be applied, cost or weight limit).
9. Repeat the process using combined stresses, as appropriate and within the equipment capabilities (temperature, vibration, power supply voltage, etc.).

The selection of stresses to be applied, singly or in combination, is based upon experience and on the hardware being tested, and not on specifications or standards.

HALT and production testing

HALT does not only provide evidence on how to make designs more robust. It also provides the information necessary to optimize stress screens for manufacturing. The basic difference between the objectives of accelerated test in development and in manufacturing is that, whilst we try to cause all development test items to fail in order to learn how to improve the design, we must try to avoid damaging good manufactured items, whilst causing weak or defective ones to fail so that they can be corrected or segregated. The knowledge that we gain by applying the full HALT sequence, including the design ruggedization, can be used to design a stress test regime that is optimized for the product, and which is far more effective than conventional production testing. This is called *highly accelerated stress screening* (HASS). *HASS provides the same benefits in manufacturing as HALT does in development, in greatly increasing the effectiveness of manufacturing screens whilst reducing test cost and time.* We will describe manufacturing testing in detail in Chapter 13.

HALT and HASS represent an integrated approach to testing to ensure that both the design and the manufacturing processes will generate highly reliable products, at minimum cost and time. Conventional, separate, approaches to development and manufacturing tests do not assure this integration, and therefore can result in much lower reliability and higher costs.

DoE or HALT?

Statistical experiments and HALT are complementary approaches in development testing. Table 11.1 gives some guidance on which approach to select for particular situations.

Note that these are by no means clear-cut criteria, and there will often be shades of grey between them. We must decide on the most appropriate method or combination of methods in relation to all of the factors: risks, knowledge, costs, time.

Table 11.1 DoE/HALT Selection

Important Variables, Effects, etc.	DoE/HALT?
Parameters: electrical, dimensions, etc.	DoE
Effects on measured performance parameters, yields	DoE
Stress: temperature, vibration, etc.	HALT
Effects on reliability/durability	HALT
Several uncertain variables	DoE
Not enough items available for DoE	HALT
Not enough time available for DoE	HALT

FAILURE REPORTING, ANALYSIS AND CORRECTIVE ACTION SYSTEMS (FRACAS)

FRACAS is an apt acronym for the task of failure reporting, analysis and corrective action. It is essential that all failures which occur during development testing are carefully reported and investigated. It can be very tempting to categorize a failure as irrelevant, or not likely to cause problems in service, especially when engineers are working to tight schedules and do not want to be delayed by filling in failure reports. However, time and costs will nearly always be saved in the long run if the first occurrence of every failure mode is treated as a problem to be investigated and corrected. Failure modes which affect reliability in service can often be tracked back to incidents during development testing, when no corrective action was taken.

A failure review board should be set up with the task of assessing failures, instigating and monitoring corrective action, and monitoring reliability growth. An important part of the board's task is to ensure that the corrective action is effective in preventing any recurrence of failure. The board should consist of:

—The project reliability engineer
—The designer
—Others who might be able to help with the solutions, such as the quality engineer, production or test engineer

The failure review board should operate as a team which works together to solve problems, not as a forum to argue about blame or to consign failure reports to the 'random, no action required' category. Its recommendations should be actioned quickly or reported to the project management if the board cannot decide on immediate action, e.g. if the solution to the problem requires more resources. This approach has much in common with the quality circles method described in Chapter 13.

US MIL-HDBK-781 provides a good description of failure reporting methods. Since a consistent reporting system should be used throughout the programme MIL-STD-781 can be recommended as the basis for this.

These data should be recorded for each failure:

1. Description of failure symptoms, and effect of failure.
2. Immediate repair action taken.

3. Equipment operating time at failure (e.g. elapsed time indicator reading, mileage).
4. Operating conditions.
5. Date/time of failure.
6. Failure classification (e.g. design, manufacturing, maintenance-induced).
7. Report of investigation into failed component and reclassification, if necessary.
8. Recommended action to correct failure mode.
9. Corrective action follow-up (test results, etc.).

Failure report forms should be designed to allow these data to be included. They should also permit easy input to computer files, by inclusion of suitable coding boxes for use by the people using the forms. An example is given in Appendix 8.

Failure data analysis methods are described in the next chapter.

Corrective action effectiveness

When a change is made to a design or to a process to correct a cause of failure, it is important to repeat the test which generated the failure to ensure that the corrective action is effective. Corrective action sometimes does not work. For example, it can have the effect of transferring the problem to the next weakest item in the sequence of stress-bearing items, or the true underlying cause of failure might be more complex than initial analysis indicates. Therefore re-test is important to ensure that no new problems have been introduced and that the change has the desired effect.

Analysis of test results must take account of the expected effectiveness of corrective action. Unless the causes of a failure are very well understood, and there is total confidence that the corrective action will prevent recurrence, 100 per cent effectiveness should not be assumed.

BIBLIOGRAPHY

1. P. D. T. O'Connor, *Test Engineering*. Wiley (2001).
2. US MIL-STD-810: *Environmental Test Methods*. Available from the National Technical Information Service, Springfield, Virginia.
3. UK Defence Standard 07-55: *Environmental Testing*. HMSO.
4. ISO/IEC 60068: *Environmental Testing*.
5. US MIL-HDBK-781: *Reliability Testing for Engineering Development, Qualification and Production*. Available from the National Technical Information Service, Springfield, Virginia.
6. C. E. Harris and C. E. Crede (eds), *Shock and Vibration Handbook*. McGraw-Hill (1976).
7. D. S. Steinberg, *Vibration Analysis for Electronic Equipment*. Wiley (1973).
8. US MIL-STD-462D, *Measurement of Electromagnetic Interference Characteristics*. Available from NTIS, Springfield, Virginia.
9. IEC 61000, *Electromagnetic Compatibility*.
10. *Environmental Stress, Screening Guidelines*. Institute of Environmental Sciences and Technology (USA).
11. ISO/IEC 61163, *Reliability Stress Screening*.
12. G. Hobbs, *Accelerated Reliability Engineering: HALT and HASS*. Wiley (2001).

QUESTIONS

1. Describe the concept of integrated test planning. What are the main categories of test that should be included in an integrated test programme for a new design, and what are the prime objectives of each category?

2. List the important information that should be considered in planning a reliability test programme.

3. Identify one major reference standard providing guidance on environmental testing. Identify the major factors to be considered in setting up tests for temperature, vibration, or electromagnetic compatibility.

4. Briefly describe the concept of combined environmental reliability testing (CERT). What are the main environmental stresses you may consider in planning a CERT for (i) a domestic dishwasher electronic controller; (ii) a communications satellite electronic module; (iii) an industrial hydraulic pump?

5. State your reservations concerning the use of standard environmental test specifications in their application to the equipments in question 4.

6. What is 'accelerated testing' and what are the main advantages of this type of testing in comparison with non-accelerated tests?

7. Explain why the stresses applied in accelerated stress tests should not necessarily simulate expected in-service levels.

8. How are tests accelerated for (i) mechanical components under fatigue loading and (ii) electronic systems operating at high temperatures? Comment on the methods used for analysis for the test results for each.

9. What is "highly accelerated life testing"? Describe the main benefits claimed for this method.

10. Why is it important to ensure that all failures experienced during engineering development are reported? Describe the essential features of an effective failure reporting, analysis and corrective action system (FRACAS).

12

Analysing Reliability Data

INTRODUCTION

This chapter describes a number of techniques, further to the probability plotting methods described in Chapter 3, that can be used to analyse reliability data derived from development tests and service use, with the objectives of monitoring trends, identifying causes of unreliability, and measuring or demonstrating reliability.

Since most of the methods are based on statistical analysis, the caution given in Chapter 2 (page 73) must be heeded, and all results obtained must be judged in relation to appropriate engineering and scientific knowledge.

PARETO ANALYSIS

As a first step in reliability data analysis we can use the Pareto principle of the 'significant few and the insignificant many'. It is often found that a large proportion of failures in a product are due to a small number of causes. Therefore, if we analyse the failure data, we can determine how to solve the largest proportion of the overall reliability problem with the most economical use of resources. We can often eliminate a number of failure causes from further analysis by creating a Pareto plot of the failure data. For example, Fig. 12.1 shows failure data on a domestic washing machine, taken from warranty records. These data indicate that attention paid to the programme switch, the outlet pump, the high level switch and leaks would be likely to show the greatest payoff in warranty cost reduction. However, before committing resources it is important to make sure that the data have been fully analysed, to obtain the maximum amount of information contained therein. The data in Fig. 12.1 show the parts replaced or adjusted.

In this case further analysis of records reveals:

1. For the programme switch: 77 failures due to timer motor armature open-circuit, 18 due to timer motor end bearing stiff, 10 miscellaneous. Timer motor failures show a decreasing hazard rate during the warranty period.
2. For the outlet pump: 79 failures due to leaking shaft seal allowing water to reach motor coils, 21 others. Shaft seal leaks show an increasing hazard rate.

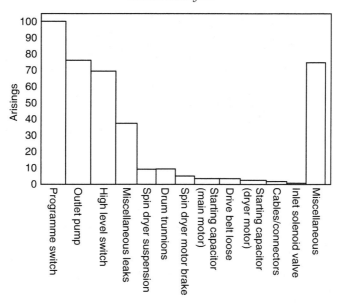

Figure 12.1 Pareto plot of failure data

3. For the high level switch: 58 failures due to failure of spot weld, allowing contact assembly to short to earth (decreasing hazard rate), 10 others.

These data reveal definite clues for corrective action. The timer motor and high level switch appear to exhibit manufacturing quality problems (decreasing hazard rate). The outlet pump shaft leak is a wear problem (increasing hazard rate). However, the leak is made more important because it damages the pump motor. Two types of corrective action might be considered: reorientation of the pump so that the shaft leak does not affect the motor coils and attention to the seal itself. Since this failure mode has an increasing rate of occurrence relative to equipment age, improving the seal would appreciably reduce the number of repair calls on older machines. Assuming that corrective action is taken on these four failure modes and that the improvements will be 80 per cent effective, future production should show a reduction in warranty period failures of about 40 per cent.

The other failure modes should also be considered, since, whilst the absolute payoff in warranty cost terms might not be so large, corrective action might be relatively simple and therefore worth while. For example, some of the starting capacitor failures on older machines were due to the fact that they were mounted on a plate, onto which the pump motor shaft leak and other leaks dripped. This caused corrosion of the capacitor bodies. Therefore, rearrangement of the capacitor mounting and investigation into the causes of leaks would both be worth considering.

This example shows the need for good data as the basis for decision-making on where to apply effort to improve reliability, by solving the important problems first. The data must be analysed to reveal as much as possible about the relative severity of problems and the likely causes. Even quite simple data, such as a brief description of the cause of failure, machine and item part number, and purchase date, are often enough to reveal the main causes of unreliability. In other applications, the failure data can be analysed in relation to contributions to down-time or by repair cost, depending upon the criteria of importance.

ACCELERATED TEST DATA ANALYSIS

Failure and life data from accelerated stress tests can be analysed using the methods described in Chapters 2–5. If the mechanism is well understood, for example material fatigue and some electronic degradation processes, then the model for the process can be applied to interpret the results and to derive reliability or life values at different stress levels. Also, the test results can be used to determine or confirm a life model. Typical stress–life models include:

Exponential Model: Life $= Ae^{-Bx}$ (12.1)

In this model there is an exponential relationship between the life (or other performance measure) and the stress variable x. Failure mechanisms of components such as film resistors and ceramic capacitors typically follow this relationship over most of their life for dependence on temperature. Some materials follow this type of relationship as a function of weathering.

Power model: Life $= A(\text{Stress})^{-x}$ (12.2)

In this model there is a simple power relationship between the performance measure and the stress variable. Failure mechanisms of components such as film resistors and ceramic capacitors typically follow this relationship over most of their life for dependence on voltage. Electric motors follow a similar form for dependence on stresses other than temperature. Mechanical creep of some materials may follow a form such as this.

Exponential power model: Life $= Ae^{-(x)^B}$ (12.3)

In this model there is an exponential relationship between the performance measure and the stress variable, raised to a power. This is a form of the Weibull life relationship, typical of fatigue failures and wear failures of components such as bearings.

Arrhenius model: Life $= (A)\dfrac{1}{T}e^{-\frac{\varsigma}{T}}$ (12.4)

Life is non-linear in the single stress variable, temperature (T). It describes many physical and chemical temperature-dependent processes, including diffusion and corrosion.

Eyring (combinedstress) linear model: Life $= Ae^{-Bx}(S)^{-N}$ (12.5)

In this relationship, two independent stress variables are present, one with a power form and the other an exponential. This form occurs for many components for temperature and voltage or temperature and humidity for restricted ranges of stress. It assumes that there is no interaction between the two stresses.

Eyring (combined stress) non-linear model: $Life = Ae^{-Bx}(S)^{-N}e^{-\frac{CX}{S}}$ $\qquad$ (12.6)

In this relationship, two dependent stress variables are present, one with a power form and the other an exponential. A third term in the relationship contains both stress variables. This form occurs for many components for temperature and voltage or temperature and humidity for high ranges of stress. The interaction term between the two stresses may take many forms.

Example 12.1 illustrates the method.

Example 12.1

An electrical insulator is rated for normal use at 12 KV. Prior tests of a sample have suggested this insulator will operate over 30 KV and a stress-life exponent of the power model was found to be $N = 5.5$. How long must one run an accelerated life test at 30 KV to demonstrate an equivalent time to 10 per cent life (based upon voltage only) if the 10 per cent life at 12 KV is desired to be at least 25 years?

Let life, $L = B(V)^{-N}$ represent the typical stress–life relationship

at 12 KV one has:
$$25 \text{ years} = B(12\,KV)^{-N}$$
at 30 KV one has:

$$L = B(30\,KV)^{-N}$$

Taking the ratio of the two equations we have:

$$\frac{25 \text{ years}}{L_{30}} = \left(\frac{12}{30}\right)^{-5.5} = 154.41$$

$$\text{Test time at } 30\,KV = \frac{25 \text{ years } (8760\,h)}{154.408} = 1418.3\,h$$

Extrapolation of accelerated test results to expected in-service conditions can be misleading if the test stresses are much higher, since different failure mechanisms might be stimulated. This is particularly the case if very high stresses, particularly combined stresses, are applied, as in HALT. It is important that the primary objective of the test is understood: whether it is to determine or confirm a life characteristic, or to help to create designs that are inherently failure free.

Life characteristics such as those listed above can only be analysed when the data represent a single (or dominant) failure mode. Therefore these methods are not appropriate for failures of assemblies or systems, when several different failure modes might be present.

Probability plotting methods (Chapter 3) can also be used for analysing such data, when sufficient data are available.

Probability plotting methods can be used to analyse data from step-stress tests. The horizontal axis then represents combined stress-time values. The probability of failure at a stress-time value can then be determined. Multiple stresses can be combined, each stress value being incremented simultaneously. Example 12.2 provides an illustration of a combined step-stress test.

Example 12.2

A rocket engine component must operate under fatigue loading conditions at high temperature. A reliability of 0.995 must be demonstrated under the extreme conditions of completely reversed cyclic (150 Hz) stress of $2 \times 10^8 \, \text{Nm}^{-2}$ at 450 °C, over a 10 min engine run-time. Design an accelerated test to demonstrate this.

The specified load and temperature stresses are fixed as the values for the start of the test. These values are maintained for 10 min, after which load is increased by $0.1 \times 10^8 \, \text{Nm}^{-2}$ and temperature by 10 °C and the test continued for 30 s. These increments are repeated after each subsequent 30 s period. Ten samples are tested to failure with results as shown in Table 12.1. The results are plotted in Fig. 12.2. This shows the line of plotted failure data lying to the right of the design point (0.5 per cent failure, 10 min). Therefore the test demonstrates compliance with the requirement, but by a small (insufficient?) margin.

Table 12.1   Results for ten samples tested to failure (Example 12.2)

			Stress	
Failure no.	Mean rank $x_i/(10+1)(\%)$	Time (min)	Load ($\text{Nm}^{-2} \times 10^8$)	Temperature (°C)
1	9.1	13.1	3.1	520
2	18	13.2	3.1	520
3	27	14.4	3.4	550
4	36	15.0	3.4	550
5	45	15.0	3.4	550
6	54	15.0	3.4	550
7	64	15.5	3.5	560
8	73	15.6	3.6	570
9	82	15.9	3.6	570
10	91	15.9	3.6	570

It is not meaningful to estimate distribution parameters or *s*-confidence limits for this type of data, since additional variables have been added. A failure-free life will probably exist in many test situations of this type. However, this will not normally be detectable from curvature of the plotted data, since step-stress failure data plot as a steeper slope than ordinary life data. Normally the design would provide for a failure-free stress life greater than the design life. The step-stress test can then be used to confirm economically that the design is correct.

Methods for analysing accelerated test data are described in Reference 1.

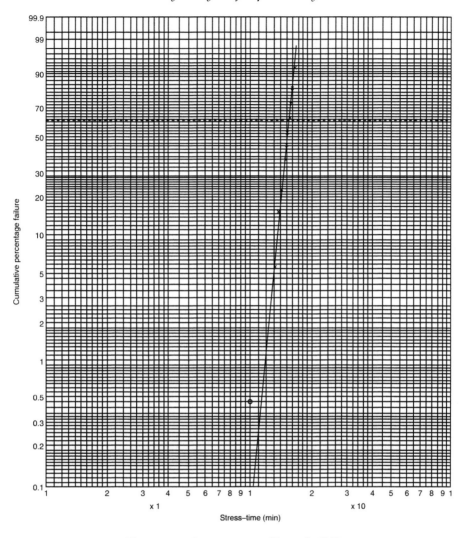

Figure 12.2 Step-stress test (Example 12.2)

RELIABILITY ANALYSIS OF REPAIRABLE SYSTEMS

Chapter 3 described methods for analysing data related to the time to first failure. The distribution function of times to first failure are obviously important when we need to understand failure processes of parts which are replaced on failure, or when we are concerned with survival probability, for example, for missiles, spacecraft or underwater telephone repeaters.

However, for repairable systems, which really represent the great majority of everyday reliability experience, the distribution of times to first failures are much less important than is the *failure rate* or *rate of occurrence of failures* (ROCOF) of the system.

Any repairable system may be considered as an assembly of parts, the parts being replaced when they fail. The system can be thought of as comprising 'sockets' into which non-repairable parts are fitted. We are concerned with the pattern of successive failures of the 'sockets'. Some parts are repaired (e.g. adjusted, lubricated, tightened, etc.) to correct system failures, but we will consider first the case where the system consists only of parts that are replaced on failure (e.g. most electronic systems). Therefore, as each part fails a new part takes its place in the 'socket'. If we ignore replacement (repair) times, which are usually small in comparison with standby or operating times, and if we assume that the time to failure of any part is independent of any repair actions, then we can use the methods of event series analysis in Chapter 2 to analyse the system reliability.

Consider the data of Example 2.20. The interarrival and (chronologically ordered) arrival values between successive component failures were as shown in columns 1 and 2:

1 X_i	2 Chronological x_i	3 Ranked X_i
175	175	12
21	196	14
108	304	21
111	415	23
89	504	38
12	516	47
102	618	51
23	641	89
38	679	102
47	726	108
14	740	111
51	791	175

Example 2.20 showed that the failure rate was increasing, the interarrival values tending to become shorter. In other words, the interarrival values are not IID. If, however, we had not performed the centroid test and assumed that the data were IID, we might order the data in rank order (column 3) and plot on probability paper. These are shown plotted on Weibull paper in Fig. 12.3. The plot shows an apparently exponential component life distribution. This is obviously a misleading result, since there is clearly an increasing failure rate trend for the 'socket' when the data are studied chronologically.

This example shows how important it is for failure data to be analysed correctly, depending on whether we need to understand the reliability of a non-repairable part or of a repairable system consisting of 'sockets' into which parts are fitted. The presence of a trend when the data are ordered *chronologically* shows that times to failure are not IID, and ordering by magnitude, which implies IID, will therefore give misleading results. *Whenever failure data are reordered* all trend information is ignored. The appropriate method to apply is trend (time series) analysis.

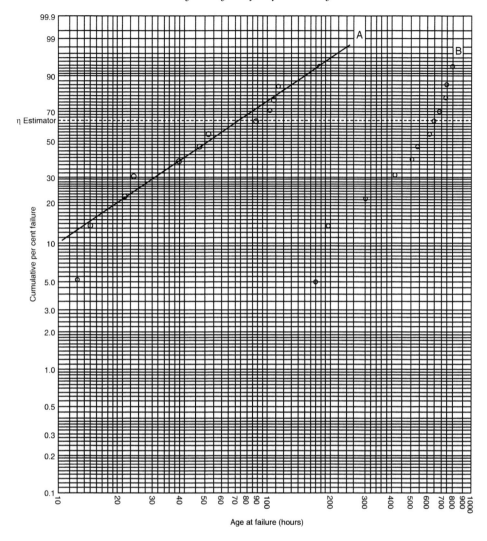

Figure 12.3 Plotted data of Example 2.20

We can derive the system reliability over a period by plotting the cumulative times to failure in chronological order (column 2) rather than in rank order. This is shown in Fig. 12.3. (points on right side) It shows the progressively increasing failure rate (though the 'socket' times to failure are not Weibull-distributed).

Multisocket systems

Now we will consider a more typical system, comprised of several parts which exhibit independent failure patterns. Each part fills a 'socket'. The failure pattern of such a system, comprising six 'sockets', is shown in Fig. 12.4.

Figure 12.4   The failure pattern of a multisocket system

Socket 1 generates a high, constant rate of system failures. Socket 2 generates an increasing rate of system failures as the system ages, and so on. The combined failure rate can be seen on the bottom line. The estimate of U for each part and for the system is shown. When U is negative, it denotes a 'happy' socket, with an increasing inter-arrival time between failures (DFR). A positive value of U indicates a 'sad' socket (IFR).

If there are no perturbations (which will be discussed below) the failure rate will tend to a constant value after most parts have been replaced at least once, regardless of the failure trends of the sockets (see page 73). This is one of the main reasons why the CFR assumption has become so widely used for systems, and why part hazard rate has been confused with failure rate. However, the time by which most parts have been replaced in a system is usually very long, well beyond the expected life of most systems.

If part times to failure (in a series system, see Chapter 6) are independently and identically exponentially distributed (IID exponential) the system will have a CFR which will be the sum of the reciprocals of the part mean times to failure, i.e.

$$\lambda_s = \sum_1^n \frac{1}{x_i}$$

The assumption of IID exponential for part times to failure within their sockets in a repairable system can be very misleading. The reasons for this are (adapted from Reference 2, with permission):

1. The most important failure modes of systems are usually caused by parts which have failure probabilities which increase with time (wearout failures).

2. Failure and repair of one part may cause damage to other parts. Therefore times between successive failures *are not necessarily independent*.

3. Repairs often do not 'renew' the system. Repairs are often imperfect or they introduce other defects leading to failures of other parts.

4. Repairs might be made by adjustment, lubrication, etc., of parts which are wearing out, thus providing a new lease of life, but not 'renewal', i.e. the system is not made as good as new.

5. Replacement parts, if they have a decreasing hazard rate, can make subsequent failure initially more likely to occur.

6. Repair personnel learn by experience, so diagnostic ability (i.e. the probability that the repair action is correct) improves with time. Generally, changes of personnel can lead to reduced diagnostic ability and therefore more reported failures.

7. Not all part failures will cause system failures.

8. Factors such as on–off cycling, different modes of use, different system operating environments or different maintenance practices are often more important than operating times in generating failure-inducing stress.

9. Reported failures are nearly always subject to human bias and emotion. What an operator or maintainer will tolerate in one situation might be reported as a failure in another, and perception of failure is conditioned by past experience, whether repair is covered by warranty, etc. Wholly objective failure data recording is very rare.

10. Failure probability is affected by scheduled maintenance or overhaul. Systems which are overhauled often display higher failure rates shortly after overhaul, due to disturbance of parts which would otherwise not have failed. If there is a post-overhaul test period before the system is returned to service, many of these failures might be repaired then. The failure data might or might not include these failures.

11. Replacement parts are not necessarily drawn from the same population as the original parts—they may be better or worse.

12. System failures might be caused by parts which individually operate within specification (i.e. do not fail) but whose combined tolerances cause the system to fail.

13. Many reported failures are not caused by part failures at all, but by events such as intermittent connections, improper use, maintainers using opportunities to replace 'suspect' parts, etc.

14. Within a system not all parts operate to the overall system cycle.

Any practical person could add to this list from his or her own experience. The factors listed above often predominate in systems to be modelled and in collected reliability data. Large data-collection systems, in which failure reports might be coded and analysed remotely from the work locations, are usually most at fault in perpetrating the analytical errors described. Such data systems might generate 'MTBFs' for systems and for parts by merely counting total reported failures and dividing into total operating time. For example, MTBFs in flying hours are quoted for aircraft electronic equipment, when the equipment only operates for part of the flight, or MTBFs in hours are quoted for valves, ignoring whether they are normally closed, normally open or how often they are opened and closed. These data are often used for reliability predictions for new systems (see Chapter 6), thus adding insult to injury.

A CFR is often a practicable and measurable first-order assumption, particularly when data are not sufficient to allow more detailed analysis.

The effect of successive repairs on the reliability of an ageing system are shown vividly in the next example (from Reference 2).

Example 12.3 (Reprinted from Reference 2 by courtesy of Marcel Dekker, Inc.)

Data on the miles between major failures (interarrival values) of bus engines are shown plotted in Fig. 12.5. These show the miles between first, second, ..., fifth major failure. Note that the interarrival mileages to the first failures (X_i) are nearly *s*-normally distributed. Successive interarrival times (second, third, fourth, fifth failures) show a tendency to being exponentially distributed. Nevertheless, the results show clearly that the reliability decreases with successive repairs, since the mean of the interarrival distances is progressively reduced:

Failure No.	$\bar{X}_i$ miles
1	94 000
2	70 000
3	54 000
4	41 000
5	33 000

The importance of this result lies in the evidence that:

1. Repair does not return the engines to an 'as new' condition.
2. Successive X_is are not IID exponential.
3. The failure rate tends to a constant value only after nearly all engines have been repaired several times. Even after five repairs the steady state has not been reached.
4. Despite the *appearance* of 'exponentiality' after several failures, replacement or more effective overhaul appears to be necessary.

CUSUM CHARTS

The 'cumulative sum', or CUSUM, chart is an effective graphical technique for monitoring trends in quality control and reliability. The principle is that, instead of monitoring the measured value of interest (parameter value, success ratio), we plot the divergence, plus or minus, from the target value. The method is the same as the scoring principle in golf, in which the above or below par score replaces the stroke count. The method enables us to report progress simply and in a way that is very easily comprehended.

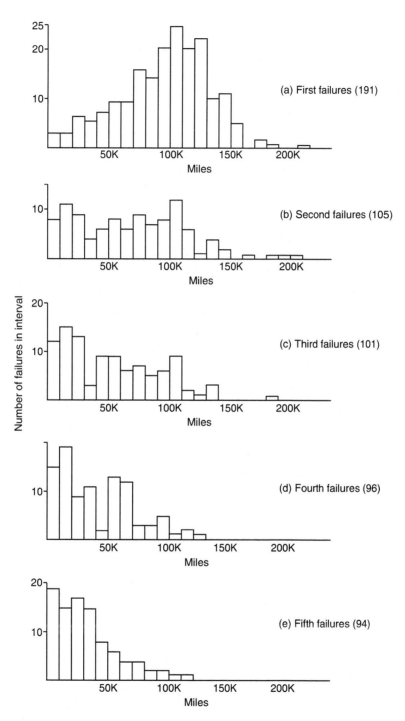

Figure 12.5 Bus engine failure data (a) First failures (191), (b) Second failures (105), (c) Third failures (101), (d) Fourth failures (96), (e) Fifth failures (94)

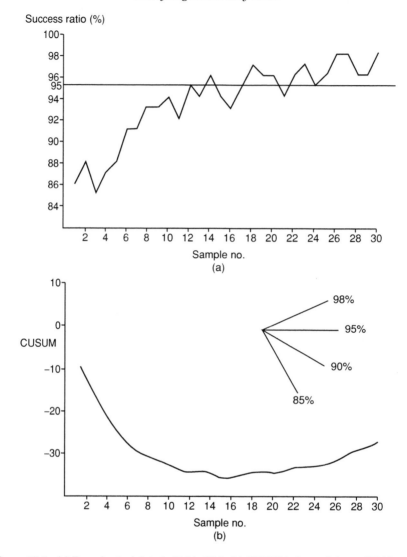

Figure 12.6 (a) Run chart of data in Table 12.2. (b) CUSUM chart of data of Table 12.2

The CUSUM chart also provides a sensitive indication of trends and changes. Instead of indicating measured values against the sample number, the plot shows the CUSUM, and the slope provides a sensitive indicator of the trend, and of points at which the trend changes.

Table 12.2 shows data from a reliability test on a one-shot item. Batches of one hundred are tested, and the target success ratio is 0.95. Figure 12.6 (a) shows the results plotted on a conventional run chart. Figure 12.6 (b) shows the same data plotted on a CUSUM chart, with the CUSUM values calculated as shown in Table 12.2.

The CUSUM can be restarted with a new target value if a changed, presumably improved, process average is attained. Decisions on when to restart, the sample size to take, and scaling of the axes will depend upon particular circumstances.

Table 12.2 Reliability test data Target = 95% (T)

Sample, i	x_i	$x_i - T$	CUSUM $\Sigma(x_i - T)$
1	86	−9	−9
2	88	−7	−16
3	85	−10	−26
4	87	−8	−34
5	88	−7	−41
6	91	−4	−45
7	91	−4	−49
8	93	−2	−51
9	93	−2	−53
10	94	−1	−54
11	92	−3	−57
12	95	0	−57
13	94	−1	−58
14	96	1	−57
15	94	−1	−58
16	93	−2	−60
17	95	0	−60
18	97	2	−58
19	96	1	−57
20	96	1	−56
21	94	−1	−57
22	96	1	−56
23	97	2	−54
24	95	0	−54
25	96	1	−53
26	97	2	−51
27	98	3	−48
28	98	3	−45
29	96	1	−44
30	98	3	−41

Guidance on the use of CUSUM charts is given in Reference 3, and in good books on statistical process control, such as those listed in the Bibliography for Chapter 13.

EXPLORATORY DATA ANALYSIS AND PROPORTIONAL HAZARDS MODELLING

Exploratory data analysis is a simple graphical technique for searching for connections between time series data and explanatory factors. In the reliability context, the failure data are plotted as a time series chart, along with the other information. For example, overhaul intervals, seasonal changes, or different operating patterns can be shown on the chart. Figure 12.7 shows failure data plotted against time between scheduled overhauls. There is a clear pattern of clustering of failures shortly after

Figure 12.7 Time series chart: failure *vs* time (overhaul interval 1000 h)

each overhaul, indicating that the overhaul is actually adversely affecting reliability. In this case, further investigation would be necessary to determine the reasons for this, for example, the quality of the overhaul work might be inadequate. Another feature that shows up is a tendency for failures to occur in clusters of two or more. This seems to indicate that failures are often not diagnosed or repaired correctly the first time.

This method of presenting data can be very useful for showing up causes of unreliability in systems such as vehicle fleets, process plant, etc. The data can be shown separately for each item, or by category of user, etc., depending on the situation, and analysed for connections or correlations between failures and the explanatory factors.

Proportional hazards modelling (PHM) is a mathematical extension of EDA. The basic proportional hazards model is of the form

$$\lambda(t; Z_1, Z_2, \ldots, Z_k) = \lambda_0(t) \exp\left(\beta_1 Z_1 + \beta_2 Z_2 + \ldots \beta_k Z_k\right) \tag{12.7}$$

where $\lambda(t; Z_1, Z_2, \ldots, Z_k)$ represents the hazard rate at time t, $\lambda_0(t)$ is the baseline hazard rate function, $Z_1, Z_2, \ldots, Z_k$ are the explanatory factors (or *covariates*), and $\beta_1, \beta_2, \ldots, \beta_k$ are the model parameters.

In the proportional hazards model, the covariates are assumed to have multiplicative effects on the total hazard rate. In standard regression analysis or analysis of variance the effects are assumed to be additive. The multiplicative assumption is realistic, for example, when a system with several failure modes is subject to different stress levels, the stress having similar effects on most of the failure modes. The proportional hazards approach can be applied to failure data from repairable and non-repairable systems.

The theoretical basis of the method is described in Reference 4. The derivation of the model parameters requires the use of advanced statistical software, as the analysis is based on iterative methods. This limits application of the technique to teams with specialist knowledge and access to the appropriate software.

RELIABILITY DEMONSTRATION

It is often necessary to measure the reliability of equipment and systems during development, production and in use. Demonstration of reliability might be required as part of a development and production contract, or prior to release to production, to ensure that the requirements have been met. Two basic forms of reliability measurement are used. A sample of equipments may be subjected to a formal reliability test, with the conditions specified in detail. Reliability may also be monitored during development and use, as test and utilization proceed, without tests being set up specifically for reliability measurement. Both approaches have common features, such as the need to define failures and to collect and analyse data, and there are advantages in practice if both approaches are used as parts of an integrated reliability programme (see Chapter 15). This section describes standard methods of test and analysis which are used to demonstrate compliance with reliability requirements.

The standard methods are not substitutes for the statistical analysis methods described in Chapter 3. The standard methods may be referenced in procurement contracts, particularly for government equipment, but they may not provide the statistical engineering insights given, for example, by Weibull probability plotting of failure data. Therefore the standards should be seen as complementary to the statistical engineering methods and useful (or mandatory) for demonstrating and monitoring reliability of products which are past the development phase.

Probability ratio sequential test (PRST) (US MIL-HDBK-781)

The best known standard method for formal reliability demonstration testing for repairable equipment which operates for periods of time, such as electronic equipment, motors, etc., is US MIL-HDBK-781: *Reliability Testing for Engineering Development, Qualification and Production* (Reference 4). It provides details of test methods and environments, as well as reliability growth monitoring methods.

MIL HDBK-781 testing is based on *probability ratio sequential testing* (PRST), the results of which (failures and test time) are plotted as in Fig. 12.8. Testing continues until the 'staircase' plot of failures versus time crosses a decision line. The reject line (dotted) indicates the boundary beyond which the equipment will have failed to meet the test criteria. Crossing the accept line denotes that the test criteria have been met. The decision lines are truncated to provide a reasonable maximum test time. Test time is stated as multiples of the specified MTBF. British and international standards for reliability demonstration are based on MIL-HDBK-781 (see Bibliography).

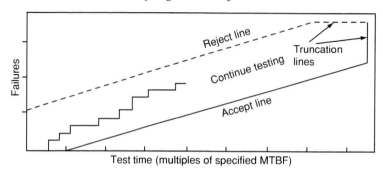

Figure 12.8 Typical probability ratio sequential test (PRST) plan

Test plans

MIL-HDBK-781 contains a number of test plans, which allow a choice to be made between the statistical risks involved (i.e. the risk of rejecting an equipment with true reliability higher than specified or of accepting an equipment with true reliability lower than specified) and the ratio of minimum acceptable to target reliability. The risk of a good equipment (or batch) being rejected is called the *producer's risk* and is denoted by α. The risk of a bad equipment (or batch) being accepted is called the *consumer's risk*, denoted by β. MIL-HDBK-781 test plans are based upon the assumption of a constant failure rate, so MTBF is used as the reliability index. Therefore MIL-HDBK-781 tests are appropriate for equipment where a constant failure rate is likely to be encountered, such as fairly complex maintained electronic equipment, after an initial burn-in period. Such equipment in fact was the original justification for the development of the precursor to MIL-HDBK-781, the AGREE report (see page 11). If predominating failure modes do not occur at a constant rate, tests based on the methods described in Chapter 2 should be used. In any case it is a good idea to test the failure data for trend, as described in Chapter 2.

The criteria used in MIL-HDBK-781 are:

1. Upper test MTBF, θ_0. This is the MTBF level considered 'acceptable'.
2. Lower test MTBF, θ_1. This is the specified, or contractually agreed, minimum MTBF to be demonstrated.
3. Design ratio, $d = \theta_0/\theta_1$.
4. Producer's risk, α (the probability that equipment with MTBF higher than θ_1 will be rejected).
5. Consumer's risk β (the probability that equipment with MTBF lower than θ_0 will be accepted).

The PRST plans available in MIL-HDBK-781 are shown below in Table 12.3. In addition, a number of fixed-length test plans are included (plans IX–XVI), in which testing is required to be continued for a fixed multiple of design MTBF. These are listed in Table 12.4. A further test plan (XVII) is provided, for production reliability acceptance testing (PRAT), when all production items are to be tested. The

Table 12.3 MIL-HDBK-781 PRST plans

Test plan	Decision risks (%)		Design ratio, $d = \theta_0/\theta_1$
	α	β	
I	10	10	1.5
II	20	20	1.5
III	10	10	2.0
IV	20	20	2.0
V	10	10	3.0
VI	20	20	3.0
VII[a]	30	30	1.5
VIII[a]	30	30	2.0

[a]Test plans VII and VIII are known as short-run high risk PRST plans.

Table 12.4 MIL-HDBK-781 fixed length test plans

Test plan	Decision risks (%)		Design ratio, d	Test duration, $X\theta_1$	Reject $\geq$ failures	Accept $\leq$ failures
	α	β				
IX	10	10	1.5	45.0	37	36
X	20	20	1.5	21.1	18	17
XI	10	10	2.0	18.8	14	13
XII	20	20	2.0	7.8	6	5
XIII	30	30	2.0	3.7	3	2
XIV	10	10	3.0	9.3	6	5
XV	20	20	3.0	4.3	3	2
XVI	30	30	3.0	1.1	1	0

plan is based on test plan III. The test time is not truncated by a multiple of MTBF but depends upon the number of equipments produced.

Statistical basis for PRST plans

MIL-HDBK-781 PRST testing is based on the assumption of a constant failure rate. The decision risks are based upon the risks that the estimated MTBF will not be more than the upper test MTBF (for rejection), or not less than the lower test MTBF (for acceptance).

We thus set up two null hypotheses:

For H_0: $\hat{\theta} \leq \theta_0$

For H_1: $\hat{\theta} \geq \theta_1$

The probability of accepting H_0 is $(1 - \alpha)$, if $\hat{\theta} = \theta_1$; the probability of accepting H_1 is β, if $\hat{\theta} = \theta_0$. The time at which the ith failure occurs is given by the exponential distribution function $f(t_i) = (1/\theta)\exp(-t_i/\theta)$. The *sequential probability ratio*, or ratio of the expected number of failures given $\theta = \theta_0$ or θ_1, is

$$\prod_{i=1}^{n} \frac{(1/\theta_1)\exp(-t_i/\theta_1)}{(1/\theta_0)\exp(-t_i/\theta_0)} \qquad\qquad (12.8)$$

where n is the number of failures.

The upper and lower boundaries of any sequential test plan specified in terms of θ_0, θ_1, α and β can be derived from the sequential probability ratio. However, arbitrary truncation rules are set in MIL-HDBK-781 to ensure that test decisions will be made in a reasonable time. The truncation alters the accept and reject probabilities somewhat compared with the values for a non-truncated test, and the α and β rules given in MIL-HDBK-781 are therefore approximations. The exact values can be determined from the *operating characteristic* (OC) curve appropriate to the test plan and are given in MIL- HDBK-781. The OC curves are described in the next section.

Operating characteristic curves and expected test time curves

An operating characteristic (OC) curve can be derived for any sequential test plan to show the probability of acceptance (or rejection) for different values of true MTBF. Similarly, the s-expected test time (ETT—time to reach an accept or reject decision) for any value of θ can be derived. OC and ETT curves are given in MIL-HDBK-781 for the specified test plans. Typical curves are shown in Figs 12.9 and 12.10.

Selection of test criteria

Selection of which test plans to use depends upon the degrees of risk which are acceptable and upon the cost of testing. For example, during development of a new equipment, when the MTBF likely to be achieved might be uncertain, a test plan with 20 per cent risks may be selected. Later testing, such as production batch acceptance testing, may use 10 per cent risks. A higher design ratio would also be appropriate for early development reliability testing. The higher the risks (i.e. the

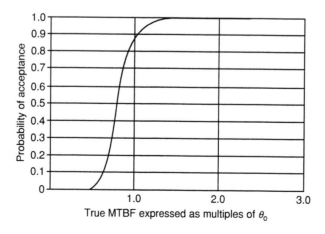

Figure 12.9 Operating characteristic (OC) curve. Test plan 1: $\alpha = 10\%$, $\beta = 10\%$ and $d = 1.5$

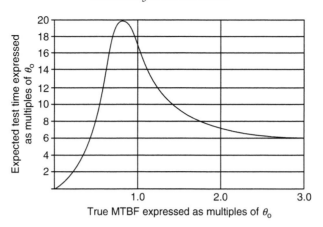

Figure 12.10 Expected test time (ETT) curve. Test plan 1: $\alpha = 10\%$, $\beta = 10\%$ and $d = 1.5$

higher the values of α and β) and the lower the design ratio, the longer will be the s-expected test duration and therefore the expected cost.

The design MTBF should be based upon reliability prediction, development testing and previous experience. MIL-HDBK-781 requires that a reliability prediction be performed at an early stage in the development programme and updated as development proceeds. We discussed the uncertainties of reliability prediction in Chapter 6. However, this should only apply to the first reliability test, since the results of this can be used for setting criteria for subsequent tests.

The lower test MTBF may be a figure specified in a contract, as is often the case with military equipment, or it may be an internally generated target, based upon past experience or an assessment of market requirements.

Test sample size

MIL-HDBK-781 provides recommended sample sizes for reliability testing. For a normal development programme, early reliability testing (qualification testing) should be carried out on at least two equipments. For production reliability acceptance testing the sample size should be based upon the production rate, the complexity of the equipment to be tested and the cost of testing. Normally at least three equipments per production lot should be tested.

Burn-in

If equipment is burned-in prior to being submitted to a production reliability acceptance test, MIL-HDBK-781 requires that all production equipments are given the same burn-in prior to delivery.

Test environmental conditions

The test conditions specified in MIL-HDBK-781 are based upon the expected conditions of use most likely to generate failure. The conditions used are called *combined environment reliability testing* (CERT), and normally include:

1. Operating stress (usually electrical stress).
2. Temperature cycling.
3. Vibration cycling (fixed frequency or swept frequency).
4. Humidity cycling.

The environmental stresses (type, severity, cyclic pattern) to be used for a particular test depend upon the anticipated usage environmental conditions. MIL-HDBK-781 describes typical conditions, including the analyses necessary to establish these. The equipment's environmental specification is usually derived from the 'mission profile', which is often specified for military equipment. Where it is not specified, typical or worst case profiles should be established by analysis and measurement. For example, if the equipment will be subjected to shock or vibration during transport or use, the conditions should be measured and used as the basis for planning the environmental test conditions. A typical CERT test profile was shown in Fig. 11.4. The reliability test operating conditions are not usually worst case or design limit values. Environmental qualification tests (see Chapter 11) should have been performed prior to reliability demonstration to prove the ability of the design to withstand these conditions. The purpose of PRST is to show, within specified levels of risk, that the equipment's MTBF is greater than the lower test MTBF and possibly also to provide contractual milestones. Therefore accelerated stresses are not applied, since the objective is to measure the reliability, not to stimulate failure.

Practical problems of PRST

Reliability demonstration testing using PRST is subject to severe practical problems and limitations, which cause it to be a controversial method. We have already covered one fundamental limitation: the assumption of a constant failure rate. However, it is also based upon the implication that MTBF is an inherent parameter of a system which can be experimentally demonstrated, albeit within s-confidence limits. In fact, reliability measurement is subject to the same fundamental constraint as reliability prediction: *reliability is not an inherent physical property of a system, as is mass or electric current.* The mass or power consumption of a system is measurable (also within statistical bounds, if necessary). Anyone could repeat the measurement with any copy of the system and would expect to measure the same values. However, if we measure the MTBF of a system in one test, *it is unlikely* that another test will demonstrate the same MTBF, quite apart from considerations of purely statistical variability. In fact there is no logical or physical reason to expect repeatability of such experiments. This can be illustrated by an example.

Suppose a computer is subjected to a PRST reliability demonstration. Four computers are tested for 400 hours and show failure patterns as follows:

No. 1   2 memory device failures (20 h, 48 h)
        1 connector intermittent (150 h)
        1 capacitor short circuit (60 h)

No. 2   1 open-circuit PCB track (40 h)
        1 IC socket failure (200 h)

No. 3   No failures

No.4   1 shorted connector (trapped on assembly) (0 h)

Total failures: 6
Total running time: 1600 h. Observed MTBF $(\hat{\theta}) = 267$ h

Note that these failures are quite typical. However, if the experiment were repeated with another four computers, there would be no reason to expect the same number or pattern of failures. If the same four computers were tested for another 1600 hours the pattern of failures would almost certainly be different. The pattern of failures and their likelihood can be influenced by quality control of manufacture and repair. Therefore the MTBF measured in this test is really no more than historical data, related to those four computers over that period of their lives. It does not predict the MTBF of other computers or of those four over a subsequent period, any more than four sales in one day would be a prediction of the next day's sales. If any design or process changes are made as a result of the test, forecasting becomes even more uncertain.

Of course, if a large number of computers were tested we would be able to extrapolate the results with rather greater credibility and to monitor trends (e.g. average failures per computer). However, PRST can seldom be extended to such large quantities because of the costs involved.

PRST is often criticized on the grounds that in-service experience of MTBF is very different to the demonstrated figure. From the discussion above this should not surprise anyone. In addition, in-service conditions are almost always very different to the environments of MIL-HDBK-781 testing, despite attempts to simulate realistic conditions in CERT.

PRST is not consistent with the reliability test philosophy described in the last chapter, since the objective is to count failures and to hope that few occur. An effective reliability test programme should aim at generating failures, *since they provide information on how to improve the product*. Failure-counting should be a secondary consideration to failure analysis and corrective action. Also, a reliability test should not be terminated solely because more than a predetermined number of failures occur. PRST is very expensive, and the benefit to the product in terms of improved reliability is sometimes questionable.

Reliability demonstration for one-shot items

For equipments which operate only once, or cyclically, such as pyrotechnic devices, missiles, fire warning systems and switchgear, the sequential method of testing based on operating time may be inappropriate. Statistical acceptance sampling method can be used for such items, as described in Chapter 13. Alternatively, a MIL-HDBK-781 test could be adapted for items which operate cyclically, using a baseline of mean cycles to failure, or MTBF assuming a given cycling rate.

COMBINING RESULTS USING BAYESIAN STATISTICS

It can be argued that the result of a reliability demonstration test is not the only information available on a product, but that information is available prior to the start of the test, from component and subassembly tests, previous tests on the product and even intuition based upon experience. Why should this information not be used to supplement the formal test result? Bayes theorem (Chapter 2) states (Eqn 2.9):

$$P(B|A) = \frac{P(A|B)P(B)}{P(A)}$$

enabling us to combine such probabilities. Equation 2.9 can be extended to cover probability distributions:

$$p(\lambda|\phi) = \frac{f(\phi|\lambda)p(\lambda)}{f(\phi)} \tag{12.9}$$

where λ is a continuous variable and ϕ represents the new observed data: $p(\lambda)$ is the *prior* distribution of λ; $p(\lambda|\phi)$ is the *posterior* distribution of λ, given ϕ; and $f(\phi|\lambda)$ is the sampling distribution of ϕ, given λ.

Let λ denote failure rate and t denote successful test time. Let the density function for λ be gamma distributed with

$$p(\lambda) = \frac{t}{\Gamma(a)}(\lambda t)^{a-1}\exp(-\lambda t)$$

If the prior parameters are a_0, t_0, then the prior mean failure rate is $\mu = a_0/t_0$ and the prior variance is $\sigma^2 = a_0/t_0^2$ (appropriate symbol changes in Eqn 2.28). The posterior will also be gamma distributed with parameters a_1 and t_1 where $a_1 = a_0 + n$ and $t_1 = t_0 + t$ and n is the number of events in the interval from 0 to t. The *s*-confidence limits on the posterior mean are $[\chi^2_{\alpha,\beta}(\nu = 2a_1)]/2t_1$.

Example 12.4

The prior estimate of the failure rate of an item is 0.02, with a standard deviation of 0.01. A reliability demonstration test results in $n = 14$ failures in $t = 500\,\text{h}$. What is the posterior estimate of failure rate and the 90 per cent lower *s*-confidence limit?

The prior mean failure rate is

$$\mu = \frac{a_0}{t_0} = 0.02\,\text{h}^{-1}$$

$$\sigma^2 = \frac{a_0}{t_0^2} = 10^{-4}\,\text{h}^{-2}$$

Therefore,

$$a_0 = \frac{\mu^2}{\sigma^2} = 4.0 \text{ failures}$$

$$t_0 = \frac{\mu}{\sigma^2} = \frac{0.02}{10^{-4}} = 200 \text{ h}$$

$$a_1 = 4 + 14 = 18 \text{ failures}$$

$$t_1 = 200 + 500 = 700 \text{ h}$$

The posterior estimate for failure rate is

$$\lambda_1 = \frac{a_1}{t_1} = \frac{18}{700} = 0.0257 \text{ h}^{-1}$$

This compares with the traditional estimate of the failure rate from the test result of $14/500 = 0.028 \text{ h}^{-1}$.

The 90 per cent lower s-confidence limit on the mean is

$$\frac{\chi^2_{10}(\nu = 2 \times 18)}{2t_1} = \frac{26}{2 \times 700} = 0.0186 \text{ h}^{-1}$$

compared with the traditional estimate (from Eqn 2.43) of

$$\frac{1}{\lambda_1} = \frac{2 \times 500}{\chi^2_{10}(\nu = 2 \times 14 + 1)}$$

$$\lambda_1 = 0.0198 \text{ h}^{-1}$$

In Example 12.4 use of the prior information has resulted in a failure rate estimate lower than that given by the test, and closer s-confidence limits.

The Bayesian approach is very controversial in reliability engineering, particularly as it has been argued that it provides a justification for less reliability testing. Choosing a prior distribution based on subjective judgement or other test experience can also be very contentious. Combining subassembly test results in this way also ignores the possibility of interface problems. The Bayesian approach is not normally recommended and it has not been approved in any national standards.

NON-PARAMETRIC METHODS

Non-parametric statistical techniques (see page 65) can be applied to reliability measurement. They are arithmetically very simple and so can be useful as quick tests in advance of more detailed analysis, particularly when no assumption is made of the underlying failure distribution.

The C-rank method

If n items are tested and k fail, the reliability of the sample is

$$R_C \approx 1 - [C\text{-rank of the } (k+1)\text{th ordered value in } (n+1)] \qquad (12.10)$$

where C denotes the confidence level required, using the appropriate median rank table (Appendix 6).

Example 12.5

Twenty items were subjected to a 100 h test in which three failed. What is the reliability at the 50 and 95 per cent lower confidence levels?

$$k + 1 = 3 + 1 = 4$$

$$n + 1 = 20 + 1 = 21$$

From Appendix 6, from the median and 95 per cent rank tables, the C-rank of four items in a sample of 21 is:

At 50%:   0.172   (i.e. $R_{50} \approx 0.828$)

At 95%:   0.329   (i.e. $R_{95} \approx 0.671$)

(cf. $17/20 = 0.85$ for $\hat{R}$)

The success-run method

When tests are run without failure, reliability can be estimated using the equation

$$R_C \approx (1 - C)^{1/(n+1)} \qquad (12.11)$$

Example 12.6

Twenty items are tested, without failure. What is the 90 per cent lower confidence limit of reliability?

$$R_{0.9} \approx (1 - 0.9)^{1/21}$$
$$= 0.897$$

RELIABILITY GROWTH MONITORING

The Duane method

It is common for new products to be less reliable during early development than later in the programme, when improvements have been incorporated as a result of failures observed and corrected. Similarly, products in service often display reliability growth. This was first analysed by J. T. Duane, who derived an empirical relationship based upon observation of the MTBF improvement of a range of items used on aircraft. Duane observed that the cumulative MTBF θ_c (total time divided by total failures) plotted against total time on log–log paper gave a straight line. The slope (α) gave an indication of reliability (MTBF) growth, i.e.

$$\log \theta_c = \log \theta_0 + \alpha(\log T - \log T_0)$$

where θ_0 is the cumulative MTBF at the start of the monitoring period T_0. Therefore,

$$\theta_c = \theta_0 \left(\frac{T}{T_0}\right)^{\alpha} \tag{12.12}$$

The relationship is shown plotted in Fig. 12.11.

The slope α gives an indication of the rate of MTBF growth and hence the effectiveness of the reliability programme in correcting failure modes. Duane observed that typically α ranged between 0.2 and 0.4, and that the value was correlated with the intensity of the effort on reliability improvement.

The Duane method is applicable to a population with a number of failure modes which are progressively corrected, and in which a number of items contribute different running times to the total time. Therefore it is not appropriate for monitoring early development testing, and it is common for early test results to show a poor fit to the Duane model. The method is also not consistent with the use of accelerated tests during development, since the objective of these is to force failures, not to generate reliability statistics, as described in Chapter 11.

We can derive the instantaneous MTBF θ_i of the population by differentiation of Eqn (12.12)

$$\theta_c = \frac{T}{n}$$

where n is the number of failures. Therefore,

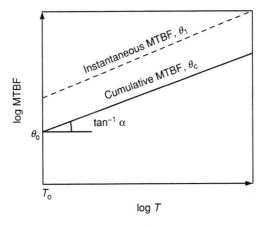

Figure 12.11 Duane reliability growth

$$n = \frac{T}{\theta_c} = \frac{T}{\theta_0 (T/T_0)^\alpha}$$

$$= T^{(1-\alpha)} \left(\frac{T_0^\alpha}{\theta_0} \right)$$

(T_0^α / θ_0) is a constant. Differentiation gives

$$\frac{dn}{dT} = (1 - \alpha) T^{-\alpha} \left(\frac{T_0^\alpha}{\theta_0} \right) = (1 - \alpha) \left(\frac{T_0}{T} \right)^\alpha \frac{1}{\theta_0}$$

$$= \frac{1 - \alpha}{\theta_c}$$

$$\frac{dn}{dT} = \frac{1}{\theta_i}$$

So

$$\theta_i = \frac{\theta_c}{1 - \alpha} \tag{12.13}$$

θ_i is shown in Fig. 12.11. The plot of θ_i is parallel to that for θ_c. A reliability monitoring programme may be directed towards a target either for cumulative or instantaneous MTBF.

After the end of a development programme in which MTBF growth is being managed, the anticipated MTBF of production items is θ_i, measured at the end of the programme. This assumes that the development testing accurately simulated the expected in-use stresses of the production items and that the standard of items being tested at the end of the development programme fully represents production items. Of course, these assumptions are often not valid and extrapolations of reliability values from one set of conditions to another must always be considered to be tentative approximations. Nevertheless, the empirical Duane method provides a reasonable approach to monitoring and planning MTBF growth for complex systems.

The Duane method can also be used in principle, to assess the amount of test time required to attain a target MTBF. If the MTBF is known at some early stage, the test time required can be estimated if a value is assumed for α. The value chosen must be related to the expected effectiveness of the programme in detecting and correcting causes of failure. Knowledge of the effectiveness of past reliability improvement programmes operated by the organization can provide guidance in selecting a value for α. The following may be used as a guide:

$\alpha = 0.4 - 0.6$. Programme dedicated to the elimination of failure modes as a top priority. Use of accelerated (overstress) tests. Immediate analysis and effective corrective action for all failures.

$\alpha = 0.3 - 0.4$. Priority attention to reliability improvement. Normal (typical expected stresses) environment test. Well-managed analysis and corrective action for important failure modes.

$\alpha = 0.2$. Routine attention to reliability improvement. Testing without applied environmental stress. Corrective action taken for important failure modes.

$\alpha = 0.2-0$. No priority given to reliability improvement. Failure data not analysed. Corrective action taken for important failure modes, but with low priority.

Example 12.7

The first reliability qualification test on a new electronic test equipment generates 11 failures in 600 h, with no one type of failure predominating. The requirement set for the production standard equipment is an MTBF of not less than 500 h in service. How much more testing should be planned, assuming values for α of 0.3 and 0.5?

$$\hat{\theta}_0 = \frac{600}{11} = 54.4\,\text{h}$$

When $\theta_i = 500$,

$$\theta_c = 500(1 - \alpha)$$

$$\begin{cases} = 350 \text{ (for } \alpha = 0.3) \\ = 250 \text{ (for } \alpha = 0.5) \end{cases}$$

Using $\theta_0 = 54.4$, from Eqn (12.12),

$$\theta_c = \theta_0 \left(\frac{T}{T_0}\right)^{\alpha}$$

$$T = T_0 \left(\frac{\theta_c}{\theta_0}\right)^{1/\alpha}$$

$$= 600 \left(\frac{350}{54.4}\right)^{1/0.3} = 297\ 200\,\text{h} \quad \text{(for } \alpha = 0.3)$$

$$= 600 \left(\frac{250}{54.4}\right)^{1/0.5} = 12\ 670\,\text{h} \quad \text{(for } \alpha = 0.5)$$

Graphical construction can be used to derive the same result, as shown in Fig 12.12 [$\tan^{-1}(0.3) = 17°$; $\tan^{-1}(0.5) = 27°$; $\theta_i = \theta_c/(1 - \alpha)$].

Obviously nearly 300 000 h of testing is unrealistic, and therefore in this case a value for α of 0.5 would have to be the objective to achieve the MTBF requirement of 500 h in a further $(12\ 670-600) \approx 12\ 000\,\text{h}$ of testing.

Example 12.7 shows that the results of a Duane analysis are very sensitive to the starting assumptions. If θ_0 was 54.5 h at $T_0 = 200\,\text{h}$, the test time required for a 500 h MTBF would be 4200 h. The initial reliability figure is usually uncertain, since data at the early stage of the programme are limited. It might be more appropriate to use a starting reliability based upon a combination of data and engineering judgement. If in the previous example immediate corrective action was being taken to remove

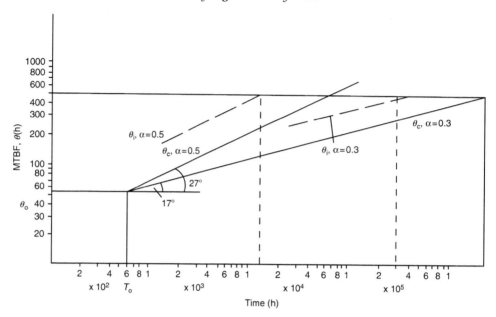

Figure 12.12 Duane plot for Example 12.7

some of the causes of earlier failures, a higher value of θ_0 could have been used. It is important to monitor early reliability growth and to adjust the plan accordingly as test results build up.

The Duane model is criticized as being empirical and subject to wide variation. It is also argued that reliability improvement in development is not usually progressive but occurs in steps as modifications are made. However, the model is simple to use and it can provide a useful planning and monitoring method for reliability growth. Difficulties can arise when results from different types of test must be included, or when corrective action is designed but not applied to all models in the test programme. These can be overcome by common-sense approaches. A more fundamental objection arises from the problem of quantifying and extrapolating reliability data. The comments made earlier about the realism of reliability demonstration testing apply equally to reliability growth measurement. As with any other failure data, trend tests as described in Chapter 2 should be performed to ascertain whether the assumption of a constant failure rate is valid.

Other reliability growth models are also used, some of which are described in MIL-HDBK-781, which also describes the management aspects of reliability growth monitoring. Reliability growth monitoring for one-shot items can be performed similarly by plotting cumulative success rate. Statistical tests for MTBF or success rate changes can also be used to confirm reliability growth, as described in Chapter 2. Example 12.8 shows a typical reliability growth plan and record of achievement.

Example 12.8

Reliability growth plan:

Office Copier Mk 4

Specification:

In-use call rate: 2 per year max. (at end of development)
              1 per year max. (after first year)

Average copies per machine per year: 40 000

Assumptions:

$\theta_0 = 1000$ copies per failure at 10 000 copies on prototypes

$\alpha \begin{cases} = 0.5 \text{ during development} \\ = 0.3 \text{ in service} \end{cases}$

Notes to Duane plot (Fig. 12.13)

1. Prototype reliability demonstration: models 2 and 3: 10 000 copies each
2. First interim reliability demonstration: models 6–8, 10: 10 000 copies each
3,5. Accelerated and ageing tests (data not included in θ_c)
4. Second interim reliability demonstration: models 8, 10, 12, 13: 10 000 copies each
6. Final reliability demonstration: models 12, 13: 10 000 copies each. Models 8, 10: 20 000 copies each

Reliability demonstration test results give values for θ_i.

   Note that in Example 12.8 the accelerated stress test results are plotted separately, so that the failures during these tests will not be accumulated with those encountered during tests in the normal operating environment. Therefore, accelerated stress failure data will be obtained to enable potential in-use failure modes to be highlighted, without confusing the picture as far as measured reliability achievement is concerned. The improvement from the first to the second accelerated stress test has a higher Duane slope than the main reliability growth line, indicating effective improvement. The example also includes a longevity test, to show up potential wearout failure modes, by having two of the first reliability demonstration units continue to undergo test in the second interim and final reliability demonstrations. These units will have been modified between tests to include all design improvements shown to be necessary. A lower value for α is assumed for the in-service phase, as improvements are more difficult to implement once production has started.

The $M(t)$ method

The $M(t)$ method of plotting failure data is a simple and effective way of monitoring reliability changes over time. It is most suitable for analysing the reliability performance of equipment in service. Figure 12.14(a) shows a typical situation in which new equipments are introduced to service over a period (calendar time), and suffer failures which incur down time (repair, delays, etc.). Figure 12.14(b) shows the same population, but now with operating time (calendar time minus down time, or operating hours, as appropriate) as the horizontal scale. The *population graph* (Fig. 12.14(c)) shows the systems at risk as a function of operating time.

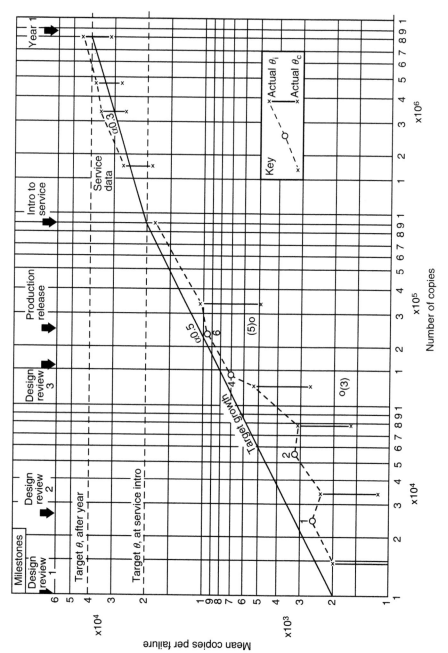

Figure 12.13 Duane plot for Example 12.8

$M(t)$ is calculated at each failure from the formula

$$M(t_i) = M(t_{i-1}) + 1/N(t_i) \tag{12.14}$$

where $M(t_i)$ is the value of M at operating time t_i
$M(t_{i-1})$ is the preceding value of M
$N(t_i)$ is the number of equipments in service at operating time t_i

$M(t)$ is the mean accumulated number of failures as a function of operating time.
$M(t_i)$ can also be calculated for groups of failures occurring over intervals, by replacing $1/N(t_i)$ with $\Delta r(t_i)/N(t_i)$, where $\Delta r(t_i)$ is the failures that have occurred in the time interval between t_{i-1} and t_i.

Figure 12.14(d) is the $M(t)$ graph. The slope of the line indicates the proportion per time unit failing at that time, or the failure intensity. Reliability improvement (either due to a decreasing instantaneous failure rate pattern or to reliability improvement actions) will reduce the slope. A straight line indicates a constant (random) pattern. An increasing slope indicates an increasing pattern, and v.v. Changes in slope indicate changing trends: for example, Fig. 12.14(d) is typical of an early 'infant mortality' period, caused probably by manufacturing problems, followed by a constant failure intensity. The failure intensity over any period can be calculated by measuring the slope. The proportions failing over a period can be determined by reading against the $M(t)$ scale.

The $M(t)$ method can be used to monitor reliability trends such as the effectiveness of improvement actions. Example 12.9 illustrates this.

Example 12.9

Table 12.5 shows data on systems in service. Figure 12.15a and 12.15b show $\Delta r(t)$ and $N(t)$ respectively, and fig. 12.15c is the $M(t)$ plot.

From the $M(t)$ graph we can determine that:

Table 12.5 Service data

Units at risk ($N(t)$)	Operating time t (hours)	Failures ($\Delta r(t)$)	$\Delta r(t)/N(t)$	$M(t)$
105	0	0	0.00	0.00
105	400	0	0.00	0.00
105	600	4	0.04	0.04
85	800	10	0.12	0.16
65	1000	17	0.26	0.42
45	1200	22	0.49	0.91
35	1400	18	0.51	1.42
25	1600	5	0.20	1.62
15	1800	3	0.20	1.82
5	2000	1	0.20	2.02

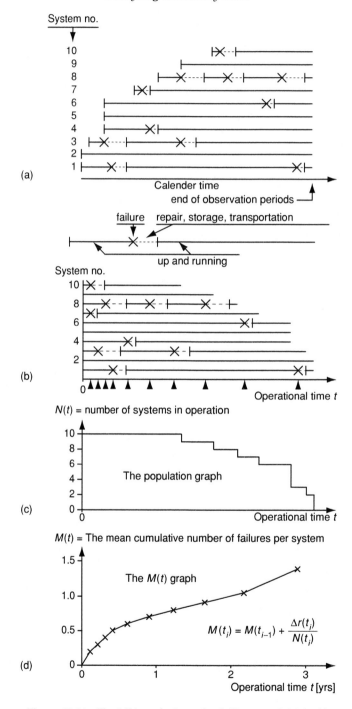

Figure 12.14   The $M(t)$ analysis method (Courtesy J. Møltoft)

Figure 12.15 *M*(*t*) graph of Table 12.5

1. Over the first 400 h no failures occur, but then failures occur at a fairly high and increasing intensity thereafter. This could indicate a failure-free period for the population.
2. The failure intensity continues at a roughly constant level (about 2.5 failures/ unit/1000 h), until 1400 h, when it drops to about 1 failure/unit/1000 h. This might be due to the weeding out of the flaws (latent defects) in the systems, modification to units in service, improved maintenance, etc.
3. By extrapolating the asymptotic part of the curve back to the *M*(*t*) axis, we see that on average each system has failed and been repaired about twice.

The *M*(*t*) method can be useful for identifying and interpreting failure trends. It can also be used for evaluating logistics and warranty policies. For example, if the horizontal scale represents calendar time, the expected number of failures over selected periods can be determined by reading the intercepts on the $M(t)$ scale (e.g. in Example 12.9 a linear extrapolation of the final slope (the dotted line) would indicate about 1 failure/unit/1000 h of the matured systems and about 12 per cent flaws). The method is described in Reference 8.

Reliability growth estimation by failure data analysis

Reliability growth can be estimated by considering the failure data and the planned corrective action. No empirical model is used and the method takes direct account of what is known and planned, so it can be easier to sell. However, it can only be applied when sufficient data are available, well into a development programme or when the product is in service.

If we know that 20 per cent of failures are caused by failure modes for which corrective action is planned and we are sure that the changes will be effective, we can simply estimate that the improvement in failure rate will be 20 per cent. Alternatively, we could assign an effectiveness value to the changes, say 80 per cent, in which case the failure rate improvement will be 16 per cent.

This approach should be used whenever failure data and failure investigations are comprehensive enough to support a Pareto analysis, as described earlier. The method can be used in conjunction with a Duane plot. If known failure modes can be corrected, reliability growth can be anticipated. However, if reliability is below target and no corrective action is planned, the reliability growth forecast will not have much meaning.

MAKING RELIABILITY GROW

In this chapter we have covered methods for measuring reliability achievement and growth. Of course it is not enough just to measure performance. Effort must be directed towards achieving the objective. In reliability engineering this means taking positive action to unearth design and production shortfalls which can lead to failure, to correct these deficiencies and to prove that the changes are effective. In earlier chapters we have covered the methods of stress analysis, design review, testing and failure data analysis which can be used to ensure a product's reliability. To make these activities as effective as possible, it is necessary to ensure that they are all directed towards the reliability achievement which has been specified.

There is a dilemma in operating such a programme. There will be a natural tendency to try to demonstrate that the reliability requirements have been met. This can lead to reluctance to induce failures by increasing the severity of tests and to a temptation to classify failures as non-relevant or unlikely to recur. On the other hand, reliability growth is maximized by deliberate and aggressive stress-testing, analysis and corrective action as described in Chapter 11. Therefore, the objective should be to stimulate failures during development testing and not to devise tests and failure-reporting methods which will maximize the chances of demonstrating that a specification has been satisfied. Such an open and honest programme makes high demands on teamwork and integrity, and emphasizes the importance of the project manager understanding the aims and being in control of the reliability programme.* The reliability milestones should be stated early in programme and achievement should be monitored against this statement.

Test, analyse and fix

Reliability growth programmes as described above have come to be known as test, analyse and fix (TAAF) programmes. It is very important in such programmes that:

1. All failures are analysed fully, and action taken in design or production to ensure that they should not recur. No failure should be dismissed as being 'random' or 'non-relevant' during this stage, unless it can be demonstrated conclusively that such a failure cannot occur on production units in service.
2. Corrective action must be taken as soon as possible on all units in the development programme. This might mean that drawings have to be altered more often, and can cause programme delays. However, if faults are not corrected reliability growth will be delayed, potential failure modes at the 'next weakest link' may not be highlighted, and the effectiveness of the corrective action will not be adequately tested.

Action on failures should be based on a disciplined FRACAS (Chapter 11 page 333) and Appendix 8.

Whenever failures occur, the investigation should refer back to the reliability predictions, stress analyses and FMECAs to determine if the analyses were correct. Discrepancies should be noted and corrected to aid future work of this type.

Reliability growth in service

The same principles as described above should be applied to reliability growth in service. However, there are three main reasons why in-service reliability growth is more difficult to achieve than during the development phase.

1. Failure data are often more difficult to obtain. Warranty or service contract repair reports are a valuable source of reliability data, but they are often harder to control, investigation can be more difficult with equipment in the users' hands and data often terminate at the end of the warranty period. Some companies make arrangements with selected dealers to provide comprehensive service data. Military and other government customers often operate their own in-use failure data systems. However, in-use data very rarely match the needs of a reliability growth programme.
2. It is much more difficult and much more expensive to modify delivered equipment or to make changes once production has started.
3. A product's reputation is made by its early performance. Reliance on reliability growth in use can be very expensive in terms of warranty costs, reputation and markets.

*The politics of test planning to ensure accept decisions for contractual or incentive purposes is an aspect of reliability programme management which will not be covered here.

Nevertheless, a new product will often have in-service reliability problems despite an intensive development programme. Therefore reliability data must be collected and analysed, and improvements designed and implemented. Most products which deserve a reliability programme have a long life in service and undergo further evolutionary development, and therefore a continuing reliability improvement programme is justifiable. When evolutionary development takes place in-use data can be a valuable supplement to further reliability test data, and can be used to help plan the follow-on development and test programme. The FRACAS described in Appendix 8 can be used for in-service failures.

Another source of reliability data is that from production test and inspection. Many products are tested at the end of the production line and this includes burn-in for many types of electronic equipment. Whilst data from production test and inspection are collected primarily to monitor production quality costs and vendor performance, they can be a useful supplement to in-use reliability data. Also, as the data collection and the product are still under the manufacturer's control, faster feedback and corrective action can be accomplished.

The manufacturer can run further tests of production equipment to verify that reliability and quality standards are being maintained. Such tests are often stipulated as necessary for batch release in government production contracts. As in-house tests are under the manufacturer's control, they can provide early warning of incipient problems and can help to verify that the reliability of production units is being maintained or improved.

BIBLIOGRAPHY

1. W. Nelson, *Accelerated Testing: Statistical Models, Test Plans and Data Analysis*. Wiley (1989).
2. H. Ascher and H. Feingold, *Repairable Systems Reliability*. Dekker (1984).
3. British Standard BS 5703: *Guide to Data Analysis and Quality Control using Cusum Techniques*. British Standards Institute, London.
4. J. D. Kalbfleisch and R. L. Prentice, *The Statistical Analysis of Failure Time Data*. Wiley (1980).
5. US MIL-HDBK-781: *Reliability Testing for Equipment Development, Qualification and Production*. Available from the National Technical Information Service, Springfield, Virginia.
6. British Standard, BS 5760: *Reliability of Systems, Equipments and Components*, Part 2. British Standards Institution, London.
7. ISO IEC 60605: *Equipment Reliability Testing*. International Standards Organisation Geneva.
8. J. Miltoft, *Reliability Engineering Based on Field Information: The Way Ahead*. Quality and Reliability Engineering International Vol. 10 399–409 (1994).

QUESTIONS

1. (a) Explain why and under what circumstances it might be valid to assume the exponential distribution for interfailure times of a complex repairable system even though it may contain 'wear-out' components.

   (b) Such a system has, on test, accumulated 1053 h of running during which there have been two failures. Estimate the MTBF of the system, and its lower 90 per cent confidence limit.

(c) On the assumption that no more failures occur, how much more testing is required to demonstrate with 90 per cent confidence that the true MTBF is not less than 500 h? Comment on the implications of your answer.

2. Identify at least three models of stress–life relationships and provide examples when these might occur.

3. Consider the following data.

   a. Does the stress–life relationship follow an expected log–log relationship?

Stress (V)	Time to fail (h)
25	5.6
50	10
70	14.3
90	27.8

   b. Give reasons why a log stress vs. log life relationship might not always be linear at high stress and at low stress.

4. Explain why a simple power law of life vs. stress is unlikely to apply to an assembly that consists of several different stressed components.

5. You desire to estimate the accelerating influences of a corrosive solution through an accelerated life test. The concentration of the solution will be the accelerating factor and four levels will be employed in the test. Metallic samples will be soaked in the various concentrations for 10 per cent of the day and then allowed to air dry at room temperature. Describe how to approach this test and use the results in the future.

6. Let S measure the time-dependent strength of a material under test and S_0 is the initial strength as measured at the beginning of test. Prior tests and evaluations have suggested that the following stress–time relationship appears to work.

$$S^2 = (S_0)^2 - 2Ct \ (t = \text{time})$$

If the strength declines 20 per cent after exactly three weeks of accelerated test, what is the expected time to failure at this condition, when the failure is defined as a decline of strength of 50 per cent?

7. Let the functional relationship of question 6 become $S^4 = (S_0)^4 - 4Ct$ instead (this is a typical degradation curve). Redo the analysis of question 6.

8. The following table of data was generated from a valid accelerated life test. The failure definition employed was the time to 10 per cent failures of a sample operated at each condition. If the basic model below is correct, what are the values of the three unknowns E_o, N and B for the typical formula shown below?

$$\text{Life} = B(V)^{-N} e E_a / KT$$

Operating temp (°C)	Operating volts	Life (h)
150	70	10
150	50	14.3
150	25	44.7
125	70	27.8
125	50	46.4
125	25	199.5
100	70	117.1
100	50	188.9

9. Explain why the times between successive failures of a repairable system might not be independently and identically distributed (IID).

10. Question 3 in Chapter 3 describes the behaviour of a component in a 'socket' of a repairable system. Referring again to that question, suppose you have been given the additional information that machine B was put into service when machine A had accumulated 500 h, machine C when machine A had accumulated 1000 h, machine D when machine A had accumulated 1500 h, and machine E when machine A had accumulated 2000 h.

    (a) Use this additional information about the *sequencing* of failures to calculate the trend statistic (Eqn 2.50), and hence judge whether, as far as this socket is concerned, the system is 'happy', 'sad' or indeterminate (IID) in terms of Fig. 12.4.

    (b) Repeat the exercise, but splitting the data to deal separately with (i) the first eight *sequenced* failures. and (ii) the second eight. What do these results tell you about the dangers of assuming IID failures when the assumption may not be valid?

11. Explain the principles of probability ratio sequential testing (PRST) to demonstrate MTBF. What are the main limitations of the method.

12. A prototype of a repairable system was subjected to a test programme where engineering action is supposedly taken to eliminate causes of failure as they occur. The first 500 h of running gave failures at 12, 36, 80, 120, 200, 360, 400, 440 and 480 hours.

    (a) Use a Duane plot to discover whether reliability growth is occurring.

    (b) Calculate the trend statistic (Eqn. 2.50) and see whether it gives results consistent with (a).

13. In an investigation into cracking of brake discs on a locomotive, a proportional hazards analysis was undertaken on a sample containing 205 failures and 905 censorings. (A failure was removal of an axle because a crack had propagated to such an extent that replacement was needed to avoid any possibility of fracture, a censoring was removal of an axle for any other reason.) Referring to Eqn (12.7), the covariates were:

    Z_1 = region of operation (0 for Eastern region, 1 for Western region)

Z_2 = braking system (0 for type A, 1 for type B)

Z_3 = disc material (0 for material X, 1 for material Y).

The data were analysed using computer methods (the only practicable way) to give the following coefficients: $\beta_1 = 0.39$, $\beta_2 = 0.72$, $\beta_3 = 0.95$.

(a) At any given age, what is the ratio between the hazard functions of axles running on the two regions?

(b) What is the ratio for the two braking systems?

(c) What is the ratio for the two disc materials?

[This example is based on real data first reported by Newton and Walley in 1983. The analysis is further developed in Bendell, Walley, Wightman and Wood (1986), 'Proportional hazards modelling in reliability analysis—an application to brake disks on high-speed trains', *Quality and Reliability International*, **2**, 42–52.]

14. The following data give the times between successive failures of an aircraft air-conditioning unit: 48, 29, 502, 12, 70, 21, 29, 386, 59, 27, 153, 26 and 326 hours. When a unit fails it is repaired *in situ*. Repairs can be assumed to be instantaneous.

    (a) Examine the distribution of failure times, treating the unit as a component on the aircraft.

    (b) Examine the trend of failures, viewing the air-conditioning system itself as a repairable system.

    (c) Describe clearly the conclusions to be drawn from both these analyses.

15. Two prototypes of a newly designed VHF communications set have been manufactured. Each has been subjected to a life-test as follows:
    Prototype A—had failures at 37, 53, 102, 230 and 480 h running; withdrawn from test at 600 h
    Prototype B—started test when A was withdrawn (only one test rig was available), and had failures after 55, 290, 310, 780 and 1220 h; is still running, having accumulated 1700 h.

    (a) On the assumption of random failures, estimate the failure rate and the probability of surviving a 100 h mission without failure.

    (b) Revise your answers to (a) if you apply a reliability growth model to the data, assuming that all modifications found necessary during the test on prototype A were applied to prototype B.

16. Calculate the overall mean time between failures for the data in Question 10. Produce a CUSUM chart of the data, plotting $\Sigma(t_i - T)$ against i where t_i is the elapsed time between the ith and the $(i - 1)$th sequenced failures and T is the expected time since the previous failure based on the overall MTBF.

    From this plot, identifying when any change to the MTBF might have occurred, and estimate the MTBF before and after the change.

    Consider whether this plot shows the situation more clearly than a straightforward trend plot of cumulative failures against cumulative time.

13

Reliability in Manufacture

INTRODUCTION

It is common knowledge that a well-designed product can be unreliable in service because of poor quality of production. Control of production quality is therefore an indispensable facet of an effective reliability effort. This involves controlling and minimizing variability and identifying and solving problems.

Human operations, particularly repetitive, boring or unpleasant tasks, are frequent sources of variability. Automation of such tasks therefore usually leads to quality improvement. Typical examples are paint spraying, welding by robots in automobile production, component placement and soldering in electronic production and CNC machining.

Variability can never be completely eliminated, since there will nearly always be some human operations, and automatic processes are not without variation. A reliable design should cater for expected production variation, so designers must be made aware of the variability inherent in the manufacturing processes to be used.

The production quality team should use the information provided by design analyses, FMECAs and reliability tests. A reliable and easily maintained design will be cheaper to produce, in terms of reduced costs of scrap and rework.

The integration of reliability and manufacturing quality programmes is covered in more detail in Chapter 15.

CONTROL OF PRODUCTION VARIABILITY

The main cause of production-induced unreliability, as well as rework and scrap, is the variability inherent in production processes. In principle, a correct design, correctly manufactured, should not fail in normal use. However, all manual processes are variable. Automatic processes are also variable, but the variability is usually easier to control. Bought-in components and materials also have variable properties. Production quality control (QC) is primarily concerned with measuring, controlling and minimizing these variations in the most cost-effective way.

Statistical process control (SPC) is the term used for the measurement and control of production variability. In SPC, QC people rely heavily on the s-normal distribution.

However, the comments in chapter 2 (pages 48 and 73) should be noted: conventional SPC often ignores the realities discussed.

Process capability

If a product has a tolerance or specification width, and it is to be produced by a process which generates variation of the parameter of interest, it is obviously important that the process variation is less than the tolerance. The ratio of the tolerance to the process variation is called the *process capability*, and it is expressed as

$$C_p = \frac{\text{Tolerance width}(T)}{\text{Process } 3\sigma \text{ limits}}$$

A process capability index of 1 will generate, in theory for *s*-normal variation, approximately 0.15 per cent out of tolerance product, at each extreme (Fig. 13.1). A process capability index of greater than 1.33 will theoretically generate about 0.005 per cent out of tolerance product, or practically 100 per cent yield.

C_p values assume that the specification centre and the process mean coincide. To allow for the fact that this is not necessarily the case, an alternative index, C_{p_k}, is used, where

$$C_{p_k} = (1 - K)C_{p'}$$

and

$$K = \frac{D - \bar{x}}{T/2} \quad (\text{if } D > \bar{x}, \text{ otherwise use } \bar{x} - D)$$

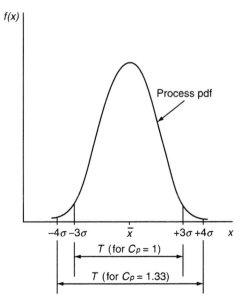

Figure 13.1 Process capability, C_p

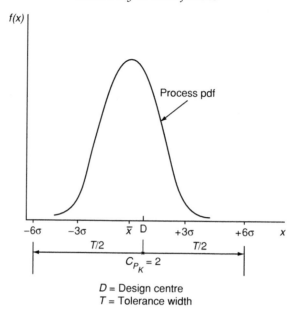

Figure 13.2   Process capability C_{p_k}

D being the design centre, $\bar{x}$ the process mean, and T the tolerance width.

Figure 13.2 shows examples of C_{p_k}. Ideally $C_p = C_{p_k}$. Modern production quality requirements typically demand C_{p_k} values of 2 or even higher, to provide high assurance of consistent performance. The "six sigma" approach (page 446) extends the concept even further.

Use of the process capability index assumes that the process is s-normally distributed far into the tails and is stationary. Any systematic divergence, due for example to set-up errors, movement of the process mean during the manufacturing cycle, or other causes, could significantly affect the output. Therefore the use of the capability index to characterize a production process is appropriate only for processes which are under statistical control, i.e. when there are no *special causes* of variation such as those just mentioned, only *common causes* (Chapter 2 page 52). Common cause variation is the random variation inherent in the process, when it is under *statistical control*.

The necessary steps to be taken when setting up a production process are:

1. Using the information from the product and process design studies and experiments, determine the required tolerance.
2. Obtain information on the process variability, either from previous production or by performing experiments.
3. Evaluate the process capability index.
4. If the process capability index is high enough, start production, and monitor using statistical control methods, as described below.
5. If C_p/C_{p_k} is too low, investigate the causes of variability, and reduce them, before starting production (see later).

Process control charts

Process control charts are used to ensure that the process is under statistical control, and to indicate when special causes of variation exist. In principle, an in-control process will generate a random fluctuation about the mean value. Any trend, or continuous performance away from the mean, indicates a special cause of variation.

Figure 13.3 is an example of a process control chart. As measurements are made the values are marked as points on the control chart against the sample number on the horizontal scale. The data plotted can be individual values or sample averages; when sample averages are plotted the chart is called an $\bar{X}$ *chart*. The $\bar{X}$ chart shows very clearly how the process value varies. Upper and lower control limits are drawn on the chart, as shown, to indicate when the process has exceeded preset limits.

The control limits on an $\bar{X}$ chart are based on the tolerance required of the process. Warning limits are also used. These are set within the control limits to provide a warning that the process might require adjustment. They are based on the process capability, and could be the process 3σ values ($C_{p_k} = 1.0$), or higher. Usually 2 or more sample points must fall outside the warning limits before action is taken.

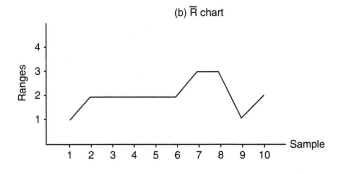

Figure 13.3 Process control charts

However, any point falling outside the control limit indicates a need for immediate investigation and corrective action.

Figure 13.3(b) is a *range chart* ($\bar{R}$ chart). The plotted points show the range of values within the sample. The chart indicates the repeatability of the process.

$\bar{X}$ and $\bar{R}$ charts (also called *Shewhart charts*) are the basic tools of SPC for control of manufacturing processes. Their ease of use and effectiveness make them very suitable for use by production operators for controlling their own work, and therefore they are commonly used in operator control and quality circles (see later in this chapter). Computer programs are available which produce $\bar{X}$ and $\bar{R}$ charts automatically when process data are input. Integrated measurement and control systems provide for direct input of measured values to the control chart program, or include SPC capabilities (analysis and graphics).

Statistical process control is applicable to relatively long, stable production runs, so that the process capability can be evaluated and monitored with reasonable statistical and engineering confidence. Methods have, however, been developed for batch production involving smaller quantities. Other types of control chart have also been developed, including variations on the basic Shewhart charts, and non-statistical graphical methods. These are all described in References 1 to 8.

The most effective application of SPC is the detection of special causes of variation, to enable process improvements to be made. Statistical finesse and precision are not usually important or essential. The methods must be applied carefully, and selected and adapted for the particular processes. Personnel involved must be adequately trained and motivated, and the methods and criteria must be refined as experience develops.

It is important to apply SPC to the processes that influence the quality of the product, not just to the final output parameter, whenever such upstream process variables can be controlled. For example, if the final dimension of an item is affected by the variation in more than one process, these should be statistically monitored and controlled, not just the final dimension. Applying SPC only to the final dimension might not indicate the causes of variation, and so might not provide effective and timely control of the processes.

CONTROL OF HUMAN VARIATION

Several methods have been developed for controlling the variability inherent in human operations in manufacturing, and these are well documented in the references on quality assurance. Psychological approaches, such as improving motivation by better work organization, exhortation and training, have been used since early industrialization, particularly since the 1940s. These were supported by the development of statistical methods, described earlier.

Inspection

One way of monitoring and controlling human performance is by independent inspection. This was the standard QC approach until the 1950s, and is still used to some extent. An inspector can be made independent of production requirements and can be given authority to reject work for correction or scrap. However, inspection is subject to three major drawbacks:

1. Inspectors are not perfect; they can fail to detect defects. On some tasks, such as inspecting large numbers of solder joints or small assemblies, inspection can be a very poor screen, missing 10 to 50 per cent of defects. Inspector performance is also as variable as any other human process.
2. Independent inspection can reduce the motivation of production people to produce high quality work. They will be concerned with production quantity, relying on inspection to detect defects.
3. Inspection is expensive. It is essentially non-productive, the staff employed are often more highly paid than production people and output is delayed while inspection takes place. Probably worse, independent inspection can result in an overlarge QC department, unresponsive to the needs of the organization.

These drawbacks, particularly the last, have led increasingly to the introduction of operator control of quality, described below. Automatic inspection aids and systems have also been developed, including computerized optical comparators and automatic gauging systems.

Operator control

Under operator control, the production worker is responsible for monitoring and controlling his or her own performance. For example, in a machining operation the operator will measure the finished article, log the results and monitor performance on SPC charts. Inspection becomes part of the production operation and worker motivation is increased. The production people must obviously be trained in inspection, measurement and SPC methods, but this is usually found to present few problems. Operator control becomes even more relevant with the increasing use of production machinery which includes measuring facilities, such as self-monitoring computer numerically controlled (CNC) machines.

A variation of operator control is to have production workers inspect the work of preceding workers in the manufacturing sequence, before starting their production task. This provides the advantages of independent inspection, whilst maintaining the advantages of an integrated approach.

ACCEPTANCE SAMPLING

Acceptance sampling provides a method for deciding whether to accept a particular production lot, based upon measurements of samples drawn at random from the lot. Sampling can be by *attributes* or by *variables*. Criteria are set for the allowable proportion defective, and the sampling risks.

Sampling by attributes

Sampling by attributes is applicable to go/no-go tests, using binomial and Poisson statistics. Sampling by attributes is covered in standard plans such as ANSI/ASQZ1-

4 and BS 6001 (References 9 and 10). These give accept and reject criteria for various sampling plans, based upon sample size and risk levels. The main criterion is the *acceptable quality level* (AQL), defined as the maximum percentage defective which can be accepted as a process average. The tables in the standards give accept and reject criteria for stated AQLs, related to sample size, and for 'tightened', 'normal' and 'reduced' inspection. These inspection levels relate to the consumer's risk in accepting a lot with a percentage defective higher than the AQL. Table 13.1 shows a typical sampling plan. Some sampling plans are based upon the *lot tolerance percentage defective* (LTPD). The plans provide the minimum sample size to assure, with given risk, that a lot with a percentage defective equal to or more than the specified LTPD will be rejected. LTPD tests give lower consumers' risks that substandard lots will be accepted. LTPD sampling plans are shown in Table 13.2.

For any attribute sampling plan, an *operating characteristic* curve can be derived. The OC curve shows the power of the sampling plan in rejecting lots with a given percentage defective. For example, Fig. 13.4 shows OC curves for single sampling plans for 10 per cent samples drawn from lots of 100, 200 and 1000, when one or more defectives in the sample will lead to rejection (acceptance number = 0). If the lot contains, say, 2 per cent defective the probability of acceptance will be 10 per cent for a lot of 1000, 65 per cent for a lot of 200, and 80 per cent for a lot of 100. Therefore the lot size is very important in selecting a sampling plan.

Double sampling plans are also used. In these the reject decision can be deferred pending the inspection of a second sample. Inspection of the second sample is required if the number of defectives in the first sample is greater than allowable for immediate acceptance but less than the value set for immediate rejection. Tables and OC curves for double (and multiple) sampling plans are also provided in the references quoted above.

Sampling by variables

Sampling by variables involves using actual measured values rather than individual attribute ('good or bad') data. The methods are based upon use of the s-normal distribution. They are described in References 1 to 8.

Sampling by variables is not as popular as sampling by attributes, since it is a more complex method. However, it can be useful when a particular production variable is important enough to warrant extra control.

General comments on sampling

Whilst standard sampling plans can provide some assurance that the proportion defective is below a specified figure, they do not provide the high assurance necessary for many modern products. For example, if an electronic assembly consists of 100 components, all with an AQL of 0.1 per cent, there is on average a probability of about 0.9 that an assembly will be free of defective components. If 10 000 assemblies are produced, about 1000 on average will be defective. The costs involved in diagnosis and repair or scrap during manufacture would obviously be very high. With such quantities typical of much modern manufacturing, higher assurance than can realistically be obtained from statistical sampling plans is

Table 13.1 Master table for normal inspection-single sampling (MIL-STD-105D, Table II-A)

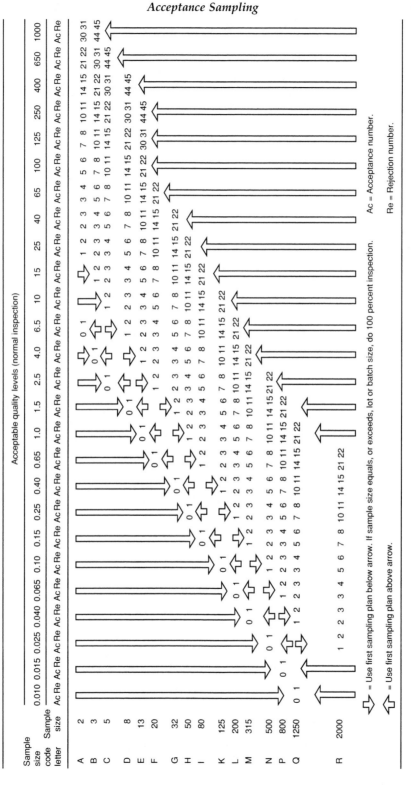

Table 13.2 LTPD sampling plans.[a] Minimum size of sample to be tested to assure, with 90 per cent confidence, that a lot having percentage defective equal to the specified LTPD will not be accepted (single sample)

Acceptance number (c) (r = c + 1)	Max. percentage defective (LTPD) or λ												
	30	20	15	10	7	5	3	2	1.5	1	0.7	0.5	0.3
	Minimum sample sizes (for device-hours required for life test, multiply by 1000)												
0	8	11	15	22	32	45	76	116	153	231	328	461	767
	(0.64)	(0.46)	(0.34)	(0.23)	(0.16)	(0.11)	(0.07)	(0.04)	(0.03)	(0.02)	(0.02)	(0.01)	(0.007)
1	13	18	25	38	55	77	129	195	258	390	555	778	1296
	(2.7)	(2.0)	(1.4)	(0.94)	(0.65)	(0.46)	(0.28)	(0.18)	(0.14)	(0.09)	(0.06)	(0.045)	(0.027)
2	18	25	34	52	75	105	176	266	354	533	759	1065	1773
	(4.5)	(3.4)	(2.24)	(1.6)	(1.1)	(0.78)	(0.47)	(0.31)	(0.23)	(0.15)	(0.11)	(0.080)	(0.045)
3	22	32	43	65	94	132	221	333	444	668	953	1337	2226
	(6.2)	(4.4)	(3.2)	(2.1)	(1.5)	(1.0)	(0.62)	(0.41)	(0.31)	(0.20)	(0.14)	(0.10)	(0.062)
4	27	38	52	78	113	158	265	398	531	798	1140	1599	2663
	(7.3)	(5.3)	(3.9)	(2.6)	(1.8)	(1.3)	(0.75)	(0.50)	(0.37)	(0.25)	(0.17)	(0.12)	(0.074)
5	31	45	60	91	131	184	308	462	617	927	1323	1855	3090
	(8.4)	(6.0)	(4.4)	(2.9)	(2.0)	(1.4)	(0.85)	(0.57)	(0.42)	(0.28)	(0.20)	(0.14)	(0.085)
6	35	51	68	104	149	209	349	528	700	1054	1503	2107	3509
	(9.4)	(6.6)	(4.9)	(3.2)	(2.2)	(1.6)	(0.94)	(0.62)	(0.47)	(0.31)	(0.22)	(0.155)	(0.093)
7	39	57	77	116	186	234	390	589	783	1178	1680	2355	3922
	(10.2)	(7.2)	(5.3)	(3.5)	(2.4)	(1.7)	(1.0)	(0.57)	(0.51)	(0.34)	(0.24)	(0.17)	(0.101)

[a]MIL-S-19500 and MIL-M-38510. Sample sizes are based on the Poisson exponential binomial limit. The minimum quality (approximate AQL) required to accept (on the average) 19 of 20 lots is shown in parentheses for information only. The life test failure rate, λ, shall be defined as the LTPD per 1000 h.

Figure 13.4 Operating characteristic (OC) curves for single sampling plans (10 per cent sample, acceptance number = 0)

obviously necessary. Also, standard QC sampling plans often do not provide assurance of long-term reliability.

Electronic component manufacturers sometimes quote quality levels in parts defective per million (p.p.m.). Typical figures for integrated circuits are 5 to 30 p.p.m. and lower for simpler components such as transitors, resistors and passive components.

All statistical sampling methods rely on the inspection or testing of samples which are drawn at random from the manufacturing batches, then using the mathematics of probability theory to make assertions about the quality of the batches. The sampling plans are based upon the idea of balancing the cost of test or inspection against minimizing the probability of the batch being accepted with an actual defective proportion higher than the AQL or LTPD.

However, optimizing the cost of test or inspection is not an appropriate objective. The logically correct objective in any test and inspection situation is to minimize the total cost of manufacture and support. When analysed from this viewpoint, the only correct decision is either to perform 100 per cent or zero test/inspection. There is no theoretical sample size between these extremes that will satisfy the criterion of total cost minimization. In addition, most modern manufacturing processes, particularly at the level of components, generate such small proportions defective (typically a few per million) that the standard statistical sampling methods such as AQL and LTPD cannot discriminate whether or not a batch is 'acceptable'.

The fundamental illogic of statistical acceptance sampling was first explained by Deming (Reference 11). If the cost of test or inspection of one item is k_1 the cost of a later failure caused by not inspecting or testing is k_2, and the average proportion defective is p, then if p is less than k_1/k_2 the correct (lowest total cost) strategy is not to test any. If p is greater than k_1/k_2 the correct strategy is to test all. This explanation represents the simplest case, but the principle is applicable generally: there is no alternative theoretically optimum sample size to test or inspect. The logic holds for inspection or test at any stage, whether of components entering a factory or of assembled products leaving.

For example, if an item costs \$50 to test at the end of production, and the average cost of failure in service is \$1000 (warranty, repair, spares, reputation), then $k_1/k_2 = 0.05$. So long as we can be confident that the production processes can ensure that fewer than 5 per cent will have defects that will cause failures in service, then the lowest cost policy is not to test or inspect any.

The logic of 0 or 100 per cent test or inspection is correct in stable conditions, that is, p, k_1 and k_2 are known and are relatively constant. Of course this is often not the case, but so long as we know that p is either much larger or much smaller than k_1/k_2 we can still base our test and inspection decisions on this criterion. If we are not sure, and particularly if the expected value of p can approach k_1/k_2, we should test/inspect 100 per cent.

There are some situations in which sampling is appropriate. In any production operation where the value of p is lower than the breakeven point but is uncertain, or might vary so that it approaches or exceeds this, then by testing or inspecting samples we might be able to detect such deviations and take corrective action or switch to 100 per cent inspection/test. It is important to note, however, that there can be no calculated optimum statistical sampling plan, since we do not know whether p changes or by how much it might. The amount and frequency of sampling can be determined only by practical thinking in relation to the processes, costs and risks involved. For example, if the production line in the example above produces items that are on average only 0.01 per cent defective, at a rate of 1000/week, we might decide to inspect or test 10/week as an audit, because 10 items can be dealt with or fitted into a test chamber with minimum interruption to production and delivery.

Items that operate or are used only once, such as rivets or locking fasteners, airbag deployment systems and pressure bursting discs can be tested only on a sample basis, since 100 per cent testing of production items is obviously not feasible. The optimum sample plan is still not statistically calculable, however, since the proportions defective are usually much lower than can be detected by any sample, and it will nearly always be highly uncertain and variable.

For these reasons statistical sampling is very little used nowadays.

IMPROVING THE PROCESS

When a production process has been started, and is under statistical control, it is likely still to produce an output with some variation, even if this is well within the allowable tolerance. Also, occasional special causes might lead to out-of-tolerance or otherwise defective items. It is important that steps are taken to improve variation and yield, even when these appear to be at satisfactory levels. Continuous improvement nearly always leads to reduced costs, higher productivity, and higher reliability. The concept of continuous improvement was first put forward by W. E. Deming (Reference 11), and taken up enthusiastically in Japan, where it is called *Kaizen*.

The idea of the quality loss function, due to Taguchi (page 147), also provides economic justification for continuous process improvement.

Methods that are available to generate process improvement are described below, and in References 3–7 and 12–17.

Simple charts

A variety of simple charting techniques can be used to help to identify and solve process variability problems. The Pareto chart (page 336) is often used as the starting point to identify the most important problems and the most likely causes. Where problems can be distributed over an area, for example defective solder joints on electronic assemblies, or defects in surface treatments, the *measles chart* is a useful aid. This consists simply of a diagram of the item, on which the locations of defects are marked as they are identified. Eventually a pattern of defect locations builds up, and this can indicate appropriate corrective action. For example, if solder defects cluster at one part of a PCB, this might be due to incorrect adjustment of the solder system.

The *cause-and-effect diagram* was invented by K. Ishikawa (Reference 15) as an aid to structuring and recording problem-solving and process improvement efforts. The diagram is also called a *fishbone*, or Ishikawa, diagram. The main problem is indicated on a horizontal line, and possible causes are shown as branches, which in turn can have sub-causes, indicated by sub-branches, and so on. An example is shown in Fig. 13.5.

Control charts

When process control charts are in use, they should be monitored continuously for trends that might indicate special causes of variation, so that the causes can be eliminated. Trends can be a continuous run high or low on the chart, or any cyclic pattern. A continuous high or low trend indicates a need for process or measuring adjustment. A cyclic trend might be caused by temperature fluctuations, process drifts between settings, operator changeover, change of material, etc. Therefore it is important to record supporting data on the SPC chart, such as time and date, to help with the identification of causes. When a process is being run on different machines the SPC charts of the separate processes should be compared, and all significant differences investigated.

Multi-vari charts

A *multi-vari chart* is a graphical method for identifying the major causes of variation in a process. Multi-vari charts can be used for process development and for problem solving, and they can be very effective in reducing the number of variables to include in a statistical experiment.

Multi-vari charts show whether the major causes of variation are spatial, cyclic or temporal. The principle of operation is very simple: the parameter being monitored is measured in different positions (e.g. locations for measurement of a dimension, hardness, etc.), at different points in the production cycle (e.g. batch number from

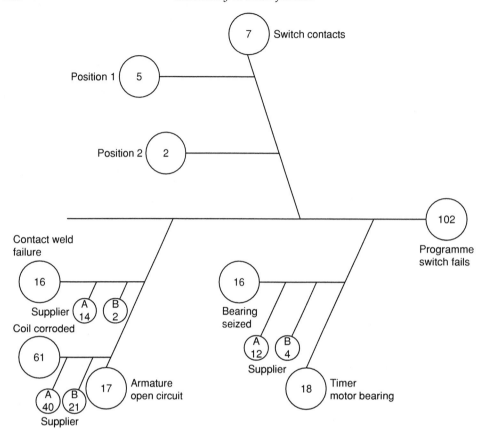

Figure 13.5 Cause and effect diagram

tool change), and at different times. The results are plotted as shown in Fig. 13.6, which shows a machined dimension plotted against two measurement locations, e.g. diameters at each end of a shaft, plotted against batch number from set-up. It shows that batch-to-batch variation is the most significant cause, with a significant pattern of end-to-end variation (taper). This information would then be used to seek the reasons for the major cause, if necessary by running further experiments. Finally, statistical experiments can be run to refine the process further, particularly if interactions are significant. The multi-vari method is described in Reference 16.

Statistical methods

The methods for analysis of variation, described in Chapter 5, can be used just as effectively for variation reduction in production processes. They should be used for process improvement, in the same way as for product and process initial design. If a particular process has been the subject of such experiments during development, then the results can be used to guide studies for further improvement.

The methods described above can also be used to identify the major causes of variation, prior to setting up statistical experiments. In this way the number of

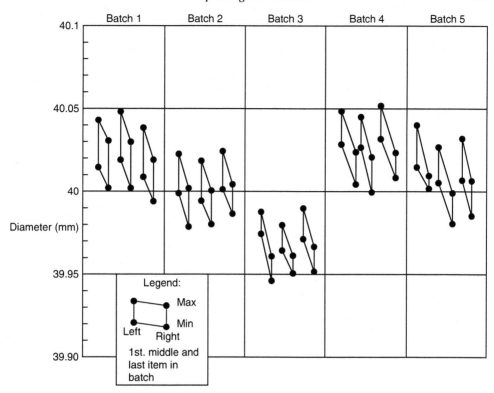

Figure 13.6 Multi-vari chart

variables to be investigated in the statistical experiment can be reduced, leading to cost savings.

'Zero defects'

The *'zero defects'* (ZD) approach to QC was developed in the United States in the 1960s. ZD is based very much upon setting QC targets, publicizing results and exhortation by award presentations and poster campaigns. Successes were claimed for ZD but it has always been controversial. The motivational basis is evangelical and emotional, and the initial enthusiasm is hard to sustain. There are few managers who can set up and maintain a ZD programme as originally publicized, and consequently the approach is now seldom used.

Quality circles

The quality circles movement started in Japan in the 1950s, and is now used world wide. The idea is largely based on Drucker's management teaching, developed and taught for QC application by W. E. Deming and K. Ishikawa. It uses the methods of operator control, consistent with Drucker's teaching that the most effective management is that nearest to the action; this is combined with basic SPC and problem solving methods to identify and correct problems at their sources. The operator is often the person most likely to understand the problems of the process he or she operates, and how to solve them. However, the individual operator does not usually have the authority or the motivation to make changes. Also, he or she might not be able to influence problems elsewhere in the manufacturing system. The quality circles system gives workers this knowledge and influence, by organizing them into small work groups, trained to monitor quality performance, to analyse problems and to recommend solutions to management.

Quality circle teams manage themselves, select their leaders and members, and the problems to be addressed. They introduce the improvements if the methods are under their control. If not, they recommend the solutions to management, who must respond positively.

It is therefore a very different approach to that of ZD, since it introduces quality motivation as a normal working practice at the individual and team level, rather than by exhortation from above. Whilst management must be closely involved and supportive, it does not have to be continually active and visible in the way ZD requires.

Quality circles are taught to use analytical techniques to help to identify problems and generate solutions. These are called the *seven tools of quality*. The seven tools are:

1. Brainstorm, to identify and prioritize problems.
2. Data collection.
3. Data analysis methods, including measles charts, trend charts and regression analysis.
4. Pareto chart.
5. Histogram.
6. Cause-and-effect (or Ishikawa) diagram.
7. Statistical process control (SPC) chart.

For example, the team would be trained to interpret SPC charts to identify special causes of variation, and to use cause-and-effect diagrams. The cause-and-effect diagram is used by the team leader, usually on a flip chart, to put on view the problem being addressed and the ideas and solutions that are generated by the team, during the 'brainstorming' stage.

Quality circles must be organized with care and with the right training, and must have full support from senior and middle management. In particular, quality circles recommendations must be carefully assessed and actioned whenever they will be effective, or good reasons given for not following the recommendation.

The concept is really straightforward, enlightened management applied to the quality control problem, and in retrospect it might seem surprising that it has taken so long to be developed. It has proved to be highly successful in motivating people to produce better quality, and has been part of the foundation of the Japanese industrial revolution since the Second World War. The quality circles approach can be very effective when there is no formal quality control organization, for example in a small company.

The quality circles approach to quality improvement is described fully in Reference 17.

QUALITY CONTROL IN ELECTRONICS PRODUCTION

Test methods

Electronic equipment production is characterized by very distinct assembly stages, and specialist test equipment has been developed for each of these. Since electronic production is so ubiquitous, and since test methods can greatly affect quality costs and reliability, it is appropriate to consider test methods for electronics in this context. Test equipment for electronic production falls into the following main categories:

1. Manual test equipment, which includes basic instruments such as 'AVO' meters (amps, volts, ohms) or digital multimeters (DMMs), oscilloscopes, spectrum analysers, waveform generators, and logic analysers, as well as special instruments, such as radio frequency testers, distortion meters, high voltage testers, optical signal testers, etc. Computer-based testing uses software that enables PCs to emulate test equipment.
2. Automatic Test Equipment

Automatic test equipment (ATE) is used for testing manufactured circuits, and also for in-service fault-finding and for testing repaired units. The main types of ATE for assembly testing are:

Vision systems

Vision systems refer generically to inspection systems that acquire an image and then analyse it. They do not actually test circuits, but they have become part of many production test sequences because of the great difficulty of human inspection of the large numbers of components, solder connections and tracks on modern circuits. *Automatic optical inspection* (AOI) machines are capable of scanning a manufactured circuit and identifying anomalies such as damaged, misplaced or missing components, faulty solder joints, solder spills across conductors, etc. X-ray systems (AXI) are also used, to enable inspection of otherwise hidden aspects such as BGA solder attachments and internal component problems. Other technologies, such as infrared and laser scanning, are also used.

In-circuit testers (ICT), manufacturing defects analysers (MDA)

ICT tests the functions of components within circuits, on loaded circuit boards. It does not test the circuit function. The ICT machine accesses the components, one at a time, via a test fixture (sometimes referred to as a 'bed of nails' fixture), which consists of a large number of spring loaded contact pins, spaced to make contact with the appropriate test points for each component, for example the two ends of a resistor or the pin connections for an IC. ICT does not test circuit-level aspects such as tolerance mismatch, timing, interference, etc. MDAs are similar but lower cost machines, with capabilities to detect only manufacturing-induced defects such as opens, shorts, and missing components: justification for their use instead of ICT is the fact that, in most modern electronics assembly, such defects are relatively more common than faulty components. *Flying probe testers* (also called *fixtureless testers*) perform the same functions, but access test points on the circuit using probes that are rapidly moved between points, driven by a programmed high-speed positioning system. The advantage over ICT/MDA is the fact that the probe programming is much less expensive and more adaptable to circuit changes than are expensive multipin ICT/MDA adaptors, which must be designed and built for each circuit to be tested. They can also gain access to test points that are difficult to access via a bed-of-nails adaptor.

Functional testers (FT)

Functional testers access the circuit, at the circuit board or assembly level, via the input and output connectors or via a smaller number of spring loaded probes. They perform a functional test, as driven by the test software. Functional testers usually include facilities for diagnosing the location of causes of incorrect function. There is a wide range, from low-cost bench-top ATE for use in development labs, in relatively low complexity/low production rate manufacture, in-service tests and in repair shops, to very large, high-speed high-capability systems. The modern trend is for production ATE to be specialized, and focussed at defined technology areas, such as computing, signal processing, etc. Some ATE for circuit testing during manufacture includes combined ICT and FT.

Electronics testing is a very fast-moving technology, driven by the advances in circuit performance, packaging and connection technology and production economics. Reference 18 provides an introduction.

It is very important to design circuits to be testable by providing test access points and by including additional circuitry, or it will not be possible to test all functions or to diagnose the causes of certain failures. Design for testability can have a major impact on quality costs and on future maintenance costs. These aspects are covered in more detail in Chapters 9 and 14 and in Reference 18.

The optimum test strategy for a particular electronic product will depend upon factors such as component quality, quality of the assembly processes, test equipment capability, costs of diagnosis and repair at each stage, and production throughput. Generally, the earlier a defect is detected, the lower will be the cost. A common rule of thumb states that the cost of detecting and correcting a defect increases by a factor of 10 at each test stage (component/PCB, ICT, functional test stages, in-service). Therefore the test strategy must be based on detecting and correcting defects as early as practicable in the production cycle, but defect probability, detection probability and test cost must also be considered.

The many variables involved in electronic equipment testing can be assessed by using computer models of the test options, under the range of assumed input conditions. Figure 13.7 shows a typical test flow arrangement.

Test results must be monitored continuously to ensure that the process is optimised in relation to total costs, including in-service costs. There will inevitably be variation over time. It is important to analyse the causes of failures at later test stages, to determine whether they could have been detected and corrected earlier.

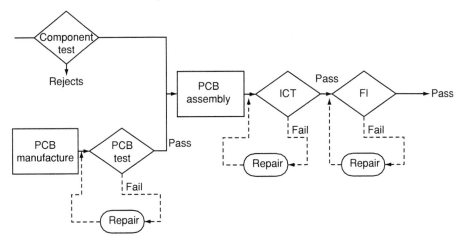

Figure 13.7 Electronic equipment test strategy

Throughout, the causes of defects must be rapidly detected and eliminated, and this requires a very responsive data collection and analysis system (see page 396). Some ATE systems provide data logging and analysis, and data networking between test stations is also available.

References 18 and 19 provide introductions to electronic test methods and economics.

Reliability of connections

Complete opens and shorts will nearly always be detected during functional test, but open-circuit and intermittent failures can occur in use due to corrosion or fatigue of joints which pass initial visual and functional tests. The main points to watch to ensure solder joint reliability are:

1. *Process control.* Solder temperature, solder mix (when this is variable, e.g. in wave solder machines), soldering time, fluxes, PCB cleaning.
2. *Component mounting.* The solder joint should not be used to provide structural support, particularly in equipment subject to vibration or shock, or for components of relatively large mass, such as trimpots and some capacitors.
3. *Component preparation.* Component connections must be clean and wettable by the solder. Components which have been stored unpackaged for more than a few days or packaged for more than six months require special attention, as oxide formation may inhibit solderability, particularly on automatic soldering systems, where all joints are subject to the same time in the solder wave or oven. If necessary, such components should have their leads cleaned and retinned prior to assembly. Components should be subject to sampling tests for solderability, as near in time to the assembly stage as practicable.
4. *Solder joint inspection.* Inspectors performing visual inspection of solder joints are typically about 80 per cent effective in seeing joints which do not pass visual inspection standards. Also, it is possible to have unreliable joints which meet

appearance criteria. If automatic testing for opens and shorts is used instead of 100 per cent visual inspection, it will not show up marginal joints which can fail later.

Surface mounted devices, as described in Chapter 9, present particular problems, since the solder connections are so much smaller and more closely spaced. Manual soldering is not practicable, so automatic placement and soldering systems must be used. Visual inspection is difficult and this has resulted in the development of semi-automatic and automated optical inspection systems, though these cannot be considered to be totally reliable. SMD solderability and soldering must be very carefully controlled to minimize the creation of defective joints.

STRESS SCREENING

Stress screening, or *environmental stress screening* (ESS), is the application of stresses that will cause defective production items which pass other tests to fail on test, while not damaging or reducing the useful life of good ones. It is therefore a method for improving reliability and durability in service. Other terms are sometimes used for the process, the commonest being *burn-in*, particularly for electronic components and systems, for which the stresses usually applied are high temperature and electrical stress (current, voltage). The stress levels and durations to be applied must be determined in relation to the main failure-generating processes and the manufacturing processes that could cause weaknesses. Stress screening is normally a 100 per cent test, i.e., all manufactured items are tested. Stress screening is applied mainly to electronic components and assemblies, but it should be considered for non-electronic items also, for example precision mechanisms (temperature, vibration) and high pressure tests for pneumatics and hydraulics to check for leaks or other weaknesses.

ESS guidelines have been developed for electronic components and systems. The US Navy has published guidelines (NAVMAT P-9492), and the US DOD published MIL-STD-2164 (ESS Guidelines for Electronics Production), but these were inflexible and the stress levels specified were not severe (typically temperature cycling between 20 °C to 60 °C for 8 h, with random or fixed frequency vibration in the range 20–2000 Hz, and the equipment not powered or monitored). The US Institute for Environmental Sciences and Technology (IEST) developed more detailed guidelines in 1990 (Environmental Stress Screening of Electronic Hardware (ESSEH)), to cover both development and manufacturing tests. These recommended stress regimes similar to the military ESS guidelines. The details were based to a large extent on industry feedback of the perceived effectiveness of the methods that had been used up to the preparation of the guidelines, so they represented past experience, particularly of military equipment manufacture. They were not based upon the results of development testing on a product-by-product basis, or on any scientific study of technologies and failure mechanisms.

Highly accelerated stress screening

Highly accelerated stress screening (HASS) is an extension of the HALT principle, as described in Chapter 11, that makes use of very high combined stresses. No 'guidelines' are published to recommend particular stresses and durations. Instead, the stresses, cycles and durations are determined separately for each product (or group of similar products) by applying HALT during development. HALT shows up the product weak points, which are then strengthened as far as practicable so that failures will occur only well beyond the envelope of expected in-service combined stresses. The stresses that are then applied during HASS are higher than the operating limit, and extend into the lower tail of the distribution of the permanent failure limit. It is essential that the equipment under test is operated and monitored throughout.

The stresses applied in HASS, like those applied in HALT, *are not designed to represent worst-case service conditions*. They are designed to precipitate failures which would otherwise occur in service. This is why they can be developed only by applying HALT in development, and why they must be specific to each product design. Because the stress levels are so high, they cannot be applied safely to any design that has not been ruggedized through HALT.

Obviously the determination of this stress–time combination cannot be exact, because of the uncertainty inherent in the distribution of strength. However, by exploring the product's behaviour on test, we can determine appropriate stress levels and durations. The durations will be short, since the stress application rates are very high and there is usually no benefit to be gained by, for example, operating at constant high or low temperatures for longer than it takes for them to stabilize. Also, only a few stress cycles will be necessary, typically one to four.

When stresses above the operating limit are applied, it will not be possible to perform functional tests. Therefore the stresses must then be reduced to levels below the operating limit. The functional test will then show which items have been caused to fail, and which have survived. The screening process is therefore in two stages: the *precipitation screen* followed by the *detection screen*, as shown in Fig. 13.8.

The total test time for HASS is much less than for conventional ESS: *a few minutes vs. many hours*. HASS is also much more effective at detecting defects. *Therefore it is far more cost effective, by orders of magnitude*. HASS is applied using the same facilities, particularly environmental chambers, as used for HALT. Since the test times are so short, it is often convenient to use the same facilities during development and in production, leading to further savings in relation to test and monitoring equipment, interfaces, etc. The HASS concept can be applied to any type of product or technology. It is by no means limited to electronic assemblies. If the design can be improved by HALT, as described in Chapter 11, then in principle manufacturing quality can be improved by HASS. The HASS approach to manufacturing stress screening is described fully in Reference 20.

If ESS/HASS shows up very few defects, it is either insufficiently severe or the product being screened is already highly reliable. Obviously the latter situation is preferable, and all failures during screening should be analysed to determine if QC methods should have prevented them or discovered them earlier, particularly prior

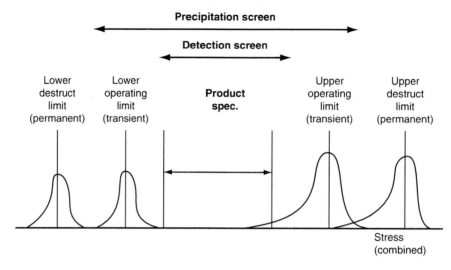

Figure 13.8   HASS

to assembly. At this stage, repairs are expensive, so eliminating the need for them by using high quality components and processes during production is a worthwhile objective.

Screening is expensive, so its application must be carefully monitored. Failure data should be analysed continuously to enable the process to be optimized in terms of operating conditions and duration. Times between failures should be logged to enable the process to be optimized.

Screening must be considered in the development of the production test strategy. The costs and effects in terms of reduced failure costs in service and the relationships with other test methods and QC methods must be assessed as part of the integrated quality assurance approach. Much depends upon the reliability requirement and the costs of failures in service. Also, screening should ensure that manufacturing quality problems are detected before shipment, so that they can be corrected before they affect much production output. With large-scale production, particularly of commercial and domestic equipment, screening of samples is sometimes applied for this purpose.

Reference 21 describes methods and analytical techniques for screening.

PRODUCTION FAILURE REPORTING ANALYSIS AND CORRECTIVE ACTION SYSTEM (FRACAS)

Failure reporting and analysis is an important part of the QA function. The system must provide for:

1. Reporting of all production test and inspection failures with sufficient detail to enable investigation and corrective action to be taken.
2. Reporting the results of investigation and action.

3. Analysis of failure patterns and trends, and reporting on these.
4. Continuous improvement ("*Kaizen*") by removal of causes of failures.

The FRACAS principles (Chapter 11, page 333) are equally applicable to production failures.

The data system should be computerized for economy and accuracy. Modern ATE sometimes includes direct test data recording and inputting to a central system by networking. Manual input to computer terminals connected to the network also speeds up the system, but data input using paper forms is still common. The data analysis system should provide Pareto analysis, probability plots and trend analyses for management reporting.

Production defect data reporting and analysis must be very quick to be effective. Trends should be analysed daily, or weekly at most, particularly for high rates of production, to enable timely corrective action to be taken. Production problems usually arise quickly, and the system must be able to detect a bad batch of components or an improperly adjusted process as soon as possible. Many problems become immediately apparent without the aid of a data analysis system, but a change of, say, 50 per cent in the proportion of a particular component in a system failing on test might not be noticed otherwise. The data analysis system is also necessary for indicating areas for priority action, using the Pareto principle of concentrating action on the few problem areas that contribute the most to quality costs. For this purpose longer term analysis, say monthly, is necessary.

Defective components should not be scrapped immediately, but should be labelled and stored for a period, say 1 to 2 months, so that they are available for more detailed investigation if necessary.

Production defect data should not be analysed in isolation by people whose task is primarily data management. The people involved (production, supervisors, QC engineers, test operators, etc.) must participate to ensure that the data are interpreted by those involved and that practical results are derived. The quality circles approach provides very effectively for this.

Production defect data are important for highlighting possible in-service reliability problems. Many in-service failure modes manifest themselves during production inspection and test. For example, if a component or process generates failures on the final functional test, and these are corrected before delivery, it is possible that the failure mechanism exists in products which pass test and are shipped. Metal surface protection and soldering processes present such risks, as can almost any component in electronic production. Therefore production defects should always be analysed to determine the likely effects on reliability and on external failure costs, as well as on internal production quality costs.

CONCLUSIONS

The modern approach to production quality control and improvement is based on the use of statistical methods and on organizing, motivating and training production people at all levels to work for continuously improving performance, of people and of processes. A very close link must exist between design and development of the

product and of the production processes, and the criteria and methods to be used to
control the processes. This integrated approach to management of the design and
production processes is described in more detail in Chapter 15, and also in References 11, 14 and 22. The journals of the main quality assurance professional societies
(e.g. References 23 and 24) also provide information on new developments.

BIBLIOGRAPHY

1. J. M. Juran, *Quality Control Handbook*, 3rd edn. McGraw-Hill (1974).
2. A. V. Feigenbaum, *Total Quality Control* 3rd edn. McGraw-Hill (1983).
3. J. M. Juran, F. M. Gryna, *Quality Planning and Analysis* 2nd edn. McGraw-Hill (1980).
4. E. L. Grant, R. S. Leavenworth, *Statistical Quality Control* 5th edn. McGraw-Hill Kogakusha Ltd (1980).
5. D. C. Montgomery, *Introduction to Statistical Quality Control* 2nd edn. Wiley (1991).
6. J. S. Oakland, R. F. Followell, *Statistical Process Control, a Practical Guide* 2nd edn. Heinemann (1990).
7. T. P. Ryan, *Statistical Methods for Quality Improvement*. Wiley–Interscience (1989).
8. A. J. Duncan, *Quality Control and Industrial Statistics*, 5th edn. Irwin (1986).
9. ANSI/ASQ Z1–4, *Sampling Procedures and Tables for Inspection by Attributes*.
10. British Standard BS 6001, *Sampling Procedures and Tables for Inspection by Attributes*.
11. W. E. Deming, *Out of the Crisis*. MIT University Press (1987).
12. M. Imai, *Gemba Kaizen*. McGraw-Hill (1997).
13. B. Thomas, *The Human Dimension of Quality*. McGraw-Hill (1995).
14. B. Bergman and B. Klefsjö, *Quality: from Customer Needs to Customer Satisfaction*. McGraw-Hill (1994).
15. K. Ishikawa, *Guide to Quality Control*. Chapman and Hall (1991).
16. K. R. Bhote, *World Class Quality*. American Management Association (1988).
17. D. C. Hutchins, *Quality Circles Handbook*. Pitman (1985).
18. P. D. T. O'Connor, *Test Engineering*. Wiley (2001).
19. B. Davis, *The Economics of Automatic Test Equipment* 2nd edn. McGraw-Hill (1994).
20. G. Hobbs, *Accelerated Reliability Engineering: HALT and HASS*. Wiley (2001).
21. F. Jensen, N. E. Peterson, *Burn-in: An Engineering Approach to the Design and Analysis of Burn-in Procedures*. Wiley (1983).
22. M. Groocock, *The Chain of Quality*. Wiley (1986).
23. *Quality Progress*, Journal of the American Society for Quality.
24. *Quality Assurance*, Journal of the Institute of Quality Assurance (UK).

QUESTIONS

1. A machined dimension on a component is specified as 12.50 mm ± 0.10 mm. A
   preliminary series of 10 samples, each of 5 components, is taken from the process,
   with measurements of the dimension as follows:

Sample		Dimensions (mm)			
1	12.55	12.51	12.48	12.55	12.46
2	12.54	12.56	12.51	12.54	12.47
3	12.53	12.46	12.49	12.45	12.50
4	12.55	12.55	12.49	12.55	12.47
5	12.49	12.52	12.49	12.48	12.48

6	12.51	12.54	12.51	12.52	12.45
7	12.53	12.52	12.49	12.46	12.50
8	12.50	12.55	12.52	12.44	12.46
9	12.48	12.52	12.54	12.49	12.50
10	12.50	12.50	12.49	12.54	12.54

(a) Use the averages and ranges for each sample to assess the capability of the process (calculated C_p and C_{p_k}).

Use the relationship (taken from BS 5700) that, for samples of size 5, standard deviation = average range $\times$ 2.326.

(b) Suggest any action that may be necessary.

2. Sketch the two charts used for statistical process control. What is the primary objective of the method?

3. Explain why statistical acceptance sampling is not an effective method for monitoring processes quality.

4. A particular component is used in large quantities in an assembly. It costs very little, but some protection against defective components is needed as they can be easily detected only after they have been built-in to an expensive assembly, which is then scrap if it contains a defective component. 100 per cent inspection of the components is not possible as testing at the component level would be destructive, but the idea of acceptance sampling is attractive.

(a) How would you decide on the AQL to use?

(b) If the AQL was 0.4 per cent defective and these components were supplied in batches of 2500, what sampling plan would you select from Table 13.1? (Batches of 2500 require sample size code K for general inspection, level II.)

(c) What would you do if you detected a single defective component in the sample?

(d) What would you do if this plan caused a batch to be rejected?

5. A manufacturing line produces car radios. The cost of the final test is $15, and the average proportion found to be not working is 0.007, though on some batches it is as high as 0.02. If the total cost of selling a non-working radio (replacement, administration, etc.) is $400, comment on the continued application of the test.

6. List the '7 tools of quality' as applied in the Quality Circles approach. Explain how they are applied by a Quality Circles team.

7. Briefly describe the three main approaches used for automatic inspection and test of modern electronic circuits. Sketch a typical inspection/test flow for a line producing circuit assemblies.

8. Describe the stresses that are typically applied to electronic hardware during environmental stress screening (ESS) in manufacturing.

9. (a) Define the main advantages and disadvantages of applying environmental stress screening (ESS) to electronic assemblies in production.

   (b) Discuss why it would be wrong to impose an ESS using MIL-HDBK-781 type environmental profiles.

10. How does highly accelerated stress screening (HASS) differ from conventional ESS? How is it related to HALT in development?

11. (a) Explain the difference between reducing failure rate by "burn-in" and by reliability growth in service.

   (b) Assuming that you are responsible for the reliability of a complex electronic system that is about to go into production, outline the way that you would set up "burn-in" testing for purchased components, sub-assemblies and the completed product, in each case stating the purpose of the test and the criteria you would consider in deciding its duration.

14

Maintainability, Maintenance and Availability

INTRODUCTION

Most engineered systems are maintained, i.e. they are repaired when they fail, and work is performed on them to keep them operating. The ease with which repairs and other maintenance work can be carried out determines a system's maintainability.

Maintained systems may be subject to *corrective* and *preventive* maintenance (CM and PM). Corrective maintenance includes all action to return a system from a failed to an operating or available state. The amount of corrective maintenance is therefore determined by reliability. Corrective maintenance action usually cannot be planned; we must repair failures when they occur, though sometimes repairs can be deferred.

Corrective maintenance can be quantified as the *mean time to repair* (MTTR). The time to repair, however, includes several activities, usually divided into three groups:

1. Preparation time: finding the person for the job, travel, obtaining tools and test equipment, etc.
2. Active maintenance time: actually doing the job.
3. Delay time (logistics time): waiting for spares, etc., once the job has been started.

Active maintenance time includes time for studying repair charts, etc., before the actual repair is started, and time spent in verifying that the repair is satisfactory. It might also include time for post-repair documentation when this must be completed before the equipment can be made available, e.g. on aircraft. Corrective maintenance is also specified as a *mean active corrective maintenance time* (MACMT), since it is only the active time (excluding documentation) that the designer can influence.

Preventive maintenance seeks to retain the system in an operational or available state by preventing failures from occurring. This can be by servicing, such as cleaning and lubrication, or by inspection to find and rectify incipient failures, e.g. by crack detection or calibration. Preventive maintenance affects reliability directly. It is planned and should be performed when we want it to be. Preventive maintenance is measured by the time taken to perform the specified maintenance tasks and their specified frequency.

Maintainability affects availability directly. The time taken to repair failures and to carry out routine preventive maintenance removes the system from the available state. There is thus a close relationship between reliability and maintainability, one affecting the other and both affecting availability and costs. In the steady state, i.e. after any transient behaviour has settled down and assuming that maintenance actions occur at a constant rate:

$$\text{Availability} = \frac{\text{MTBF}}{\text{MTBF} + \text{MTTR}} \times \frac{\text{PM cycle}}{\text{PM cycle} + \text{PM time (total)}}$$

where PM cycle = time during which all PM tasks are performed.
Availability analysis and prediction were covered in Chapter 6.

The maintainability of a system is clearly governed by the design. The design determines features such as accessibility, ease of test, diagnosis and repair and requirements for calibration, lubrication and other preventive maintenance actions.

This chapter describes how maintainability can be optimized by design, and how it can be predicted and measured. It also shows how plans for preventive maintenance can be optimized in relation to reliability, to minimize downtime and costs.

MAINTENANCE TIME DISTRIBUTIONS

Maintenance times tend to be lognormally distributed (see Fig. 14.1). This has been shown by analysis of data. It also fits our experience and intuition that, for a task or group of tasks, there are occasions when the work is performed rather quickly, but it is relatively unlikely that the work will be done in much less time than usual, whereas it is relatively more likely that problems will occur which will cause the work to take much longer than usual.

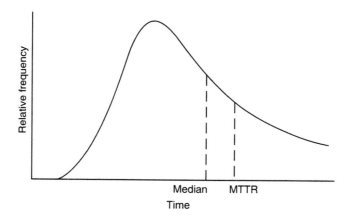

Figure 14.1   The lognormal distribution of maintenance times

In addition to the job-to-job variability, leading typically to a lognormal distribution of repair times, there is also variability due to learning. Depending upon how data are collected, this variability might be included in the job-to-job variability, e.g. if technicians of different experience are being used simultaneously. However, both the mean time and the variance should reduce with experience and training.

The properties of the lognormal distribution are described in Chapter 2. Lognormal or Weibull probability paper can be used for plotting maintenance time data (Chapter 3).

PREVENTIVE MAINTENANCE STRATEGY

The effectiveness and economy of preventive maintenance can be maximized by taking account of the time-to-failure distributions of the maintained parts and of the failure rate trend of the system.

In general, if a part has a decreasing hazard rate, any replacement will increase the probability of failure. If the hazard rate is constant, replacement will make no difference to the failure probability. If a part has an increasing hazard rate, then scheduled replacement at any time will in theory improve reliability of the system. However, if the part has a failure-free life (Weibull $\gamma > 0$), then replacement before this time will ensure that failures do not occur. These situations are shown in Fig. 14.2.

These are theoretical considerations. They assume that the replacement action does not introduce any other defects and that the time-to-failure distributions are exactly defined. As explained in Chapters 2 and 6, these assumptions must not be made without question. However, it is obviously of prime importance to take account of the time-to-failure distributions of parts in planning a preventive maintenance strategy.

In addition to the effect of replacement on reliability as theoretically determined by considering the time-to-failure distributions of the replaced parts, we must also take account of the effects of the maintenance action on reliability. For example, data might show that a high pressure hydraulic hose has an increasing hazard rate after a failure-free life, in terms of hose leaks. A sensible maintenance policy might therefore be to replace the hose after, say, 80 per cent of the failure-free life. However, if the replacement action increases the probability of hydraulic leaks from the hose end connectors, it might be more economical to replace hoses on failure.

The effects of failures, both in terms of effects on the system and of costs of downtime and repair, must also be considered. In the hydraulic hose example, for instance, a hose leak might be serious if severe loss of fluid results, but end connector leaks might generally be only slight, not affecting performance or safety. A good example of replacement strategy being optimized from the cost point of view is the replacement of incandescent and fluorescent light units. It is cheaper to replace all units at a scheduled time before an expected proportion will have failed, rather than to replace each unit on failure, in large installations such as offices and street lights. However, at home we would replace only on failure.

In order to optimize preventive replacement, it is therefore necessary to know the following for each part:

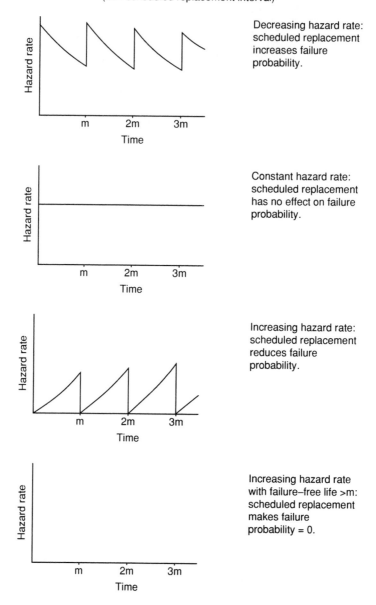

Figure 14.2 Theoretical reliability and scheduled replacement relationships

1. The time-to-failure distribution parameters for the main failure modes.
2. The effects of all failure modes.
3. The cost of failure.
4. The cost of scheduled replacement.
5. The likely effect of maintenance on reliability.

We have considered so far parts which do not give any warning of the onset of failure. If incipient failure can be detected, e.g. by inspection, non-destructive testing, etc., we must also consider:

6. The rate at which defects propagate to cause failure.
7. The cost of inspection or test.

Note that, from 2, an FMECA is therefore an essential input to maintenance planning.

This systematic approach to maintenance planning, taking account of reliability aspects, is called *reliability centred maintenance* (RCM). Figure 14.3 shows the basic logic of the method. RCM is widely applied, for example on aircraft, factory systems, etc. It is described in Reference 2.

Example 14.1

A flexible cable on a robot assembly line has a time-to-failure distribution which is Weibull, with $\gamma = 150\,\text{h}, \beta = 1.7$ and $\eta = 300\,\text{h}$. If failure occurs whilst in use the

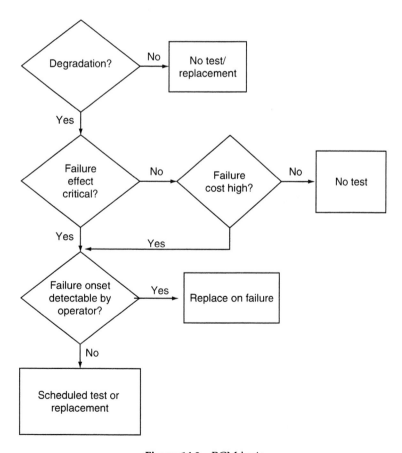

Figure 14.3   RCM logic

cost of stopping the line and replacing the cable is $5000. The cost of replace-
ment during scheduled maintenance is $500. If the line runs for 5000 hours
a year and scheduled maintenance takes place every week (100 hours), what
would be the annual expected cost of replacement at one-weekly or two-weekly
intervals?

With no scheduled replacement the probability of a failure occurring in t hours
will be

$$1 - \exp\left[-\left(\frac{t - 150}{300}\right)^{1.7}\right]$$

(2.32)

With scheduled replacement after m hours, the scheduled maintenance cost in
5000 h will be

$$\frac{5000}{m} \times 500 = \frac{2.5 \times 10^6}{m}$$

and the expected failure cost in each scheduled replacement interval will be (assuming not more than one failure in any replacement interval):

$$5000\left\{1 - \exp\left[1 - \left(\frac{m - 150}{300}\right)^{1.7}\right]\right\}$$

Then the total cost per year

$$C = \frac{2.5 \times 10^6}{m} + \frac{5000 \times 5000}{m}\left\{1 - \exp\left[-\left(\frac{m - 150}{300}\right)^{1.7}\right]\right\}$$

Results are as follows:

m	No. of scheduled replacements	Expected failures	C
100	50	0	$25 000
200	25	1.2	$18 304
400	12	6.5	$38 735

Therefore the optimum policy might be to replace the cables at alternate scheduled
maintenance intervals, taking a slight risk of failure. (Note that the example assumes
that not more than one failure occurs in any scheduled maintenance interval. If m is
only a little more than γ this is a reasonable assumption.)

A more complete analysis could be performed using Monte Carlo simulation
(Chapter 6 page 190). We could then take account of more detailed maintenance
policies; e.g. it might be decided that if a cable had been replaced due to failure

shortly before a scheduled maintenance period, it would not be replaced at that time.

Practical implications

The time-to-failure patterns of components in a system therefore largely dictate the optimum maintenance policy. Generally, since electronic components do not wear out, scheduled tests and replacements do not improve reliability. Indeed they are more likely to induce failures (real or reported). Electronic equipment should only be subjected to periodic test or calibration when drifts in parameters or other failures can cause the equipment to operate outside specification without the user being aware. Built-in test and autocalibration can reduce or eliminate the need for periodic test.

Mechanical equipment subject to wear, corrosion, fatigue, etc., should be considered for preventive maintenance.

FMECA AND FTA IN MAINTENANCE PLANNING

The FMECA is an important prerequisite to effective maintenance planning and maintainability analysis. As shown earlier, the effects of failure modes (costs, safety implications, detectability) must be considered in determining scheduled maintenance requirements. The FMECA is also a very useful input for preparation of diagnostic procedures and checklists, since the likely causes of failure symptoms can be traced back using the FMECA results. When a fault tree analysis (FTA) has been performed it can also be used for this purpose.

MAINTENANCE SCHEDULES

When any scheduled maintenance activity is determined to be necessary, we must also determine the most suitable intervals between its performance. Maintenance schedules should be based upon the most appropriate time or other base. These can include:

—Road and rail vehicles: distance travelled.
—Aircraft: hours flown, numbers of takeoff/landing cycles.
—Electronic equipment: hours run, numbers of on/off cycles.
—Fixed systems (radars, rail infrastructure, etc.): calendar time.

The most appropriate base is the one that best accounts for the equipment's utilization in terms of the causes of degradation (wear, fatigue, parameter change, etc.), and is measured. For example, we measure the distance travelled by our cars, and most degradation is related to this. On the other hand, there is no point in setting a calibration schedule based upon running time for a measuring instrument unless an automatic or manual record of its utilization were maintained, so these are usually calendar-based.

TECHNOLOGY ASPECTS

Mechanical

Monitoring methods are used to provide periodic or continuous indications of the condition of mechanical components and systems. These include:

—Non-destructive test (NDT) for detection of fatigue cracks.
—Temperature and vibration monitors on bearings, gears, engines, etc.
—Oil analysis, to detect signs of wear or breakup in lubrication and hydraulic systems.

Electronic and electrical

Electronic components and assemblies generally do not degrade in service, so long as they are protected from environments such as corrosion. Electronic components and connections do not suffer from wear or fatigue, except as discussed in Chapter 9, so there is very seldom a pronounced 'wearout' phase in which failures become more likely or frequent. Therefore, apart from calibration for items like measuring instruments, scheduled tests are seldom appropriate.

"No fault found"

A large proportion of the reported failures of many electronic systems are not confirmed on later test. These occurrences are called *no fault found* (NFF) or *re-test OK* (RTOK) faults. There are several causes of these, including:

—Intermittent failures, such as components that fail under certain conditions (temperature, etc.), intermittent opens on conductor tracks or solder connections, etc.
—Tolerance effects, which can cause a unit to operate correctly in one system or environment but not in another.
—Connector troubles. The failure seems to be cleared by replacing a unit, when in fact it was caused by the connector that is disturbed to replace it.
—Built-in test (BIT) systems which falsely indicate failures that have not actually occurred (see below).
—Failures that have not been correctly diagnosed and repaired, so that the symptoms recur.
—Inconsistent test criteria between the in-service test and the test applied during diagnosis elsewhere such as the repair depot.
—Human error or inexperience.
—In some systems, the diagnosis of which item (card, box) has failed might be ambiguous, so more than one is replaced even though only one has failed. Sometimes it is quicker and easier for the technician to replace multiple items,

rather than trying to find out which has failed. In these cases multiple units are sent for diagnosis and repair, resulting in a proportion being classified NFF. (Returning multiple units can sometimes be justified economically. For example, it might be appropriate to spend the minimum time diagnosing the cause of a problem on a system such as an aircraft or oil rig, in order to return the system to operation as soon as possible.)

The proportion of reported failures that are caused in these ways can be very high, often over 50 per cent and sometimes up to 80 per cent. This can generate high costs, in relation to warranty, spares, support, test facilities, etc. NFF rates can be minimized by effective management of the design in relation to in-service test, and of the diagnosis and repair operations. Stress screening of repaired items can also reduce the proportion of failures that are not correctly diagnosed and repaired.

Software

As discussed in Chapter 10 software does not fail in the ways that hardware can, so there is no 'maintenance'. If it is found to be necessary to change a program for any reason (change in system requirements, correction of a software error), this is really redesign of the program, not repair. So long as the change is made to all copies in use, they will all work identically, and will continue to do so.

Built-in test (BIT)

Complex electronic systems such as laboratory instruments, avionics and process control systems now frequently include built-in test (BIT) facilities. BIT consists of additional hardware (and often software) which is used for carrying out functional test on the system. BIT might be designed to be activated by the operator, or it might monitor the system continuously or at set intervals.

BIT can be very effective in increasing system availability and user confidence in the system. However, BIT inevitably adds complexity and cost and can therefore increase the probability of failure. Additional sensors might be needed as well as BIT circuitry and displays. In microprocessor-controlled systems BIT can be largely implemented in software.

BIT can also adversely affect apparent reliability by falsely indicating that the system is in a failed condition. This can be caused by failures within the BIT, such as failures of sensors, connections, or other components. BIT should therefore be kept simple, and limited to monitoring of essential functions which cannot otherwise be easily monitored.

It is important to optimize the design of BIT in relation to reliability, availability and cost. Sometimes BIT performance is specified (e.g. '90 per cent of failures must be detected and correctly diagnosed by BIT'). An FMECA can be useful in checking designs against BIT requirements since BIT detection can be assessed against all the important failure modes identified.

CALIBRATION

Calibration is the regular check or test of equipment used for measuring physical parameters, by making comparisons against standard sources. Calibration is applied to basic measurement tools such as micrometers, gauges, weights and torque wrenches, and to transducers and instruments that measure parameters such as flow rate, electrical potential, current and resistance, frequency, etc. Therefore calibration may involve simple comparisons (weight, timekeeping, etc.) or more complex testing (radio frequency, engine torque, etc.).

Whether an item needs to be calibrated or not depends primarily upon its application, and also upon whether or not inaccuracy would be apparent during normal use. Any instrument used in manufacturing should be calibrated, to ensure that correct measurements are being made. Calibration is often a legal requirement, for example of measuring systems in food or pharmaceutical production and packaging, retailing, safety–critical processes, etc.

Calibration requirements and systems are described in Reference 12.

MAINTAINABILITY PREDICTION

Maintainability prediction is the estimation of the maintenance workload which will be imposed by scheduled and unscheduled maintenance. A standard method used for this work is US MIL-HDBK-472 (Reference 9), which contains four methods for predicting the mean time to repair (MTTR) of a system. Method II is the most frequently used. This is based simply on summing the products of the expected repair times of the individual failure modes and dividing by the sum of the individual failure rates, i.e.:

$$\text{MTTR} = \frac{\Sigma(\lambda t_r)}{\Sigma\lambda} \tag{14.1}$$

The same approach is used for predicting the mean preventive maintenance time, with λ replaced by the frequency of occurrence of the preventive maintenance action.

MIL-HDBK-472 describes the methods to be used for predicting individual task times, based upon design considerations such as accessibility, skill levels required, etc. It also describes the procedures for calculating and documenting the analysis, and for selection of maintenance tasks when a sampling basis is to be used (method III), rather than by considering all maintenance activities, which is impracticable on complex systems.

MAINTAINABILITY DEMONSTRATION

A standard approach to maintainability demonstration is MIL-HDBK-470 (Reference 10). The technique is the same as for maintainability prediction using method III of MIL-HDBK-472, except that the individual task times are measured rather than estimated from the design. Selection of task times to be demonstrated might be by agreement or by random selection from a list of maintenance activities.

DESIGN FOR MAINTAINABILITY

It is obviously important that maintained systems are designed so that maintenance tasks are easily performed, and that the skill levels required for diagnosis, repair and scheduled maintenance are not too high, considering the experience and training of likely maintenance personnel and users. Features such as ease of access and handling, the use of standard tools and equipment rather than specials, and the elimination of the need for delicate adjustment or calibration are desirable in maintained systems. As far as is practicable, the need for scheduled maintenance should be eliminated. Whilst the designer has no control over the performance of maintenance people, he or she can directly affect the inherent maintainability of a system.

Design rules and checklists should include guidance, based on experience of the relevant systems, to aid design for maintainability and to guide design review teams.

Design for maintainability is closely related to design for ease of production. If a product is easy to assemble and test maintenance will usually be easier. Design for testability of electronic circuits is particularly important in this respect, since circuit testability can greatly affect the ease and accuracy of failure diagnosis, and thus maintenance and logistics costs. Design of electronic equipment for testability is covered in more detail in Chapter 9.

Interchangeability is another important aspect of design for ease of maintenance of repairable systems. Replaceable components and assemblies must be designed so that no adjustment or recalibration is necessary after replacement. Interface tolerances must be specified to ensure that replacement units are interchangeable.

INTEGRATED LOGISTIC SUPPORT

Integrated logistic support (ILS) is a concept developed by the military, in which all aspects of design and of support and maintenance planning are brought together, to ensure that the design and the support system are optimized. Operational effectiveness, availability, and total costs of deployment and support are all considered. The approach is described in US-MIL-HDBK 1388 (Reference 13).

ILS, and the associated logistic support analysis (LSA), require inputs of reliability and maintainability data and forecasts, as well as data on costs, weights, special tools and test equipment, training requirements, etc. MIL-HDBK-1388 requires that all analyses are computerized, and lays down standard input and output formats. Several commercial computer programs have been developed for the tasks.

ILS/LSA outputs are obviously very sensitive to the accuracy of the inputs. In particular, reliability forecasts can be highly uncertain, as explained in Chapter 6. Therefore such analyses, and decisions based upon them, should take full account of these uncertainties.

BIBLIOGRAPHY

1. J. D. Patton, *Maintainability and Maintenance Management*. Instrument Society of America, Pittsburgh (1980).
2. A. S. Goldman and T. B. Slattery, *Maintainability: A Major Element of Systems Effectiveness*. Wiley (1964).
3. B. S. Blanchard and E. E. Lowery, *Maintainability Principles and Practices*. McGraw-Hill (1969).
4. C. E. Cunningham and W. Cox, *Applied Maintainability Engineering*. Wiley (1972).
5. J. Moubray, *Reliability Centred Maintenance*. Butterworth-Heinemann (1991).
6. A. K. S. Jardine, *Maintenance, Replacement and Reliability*. Pitman (1973).
7. D. J. Smith, *Reliability and Maintainability in Perspective*, 2nd edn. Macmillan (1985).
8. J. E. Arsenault and J. A. Roberts, *Reliability and Maintainability of Electronic Systems*. Computer Science Press (1980).
9. US MIL-HDBK-472: *Maintainability Prediction*. Available from the National Technical Information Service, Springfield, Virginia.
10. US MIL-HDBK-470: *Maintainability Program Requirements*. Available from the National Technical Information Service, Springfield, Virginia.
11. UK Defence Standard 00–40: *The Management of Reliability and Maintainability*. HMSO.
12. A. S. Morris, *Measurement and Calibration Requirements for Quality Assurance to ISO9000*. J.Wiley (1997).
13. US MIL-HDBK-1388: *Integrated Logistic Support*. Available from the National Technical Information Service, Springfield, Virginia.

QUESTIONS

1. Define and explain the following terms: (a) availability; (b) maintainability; (c) condition monitoring.

2. (a) Explain circumstances in which it is possible to improve maintainability to counteract poor reliability, and also identify circumstances where this approach is unrewarding.

   (b) The expression for steady-state availability of a single element is:

   $$A(t) = \frac{\mu}{\lambda + \mu}$$

   Explain what is meant by a steady state. State the meanings of the symbols μ and λ, and the assumptions inherent in the expression.

3. Describe the concept of reliability centred maintenance (RCM), including the key inputs to the analysis.

4. Recommend a planned replacement policy for the pumps in question 4 in Chapter 3.

5. Detail the time-related activities you may consider when analysing a maintenance-related task. (Hint: Break the time down into active (or uptime) and inactive (or downtime)).

6. The median (50th percentile) active time to restore/repair a system after failure, using specified procedures and resources, is not to be more than 4.5 h. The maximum 15 per cent active restore/repair time should not be more than 13.5 h (i.e. mean repair time = 5.7 h). Criticize the statement given and deduce what would be a realistic set of consistent numbers. You should use the following equations related to the log-normal distribution:

$$\sigma = [2(\ln t_{MART} - \ln t_m)]^{1/2} \text{ and } \sigma Z_\alpha = \ln t_\alpha - \ln t_m,$$

Where t_{MART} is the mean time to repair, t_m the median time to repair, and t_α the 'maximum' time to repair, evaluated at the $(1 - \alpha)$ percentile point on the distribution. Z_α is the standardized normal deviate found in Appendix 1, and σ is the standard deviation of the distribution.

7. Given the removal and replacement time data in the table, calculate t_{MART} (median time to repair)

Part identity	Quantity	Failure rate $\times 10^{-4}(h)$	Total M task time (h)
Bolts	3	0.46	0.20
Earth strap	1	0.12	0.10
Power lead	1	0.36	0.36
Signal lead	1	1.16	0.10
Cover plate	1	1.05	0.26
Brushes	2	23.6	0.35

15

Reliability Management

CORPORATE POLICY FOR RELIABILITY

A really effective reliability function can exist only in an organization where the achievement of high reliability is recognized as part of the corporate strategy and is given top management attention. If these conditions are not fulfilled, and if it receives only lip service, reliability effort will be cut back whenever cost or time pressures arise. Reliability staff will suffer low morale and will not be accepted as part of project teams. Therefore, quality and reliability awareness and direction must start at the top and must permeate all functions and levels where reliability can be affected. It is significant that in Japan the government has made quality a national objective, with a requirement that a product attains set quality standards as a condition of being granted an export licence. It is common in Japanese corporations for reliability training to start at senior executive level and to be extended downwards to include every designer. The effects of this high level attention have been dramatic in leading Japanese industry to the forefront in many areas of technological competition, and many Western companies now adopt this approach.

Several factors of modern industrial business make such high level awareness essential. The high costs of repairs under warranty, and of those borne by the user, even for relatively simple items such as domestic electronic and electrical equipment, make reliability a high value property. Other less easily quantifiable effects of reliability, such as customer goodwill and product reputation, and the relative reliability of competing products, are also important in determining market penetration and retention.

INTEGRATED RELIABILITY PROGRAMMES

The reliability effort should always be treated as an integral part of the product development and not as a parallel activity unresponsive to the rest of the development programme. This is the major justification for placing responsibility for reliability with the project manager. Whilst specialist reliability services and support can be provided from a central department in a matrix management structure, the responsibility for reliability achievement must not be taken away from the project manager,

ASPECTS TO CONSIDER	TASKS	METHODS

Project/contract requirements
Markets
Competition
Technology, knowledge
Risks, costs

SPECIFICATION
Performance, cost
Reliability/durability/safety

QFD (7)

Specification Review

Project/contract requirements (15)
Environments, stresses (7,8,9)
Variation (2,4,5)
Tools, methods
Skills, training
Suppliers
Value/cost

DESIGN
Create (product, processes)
Analyse (performance, *reliability*)
Refine/improve

Synthesis, models
CAD/CAE/EDA (schematics, etc.)
FEA, CFD Simulation (EDA, Monte Carlo) ⎫(7)
FMEA, FTA, etc.
DoE, Taguchi (5)
Maintenance, RCM (14)

Design Review (7)

Project requirements (15)
Environments, stresses (7,8,9)
Variation (2,4,5)
Methods (11)
Test items, numbers, levels (11)
Manufacturing (13)
Facilities
Suppliers

DEVELOPMENT TEST PLAN
Create

Performance, conformity
HALT(11)
DoE, Taguchi (5)
Standard methods (11)
FRACAS (11)
Data analysis (12)
Maintenance, RCM (14)

PERFORMANCE/ CONFORMITY TESTS

RELIABILITY/ DURABILITY TESTS

Design review (7)

FINAL DESIGN

PRODUCTION

Notes:
1. Items in italics are reliability specific aspects.
2. Figures in brackets indicate relevant chapters.
3. Shaded boxes indicate processes that are usually iterative.
4. Dotted lines indicate data feedback (FRACAS).

Figure 15.1 Reliability Programme Flow (design/development)

who is the only person who can ensure that the right balance is struck in allocating resources and time between the various competing aspects of product development.

The elements of a comprehensive and integrated reliability programme are shown, related to the overall development, production and in-service programme, in Figs 15.1

and 15.2. These show the continuous feedback of information, so that design iteration can be most effective.

Since production quality will be the final determinant of reliability, quality control is an integral part of the reliability programme. Quality control cannot make up for design shortfalls, but poor quality can negate much of the reliability effort. The quality control effort must be responsive to the reliability requirement and must not be directed only at reducing production costs and the passing of a final test or inspection. Quality control can be made to contribute most effectively to the reliability effort if:

1. Quality procedures, such as test and inspection criteria, are related to factors which can affect reliability, and not only to form and function. Examples are

Notes:

1. Items in italics directly influence reliability
2. Figures in brackets indicate relevant chapters.
3. Shaded boxes indicate processes that are usually iterative.
4. Dotted lines indicate data feedback (FRACAS)

Figure 15.2   Reliability Programme Flow (production, in-service)

tolerances, inspection for flaws which can cause weakening, and the need for adequate screening when appropriate.
2. Quality control test and inspection data are integrated with the other reliability data.
3. Quality control personnel are trained to recognize the relevance of their work to reliability, and trained and motivated to contribute.

An integrated reliability programme must be disciplined. Whilst creative work such as design is usually most effective when not constrained by too many rules and guidelines, the reliability (and quality) effort must be tightly controlled and supported by mandatory procedures. The disciplines of design analysis, test, reporting, failure analysis and corrective action must be strictly imposed, since any relaxation can result in a reduction of reliability, without any reduction in the cost of the programme. There will always be pressure to relax the severity of design analyses or to classify a failure as non-relevant if doubt exists, but this must be resisted. The most effective way to ensure this is to have the agreed reliability programe activities written down as mandatory procedures, with defined responsibilities for completing and reporting all tasks, and to check by audit and during programme reviews that they have been carried out.

RELIABILITY AND COSTS

Achieving high reliability is expensive, particularly when the product is complex or involves relatively untried technology. The techniques described in earlier chapters require the resources of trained engineers, management time, test equipment and products for testing, and it often appears difficult to justify the necessary expenditure in the quest for an inexact quantity such as reliability. It can be tempting to trust to good basic design and production and to dispense with specific reliability effort, or to provide just sufficient effort to placate a customer who insists upon a visible programme without interfering with the 'real' development activity. However, experience points to the fact that all well-managed reliability efforts pay off. There are usually practical limits to how much can be spent on reliability during a development programme. However, the author is unaware of any programme in which experience indicated that too much effort was devoted to reliability or that the law of diminishing returns was observed to be operating to a degree which indicated that the programme was saturated. This is mainly due to the fact that nearly every failure mode experienced in service is worth discovering and correcting during development, owing to the very large cost disparity between corrective action taken during development and similar action (or the cost of living with the failure mode) once the equipment is in service. The earlier in a development programme that the failure mode is identified and corrected the cheaper it will be, and so the reliability effort must be instituted at the outset and as many failure modes as possible eliminated during the early analysis, review and test phases. Likewise, it is nearly always less costly to correct causes of production defects than to live with the consequences in terms of production costs and unreliability.

It is dangerous to generalize about the cost of achieving a given reliability value, or of the effect on reliability of stated levels of expenditure on reliability programme activities. Some texts show a relationship as in Fig. 1.7 (page 17) with the lowest total cost (life cycle cost—LCC) indicating the 'optimum reliability' (or quality) point. However, this is a very misleading picture. The direct failure costs can usually be estimated fairly accurately, related to assumed reliability levels and yields of production processes, but the cost of achieving these levels is much more difficult to forecast. In fact, as described above, the relationship is more likely to be a decreasing one, so that the optimum quality and reliability is in fact 100 per cent, as shown in Fig. 1.8 (page 18).

Several standard references on quality management suggest considering costs under three headings, so that they can be identified, measured and controlled. These *quality costs* are the costs of all activities specifically directed at reliability and quality control, and the costs of failure. Quality costs are usually considered in three categories:

1. Prevention costs
2. Appraisal costs
3. Failure costs

Prevention costs are those related to activities which prevent failures occurring. These include reliability efforts, quality control of bought-in components and materials, training and management.

Appraisal costs are those related to test and measurement, process control and quality audit.

Failure costs are the actual costs of failure. Internal failure costs are those incurred during manufacture. These cover scrap and rework costs (including costs of related work in progress, space requirements for scrap and rework, associated documentation, and related overheads). Failure costs also include external or post-delivery failure costs, such as warranty costs; these are the costs of unreliability.

Obviously it is necessary to minimize the sum of quality and reliability costs over a suitably long period. Therefore the immediate costs of prevention and appraisal must be related to the anticipated effects on failure costs, which might be affected over several years. Investment analysis related to Q & R is an uncertain business, because of the impossibility of accurately predicting and quantifying the results. Therefore the analysis should be performed using a range of assumptions to determine the sensitivity of the results to assumed effects, such as the yield at test stages and reliability in service.

For example, two similar products, developed with similar budgets, may have markedly different reliabilities, due to differences in quality control in production, differences in the quality of the initial design or differences in the way the reliability aspects of the development programme were managed. It is even harder to say by how much a particular reliability activity will affect reliability. $20 000 spent on FMECA might make a large or a negligible difference to achieved reliability, depending upon whether the failure modes uncovered would have manifested themselves and been corrected during the development phase, or the extent to which the initial design was free of shortcomings.

The value gained from a reliability programme must, to a large extent, be a subjective judgement based upon experience and related to the way the programme is managed. The reliability programme will usually be constrained by the resources which can be usefully applied within the development time-scale. Allocation of resources to reliability programme activities should be based upon an assessment of the risks. For a complex new design, design analysis must be thorough and searching, and performed early in the programme. For a relatively simple adaptation of an existing product, less emphasis may be placed on analysis. In both cases the test programme should be related to the reliability requirement, the risks assessed in achieving it and the costs of non-achievement. The two most important features of the programme are

(1) the statement of the reliability aim in such a way that it is understood, feasible, mandatory and demonstrable,

(2) dedicated, integrated management of the programme.

Provided these two features are present, the exact balance of resources between activities will not be critical, and will also depend upon the type of product. A strong test–analyse-fix programme can make up for deficiencies in design analysis, albeit probably at higher cost; an excellent design team, well controlled and supported by good design rules, can reduce the need for testing. The reliability programme for an electronic equipment will not be the same as for a power station. As a general rule, all the reliability programme activities described in this book are worth applying insofar as they are appropriate to the product, and the earlier in the programme they can be applied the more cost-effective they are likely to prove.

In a well-integrated design, development and production effort, with all contributing to the achievement of high quality and reliability, and supported by effective management and training, it is not possible to isolate the costs of reliability and quality effort. The most realistic and effective approach is to consider all such effort as investments to enhance product performance and excellence, and not to try to classify or analyse them as though they were burdens.

Costs of unreliability

The costs of unreliability in service should be evaluated early in the development phase, so that the effort on reliability can be justified and requirements can be set, related to expected costs. The analysis of unreliability costs takes different forms, depending on the type of development programme and how the product is maintained. Two examples are given to illustrate these differences in typical programmes.

Example 15.1

A commercial electronic communication equipment is to be developed as a risk venture. The product will be sold outright, with a 2 year parts and labour warranty. Outline the LCC analysis approach and comment on the support policy options.

The analysis must take account of direct and indirect costs. The direct costs can be related directly to failure rate (or removal rate, which is likely to be higher).

The *direct costs* are:

1. Warranty repair costs.
   The annual warranty repair cost will be:

   (Number of warranted units in use) × (annual call rate per unit) × (cost per call).

   The number of warranted units will be obtained from market projections. The call rate will be related to MTBF and expected utilization.
2. Spares production and inventory costs for warranty support.
   Spares costs: to be determined by analysis (e.g. Poisson model, simulation) using inputs of call rate, proportion of calls requiring spares, spares costs, probability levels of having spares in stock, repair time to have spares back in stock, repair and stockholding costs.
3. Net of profits on post-warranty repairs and spares.
   Annual profit on post-warranty spares and repairs: analysis to be similar to warranty costs analysis, but related to post-warranty equipment utilization.

Indirect costs (not directly related to failure or removal rate):

1. Service organization (training manuals, overheads). (Warranty period contribution).
2. Product reputation.

These costs cannot be derived directly. A service organization will be required in any case and its performance will affect the product's reputation. However, a part of its costs will be related to servicing the warranty. A parametric estimate should be made under these headings, for example:

   Service organization: 50 per cent of annual warranty cost in first 2 years, 25 per cent thereafter.
   Product reputation: agreed function of call rate.

Since these costs will accrue at different rates during the years following launch, they must all be evaluated for, say, the first 5 years. The unreliability costs progression should then be plotted (Fig. 15.3) to show the relationship between cost and reliability.

The net present values of unreliability cost should then be used as the basis for planning the expenditure on the reliability programme.

This situation would be worth analysing from the point of view of what support policy might show the lowest cost for varying call rates. For example, a very low call rate might make a 'direct exchange, no repair' policy cost-effective, or might make a longer warranty period worth considering, to enhance the product's reputation. Direct exchange would result in service department savings, but a higher spares cost.

Example 15.2

A cabin air-conditioning unit is to be developed for a military aircraft. The planned deployment is 400 aircraft, each utilizing one unit. The failure rate of a similar unit in

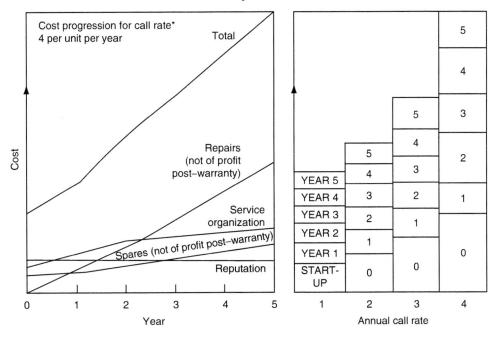

Figure 15.3 Reliability cost progression (Example 15.1)

a previous application was 2.4 per 1000 flight hours, and the average flight time per aircraft per year is 350 h. It is decided to carry out an additional reliability test programme with the objective of forcing the MTBF to 1000 h for the new unit. Assuming a Duane growth slope of 0.3, a cost per test hour of $80 and a starting MTBF of 417 h (1000/2.4) after 1500 h of development testing, comment on the cost-effectiveness of the reliability growth programme, taking into account the first 4 years of operation. The cost of a unit is $6000 and the cost of replacing a unit is $900. Make the simplifying assumptions:

(a) The fleet starts at 400, i.e. there is no build-up period.
(b) Use constant money values.
(c) There will be no reliability growth in service.
(d) Consider only removals for failure.

The total cost of operating with a 417 h MTBF, over 4 years, would be:

$$\frac{400 \times 350 \times 4 \times 900}{417} = \$1\,210\,000$$

The total cost for a 1000 h MTBF is $500 000. Therefore the LCC saving on repair costs would be about $700 000.

The instantaneous MTBF θ_i is equal to $\theta_c/(1 - \alpha)$, where θ_c is the cumulative MTBF on the Duane MTBF growth plot (Chapter 12) and α is the Duane slope. Therefore,

$\theta_c = \theta_i(1 - \alpha)$

$\quad = 1000(1 - 0.3) = 700\,\text{h}$

We must determine the number of test hours required to increase the cumulative MTBF from an initial value of 417 h to 700 h, at a slope of 0.3. With $T_0 = 1500\,\text{h}$ and using the formula

$$\theta_c = \theta_0 \left(\frac{T}{T_0}\right)^\alpha \tag{12.6}$$

we obtain

$T = 8435\,\text{h}$

At a cost of \$80 per test hour, the test cost would be \$670 000. Therefore on repair cost considerations alone there would be an approximate breakeven. However, we should also consider the value of reduced spares procurement.

Note that a better approach to this analysis would be to take account of actual failures being experienced and of the action that could be taken to reduce them (page 370).

If the aircraft are to be deployed, on ten bases of 40 aircraft each, and we desired at least a 90 per cent probability that a spare unit was available when needed, assuming Poisson arisings and a 1 month repair cycle, we can calculate the base spares requirements as

$$\text{Flying hours per month per base} = \frac{40 \times 350}{12}$$

$$= 1167\,\text{h}$$

The expected monthly arising rate with a failure rate of 2.4 per 1000 h is 2.8, and with a failure rate of 1 per 1000 h is 1.17.

Referring to Fig. 2.17, the cumulative Poisson probability curves, five spare units are required per base for a 2.8 monthly arising rate, but only two per base for a 1.17 monthly arising rate. Therefore a total saving of $10(5 - 2) = 30$ units could be made in the spares buy. This represents a further saving of \$180 000.

The two examples given above involve very simple analysis and simplifying assumptions. In both cases a Monte Carlo simulation would be a more suitable approach if we needed to consider more complex dynamic effects, such as distributed repair times and costs, multiechelon repair and progressive increase in units at risk. However, simple analysis is often sufficient to indicate the magnitude of costs, and in many cases this is all that is needed, as the input variables in logistics analysis are usually somewhat imprecise, particularly failure (removal) rate values. Simple analysis is adequate if relatively gross decisions are required. However, if it is necessary to attempt to make more precise judgements or to perform sensitivity analyses, then more powerful methods such as simulation should be considered.

There are of course other costs which can be incurred as a result of a product's unreliability. Some of these are hard to quantify, such as goodwill and market share, though these can be very large in a competitive situation and where consumer organizations are quick to publicize failure. In extreme cases unreliability can lead to litigation, especially if damage or injury results. An unreliability cost often overlooked is that due to failures in production due to unreliable features of the design. A reliable product will usually be cheaper to manufacture, and the production quality cost monitoring system should be able to highlight those costs which are due to design shortfalls.

SAFETY AND PRODUCT LIABILITY

Recent product liability legislation in the United States, Europe and in other countries adds a new dimension to the importance of eliminating safety-related failure modes, as well as to the total quality assurance approach in product development and manufacture. Before product liability (PL), the law relating to risks in using a product was based upon the principle of *caveat emptor* ('let the buyer beware'). PL introduced *caveat venditor* ('let the supplier beware'). PL was an outgrowth of the Ralph Nader campaigns in the United States, and it makes the manufacturer of a product liable for injury or death sustained as a result of failure of his product. A designer can now be held liable for a failure of his design, even if the product is old and the user did not operate or maintain it correctly. Claims for death or injury in many product liability cases can only be defended successfully if the producer can demonstrate that he has taken all practical steps towards identifying and eliminating the risk, and that the injury was entirely unrelated to failure or to inadequate design or manufacture. Since these risks may extend over ten years or even indefinitely, depending upon the law in the country concerned, long-term reliability of safety-related features becomes a critical requirement. The size of the claims, liability being unlimited in the United States, necessitates top management involvement in minimizing these risks, by ensuring that the organization and resources are provided to manage and execute the quality and reliability tasks which will ensure reasonable protection. PL insurance is a new business area for the insurance companies, who naturally expect to see a suitable reliability and safety programme being operated by the manufacturers they insure.

Reference 1 provides an overview of this topic.

STANDARDS FOR RELIABILITY, QUALITY AND SAFETY

Reliability programme standard requirements have been issued by several large agencies which place development contracts on industry. The best known of these was US MIL-STD-785—*Reliability Programs for Systems and Equipments, Development and Production* (Reference 2)—which covers all development programmes for the US Department of Defense. MIL-STD-785 was supported by other military standards, handbooks and specifications, and these have been referenced in earlier chapters. This book has referred mainly to US military documents, since in many cases they are the most highly developed and best known. However, in 1995 the US Department of Defence cancelled most military standards and specifications, including MIL-STD-785, and downgraded others to handbooks (HDBK) for guidance only. Military suppliers are now required to apply "best industry practices", rather than comply with mandatory standards.

In the United Kingdom, Defence Standards 00–40 and 00–41 (Reference 3) cover reliability programme management and methods for defence equipment, and BS 5760 (Reference 4) has been published for commercial use, and can be referenced by any organization in developing contracts. ARMP-1 is the NATO standard on reliability and maintainability. Some large agencies such as NASA, major utilities and corporations issue their own reliability standards. International standards have also been issued, and some of these are described below.

These official standards generally (but not all) tend to over-emphasize documentation, quantitative analysis, and formal test. They do not reflect the integrated approach described in this book and used by many modern engineering companies. They suffer from slow response to new ideas. Therefore they are not much used outside the defence and related industries. However, it is necessary for people involved in a reliability programme, whether from the customer or supplier side, to be familiar with the appropriate standards.

ISO/IEC60300 (Dependability)

ISO/IEC60300 (Reference 5) is the international standard for *'dependability'*, which is defined as covering reliability, maintainability and safety. It describes management and methods related to these aspects of product design and development. The methods covered include reliability prediction, design analysis, reliability demonstration tests, and mathematical/statistical techniques; most of these are described in separate standards within the ISO/IEC60000 series: see homepage. Manufacturing quality aspects are not included. The methods are mostly inconsistent with modern best practice, as described in this book. In particular, the sections on reliability testing define rigid environmental and other conditions to be applied, and for pass/fail criteria based on the statistical methods described and rejected in Chapters 2 and 12.

ISO/IEC60300 has not, so far, been made the subject of audits and registration in the way that ISO9000 has (see below).

ISO9000 (Quality systems)

The international standard for quality systems, ISO9000 (Reference 6), has been developed to provide a framework for assessing the 'quality management system' which an organization operates in relation to the goods or services provided. The concept was developed from the US Military Standard for quality management, MIL-Q-9858, which was introduced in the 1950s as a means of assuring the quality of products built for the US military. However, many organizations and companies rely on ISO9000 registration to provide assurance of the quality of products and services they buy and to indicate quality of their products and services.

Registration is to the relevant standard within the ISO9000 'family'. ISO9001 is the standard applicable to organizations that design, develop and manufacture products. We will refer to 'ISO9000 registration' as a general indication.

ISO9000 does not specifically address the quality of products and services, nor does it prescribe methods for achieving quality, such as design analysis, test and quality control. It describes, in very general and rather vague terms, the 'system' that should be in place to assure quality. In principle, there is nothing in the standard to prevent an organization from producing poor quality goods or services, so long as written procedures exist and are followed. An organization with an effective quality system would normally be more likely to take corrective action and to improve processes and service, than would one which is disorganized. However, the fact of registration cannot be taken as assurance of quality. It is often stated that registered organizations can, and sometimes do, produce 'well-documented rubbish'. An alarming number of purchasing and quality managers, in industry and in the public sector, seem to be unaware of this fundamental limitation of the standard.

In the ISO9000 approach, suppliers' quality management systems (organization, procedures, etc.) are audited by independent 'third party' assessors, who assess compliance with the standard, and issue certificates of registration. Certain organizations are 'accredited' as 'certification bodies' by the appropriate national accreditation services. The justification given for third party assessment is that it removes the need for every customer to perform his own assessment of all of his suppliers. However, a matter as important as quality cannot safely be left to be assessed spasmodically by third parties, who are unlikely to have the appropriate specialist knowledge, and who cannot be members of the joint supplier–purchaser team. The *total quality management* (TQM) philosophy (see later) demands close partnership between suppliers and purchasers.

Since its inception, ISO9000 has generated considerable controversy. The effort and expense that must be expended to obtain and maintain registration tend to engender the attitude that optimal standards of quality have been achieved. The publicity that typically goes with initial certification of a business supports this belief. The objectives of the organization, and particularly of the staff directly involved in obtaining and maintaining registration, are directed at the maintenance of procedures and at audits to ensure that staff work to them. It becomes more important to work to procedures than to develop better ways of working. Some organizations have generated real improvements as a result of certification, and some consultants and certification bodies do provide good service in quality improvement.

The leading teachers of quality management all argue against the 'systems' approach to quality, and most of the world's leading companies do not rely on it. So why is the approach so widely used? The answer is partly cultural and partly coercive.

The cultural pressure derives from the tendency to believe that people perform better when told what to do, rather than when they are given freedom and the necessary skills and motivation to determine the best ways to perform their work. This belief stems from the concept of *scientific management*, as described in References 23 and 26.

The coercion to apply the standard comes from several directions. In practice, many agencies simply exclude non-registered suppliers, or demand that bidders for contracts must be registered. All contractors and their subcontractors supplying the UK

Ministry of Defence must be registered, since the MoD decided to drop its own assessments in favour of the third party approach, and the US Department of Defense has recently decided to apply ISO9000 in place of MIL-STD-Q9858. Several large companies, as well as public utilities, demand that their suppliers are registered to ISO9000 or to industry versions, such as QS9000 for automotive and TL9000 for telecommunications applications. European Community policy for public purchasing is not explicit on ISO9000, and of course there are no such conditions regarding commercial trade, but this has not prevented registration providers in the USA from advertising that ISO9000 registration is 'a condition for doing business in Europe'. It is possible to conform to the European CE mark requirements without having ISO9000 registration, though having registration permits a manufacturer greater flexibility.

Defenders of ISO9000 say that the total quality management (TQM) approach is too severe for most organisations, and that ISO9000 can provide a 'foundation' for a subsequent total quality effort. However, the foremost teachers of modern quality management all argue against this view. (It is notable that none of these serve on the national or international committees that prepare and 'update' the standard.) They point out that any organization can adopt the TQM philosophy, and that it will lead to far greater benefits than will registration to the standard, and at much lower costs. The ISO9000 approach seeks to 'standardize' methods which directly contradict the essential lessons of the modern quality and productivity revolution.

It is notable that ISO9000 is very little used in Japan, and then mainly by companies which perceive that it will provide advantages in Western markets, not because they believe that it will lead to improvements in quality. Companies that embrace TQM set standards for product and service quality, internally and from their suppliers, far in excess of the requirements of ISO9000. These are aimed at the actual quality achievements of the products and services, and at continuous improvement in these levels. Much less emphasis is placed on the 'system'.

The recent changes to ISO9000 ('ISO9000:2000'), apart from introducing requirements for 'improvement' and more detailed requirements for analysis, design, development and measurement do not deal with these fundamental criticisms.

There are many books in publication that describe ISO9000 and its application: References 7 and 8 are examples. Reference 9 presents arguments against the approach.

ISO/IEC61508 (Functional safety of electrical/electronic/programmable electronic safety-related systems)

ISO/IEC61508 (Reference 10) is a new standard to set requirements for design, development, operation and maintenance of 'safety-related' control and protection systems which are based on electrical, electronics and software technologies. A system is 'safety-related' if any failure to function correctly can present a hazard to people. Thus, systems such as railway signalling, vehicle braking, aircraft controls, fire detection, machine safety interlocks, process plant emergency controls and car airbag initiation systems would be included. The standard lays down criteria for the extent to which such systems must be analysed and tested, including the use of independent assessors, depending on the criticality of the system. It also describes a number of methods for analysing hardware and software designs.

The extent to which the methods are to be applied is determined by the required or desired *'safety integrity level'* (SIL) of the safety function, which is stated on a range from 1 to 4. SIL 4 is the highest, relating to a 'target failure measure' between 10^{-5} and 10^{-4} per demand, or 10^{-9} and 10^{-8} per hour. For SIL 1 the figures are $10^{-2} - 10^{-1}$ and $10^{-6} - 10^{-5}$. For the quantification of failure probabilities methods such as MIL-HDBK-217 (see Chapters 3 and 4) are recommended. The methods listed include 'use of well-tried components' (recommended for all SILs), 'simulation' (recommended for SIL 2, 3 and 4), and 'modularization' ('highly recommended' for all SILs). Most of the analysis and test approaches for hardware and software recommended in this book are not included in the standard.

The standard is without any practical value or merit. The methods described are inconsistent with accepted industry practices, and many of them are known only to specialist academics, presumably including the members of the drafting committee. The issuing of the standard is leading to a growth of bureaucracy, auditors and consultants, and increased costs. It is unlikely to generate any improvements in safety, for the same reasons that ISO9000 does not improve quality. There is a danger that it could actually lead to safety being jeopardized, as a result of action and resources being applied to inappropriate methods.

SPECIFYING RELIABILITY

Designs are usually based upon specifications, which define the problems the designer must overcome. Where the specification is deficient, it is possible that the unspecified parameter or feature will be traded against when compromises are needed. This is accepted without question for features which are deterministic, such as weight, size and input and output parameter values.

However, it is more difficult to specify non-deterministic features such as appearance or reliability. It is relatively easy to assess a design for conformance with deterministic features, or even for appearance. The inherent reliability of a design is not as easily apparent, since it depends upon external factors such as load cycles, the environment in which it will be used, manufacturing quality, etc.

Deterministic performance characteristics can be accurately tested and assessed with 'breadboard' or prototype models. Reliability cannot be assessed adequately from such tests. If a prototype model passes specified performance tests, and reliability has not been specified, it is likely that production will be authorized, without any reliability testing being performed. The effects of this deficiency are apparent in many domestic, commercial and military products, which enter production with irritating design deficiencies which should have been eliminated by even a cursory design review or a limited reliability test programme.

In order to ensure that reliability does not suffer in this way, the requirement must be specified. Before describing how to specify reliability adequately, we will cover some of the ways *not* to do it:

1. Do not write vague requirements, such as 'as reliable as possible', 'high reliability is to be a feature of the design', or 'the target reliability is to be 99 per cent'. Such statements do not provide assurance against reliability being compromised.

2. Do not write unrealistic requirements. 'Will not fail under the specified operation conditions' is a realistic requirement in many cases. However, an unrealistically high reliability requirement for, say, a complex electronic equipment will not be accepted as a credible design parameter, and is likely therefore to be ignored.

The reliability specification *must* contain:

1. A definition of failure related to the product's function. The definition should cover all failure modes relevant to the function.
2. A full description of the environments in which the product will be stored, transported, operated and maintained.
3. A statement of the reliability requirement, and/or a statement of failure modes and effects which are particularly critical and which must therefore have a very low (or zero) probability of occurrence. The reliability specification may be a separate document, or it may be covered in the range of other specifications (design, test, maintenance, etc.). Reference 3 covers the preparation of reliability specifications in detail.

Definition of failure

Care must be taken in defining failure to ensure that the failure criteria are unambiguous. Failure should always be related to a measurable parameter or to a clear indication. A seized bearing indicates itself clearly, but a leaking seal might or might not constitute a failure, depending upon the leak rate, or whether or not the leak can be rectified by a simple adjustment. An electronic equipment may have modes of failure which do not affect function in normal operation, but which may do so under other conditions. For example, the failure of a diode used to block transient voltage spikes may not be apparent during functional test, and will probably not affect normal function. Defects such as changes in appearance or minor degradation that do not affect function are not usually relevant to reliability. However, sometimes a perceived degradation is an indication that failure will occur and therefore such incidents can be classified as failures.

Inevitably there will be subjective variations in assessing failure, particularly when data are not obtained from controlled tests. For example, failure data from repairs carried out under warranty might differ from data on the same equipment after the end of the warranty period, and both will differ from data derived from a controlled reliability demonstration. The failure criteria in reliability specifications can go a long way to reducing the uncertainty of relating failure data to the specification and in helping the designer to understand the reliability requirement.

Environmental specifications

The environmental specification must cover all aspects of the many loads and other effects that can influence the product's strength or probability of failure. Without a clear definition of the conditions which the product will face, the designer will not be briefed on what he is designing against. Of course, aspects of the environmental specification might sometimes be taken for granted and the designer might be expected to cater for these conditions without an explicit instruction. It is generally preferable, though, to prepare a complete environmental specification for a new product, since the discipline of considering and analysing the likely usage conditions is a worthwhile exercise if it focuses attention on any aspect which might otherwise be overlooked in the design. For most design groups only a limited number of standard environmental specifications is necessary. For example, the environmental requirements and methods of test for military equipment are covered in specifications such as US MIL-STD-810 and UK Defence Standard 07–55.

The environments to be covered must include handling, transport, storage, normal use, foreseeable misuse, maintenance and any special conditions. For example, the type of test equipment likely to be used, the skill level of users and test technicians, and the conditions under which testing might be performed should be stated if these factors might affect the observed reliability.

Stating the reliability requirement

The reliability requirement should be stated in a way which can be verified, and which makes sense relative to the use of the product. The simplest requirement to state is that no failure will occur under the stated conditions. The requirement should not include statements on the s-confidence levels of the measured reliability. The requirement relates to the population; s-confidence levels apply to the results of tests or other limited data. s-Confidence limits may be used for pass/fail decision-making and test-planning, but they should not be included in the requirement. The designer is concerned with reliability, but not with planning the reliability tests.

Levels of reliability less than unity can be stated as a success ratio, or as a life. For 'one-shot' items the success ratio is the only relevant criterion. For example, a missile solid propellant gas generator might be required to have a success ratio of 99.5 per cent after a stated period in specified storage conditions. The definition of success in terms of pressure rise, burning time and pressure variations in a specified environment would have to be stated. This specification could then form the basis for batch acceptance testing, with the test plan selected to provide the required s-confidence.

Reliability specifications based on life parameters must be framed in relation to the appropriate life distributions. Two common parameters used are MTBF, when a constant failure rate is assumed, and B-life, related to Weibull life distributions. MTBFs should not be specified if a constant failure rate assumption cannot be justified. This assumption can usually be made for complex, repairable systems (see Chapter 6), e.g. complex domestic equipment and military electronic systems. Otherwise a B-life (e.g. B_{10}, the life by which not more than 10 per cent of the population will be expected to have failed) should be specified.

Specified life parameters must clearly state the life characteristic. For example, the life of a switch, a sequence valve or a video-cassette recorder cannot be usefully stated merely as a number of hours. The life must be related to the duty cycle (in these cases switch reversals and frequency, sequencing operations and frequency, and anticipated operating cycles on record, playback and switch on/off). The life parameter may be stated as some time-dependent function, e.g. miles travelled, switching cycles, load reversals, or it may be stated as a time, with a stipulated operating cycle.

CONTRACTING FOR RELIABILITY ACHIEVEMENT

Users of equipment which can have high unreliability costs have for some time imposed contractual conditions relative to reliability. Of course, every product warranty is a type of reliability contract. However, contracts which stipulate specific incentives or penalties related to reliability achievement have been developed, mainly by the military, but also by other major equipment users such as airlines and public utilities.

The most common form of reliability contract is one which ties an incentive or penalty to a reliability demonstration. The demonstration may either be a formal test (e.g. MIL-HDBK-781) or may be based upon the user's experience. In either case, careful definition of what constitutes a relevant failure is necessary, and a procedure for failure classification must be agreed. If the contract is based only on incentive payments, it can be agreed that the customer will classify failures and determine the award, since no penalty is involved. A well-known form of reliability incentive contract is that used for spacecraft, whereby the customer pays an incentive fee for successful operation for up to, say, 2 years.

Incentive payments have advantages over incentive/penalty arrangements. It is important to create a positive motivation, rather than a framework which can result in argument or litigation, and incentives are preferable in this respect. Also, an incentive is easier to negotiate, as it is likely to be accepted as offered. Incentive payments can be structured so that, whilst they represent a relatively small percentage of the customer's saving due to increased reliability, they provide a substantial increase in profit to the supplier. The receipt of an incentive fee has significant indirect advantages, as a morale booster and as a point worth quoting in future bid situations. A typical award fee structure is shown in Fig. 15.4.

When planning incentive contracts it is necessary to ensure that other performance aspects are sufficiently well specified and, if appropriate, also covered by financial provisions such as incentives or guarantees, so that the supplier is not motivated to aim for the reliability incentive at the expense of other features. Incentive contracting requires careful planning so that the supplier's motivation is

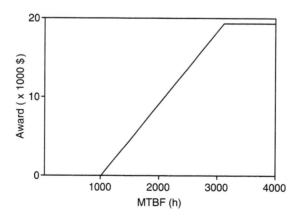

Figure 15.4   Reliability incentive structure

aligned with the customer's requirements. The parameter values selected must provide a realistic challenge and the fee must be high enough to make extra effort worth while.

Reliability improvement warranty

A type of reliability incentive contract which has recently attracted a lot of attention is reliability improvement warranty (RIW). It has been employed by the US Department of Defense and others, such as airlines. An RIW contract requires that the supplier carries out all repairs and provides all spares needed, for a fixed period, typically 4 years, for a once-off fee. The supplier's motivation is then to maximize his profit by ensuring that the repair arising rate is minimized. The customer benefits by not having to be involved in monitoring the reliability programme and not having to administer the supply and repair work. Guidelines have been developed for RIW contracting, based upon experience. The main points which need to be considered are:

1. RIW contracts should only be applied to products where there is not a high development risk, and where the utilization will be reasonably stable and centralized, so that both parties to the contract can agree on the likely reliability achievement. The product should be manufactured and used in quantity, to ensure even work flow.
2. The contract fee must provide a good chance for the supplier to make a high profit, and yet contain a reasonable risk element. There is no advantage to either side in an RIW contract which bankrupts the supplier.
3. Real difficulties can arise in administering RIW contracts in conjunction with other, conventional, repair policies. For example, hardware subject to RIW needs special handling and marking, to ensure that units are not opened or repaired by anyone other than the RIW contractor. Personnel involved must be trained, and procedures must be written to cover the operation of the RIW contract. For example, arrangements have to be made for dealing with units which are reported defective, but which are shown to have no failure on investigation.
4. The contractor must be given freedom to modify the equipment to improve its reliability (which he would do at his own expense). However, the customer may wish to have some control over modifications, as they may affect interchangeability or performance. The contract must define the arrangements for notification, approval and incorporation of changes.
5. The contract should stipulate the way in which the equipment will be used and maintained, if there is possibility that these factors can affect reliability.
6. The equipment should be put into service immediately when delivered, and failures must be reported promptly.

RIW contracts sometimes include an MTBF option, which provides for a further incentive payment related to reliability achievement.

RIW contracts can be very rewarding to both parties, but there are pitfalls to be avoided. Careful planning and management, coupled with goodwill and willingness to compromise, are essential to their success.

Total service contracts

A total service supply contract is one in which the supplier is required to provide the system, as well as all of the support. The purchaser does not specify a quantity of systems, but a level of availability. Recent railway rolling stock contracts in Europe epitomize the approach: the rail company specifies a timetable, and the supplier must determine and build the appropriate number of trains, provide all maintenance and other logistic support, including staffing and running the maintenance depots, spares provisioning, etc. All the train company does is operate the trains. The contracts include terms to cover failures to meet the timetable due to train failures or non-availability. Similar contracts are used for the supply of electronic instrumentation to large buyers, medical equipment to hospitals, and some military applications such as trainer aircraft.

A total supply contract places the responsibility and risk aspects of reliability firmly with the supplier, and can therefore be highly motivating. However, there can be long-term disadvantages to the purchaser. The purchaser's organization can lose the engineering knowledge that might be important in optimizing trade-offs between engineering and operational aspects, and in planning future purchases. When the interface between engineering and operation is purely financial and legal, with separate companies working to different business objectives, conflicts of interests can arise, leading to sub-optimization and inadequate co-operation. Since the support contract, once awarded, cannot be practically changed or transferred, the supplier is in a monopoly situation. The case of Railtrack in the UK, which 'outsourced' all rail and other infrastructure maintenance to contractors, and in the process lost the knowledge necessary for effective management, resulting directly in a fatal crash and a network-wide crisis due to cracked rails, provides a stark warning of the dangers of the approach. It is interesting that, by contrast, the airlines continue to perform their own maintenance, and the commercial aircraft manufacturers concentrate on the business they know best.

MANAGING LOWER-LEVEL SUPPLIERS

Lower-level suppliers can have a major influence on the reliability of systems. It is quite common that 80 per cent or more of failures of systems such as trains, aircraft, ships, factory and infrastructure systems, etc. can be 'bought' from the lower-level suppliers. In smaller systems, such as machines and electronic equipment, items such as engines, hydraulic pumps and valves, power supplies, displays, etc. are nearly always bought from specialist suppliers, and their failures contribute to system unreliability. Therefore it is essential that the project reliability effort is directed as much to these suppliers as to internal design, development and manufacturing.

To ensure that lower-level suppliers make the best contributions to system reliability, the following guidelines should be applied:

1. Rely on the existing commercial laws that govern trading relationships to provide assurance. In all cases this provides for redress if products or services fail to achieve the performance specified or implied. Therefore, if failures do occur, action to improve or other appropriate action can be demanded. In many cases

warranty terms can also be exercised. However, if the contract stipulates a value of reliability to be achieved, such as MTBF or maximum proportion failing, then in effect failures are being invited. When failures do occur there is a tendency for discussion and argument to concentrate on statistical interpretations and other irrelevancies, rather than on engineering and management actions to prevent recurrence. We should specify success, not failure.

2. Do not rely on ISO 9000 or similar schemes to provide assurance. As explained above, these approaches provide no assurance of product or service reliability or quality.

3. Engineers should manage the selection and purchase of engineering products. It is a common practice for companies to assign this function to a specialized purchasing organization, and design engineers must submit specifications to them. This is based on the argument that engineers might not have knowledge of the business aspects that purchasing specialists do. However, only the engineers concerned can be expected to understand the engineering aspects, particularly the longer-term impact of reliability. Engineers can quickly be taught sufficient purchasing knowledge, or can be supported by purchasing experts, but purchasing people cannot be taught the engineering knowledge and experience necessary for effective selection of engineering components and subsystems.

4. Do not select suppliers on the basis of the price of the item alone. (This is point no. 4 of Deming's famous '14 points for managers': see Reference 24.) Suppliers and their products must be selected on the basis of total value to the system, over its expected life. This includes performance, reliability, support, etc. as well as price.

5. Create long-term partnerships with suppliers, rather than seek suppliers on a project-by-project basis or change suppliers for short-term advantage. In this way it becomes practicable to share information, rewards and risks, to the long-term benefit of both sides. Suppliers in such partnerships can teach application details to the system designers, and can respond more effectively to their requirements.

6. Do not write detailed specifications for items to be supplied, during the design and development phases. Rely instead on flexible general statements of requirements. This encourages engineering co-operation, and in particular the application of the suppliers' specialist knowledge to the overall design and development process. Detailed specifications should be prepared only when designs have been finalized for production, and then in co-operation with the suppliers.

These points are discussed in more detail in Reference 26.

THE RELIABILITY MANUAL

Just as most medium to large design and development organizations have internal manuals covering design practices, organizational structure, QA procedures, etc., so reliability management and engineering should be covered. Depending on the type of product and the organization adopted, reliability can be adequately covered by appropriate sections in the engineering and QA manuals, or a separate reliability manual may be necessary. In-house reliability procedures should not attempt to

Table 15.1 Reliability manual: responsibilities

Task	Reference	Department responsible		
		Prime	Resources	Support
Stress analysis (electronic)	Procedure XX	Project design	Reliability	Reliability
Reliability test	Procedure YY	Reliability	Environmental test	Project design
Reliability data	Procedure ZZ	Reliability	Reliability	QA

teach basic principles in detail, but rather should refer to appropriate standards and literature, which should of course then be made available. If reliability programme activities are described in military, national or industry standards, these should be referred to and followed when appropriate. The bibliographies at the end of each chapter of this book list the major references.

The in-house documents should cover, as a minimum, the following subjects:

1. Corporate policy for reliability.
2. Organization for reliability.
3. Reliability procedures in design (e.g. design analysis, parts derating policy, parts, materials and process selection, approval and review, critical items listing, design review).
4. Reliability test procedures.
5. Reliability data collection, analysis and action system, including data from test, warranty, etc. (See Appendix 8.)

The written procedures must state, in every case, who carries responsibility for action and who is responsible for providing the resources and capability. They must also state who provides supporting services. A section from the reliability manual may appear as shown in Table 15.1.

THE PROJECT RELIABILITY PLAN

Every project should create and work to a written reliability plan. It is normal for customer-funded development projects to include a requirement for a reliability plan to be produced.

The reliability plan should include:

1. A brief statement of the reliability requirement.
2. The organization for reliability.
3. The reliability activities that will be performed (design analysis, test, reports).
4. The timing of all major activities, in relation to the project development milestones.
5. Reliability management of suppliers.

6. The standards, specifications and internal procedures (e.g. the reliability manual) which will be used, as well as cross-references to other plans such as for safety, maintainability and quality assurance.

When a reliability plan is submitted as part of a response to a customer request for proposals (RFP) in a competitive bid situation, it is important that the plan reflects complete awareness and understanding of the requirements and competence in compliance. A response to a competitive RFP should never query the requirements, since the responses will be judged by the people who prepared the RFP. If it is considered sensible to propose any alternative approaches, these should be included in a separate annex to the main proposal, and separately costed.

RFP responses should not include references to internal procedures, but should be complete in themselves. They should be brief, and should not repeat material in the RFP.

A reliability plan prepared as part of a project development (e.g. in response to a statement of work), after a contract has been accepted, is more comprehensive than an RFP response, since it gives more detail of activities, time-scales and reporting. Since the project reliability plan usually forms part of the contract once accepted by the customer, it is important that every aspect is covered clearly and explicitly.

A well-prepared reliability plan is useful for instilling confidence in the supplier's competence to undertake the tasks, and for providing a sound reliability management plan for the project to follow.

Appendix 9 shows an example of a reliability and maintainability plan, including safety aspects.

Specification tailoring

Specification tailoring is a term used to describe the process of suggesting alternatives to the customer's specification. 'Tailoring' is often invited in RFPs and in development contracts. A typical example occurs when a customer specifies a system and requires a formal reliability demonstration. If a potential supplier can supply a system for which adequate in-service reliability records exist, the specification could be tailored by proposing that these data be used in place of the reliability demonstration. This could save the customer the considerable expense of the demonstration. Other examples might arise out of trade-off studies, which might show, for instance, that a reduced performance parameter could lead to cost savings or reliability improvement.

USE OF EXTERNAL SERVICES

The retention of staff and facilities for analysis and test and the maintenance of procedures and training can only be cost-effective when the products involve fairly intensive and continuous development. In advanced product areas such as defence and aerospace, electronic instrumentation, control and communications, vehicles, and for large manufacturers of less advanced products such as domestic equipment and less complex industrial equipment, a dedicated reliability engineering organization is necessary, even if it is not a contractual requirement. Smaller companies with less involvement in risk-type development may have as great a need for reliability engineering expertise, but not on a continuous basis. External reliability engineering services can fulfil the requirements of smaller companies by providing the specialist support and facilities when needed. Reliability engineering consultants and specialist test establishments can often be useful to larger companies also, in support of internal staff and facilities. Since they are engaged full time across a number of different types of project they should be considered whenever new problems arise. However, they should be selected carefully and integrated in the project team.

Information on these services is provided on the book's homepage.

Small companies should also be prepared to seek the help of their major customers when appropriate. This cooperative approach benefits both supplier and customer.

CUSTOMER MANAGEMENT OF RELIABILITY

When a product is being developed under a development contract, as is often the case with military and other public purchasing, the purchasing organization plays an important role in the reliability and quality programme. As has been shown, such organizations often produce standards for application to development contracts, covering topics such as reliability programme management, design analysis methods and test methods.

There is often a tendency for reliability (and quality, safety and other 'ilities') to become bureaucratized and inflexible in large purchasing organizations. It is important for such organizations to set up effective, responsive reliability management, with the main effort devoted to projects rather than to policies.

A reliability manager should be assigned to each project. The reliability manager might have other responsibilities, such as for production quality control. Project reliability management by a centralized reliability department, not responsible to the project manager, is just as likely to result in lower effectiveness, as is this approach in the contractor's organization. A central reliability department is necessary to provide general standards, training and advice, but should not be relied upon to manage reliability programmes across a range of projects. If there is a tendency for this to happen it is usually an indication that inadequate standards or training have been provided for project staff, and these problems should then be corrected.

The prime responsibilities of the purchaser in a development reliability programme are to:

1. Specify the reliability requirements (page 427).
2. Specify the standards and methods to be used.
3. Set up the financial and contractual framework (page 429).
4. Specify the reporting requirements.
5. Monitor contract performance.

The purchaser should not attempt to run the programme, since this would usurp the supplier's management responsibilities. Proper attention to the first three items above should ensure that the supplier is effectively directed and motivated, so that the purchaser has visibility of activities and progress without having to become too deeply involved.

It is usually necessary to negotiate aspects of the specification and contract. During the specification and negotiation phases it is usual for a central reliability organization to be involved, since it is important that uniform approaches are applied. Specification tailoring (page 435) is now a common feature of development contracting and this is an important aspect in the negotiation phase, requiring experience and knowledge of the situation of other contracts being operated or negotiated.

The supplier's reliability plan, prepared in response to the purchaser's requirement, should also be reviewed by the central organization, particularly for major contracts.

The contractor's reporting tasks are often specified in the statement of work (SOW). These usually include:

1. The reliability plan.
2. Design analysis reports and updates (prediction, FMECA, FTA, etc.)
3. Test reports.

Reporting should be limited to what would be useful for monitoring performance. For example, a 50-page FMECA report, tabulating every failure mode in a system, is unlikely to be useful to the purchaser. Therefore the statement of work should specify the content, format and size of reports. The detailed analyses leading to the reports should be available for specific queries or for audit.

The purchaser should observe the supplier's design reviews. Some large organizations assign staff to supplier's premises, to monitor development and to advise on problems such as interpretation of specifications. This can be very useful on major projects such as aircraft, ships and plant, particularly if the assigned staff are subsequently involved in operation and maintenance of the system.

There are many purchasers of equipment who do not specify complete systems or let total development contracts. Also, many such purchasers do not have their own reliability standards. Nevertheless, they can usually influence the reliability and availability of equipment they buy. We will use an example to illustrate how a typical purchaser might do this.

Example 15.3

A medium-sized food-processing plant is being planned by a small group of entrepreneurs. Among other things, the plant will consist of:

1. Two large continuous-feed ovens, which are catalogue items but have some modifications added by the supplier, to the purchaser's specification. These are the most expensive items in the plant. There is only one potential supplier.

2. A conveyor feed system.
3. Several standard machines (flakers, packaging machines, etc.).
4. A process control system, operated by a central computer, for which both the hardware and software will be provided by a specialist supplier to the purchaser's specification.

The major installations except item 4 will be designed and fitted by a specialist contractor; the process control system integration will be handled by the purchasers. The plant must comply with the statutory safety standards, and the group is keen that both safety and plant availability are maximized. What should they do to ensure this?

The first step is to ensure that every supplier has, as far as can be ascertained, a good reputation for reliability and service. The purchasers should survey the range of equipment available, and if possible obtain information on reliability and service from other users. Equipment and supplier selection should be based to a large extent on these factors.

For the standard machines, the warranties provided should be studied. Since plant availability is important, the purchasers should attempt to negotiate service agreements which will guarantee up-time, e.g. for guaranteed repair or replacement within 24 hours. If this is not practicable, they should consider, in conjunction with the supplier, what spares they should hold.

Since the ovens are critical items and are being modified, the purchasers should ensure that the supplier's normal warranty applies, and service support should be guaranteed as for the standard items. They should consider negotiating an extended warranty for these items.

The process control system, being a totally new development (except for the computer), should be very carefully specified, with particular attention given to reliability, safety and maintainability, as described below. Key features of the specification and contract should be:

1. Definition of safety-critical failure effects.
2. Definition of operational failure effects.
3. Validation of correct operation when installed.
4. Guaranteed support for hardware and software, covering all repairs and corrections found to be necessary.
5. Clear, comprehensive documentation (test, operating and maintenance instructions, program listings, program notes).

For this development work, the purchasers should consider invoking appropriate standards in the contract, such as BS 5760. For example, FMECA and FTA could be very valuable for this system, and the software development should be properly controlled and documented. The supplier should be required to show how those aspects of the specification and contract will be addressed, to ensure that the requirements are fully understood. A suitable consultant engineer might be employed to specify and manage this effort.

The installation contract should also cover reliability, safety and maintainability, and service.

During commissioning, all operating modes should be tested. Safety aspects should be particularly covered, by simulating as far as possible all safety-critical failure modes.

The purchasers should formulate a maintenance plan, based upon the guidelines given in Chapter 14. A consultant engineer might be employed for this work also.

Finally, the purchasers should insure themselves against the risks. They should use the record of careful risk control during development to negotiate favourable terms with their insurers.

SELECTING AND TRAINING FOR RELIABILITY

Within the reliability organization, staff are required who are familiar with the product (its design, manufacture and test) and with statistical engineering techniques. Therefore the same qualifications and experience as apply to the other engineering departments should be represented within the reliability organization. The objective should be to create a balanced organization, in which some of the staff are drawn from product engineering departments and given the necessary reliability training, and the others are specialists in the reliability engineering techniques who should receive training to familiarize them with the product. Reliability engineering should be included as part of the normal engineering staff rotation for career development purposes. By having a balanced department, and engineers in other departments with experience of reliability engineering, the reliability effort will have credibility and will make the most effective contribution.

Reliability engineers need not necessarily be specialists in particular disciplines, such as electronic circuit design or metallurgy. Rather, a more widely based experience and sufficient knowledge to understand the specialists' problems is appropriate. The reliability engineer's task is not to solve design or production problems but to help to prevent them, and to ascertain causes of failure. He or she must, therefore, be a communicator, competent to participate with the engineering specialists in the team and able to demonstrate the value and relevance of the reliability methods applied. Experience and knowledge of the product, including manufacturing, operation and maintenance, enables the reliability engineer to contribute effectively and with credibility. Therefore engineers with backgrounds in areas such as test, product support, and user maintenance should be short-listed for reliability engineering positions.

Since reliability engineering and quality control have much in common, quality control work often provides suitable experience from which to draw, provided that the QC experience has been deeper than the traditional test and inspection approach, with no design or development involvement. For those in the reliability organization providing data analysis and statistical engineering support, specialist training is relatively more important than product familiarity.

The qualities required of the reliability engineering staff obviously are equally relevant for the head of the reliability function. Since reliability engineering should involve interfaces with several other functions, including such non-engineering areas as marketing and forecasting, this position should not be viewed as the end of the line for moderately competent engineers, but rather as one in which potential top management staff can develop general talents and further insight into the overall business, as well as providing further reliability awareness at higher levels in due course.

Since reliability engineering is often suited to general as opposed to specialist engineers, it has often been made the final resting place for 'failed' engineers. Also, because there is a statistical element, statisticians have been given reliability engineering titles and responsibilities. A combination of second-rate engineers, known as such in the other departments, and mathematical statisticians masquerading as engineers, is a recipe for an ineffective reliability organization which will lack credibility. Since reliability is a relatively new branch of engineering there is often a shortage of suitable candidates, but it is usually wiser to operate undermanned than to recruit unsuitable staff, since the reputation of the reliability organization within the firm is essential to its success. It may be that an engineer who does not excel in a specialist area can, with the right training in reliability methods, make a useful contribution, but the reliability organization must not be seen to attract more of this type of staff than do other departments.

Statistical specialists can make a very significant contribution to the integrated reliability effort. Such skills are needed for design of experiments and analysis of data, and not many engineers are suitably trained and experienced. It is important that statisticians working in engineering are made aware of the 'noisy' nature of the statistics generated, as described in earlier chapters. They should be taught the main engineering and scientific principles of the problems being addressed, and integrated into the engineering teams. They also have an important role to play in training engineers to understand and use the appropriate statistical methods.

Whilst selection and training of reliability people is important, it is also necessary to train and motivate all other members of the engineering team (design, test, production, etc.). Since product failures are nearly always due to human shortcomings, in terms of lack of knowledge, skill or effort, all involved with the product must be trained so that the chances of such failures are minimized. For example, if electronics designers understand electromagnetic interference as it affects their system they are less likely to provide inadequate protection, and test engineers who understand variation will conduct more searching tests. Therefore the reliability training effort must be related to the whole team, and not just to the reliability specialists.

ORGANIZATION FOR RELIABILITY

Because several different activities contribute to the reliability of a product it is difficult to be categorical about the optimum organization to ensure effective management of reliability. Reliability is affected by design, development, production quality control, control of subcontractors and maintenance. These activities need to be coordinated, and the resources applied to them must be related to the requirements of the product. The requirements may be determined by a market assessment, by warranty cost considerations or by the customer. The amount of customer involvement in the reliability effort varies. The military and other public organizations often stipulate the activities required in detail, and demand access to design data, test records and other information, particularly when the procurement agency funds the development. At the other extreme, domestic customers are not involved in any way directly with the development and production programme. Different activities will have greater or lesser importance depending on whether the product involves innovative or complex design, or is simple and based upon considerable experience. The reliability effort also varies as the project moves through the development, production and in-use phases, so that the design department will be very much involved to begin with, but later the emphasis will shift to the production, quality control and maintenance functions. However, the design must take account of production, test and maintenance, so these downstream activities must be considered by the specification writers and designers.

Since the knowledge, skills and techniques required for the reliability engineering tasks are essentially the same as those required for safety analysis and for maintainability engineering, it is logical and effective to combine these responsibilities in the same department or project team.

Reliability management must be integrated with other project management functions, to ensure that reliability is given the appropriate attention and resources in relation to all the other project requirements and constraints.

Two main forms of reliability organization have evolved. These are described below.

Quality assurance based organization

The quality assurance (QA) based organization places responsibility for reliability with QA management, which then controls the 'quality' of design, maintenance, etc., as well as of production. This organizational form is based upon the definition of *quality as the totality of features which bear on a product's ability to satisfy the requirement*. This is the formal European (including UK) definition of quality. Consequently, in Europe the QA department or project QA manager is often responsible for all aspects of product reliability. Figure 15.5 shows a typical organization. The reliability

Figure 15.5 QA based reliability organization

engineering team interfaces mainly with the engineering departments, while quality control is mainly concerned with production. However, there is close coordination of reliability engineering and quality control, and shared functions: e.g. a common failure data collection and analysis system can be operated, covering development, production and in-use. The QA department then provides the feedback loop from in-use experience to future design and production. This form of organization is used by most manufacturers of commercial and domestic products.

Engineering based organization

In the engineering based organization, reliability is made the responsibility of the engineering manager. The QA (or quality control) manager is responsible only for controlling production quality and may report direct to the product manager or to the production manager. Figure 15.6 shows a typical organization. This type of organization is more common in the United States.

Comparison of types of organization

The QA based organization for reliability allows easier integration of some tasks that are common to design, development and production. The ability to operate a common failure data system has been mentioned. In addition, the statistical methods used to design experiments and to analyse development test and production failure data are the same, as is much of the test equipment and test methods. For example, the environmental test equipment used to perform reliability qualification tests in development might be the same as that used for production reliability acceptance tests and screening. Engineers with experience or qualifications in QA are often familiar with reliability engineering methods, as their professional associations on both sides of the Atlantic include reliability in their areas of interest. However, for products or systems where a considerable amount of innovative design is required, the engineering based organization has advantages, since more

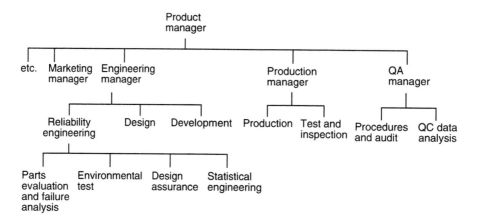

Figure 15.6 Engineering based reliability organization

of the reliability effort will have to be directed towards design assurance, such as stress analysis, design review and development testing.

The main question to be addressed in deciding upon which type of organization should be adopted is whether the split of responsibility for reliability activities inherent in the engineering based organization can be justified in relation to the amount of reliability effort considered necessary during design and development. In fact the type of organization adopted is much less important than the need to ensure integrated management of the reliability programme. So long as the engineers performing reliability activities are properly absorbed into the project team, and report to the same manager as the engineers responsible for other performance aspects, functional departmental loyalties are of secondary importance. To ensure an integrated team approach reliability engineers should be attached to and work alongside the design and other staff directly involved with the project. These engineers should have access to the departmental supporting services, such as data analysis and component evaluation, but their prime responsibility should be to the project.

MANAGING PRODUCTION QUALITY

The production department should have ultimate responsibility for the manufacturing quality of the product. It is often said that quality cannot be 'inspected in' or 'tested in' to a product. The QA department is responsible for assessing the quality of production but not for the operations which determine quality. QA thus has the same relationship to production as reliability engineering has to design and development.

In much modern production, inspection and test operations have become integrated with production operations. For example, in operator control of NC machine tools the machine operator might carry out workpiece gauging and machine calibration. Also, the costs of inspection and test can be considered to be production costs, particularly when it is not practicable to separate the functions or when, as in electronics production, the test policy can have a great impact on production costs. For these reasons, there is a trend towards routine inspection and test work being made the responsibility of production, with QA providing support services such as patrol inspection, and possibly final inspection, as well as training, calibration, etc. Determination of inspection and test policy methods and staffing should then be primarily a production responsibility, with QA in a supporting role, providing advice and ensuring that quality standards will be met.

This approach to modern quality results in much smaller QA departments than under the older system whereby production produced and passed the products to QA for inspection and test at each stage. It also obviously reduces the total cost of production (production cost plus inspection and test costs). Motivation for quality is enhanced and QA staff are better placed to contribute positively, rather than acting primarily in a policing role. The quality circles movement has also heavily influenced this trend; quality circles could not operate effectively under the old approach.

The QA department should be responsible for:

1. Setting production quality standards.
2. Monitoring production quality performance and costs.
3. QA training (SPC, motivational, etc.).
4. Specialist facilities and services.

These will be discussed in turn.

Setting production quality standards

The quality manager must decide the production quality standards to be met. These might have been set by the customer, as is increasingly the case in commercial as well as in defence equipment manufacture, in which case the quality manager is the interface with the customer on production quality matters, leaving production people to concentrate on producing. Quality standards apply to the finished product, to production processes and to bought-in materials and components. Therefore the quality manager should decide, or approve, the final inspection and test to ensure conformance. He or she should also determine such details as quality levels of components quality control of suppliers and calibration requirements for test and measuring equipment.

Monitoring production quality performance and costs

The quality manager must be satisfied that the quality objectives are being attained or that action is being taken to ensure this. These include quality cost objectives, as described earlier. Therefore the resources must be available to perform this task, as far as practicable independently of the production organization. QA staff should therefore oversee and monitor functions such as defect reporting and final conformance inspection and test. The QA department should prepare or approve quality performance and cost reports, and should monitor and assist with problem-solving. The methods described in Chapters 5 and 13 are particularly appropriate for this task.

Quality training

The quality manager is responsible for all quality control training. This is particularly important in training for operator control and quality circles, since all production people must understand and apply basic quality concepts such as simple SPC and data analysis.

Specialist facilities and services

The quality department provides facilities such as calibration services and records, vendor appraisal, component and material assessment, and defect data collection and analysis.

The assessment facilities used for testing components and materials so that their use can be approved are also the best services to use for failure investigation, since this makes the optimum use of expensive resources such as spectroscopic analysis equipment and scanning electron microscopes, and the associated specialist staff.

The joint use of these services in support of development, manufacturing and in-service is best achieved by operating an integrated approach to quality and reliability engineering.

QUALITY AUDIT

Quality audit is an independent appraisal of all of the operations, processes and management activities that can affect the quality of a product or service. The objective is to ensure that procedures are effective, that they are understood and that they are being followed. The idea of independent quality audit originated in the United States, with the objective of reducing the level of government inspection at factories. Instead of products being manufactured, then being submitted for a final customer inspection, with customer inspection of intermediate processes, the control of quality was passed to the manufacturer, with the government retaining the authority to audit the effectiveness of the suppliers' quality system. MIL-STD-Q9858 described the policy, and all US military contracts for equipment manufacture called up this standard. Quality audit became the common approach for most large defence and non-defence customers. ISO9000 has superseded these, as described earlier.

Quality audit, like financial audit, requires both internal and external audit. Internal audit is a continuing function whereby QA staff review the operations and controls, and report on discrepancies. External audit is performed by the third party assesors on a regular schedule, typically annually. However, as described earlier, this is a bureaucratic approach, not in tune with the best modern principles of design and production. Nevertheless, it might be necessary.

An effective quality audit should include review of all design, development and production, test and inspection operations, as well as associated procedures and documentation. An important aspect is the assurance that personnel know and understand their role in the quality system, including relevant procedures and responsibilities. Areas covered by the audit should include:

Organization
Procedures
Problem-reporting and corrective action
Calibration of measuring and test equipment
Quality cost monitoring
Material and process traceability

The quality manager's responsibility for audit includes all internal audit, for ensuring that the company is successful in customer or third party audits and for quality auditing of suppliers. Preparation for external audit and being subjected to it can be a very important task, and much effort is involved. The quality department should be skilled in undertaking this responsibility with minimum disruption to normal design and production work. This demands thorough knowledge of the appropriate standards and methods and the ways in which they are applied. Training is an important feature to ensure that personnel will respond correctly during audit.

Reliability aspects can be effectively included in quality audit, since there is so much common ground, particularly in relation to failure reporting and corrective action. Ideally, internal audits should be carried out by the managers responsible for the procedures and tasks concerned, or by other knowledgeable staff on their behalf. The audits should have the objective of improving processes, not merely of determining compliance, as is usually the case in ISO9000 type audits.

References 11 and 12 provide effective guidance on quality auditing.

QUALITY MANAGEMENT APPROACHES

Quality systems

The quality systems approach is epitomized by ISO9000, described earlier. It is based on the premise that if the 'system' is described and followed, then the output (products, services) will be of high quality. This can be true of machines, but it does not work for people. A good quality system (effective procedures, training, etc.) is necessary, but can provide only a baseline for achieving high quality and reliability. Despite these fundamental limitations, the quality systems approach and the ISO9000 industry that supports it are the most widely-applied in most engineering business.

'Zero defects'

Zero defects (ZD) is an approach that attempts to ensure that all processes are performed without error, by providing training and motivation to all of the people concerned. It was described in Chapter 13. It can generate improvements, but it is difficult to sustain. It is not the best approach.

'Six sigma'

The 'six sigma' approach was developed by the Motorola company in the 1980s. It is a two-pronged approach. The first is that every process should operate within 6σ limits, which implies that not more than about one operation per million will be defective or wrong. This in turn implies that every process follows the s-normal distribution. As explained in Chapter 2, this is unrealistic and misleading. The second is the application of statistical and other methods to identify opportunities for process improvements. To this end specialists are trained in the methods, gaining credentials that culminate in 'black belt' status. They are given the task of finding opportunities and driving the improvements. Top management involvement and leadership is essential.

The six sigma approach is credited with generating very large savings in several big companies. However, it is not consistent with best management practice, in that it relies to a large extent on 'experts' telling people how to do their jobs. The most effective improvements are created when the people concerned participate in their development. The six sigma approach can be applied with this principle in mind, if the 'experts' are trained and directed to work as facilitators.

The six sigma approach is described in References 14 and 15.

Quality circles

The quality circles approach to improving production quality was described in Chapter 13. It is an inherent part of the total quality management (TQM) approach described below.

Quality awards

The first national quality awards to be introduced were the Deming Awards in Japan. These are presented every year to individuals, groups and companies which achieve notable quality levels or improvements. Later the idea was followed in the USA with the Baldrige Awards, named after the then Secretary for Trade. US companies submit themselves to the assessment process, which is conducted by independent assessors, to determine whether they achieve the highest scores in a range of company categories. The award winners gain good publicity in the years that they win, and they can also generate improvements as a result of their efforts to impress the assessors.

In Europe a different approach is used: the European Foundation of Quality Management (EFQM) has produced a self-assessment guide, which companies can use to conduct their own evaluations.

Total quality management

The terms 'total quality management (TQM)' (or 'total quality control (TQC))' are often used to describe a system whereby all the activities that contribute to product quality, not just production quality control, are appraised and controlled by one manager. In this context quality is defined as the totality of features which determine a product's acceptability, and as such includes appearance, performance, reliability, support, etc.

Under this concept the QA manager has very wide authority for setting and monitoring quality standards, in this wide sense, throughout all functions of the organization. The QA manager then reports directly to the chief executive. It remains essential for line functions such as design, test and production to retain responsibility for their contributions to quality and reliability. However, the QA manager is responsible for ensuring that the total approach is coordinated, through the setting of standards, training and performance monitoring.

The total QA approach to reliability and quality can be very effective, particularly when applied to correct a situation in which quality is perceived as being lower than is required, but the reasons cannot be pinned down.

It is not easy to find people who can effectively fulfil the total QA management role. The task demands rather exceptional talents of persuasiveness and ability. It is easy for the QA manager and the organization to become dissociated from corporate realities, and the authority of the QA manager might be questioned by line departments and project managers.

The obvious solution to this is for the chief executive to undertake the responsibility. This has the supreme advantage of showing that quality and reliability are of top level concern. Functions such as design reliability and production quality control can then be integrated with design and production, and coordination of standards, training, etc., can be achieved through a chief executive's QA committee.

Only the chief executive can ensure total integration of the quality and reliability functions with the management of specifying, designing, producing and supporting the product. The increasing integration of design and production, and the pace and competitiveness of modern markets for technology-based products, demand that a fully integrated approach be used. This is to be found in many of the modern high technology companies that have grown up in the last 20 years, and in those older companies that have perceived quality and reliability as being matters too important to be left to chance or to lower levels of management. Their success has been largely due to this recognition, and to the commitment and involvement of the most senior executives.

References 16–22 and 26 describe the management principles and methods of TQM.

CHOOSING THE METHODS: STRATEGY AND TACTICS

We must apply the methods that are known to be effective in creating reliable products. We should also try to avoid methods that are misleading or counterproductive. We can consider management methods to be either strategic or tactical. Strategic principles and methods prepare the organization for the work that must be done in the future. Tactical methods are applied when necessary.

The important strategic principles and methods are:

—Top management commitment, through all functions and product phases.
—Effective organization and people.
—Effective methods, capabilities and procedures.
—Supplier selection and development.
—Training, covering management, technology, the tactical methods and consistent with the organization and procedures.
—Research.

The important tactical methods are:

—Design analysis (QFD, FMECA, stress, variation, etc., and effective use of CAE tools) (Chapters 4, 7–10).
—Test, with the emphasis on HALT (Chapter 11) and DoE (Chapter 5).
—FRACAS (Chapters 11–14).
—Production quality control and improvement (Chapter 13).
—Maintenance planning and methods (when appropriate) (Chapter 14).

The strategic methods must be applied over the long term: they do not generate benefits right away. The timescales for effective payback are typically 1–5 years, and the benefits continue to grow. However, without them the application of the tactical methods will not be as effective. Therefore they must be managed appropriately, and protected from short-term cost-cutting.

The tactical methods must be applied continuously on product development, production and in service.

The methods that should be avoided are:

—Inappropriate methods for quantifying reliability, such as prediction (Chapter 6) and demonstration (Chapter 12). These methods are sometimes referred to as 'the numbers game'.

—Reliability and quality standards, such as US MIL-STDs, ISO9000, etc. However, sometimes compliance might be necessary for contractual or other business reasons.

—Quality awards. These can be very expensive in effort and time, and the award assessors are not the customers for the product.

CONCLUSIONS: GREED, FEAR AND FREEDOM

The last 30 years or so have been marked by a revolution in reliability and quality which has quietly affected the lives of all of us. Most people are unaware of the forces involved and the methods used to provide the reliability we now so often take for granted. Of course, people are hardly ever satisfied with less than perfect reliability, and we all experience failures of products. However, a few facts should set the perspective:

1. Modern electronic equipment, despite the incredible complexity of individual parts, hardly ever fails. Thirty years ago equivalent systems built with thermionic valves or discrete electronic transistors, if feasible at all, would have been extremely unreliable.
2. Automobile manufacturers used to grow fat on the sale of spare parts for their unreliable products. Now cars are much more reliable, and reliability is a selling point in advertising.
3. Spacecraft travel to the outer planets, functioning for years with no maintenance.
4. Modern transport systems are safe and reliable.

So much for the facts. What have been the driving forces behind such performance?

There have been two forces—one to show what could be done, the other a new awareness that people want reliability and can design and produce reliable products if properly motivated. There has been a change in the balance between greed and fear in relation to reliability.

The American space programme has provided the most dramatic demonstration of what could be achieved. Here the balance between greed and fear was very heavily weighted on the side of fear. If a mission failed, particularly if there was loss of life, the future funding for NASA projects could have been drastically curtailed. Therefore everything had to work. NASA developed new techniques for reliability analysis, testing of new designs and quality control that ensured this success. We know of a few accidents, one which cost the lives of three astronauts in the capsule fire, the Apollo 13 failure which nearly ended in disaster, and the Challenger disaster; but consider the magnitude of the success in reliability terms. For example, in all the Apollo launches, over 200 rocket engines worked with only 1 failure, and these were new, complex, highly stressed systems unproven in previous applications. On the other hand, the Challenger space shuttle explosion showed how safety and reliability ultimately depend on human behaviour, and how serious can be the consequences of a lapse from standards of excellence.

The awareness that ordinary people want reliable products, and that reliability is a saleable property, has been most highly developed in Japan. The Japanese have the knack, apparently not prevalent in other countries, of deciding at a high level what needs to be done and then executing the decision right down to where the action is. They decided on quality and reliability as national priorities, and we can all see the results.

The awareness of a need for high reliability (the fear side of the balance) has also been a powerful motivator in the other fields mentioned. The new electronics gave us the opportunity to manufacture far more complex systems for everyday use, but these had to be reliable. Competition, reputation and the costs of warranties have made manufacturers aware as never before. Once national pride ('British is Best') provided motivation for quality, but sadly we do not see much of that now, and anyhow it is not enough for the new technologies and markets.

Public safety consciousness has been a powerful spur to quality and reliability in some areas, such as nuclear power and air travel, where government agencies set up and monitor reliability and safety standards. Product liability legislation is a factor which strikes fear, if not dread, into manufacturers' hearts.

Competition has provided a sharp awareness of the need for more reliable products in many consumer and industrial areas. Again, the Japanese have generally led the way, though some Western manufacturers have also been in the forefront. Now we are all trying to catch up, in quality, reliability and value, and the consumer benefits, as of course do the leading suppliers.

There has also been a drive for high reliability in military products, and in fact most of the formal methods for design reliability improvement started in the military area in the United States, so that the US military standards on reliability analysis and methods are now used in many non-military areas. The military interest in reliability stemmed from the need to reduce the enormous cost of maintaining very complex systems. Whilst the commercial world adopted many of the techniques developed by the military, there is now a swing the other way, with military buyers insisting on commercial-type methods and warranties.

The awareness that people can produce very reliable products has evolved from the debunking of the theories of scientific management. Scientific management was born in the 1930s, and taught that people will produce economically if given good conditions and if work was carefully planned. People did produce more economically, but were separated from involvement in the product. Quality of work was a separate 'scientific' function of measurement and control. Scientific management could result in higher quality, but a worker was not supposed to think, take decisions or solve problems. Scientific management led to the production line, and it dehumanized labour by forgetting that people must think and solve problems to obtain fulfilment at work.

Peter Drucker buried scientific management with publication of his book *The Practice of Management*, published in 1955 (Reference 23). Despite this, the theories of the 1930s are still taught in Western countries. Production Drucker-style involves making the worker assess and solve problems. The solution to problems is usually known at the point where the problems originate. Scientific management made it the job of management to find solutions. Drucker makes the worker interested in finding and either implementing solutions or in telling management how to solve them. This is, of course, only a small part of Drucker's teaching, but it is particularly appropriate to our subject. For example, the quality circles movement is obviously Drucker-inspired.

It was W. E. Deming who took the Drucker philosophy to the management of production, and showed how continuous improvement could be achieved while at the same time reducing costs and increasing productivity (Reference 24). Other pioneers, mentioned earlier in this book, have also made notable contributions to the drive for ever-increasing quality and reliability. The liberation of human talent at all levels in design, development, production and support, and the enhancement of those talents through training, commitment, trust and teamwork, generate levels of performance and productivity that far exceed what can be delivered by 'scientific' management.

The new philosophy of engineering management is a total one. Many companies in the West have adopted part of the philosophy, trying to combine it with cherished traditional approaches, such as functional as opposed to integrated engineering, or by making only token increases in training effort. The new philosophy of excellence requires that management inspires, supports, trusts, teaches and leads, and does not inhibit, frighten or constrain. The effects are dramatic, and the improvement never ends. The new management philosophy, as applied to engineering, is described in References 25, 26 and 27.

Quality and freedom

It is notable that no undemocratic state has been able to make any significant contribution to the reliability and quality revolution. Quality represents the essence of freedom—freedom to make decisions at work and as a consumer. Centralized bureaucratic state systems do not allow this freedom. They are stuck with scientific management, if they can even get that far. The products needed today must be complex to be competitive, whether they are toys, domestic equipment or weaponry. Modern technology has given us the ability to mass-produce complex products at low cost, but complexity is the enemy of reliability. The new techniques for controlling the reliability of design and quality of production enable us to produce complex yet reliable products, but the techniques are dependent for their success on the motivation that comes only with personal freedom.

BIBLIOGRAPHY

1. H. Abbott and M. Tyler, *Safer by Design: a Guide to the Management and Law of Designing for Product Safety.* Gower (1997).
2. US MIL-STD-785: *Reliability Programs for Systems and Equipment—Development and Production.* Available from the National Technical Information Service, Springfield, Virginia.
3. UK Defence Standard 00–40: *The Management of Reliability and Maintainability.* HMSO.
4. British Standard, BS 5760: *Reliability of Systems, Equipments and Components.* British Standards Institution, London.
5. ISO/IEC60300, *Dependability Management.* International Standards Organisation, Geneva.
6. ISO9000(2000), *Quality Systems.* International Standards Organisation, Geneva.
7. ISO/IES61508, *Functional Safety of Electrical/Electronic/Programmable Electronic Safety-related Systems.* International Standards Organisation, Geneva.
8. B. Rothery, *Standards and Certification in Europe.* Gower (1996).
9. A. J. Sayle, *Meeting ISO9000 in a TQM World,* 2nd edn. AJSL Publishing (1994).
10. J. Seddon, *In Pursuit of Quality: the Case against ISO9000.* Oak Tree Press (1997).
11. ISO/IEC61508, *Functional Safety of Electrical/Electronic/Programmable Electronic Safety-related Systems.* International Standards Organisation.
12. A. J. Sayle, *Management Audits* 3rd edn. AJSL Publishing (1997).
13. T. Conti, *Organisational Self-assessment.* Chapman and Hall (1997).
14. F. W. Breyfogle, J. M Cupello and B. Meadows, *Managing Six Sigma.* J. Wiley (2000).
15. T. Pyzdek, *The Six Sigma Handbook.* McGraw-Hill (2001).
16. J. A. Edosomwan, *Integrating Quality and Productivity Management.* Dekker (1987).
17. D. G. Raheja, *Assurance Technologies: Principles and Practices.* McGraw-Hill (1990).
18. B. Thomas, *The Human Dimension of Quality.* McGraw-Hill (1995).
19. K. Suzaki, *The New Shop Floor Management: Empowering People for Continuous Improvement.* The Free Press (1994).
20. T. Conti, *Building Total Quality.* Chapman and Hall (1993).
21. D. C. Hutchins, *In Pursuit of Quality.* Pitman (1990).
22. B. Bergman and B. Klefsjö, *Quality: from Customer Needs to Customer Satisfaction.* McGraw-Hill (1994).
23. P. F. Drucker, *The Practice of Management.* Heinemann (1955).
24. W. E. Deming, *Out of the Crisis.* MIT Press (1987).
25. D. Clausing, *Total Quality Development.* ASME Press (1994).
26. P. D. T. O'Connor, *The Practice of Engineering Management.* J. Wiley (1994).
27. P. D. T. O'Connor, *The New Management of Engineering.* www.lulu.com (2004).

QUESTIONS

1. What are the main elements of an integrated reliability programme?

2. (a) Describe briefly the main cost headings associated with achieving high quality and reliability, and the main consequential costs of failure, in development, production and use.

   (b) Explain and discuss Deming's philosophy of overall quality and reliability cost minimization.

3. (a) What are the most important aspects to be considered in preparing reliability specifications?

   (b) Write an outline reliability specification for (i) a domestic TV set, (ii) a fighter aircraft, and (iii) a gearbox bearing.

4. Discuss the ways in which reliability can be covered in procurement contracts for complex systems.

5. What are the main elements of a project reliability plan? To which other project plans should it refer?

6. What is meant by 'total quality management'? How does the concept differ from the requirements of the international standard for quality systems (ISO9000), and how does it affect reliability?

7. Your firm designs, develops and manufactures a complex consumer product which sells into a highly competitive market. The firm has recently been losing its market share and this is thought to be due to an increasing reputation for unreliable products.

   You are currently developing a new product scheduled for volume production in about 18 months' time. This product includes several new technological features and is seen very much as a 'make-or-break' product as far as the firm's future survival is concerned. The design concept is 'frozen', but little development work has taken place.
   Outline the procedures you would adopt in development and subsequent volume production to ensure the retrieval of your firm's previous reputation for high reliability.

Appendix 1. The Standard Cumulative Normal Distribution Function

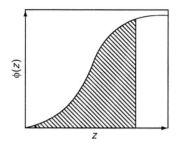

$$\Phi(z) = \frac{1}{(2\pi)^{1/2}} \int_{-\infty}^{z} \exp\left(\frac{-x^2}{2}\right) dx$$

for $0.00 \leq z \leq 4.00$

$$1 - \Phi(z) = \Phi(-z)$$

z	0.00	0.01	0.02	0.03	0.04	0.05	0.06	0.07	0.08	0.09
0.0	0.5000	0.5040	0.5080	0.5120	0.5160	0.5199	0.5239	0.5279	0.5319	0.5359
0.1	0.5398	0.5438	0.5478	0.5517	0.5557	0.5596	0.5636	0.5675	0.5714	0.5753
0.2	0.5793	0.5832	0.5871	0.5910	0.5948	0.5987	0.6026	0.6064	0.6103	0.6141
0.3	0.6179	0.6217	0.6255	0.6293	0.6331	0.6368	0.6406	0.6443	0.6480	0.6517
0.4	0.6554	0.6591	0.6628	0.6664	0.6700	0.6736	0.6772	0.6808	0.6844	0.6879
0.5	0.6915	0.6985	0.6985	0.7019	0.7054	0.7088	0.7123	0.7157	0.7190	0.7224
0.6	0.7257	0.7291	0.7324	0.7357	0.7389	0.7422	0.7454	0.7486	0.7517	0.7549
0.7	0.7580	0.7611	0.7642	0.7673	0.7703	0.7734	0.7764	0.7794	0.7823	0.7852
0.8	0.7881	0.7910	0.7939	0.7967	0.7995	0.8023	0.8051	0.8078	0.8106	0.8133
0.9	0.8159	0.8186	0.8212	0.8238	0.8264	0.8289	0.8315	0.8340	0.8365	0.8389
1.0	0.8413	0.8438	0.8461	0.8485	0.8508	0.8531	0.8554	0.8577	0.8599	0.8621
1.1	0.8643	0.8665	0.8686	0.8708	0.8729	0.8749	0.8770	0.8790	0.8810	0.8830
1.2	0.8849	0.8869	0.8888	0.8907	0.8925	0.8944	0.8962	0.8980	0.8997	0.9015
1.3	0.9032	0.9049	0.9066	0.9082	0.9099	0.9115	0.9131	0.9147	0.9162	0.9177
1.4	0.9192	0.9207	0.9222	0.9236	0.9251	0.9265	0.9279	0.9292	0.9306	0.9319
1.5	0.9332	0.9345	0.9357	0.9370	0.9382	0.9394	0.9406	0.9418	0.9430	0.9440
1.6	0.9452	0.9463	0.9474	0.9485	0.9495	0.9505	0.9515	0.9525	0.9535	0.9545
1.7	0.9554	0.9564	0.9573	0.9582	0.9591	0.9599	0.9608	0.9616	0.9625	0.9633
1.8	0.9641	0.9649	0.9656	0.9664	0.9671	0.9678	0.9686	0.9693	0.9700	0.9706
1.9	0.9713	0.9719	0.9726	0.9732	0.9738	0.9744	0.9750	0.9756	0.9762	0.9767
2.0	0.9773	0.9778	0.9783	0.9788	0.9793	0.9798	0.9803	0.9808	0.9812	0.9817
2.1	0.9821	0.9826	0.9830	0.9834	0.9838	0.9842	0.9846	0.9850	0.9854	0.9857
2.2	0.9861	0.9865	0.9868	0.9871	0.9875	0.9878	0.9881	0.9884	0.9887	0.9890
2.3	0.9893	0.9896	0.9898	0.9^2010	0.9^2061	0.9^2035	0.9^2086	0.9^2111	0.9^2134	0.9^2158
2.4	0.9^2180	0.9^2202	0.9^2224	0.9^2245	0.9^2266	0.9^2286	0.9^2305	0.9^2324	0.9^2343	0.9^2361
2.5	0.9^2379	0.9^2396	0.9^2413	0.9^2430	0.9^2446	0.9^2461	0.9^2477	0.9^2492	0.9^2506	0.9^2520
2.6	0.9^2534	0.9^2547	0.9^2560	0.9^2573	0.9^2586	0.9^2598	0.9^2609	0.9^2621	0.9^2632	0.9^2643
2.7	0.9^2653	0.9^2664	0.9^2674	0.9^2683	0.9^2693	0.9^2702	0.9^2711	0.9^2720	0.9^2728	0.9^2737
2.8	0.9^2745	0.9^2752	0.9^2760	0.9^2767	0.9^2774	0.9^2781	0.9^2788	0.9^2795	0.9^2801	0.9^2807

z	0.00	0.01	0.02	0.03	0.04	0.05	0.06	0.07	0.08	0.09
2.9	0.9^2813	0.9^2819	0.9^2825	0.9^2831	0.9^2836	0.9^2841	0.9^2846	0.9^2851	0.9^2856	0.9^2861
3.0	0.9^2865	0.9^2869	0.9^2874	0.9^2878	0.9^2882	0.9^2886	0.9^2889	0.9^2893	0.9^2897	0.9^2900
3.1	0.9^3032	0.9^3065	0.9^3096	0.9^3126	0.9^3155	0.9^3184	0.9^3211	0.9^3238	0.9^3264	0.9^3289
3.2	0.9^3313	0.9^3336	0.9^3359	0.9^3381	0.9^3402	0.9^3423	0.9^3443	0.9^3462	0.9^3481	0.9^3499
3.3	0.9^3517	0.9^3534	0.9^3550	0.9^3566	0.9^3581	0.9^3596	0.9^3610	0.9^3624	0.9^3638	0.9^3651
3.4	0.9^3663	0.9^3675	0.9^3687	0.9^3698	0.9^3709	0.9^3720	0.9^3730	0.9^3740	0.9^3749	0.9^3759
3.5	0.9^3767	0.9^3776	0.9^3784	0.9^3792	0.9^3800	0.9^3807	0.9^3815	0.9^3822	0.9^3822	0.9^3835
3.6	0.9^3841	0.9^3847	0.9^3853	0.9^3858	0.9^3864	0.9^3869	0.9^3874	0.9^3879	0.9^3883	0.9^3888
3.7	0.9^3892	0.9^3896	0.9^4004	0.9^4043	0.9^4116	0.9^4116	0.9^4150	0.9^4184	0.9^4216	0.9^4257
3.8	0.9^4277	0.9^4305	0.9^4333	0.9^4359	0.9^4385	0.9^4409	0.9^4433	0.9^4456	0.9^4478	0.9^4499
3.9	0.9^4519	0.9^4539	0.9^4557	0.9^4575	0.9^4593	0.9^4609	0.9^4625	0.9^4641	0.9^4655	0.9^4670

Appendix 2.   Values of $y = \exp(-x)$

x	e^{-x}	x	e^{-x}	x	e^{-x}	x	e^{-x}	x	e^{-x}	x	e^{-x}
0.00	1.0000	0.50	0.6065	1.00	0.3679	1.50	0.2231	2.00	0.1353	2.50	0.0821
0.01	0.9900	0.51	0.6005	1.01	0.3642	1.51	0.2209	2.01	0.1340	2.51	0.0813
0.02	0.9802	0.52	0.5945	1.02	0.3606	1.52	0.2187	2.02	0.1327	2.52	0.0805
0.03	0.9704	0.53	0.5886	1.03	0.3570	1.53	0.2165	2.03	0.1313	2.53	0.0797
0.04	0.9608	0.54	0.5827	1.04	0.3535	1.54	0.2144	2.04	0.1300	2.54	0.0789
0.05	0.9512	0.55	0.5769	1.05	0.3499	1.55	0.2122	2.05	0.1287	2.55	0.0781
0.06	0.9418	0.56	0.5712	1.06	0.3465	1.56	0.2101	2.06	0.1275	2.56	0.0773
0.07	0.9324	0.57	0.5655	1.07	0.3430	1.57	0.2080	2.07	0.1262	2.57	0.0765
0.08	0.9231	0.58	0.5599	1.08	0.3396	1.58	0.2060	2.08	0.1249	2.58	0.0758
0.09	0.9139	0.59	0.5543	1.09	0.3362	1.59	0.2039	2.09	0.1237	2.59	0.0750
0.10	0.9048	0.60	0.5488	1.10	0.3329	1.60	0.2019	2.10	0.1225	2.60	0.0743
0.11	0.8958	0.61	0.5434	1.11	0.3296	1.61	0.1999	2.11	0.1212	2.61	0.0735
0.12	0.8869	0.62	0.5379	1.12	0.3263	1.62	0.1979	2.12	0.1200	2.62	0.0728
0.13	0.8781	0.63	0.5326	1.13	0.3230	1.63	0.1959	2.13	0.1188	2.63	0.0721
0.14	0.8694	0.64	0.5273	1.14	0.3198	1.64	0.1940	2.14	0.1177	2.64	0.071
0.15	0.8607	0.65	0.5220	1.15	0.3166	1.65	0.1920	2.15	0.1165	2.65	0.0707
0.16	0.8521	0.66	0.5169	1.16	0.3135	1.66	0.1901	2.16	0.1153	2.66	0.0699
0.17	0.8437	0.67	0.5117	1.17	0.3104	1.67	0.1882	2.17	0.1142	2.67	0.0693
0.18	0.8353	0.68	0.5066	1.18	0.3073	1.68	0.1864	2.18	0.1130	2.68	0.0686
0.19	0.8270	0.69	0.5016	1.19	0.3042	1.69	0.1845	2.19	0.1119	2.69	0.0679
0.20	0.8187	0.70	0.4966	1.20	0.3012	1.70	0.1827	2.20	0.1108	2.70	0.0672
0.21	0.8106	0.71	0.4916	1.21	0.2982	1.71	0.1809	2.21	0.1097	2.71	0.0665
0.22	0.8025	0.72	0.4868	1.22	0.2952	1.72	0.1791	2.22	0.1086	2.72	0.0659
0.23	0.7945	0.73	0.4819	1.23	0.2923	1.73	0.1773	2.23	0.1075	2.73	0.0652
0.24	0.7866	0.74	0.4771	1.24	0.2894	1.74	0.1755	2.24	0.1065	2.74	0.0646
0.25	0.7788	0.75	0.4724	1.25	0.2865	1.75	0.1738	2.25	0.1054	2.75	0.0639
0.26	0.7711	0.76	0.4677	1.26	0.2837	1.76	0.1720	2.26	0.1044	2.76	0.0633
0.27	0.7634	0.77	0.4630	1.27	0.2808	1.77	0.1703	2.27	0.1033	2.77	0.0627
0.28	0.7558	0.78	0.4584	1.28	0.2780	1.78	0.1686	2.28	0.1023	2.78	0.0620
0.29	0.7483	0.79	0.4538	1.29	0.2753	1.79	0.1670	2.29	0.1013	2.79	0.0614
0.30	0.7408	0.80	0.4493	1.30	0.2725	1.80	0.1653	2.30	0.1003	2.80	0.0608
0.31	0.7334	0.81	0.4449	1.31	0.2698	1.81	0.1637	2.31	0.0993	2.81	0.0602
0.32	0.7261	0.82	0.4404	1.32	0.2671	1.82	0.1620	2.32	0.0983	2.82	0.0596
0.33	0.7189	0.83	0.4360	1.33	0.2645	1.83	0.1604	2.33	0.0973	2.83	0.0590
0.34	0.7118	0.84	0.4317	1.34	0.2618	1.84	0.1588	2.34	0.0963	2.84	0.0584
0.35	0.7047	0.85	0.4274	1.35	0.2592	1.85	0.1572	2.35	0.0954	2.85	0.0578
0.36	0.6977	0.86	0.4322	1.36	0.2567	1.86	0.1557	2.36	0.0944	2.86	0.0573
0.37	0.6907	0.87	0.4190	1.37	0.2541	1.87	0.1541	2.37	0.0935	2.87	0.0567
0.38	0.6839	0.88	0.4148	1.38	0.2516	1.88	0.1526	2.38	0.0926	2.88	0.0561
0.39	0.6771	0.89	0.4107	1.39	0.2491	1.89	0.1511	2.39	0.0916	2.89	0.0556
0.40	0.6703	0.90	0.4066	1.40	0.2466	1.90	0.1496	2.40	0.0907	2.90	0.0550
0.41	0.6637	0.91	0.4025	1.41	0.2441	1.91	0.1481	2.41	0.0898	2.91	0.0545
0.42	0.6570	0.92	0.3985	1.42	0.2417	1.92	0.1466	2.42	0.0889	2.92	0.0539
0.43	0.6505	0.93	0.3946	1.43	0.2393	1.93	0.1451	2.43	0.0880	2.93	0.0534
0.44	0.6440	0.94	0.3906	1.44	0.2369	1.94	0.1437	2.44	0.0872	2.94	0.0529
0.45	0.6376	0.95	0.3867	1.45	0.2346	1.95	0.1423	2.45	0.0863	2.95	0.0523
0.46	0.6313	0.96	0.3829	1.46	0.2322	1.96	0.1409	2.46	0.0854	2.96	0.0518
0.47	0.6250	0.97	0.3791	1.47	0.2299	1.97	0.1395	2.47	0.0846	2.97	0.0513
0.48	0.6188	0.98	0.3753	1.48	0.2277	1.98	0.1381	2.48	0.0837	2.98	0.0508
0.49	0.6126	0.99	0.3716	1.49	0.2254	1.99	0.1367	2.49	0.0829	2.99	0.0503
0.50	0.6065	1.00	0.3679	1.50	0.2231	2.00	0.1353	2.50	0.0821	3.00	0.0498

For $x < 0.01$, use the approximation $y \approx 1 - x$. If greater accuracy is needed, use $y \approx 1 - x + \frac{x^2}{2}$.

Appendix 3. Percentiles of the χ^2 Distribution

Degrees of freedom ν	α							
	0.005	0.010	0.025	0.05	0.10	0.20	0.30	0.40
1	0.0^4393	0.0^3157	0.0^3982	0.0^2393	0.0158	0.0642	0.148	0.275
2	0.0100	0.0201	0.0506	0.103	0.211	0.446	0.713	1.02
3	0.0717	0.115	0.216	0.352	0.584	1.00	1.42	1.87
4	0.207	0.297	0.484	0.711	1.06	1.65	2.19	2.75
5	0.412	0.554	0.831	1.15	1.61	2.34	3.00	3.66
6	0.676	0.872	1.24	1.64	2.20	3.07	3.83	4.57
7	0.989	1.24	1.69	2.17	2.83	3.82	4.67	5.49
8	1.34	1.65	2.18	2.73	3.49	4.59	5.53	6.42
9	1.73	2.09	2.70	3.33	4.17	5.38	6.39	7.36
10	2.16	2.56	3.25	3.94	4.87	6.18	7.27	8.30
11	2.60	3.05	3.82	4.57	5.58	6.99	8.15	9.24
12	3.07	3.57	4.40	5.23	6.30	7.81	9.03	10.2
13	3.57	4.11	5.01	5.89	7.04	8.63	9.93	11.1
14	4.07	4.66	5.63	6.57	7.79	9.47	10.8	12.1
15	4.60	5.23	6.26	7.26	8.55	10.3	11.7	13.0
16	5.14	5.81	6.91	7.96	9.31	11.2	12.6	14.0
17	5.70	6.41	7.56	8.67	10.1	12.0	13.5	14.9
18	6.26	7.01	8.23	9.39	10.9	12.9	14.4	15.9
19	6.84	7.63	8.91	10.1	11.7	13.7	15.4	16.9
20	7.43	8.26	9.59	10.9	12.4	14.6	16.3	17.8
21	8.03	8.90	10.3	11.6	13.2	15.4	17.2	18.8
22	8.64	9.54	11.0	12.3	14.0	16.3	18.1	19.7
23	9.26	10.2	11.7	13.1	14.8	17.2	19.0	20.7
24	9.89	10.9	12.4	13.8	15.7	18.1	19.9	21.7
25	10.5	11.5	13.1	14.6	16.5	18.9	20.9	22.6
26	11.2	12.2	13.8	15.4	17.3	19.8	21.8	23.6
27	11.8	12.9	14.6	16.2	18.1	20.7	22.7	24.5
28	12.5	13.6	15.3	16.9	18.9	21.6	23.6	25.5
29	13.1	14.3	16.0	17.7	19.8	22.5	24.6	26.5
30	13.8	15.0	16.8	18.5	20.6	23.4	25.5	27.4
35	17.2	18.5	20.6	22.5	24.8	27.8	30.2	32.3
40	20.7	22.2	24.4	26.5	29.1	32.3	34.9	37.1
45	24.3	25.9	28.4	30.6	33.4	36.9	39.6	42.0
50	28.0	29.7	32.4	34.8	37.7	41.4	44.3	46.9
75	47.2	49.5	52.9	56.1	59.8	64.5	68.1	71.3
100	67.3	70.1	74.2	77.9	82.4	87.9	92.1	95.8

| | | | | | | | | α | |
0.50	0.60	0.70	0.80	0.90	0.95	0.975	0.990	0.995	ν
0.455	0.708	1.07	1.64	2.71	3.84	5.02	6.63	7.88	1
1.39	1.83	2.41	3.22	4.61	5.99	7.38	9.21	10.6	2
2.37	2.95	3.67	4.64	6.25	7.81	9.35	11.3	12.8	3
3.36	4.04	4.88	5.99	7.78	9.49	11.1	13.3	14.9	4
4.35	5.13	6.06	7.29	9.24	11.1	12.8	15.1	16.7	5
5.35	6.21	7.23	8.56	10.6	12.6	14.4	16.8	18.5	6
6.35	7.28	8.38	9.80	12.0	14.1	16.0	18.5	20.3	7
7.34	8.35	9.52	11.0	13.4	15.5	17.5	20.1	22.0	8
8.34	9.41	10.7	12.2	14.7	16.9	19.0	21.7	23.6	9
9.34	10.5	11.8	13.4	16.0	18.3	20.5	23.2	25.2	10
10.3	11.5	12.9	14.6	17.3	19.7	21.9	24.7	26.8	11
11.3	12.6	14.0	15.8	18.5	21.0	23.3	26.2	28.3	12
12.3	13.6	15.1	17.0	19.8	22.4	24.7	27.7	29.8	13
13.3	14.7	16.2	18.2	21.1	23.7	26.1	29.1	31.3	14
14.3	15.7	17.3	19.3	22.3	25.0	27.5	30.6	32.8	15
15.3	16.8	18.4	20.5	23.5	26.3	28.8	32.0	34.3	16
16.3	17.8	19.5	21.6	24.8	27.6	30.2	33.4	35.7	17
17.3	18.9	20.6	22.8	26.0	28.9	31.5	34.8	37.2	18
18.3	19.9	21.7	23.9	27.2	30.1	32.9	36.2	38.6	19
19.3	21.0	22.8	25.0	28.4	31.4	34.2	37.6	40.0	20
20.3	22.0	23.9	26.2	29.6	32.7	35.5	38.9	41.4	21
21.3	23.0	24.9	27.3	30.8	33.9	36.8	40.3	42.8	22
22.3	24.1	26.0	28.4	32.0	35.2	38.1	41.6	44.2	23
23.3	25.1	27.1	29.6	33.2	36.4	39.4	43.0	45.6	24
24.3	26.1	28.2	30.7	34.4	37.7	40.6	44.3	46.9	25
25.3	27.2	29.2	31.8	35.6	38.9	41.9	45.6	48.3	26
26.3	28.2	30.3	32.9	36.7	40.1	43.2	47.0	49.6	27
27.3	29.2	31.4	34.0	37.9	41.3	44.5	48.3	51.0	28
28.3	30.3	32.5	35.1	39.1	42.6	45.7	49.6	52.3	29
29.3	31.3	33.5	36.3	40.3	43.8	47.0	50.9	53.7	30
34.3	36.5	38.9	41.8	46.1	49.8	53.2	57.3	60.3	35
39.3	41.6	44.2	47.3	51.8	55.8	59.3	63.7	66.8	40
44.3	46.8	49.5	52.7	57.5	61.7	65.4	70.0	73.2	45
49.3	51.9	54.7	58.2	63.2	67.5	71.4	76.2	79.5	50
74.3	77.5	80.9	85.1	91.1	96.2	100.8	106.4	110.3	75
99.3	102.9	106.9	111.7	118.5	124.3	129.6	135.6	140.2	100

Appendix 4.   Values of the F-distribution

(A. Hald, *Statistical Tables and Formulas*, copyright © 1952 John Wiley & Sons Inc. Reprinted by permission of John Wiley & Sons Inc.)

Tabulation of the values of $F_{0.10;\nu_1;\nu_2}$ versus ν_1 and ν_2

	Degrees of freedom for the numerator (ν_1)																	
	1	2	3	4	5	6	7	8	9	10	15	20	30	50	100	200	500	∞
1	39.9	49.5	53.6	55.8	57.2	58.2	58.9	59.4	59.9	60.2	61.2	61.7	62.3	62.7	63.0	63.2	63.3	63.3
2	8.53	9.00	9.16	9.24	9.29	9.33	9.35	9.37	9.38	9.39	9.42	9.44	9.46	9.47	9.48	9.49	9.49	9.49
3	5.54	5.46	5.39	5.34	5.31	5.28	5.27	5.25	5.24	5.23	5.20	5.18	5.17	5.15	5.14	5.14	5.14	5.13
4	4.54	4.32	4.19	4.11	4.05	4.01	3.98	3.95	3.94	3.92	3.87	3.84	3.82	3.80	3.78	3.77	3.76	3.76
5	4.06	3.78	3.62	3.52	3.45	3.40	3.37	3.34	3.32	3.30	3.24	3.21	3.17	3.15	3.13	3.12	3.11	3.10
6	3.78	3.46	3.29	3.18	3.11	3.05	3.01	2.98	2.96	2.94	2.87	2.84	2.80	2.77	2.75	2.73	2.73	2.72
7	3.59	3.26	3.07	2.96	2.88	2.83	2.78	2.75	2.72	2.70	2.63	2.59	2.56	2.52	2.50	2.48	2.48	2.47
8	3.46	3.11	2.92	2.81	2.73	2.67	2.62	2.59	2.56	2.54	2.46	2.42	2.38	2.35	2.32	2.31	2.30	2.29
9	3.36	3.01	2.81	2.69	2.61	2.55	2.51	2.47	2.44	2.42	2.34	2.30	2.25	2.22	2.19	2.17	2.17	2.16
10	3.28	2.92	2.73	2.61	2.52	2.46	2.41	2.38	2.35	2.32	2.24	2.20	2.16	2.12	2.09	2.07	2.06	2.06
11	3.23	2.86	2.66	2.54	2.45	2.39	2.34	2.30	2.27	2.25	2.17	2.12	2.08	2.04	2.00	1.99	1.98	1.97
12	3.18	2.81	2.61	2.48	2.39	2.33	2.28	2.24	2.21	2.19	2.10	2.06	2.01	1.97	1.94	1.92	1.91	1.90
13	3.14	2.76	2.56	2.43	2.35	2.28	2.23	2.20	2.16	2.14	2.05	2.01	1.96	1.92	1.88	1.86	1.85	1.85
14	3.10	2.73	2.52	2.39	2.31	2.24	2.19	2.15	2.12	2.10	2.01	1.96	1.91	1.87	1.83	1.82	1.80	1.80
15	3.07	2.70	2.49	2.36	2.27	2.21	2.16	2.12	2.09	2.06	1.97	1.92	1.87	1.83	1.79	1.77	1.76	1.76
16	3.05	2.67	2.46	2.33	2.24	2.18	2.13	2.09	2.06	2.03	1.94	1.89	1.84	1.79	1.76	1.74	1.73	1.72
17	3.03	2.64	2.44	2.31	2.22	2.15	2.10	2.06	2.03	2.00	1.91	1.86	1.81	1.76	1.73	1.71	1.69	1.69
18	3.01	2.62	2.42	2.29	2.20	2.13	2.08	2.04	2.00	1.98	1.89	1.84	1.78	1.74	1.70	1.68	1.67	1.66
19	2.99	2.61	2.40	2.27	2.18	2.11	2.06	2.02	1.98	1.96	1.86	1.81	1.76	1.71	1.67	1.65	1.64	1.63
20	2.97	2.59	2.38	2.25	2.16	2.09	2.04	2.00	1.96	1.94	1.84	1.79	1.74	1.69	1.65	1.63	1.62	1.61
22	2.95	2.56	2.35	2.22	2.13	2.06	2.01	1.97	1.93	1.90	1.81	1.76	1.70	1.65	1.61	1.59	1.58	1.57
24	2.93	2.54	2.33	2.19	2.10	2.04	1.98	1.94	1.91	1.88	1.78	1.73	1.67	1.62	1.58	1.56	1.54	1.53
26	2.91	2.52	2.31	2.17	2.08	2.01	1.96	1.92	1.88	1.86	1.76	1.71	1.65	1.59	1.55	1.53	1.51	1.50
28	2.89	2.50	2.29	2.16	2.06	2.00	1.94	1.90	1.87	1.84	1.74	1.69	1.63	1.57	1.53	1.50	1.49	1.48
30	2.88	2.49	2.28	2.14	2.05	1.98	1.93	1.88	1.85	1.82	1.72	1.67	1.61	1.55	1.51	1.48	1.47	1.46
40	2.84	2.44	2.23	2.09	2.00	1.93	1.87	1.83	1.79	1.76	1.66	1.61	1.54	1.48	1.43	1.41	1.39	1.38
50	2.81	2.41	2.20	2.06	1.97	1.90	1.84	1.80	1.76	1.73	1.63	1.57	1.50	1.44	1.39	1.36	1.34	1.33
60	2.79	2.39	2.18	2.04	1.95	1.87	1.82	1.77	1.74	1.71	1.60	1.54	1.48	1.41	1.36	1.33	1.31	1.29
80	2.77	2.37	2.15	2.02	1.92	1.85	1.79	1.75	1.71	1.68	1.57	1.51	1.44	1.38	1.32	1.28	1.26	1.24
100	2.76	2.36	2.14	2.00	1.91	1.83	1.78	1.73	1.70	1.66	1.56	1.49	1.42	1.35	1.29	1.26	1.23	1.21
200	2.73	2.33	2.11	1.97	1.88	1.80	1.75	1.70	1.66	1.63	1.52	1.46	1.38	1.31	1.24	1.20	1.17	1.14
500	2.72	2.31	2.10	1.96	1.86	1.79	1.73	1.68	1.64	1.61	1.50	1.44	1.36	1.28	1.21	1.16	1.12	1.09
∞	2.71	2.30	2.08	1.94	1.85	1.77	1.72	1.67	1.63	1.60	1.49	1.42	1.34	1.26	1.18	1.13	1.08	1.00

Degrees of freedom for the denominator (ν_2)

Appendix 4

Tabulation of the values of $F_{0.10;\nu_1;\nu_2}$ versus ν_1 and ν_2

	Degrees of freedom for the numerator (ν_1)																	
	1	*2*	*3*	*4*	*5*	*6*	*7*	*8*	*9*	*10*	*11*	*12*	*13*	*14*	*15*	*16*	*17*	*18*
1	161	200	216	225	230	234	237	239	241	242	243	244	245	245	246	246	247	247
2	18.5	19.0	19.2	19.2	19.3	19.3	19.4	19.4	19.4	19.4	19.4	19.4	19.4	19.4	19.4	19.4	19.4	19.4
3	10.1	9.55	9.28	9.12	9.01	8.94	8.89	8.85	8.81	8.79	8.76	8.74	8.73	8.71	8.70	8.69	8.68	8.67
4	7.71	6.94	6.59	6.39	6.26	6.16	6.09	6.04	6.00	5.96	5.94	5.91	5.82	5.87	5.86	5.84	5.83	5.82
5	6.61	5.79	5.41	5.19	5.05	4.95	4.88	4.82	4.77	4.74	4.70	4.08	4.66	4.64	4.62	4.60	4.59	4.58
6	5.99	5.14	4.76	4.53	4.39	4.28	4.21	4.15	4.10	4.06	4.03	4.00	3.98	3.96	3.94	3.92	3.91	390
7	5.59	4.74	4.35	4.12	3.97	3.87	3.79	3.73	3.68	3.64	3.60	3.57	3.55	3.53	3.51	3.49	3.48	3.47
8	5.32	4.46	4.07	3.84	3.69	3.58	3.50	3.44	3.39	3.35	3.31	3.28	3.26	3.24	3.22	3.20	3.19	3.17
9	5.12	4.26	3.86	3.63	3.48	3.37	3.29	3.23	3.18	3.14	3.10	3.07	3.05	3.03	3.01	2.99	2.97	2.96
10	4.96	4.10	3.71	3.48	3.33	3.22	3.14	3.07	3.02	2.98	2.94	2.91	2.89	2.86	2.85	2.83	2.81	2.80
11	4.84	2.98	3.50	3.36	3.20	3.01	2.95	2.90	2.85	2.82	2.82	2.79	2.76	2.74	2.72	2.70	2.69	2.67
12	4.75	3.89	3.49	3.26	3.11	3.00	2.91	2.85	2.80	2.75	2.72	2.69	2.66	2.64	2.62	2.60	2.58	2.57
13	4.67	3.81	3.41	3.18	3.03	2.92	2.83	2.77	2.71	2.67	2.63	2.60	2.58	2.55	2.53	2.51	2.50	2.48
14	4.60	3.74	3.34	3.11	2.96	2.85	2.76	2.70	2.65	2.60	2.57	2.53	2.51	2.48	2.46	2.44	2.43	2.41
15	4.54	3.68	3.29	3.06	2.90	2.79	2.71	2.64	2.59	2.54	2.51	2.48	2.45	2.42	2.40	2.38	2.37	2.35
16	4.49	3.63	3.24	3.01	2.85	2.74	2.66	2.59	2.54	2.49	2.46	2.42	2.40	2.37	2.35	2.33	2.32	2.30
17	4.45	3.59	3.20	2.96	2.81	2.70	2.61	2.55	2.49	2.45	2.41	2.38	2.36	2.33	2.31	2.29	2.27	2.26
18	4.41	3.55	3.16	2.93	2.77	2.66	2.58	2.51	2.46	2.41	2.37	2.34	2.31	2.29	2.27	2.25	2.23	2.22
19	4.38	3.52	3.13	2.90	2.74	2.63	2.54	2.48	2.42	2.38	2.34	2.31	2.28	2.26	2.23	2.21	2.20	2.18
20	4.35	3.49	3.10	2.87	2.71	2.60	2.51	2.45	2.39	2.35	2.31	2.28	2.25	2.22	2.20	2.18	2.17	2.15
21	4.32	3.47	3.07	2.82	2.68	2.57	2.49	2.42	2.37	2.32	2.28	2.25	2.22	2.20	2.18	2.16	2.14	2.12
22	4.30	3.44	3.05	2.84	2.66	2.55	2.46	2.40	2.34	2.30	2.26	2.23	2.20	2.17	2.15	2.13	2.11	2.10
23	4.28	3.42	3.03	2.80	2.64	2.53	2.44	2.37	2.32	2.27	2.23	2.20	2.18	2.15	2.13	2.11	2.09	2.07
24	4.26	3.40	3.01	2.78	2.62	2.51	2.42	2.36	2.30	2.25	2.21	2.18	2.15	2.13	2.11	2.09	2.07	2.05
25	4.24	3.39	2.99	2.76	2.60	2.49	2.40	2.34	2.28	2.24	2.20	2.16	2.14	2.11	2.09	2.07	2.05	2.04
26	4.23	3.37	2.98	2.74	2.59	2.47	2.39	2.32	2.27	2.22	2.18	2.15	2.12	2.09	2.07	2.05	2.03	2.02
27	4.21	3.35	2.96	2.73	2.57	2.46	2.37	2.31	2.25	2.20	2.17	2.13	2.10	2.08	2.06	2.04	2.02	2.00
28	4.20	3.34	2.95	2.71	2.56	2.45	2.36	2.29	2.24	2.19	2.15	2.12	2.09	2.06	2.04	2.02	2.00	1.99
29	4.18	3.33	2.93	2.70	2.55	2.43	2.35	2.28	2.22	2.18	2.14	2.10	2.08	2.05	2.03	2.01	1.99	1.97
30	4.17	3.32	2.92	2.69	2.53	2.42	2.33	2.27	2.21	2.16	2.13	2.09	2.06	2.04	2.01	1.99	1.98	1.96
32	4.15	3.29	2.90	2.67	2.51	2.40	2.31	2.24	2.19	2.14	2.10	2.07	2.04	2.01	1.99	1.97	1.95	1.94
34	4.13	3.28	2.88	2.65	2.49	2.38	2.29	2.23	2.17	2.12	2.08	2.05	2.02	1.99	1.97	1.95	1.93	1.92
36	4.11	3.26	2.87	2.63	2.48	2.36	2.28	2.21	2.15	2.11	2.07	2.03	2.00	1.98	1.95	1.93	1.92	1.90
38	4.10	3.24	2.85	2.62	2.46	2.35	2.26	2.19	2.14	2.09	2.05	2.02	1.99	1.96	1.94	1.92	1.90	1.88
40	4.08	3.23	2.84	2.61	2.45	2.34	2.25	2.18	2.12	2.08	2.04	2.00	1.97	1.95	1.92	1.90	1.89	1.87

Degrees of freedom for the denominator (ν_2)

							Degrees of freedom for the numerator (ν_1)											
19	*20*	*22*	*24*	*26*	*28*	*30*	*35*	*40*	*45*	*50*	*60*	*80*	*100*	*200*	*500*	∞	
248	248	249	249	249	250	250	251	251	251	252	252	252	253	254	254	254	1
19.4	19.4	19.5	19.5	19.5	19.5	19.5	19.5	19.5	19.5	19.5	19.5	19.5	19.5	19.5	19.5	19.5	2
8.67	8.66	8.65	8.64	8.63	8.62	8.62	8.60	8.59	8.59	8.58	8.57	8.56	8.55	8.54	8.53	8.53	3
5.81	5.80	5.79	5.77	5.76	5.75	5.75	5.73	5.72	5.71	5.70	5.69	5.67	5.66	5.65	5.64	5.63	4
4.57	4.56	4.54	4.53	4.52	4.50	4.50	4.48	4.46	4.45	4.44	4.43	4.41	4.41	4.30	4.37	4.37	5
3.88	3.87	3.88	3.84	3.83	3.82	3.81	3.79	3.77	3.76	3.75	3.74	3.72	3.71	3.09	3.63	3.67	6
3.46	3.44	3.43	3.41	3.40	3.39	3.38	3.36	3.34	3.33	3.32	3.30	3.20	3.29	2.25	3.24	3.23	7
3.16	3.15	3.13	3.12	3.10	3.09	3.08	3.06	3.04	3.03	3.02	3.01	2.99	2.97	2.95	2.94	2.93	8
2.95	2.94	2.92	2.90	2.89	2.87	2.86	2.84	2.83	2.81	2.80	2.79	2.77	2.76	2.23	2.72	2.71	9
2.78	2.77	2.75	2.74	2.72	2.71	2.70	2.68	2.66	2.65	2.64	2.62	2.60	2.59	2.55	2.55	2.54	10
2.66	2.65	2.63	2.61	2.59	2.58	2.57	2.55	2.53	2.52	2.51	2.49	2.47	2.46	2.43	2.42	2.40	11
2.56	2.54	2.52	2.51	2.49	2.48	2.47	2.44	2.43	2.41	2.40	2.38	2.36	2.35	2.32	2.31	2.30	12
2.47	2.46	2.44	2.42	2.41	2.39	2.38	2.36	2.34	2.33	2.31	2.30	2.27	2.26	2.23	2.22	2.21	13
2.40	2.39	2.37	2.35	2.33	2.32	2.31	2.28	2.27	2.26	2.24	2.22	2.20	2.19	2.16	2.14	2.13	14
2.34	2.33	2.31	2.29	2.27	2.26	2.25	2.22	2.20	2.19	2.18	2.16	2.14	2.12	2.10	2.08	2.07	15
2.29	2.28	2.25	2.24	2.22	2.21	2.19	2.17	2.15	2.14	2.12	2.11	2.08	2.07	2.04	2.02	2.01	16
2.24	2.23	2.21	2.19	2.17	2.16	2.15	2.12	2.10	2.09	2.08	2.06	2.03	2.02	1.09	1.97	1.96	17
2.20	2.19	2.17	2.15	2.13	2.12	2.11	2.08	2.06	2.05	2.04	2.02	1.99	1.98	1.95	1.93	1.92	18
2.17	2.16	2.13	2.11	2.10	2.08	2.07	2.05	2.03	2.01	2.00	1.98	1.96	1.94	1.91	1.89	1.88	19
2.14	2.12	2.10	2.08	2.07	2.05	2.04	2.01	1.99	1.98	1.97	1.95	1.92	1.91	1.88	1.86	1.84	20
2.11	2.10	2.07	2.05	2.04	2.02	2.01	1.98	1.96	1.95	1.94	1.92	1.89	1.88	1.84	1.82	1.81	21
2.08	2.07	2.03	2.03	2.01	2.00	1.98	1.96	1.94	1.92	1.91	1.89	1.86	1.85	1.82	1.80	1.78	22
2.06	2.05	2.02	2.00	1.99	1.97	1.96	1.93	1.91	1.90	1.88	1.85	1.84	1.82	1.79	1.77	1.76	23
2.04	2.03	2.00	1.98	1.97	1.95	1.94	1.91	1.89	1.88	1.86	1.84	1.82	1.80	1.77	1.75	1.73	24
2.02	2.01	1.98	1.96	1.95	1.93	1.92	1.89	1.87	1.85	1.84	1.82	1.80	1.78	1.75	1.73	1.71	25
2.00	1.99	1.97	1.95	1.93	1.91	1.90	1.87	1.85	1.84	1.82	1.80	1.78	1.76	1.73	1.71	1.69	26
1.99	1.97	1.95	1.93	1.91	1.90	1.88	1.86	1.84	1.82	1.81	1.79	1.76	1.74	1.71	1.69	1.67	27
1.97	1.96	1.93	1.91	1.90	1.88	1.87	1.84	1.82	1.80	1.79	1.77	1.74	1.73	1.69	1.67	1.65	28
1.96	1.94	1.92	1.90	1.88	1.87	1.85	1.83	1.81	1.79	1.77	1.75	1.73	1.71	1.67	1.65	1.64	29
1.95	1.93	1.91	1.89	1.87	1.85	1.84	1.81	1.79	1.77	1.76	1.74	1.71	1.70	1.65	1.64	1.62	30
1.92	1.91	1.88	1.86	1.85	1.83	1.82	1.79	1.77	1.75	1.74	1.71	1.69	1.67	1.63	1.61	1.59	32
1.90	1.89	1.86	1.84	1.82	1.80	1.80	1.77	1.75	1.73	1.71	1.69	1.66	1.65	1.61	1.59	1.57	34
1.88	1.87	1.85	1.82	1.81	1.79	1.78	1.75	1.73	1.71	1.69	1.67	1.64	1.62	1.59	1.56	1.55	36
1.87	1.85	1.83	1.81	1.79	1.77	1.76	1.73	1.71	1.69	1.68	1.65	1.62	1.61	1.57	1.54	1.53	38
1.85	1.84	1.81	1.79	1.77	1.76	1.74	1.72	1.69	1.67	1.66	1.64	1.61	1.50	1.55	1.53	1.51	40

Degrees of freedom for the denominator (ν_2)

					Degrees of freedom for the numerator (ν_1)													
	1	*2*	*3*	*4*	*5*	*6*	*7*	*8*	*9*	*10*	*11*	*12*	*13*	*14*	*15*	*16*	*17*	*18*
42	4.07	3.22	2.83	2.59	2.44	2.32	2.24	2.16	2.11	2.06	2.03	1.99	1.96	1.93	1.91	1.89	1.87	1.86
44	4.06	3.21	2.82	2.58	2.43	2.31	2.23	2.16	2.10	2.05	2.01	1.98	1.95	1.92	1.90	1.88	1.86	1.84
46	4.05	3.20	2.81	2.57	2.42	2.30	2.22	2.15	2.09	2.04	2.00	1.97	1.94	1.91	1.89	1.87	1.85	1.83
48	4.04	3.19	2.80	2.57	2.41	2.29	2.21	2.14	2.08	2.03	1.99	1.96	1.93	1.90	1.88	1.86	1.84	1.82
50	4.03	3.18	2.79	2.56	2.40	2.29	2.20	2.13	2.07	2.03	1.99	1.95	1.92	1.89	1.87	1.85	1.83	1.81
55	4.02	3.16	2.77	2.54	2.38	2.27	2.18	2.11	2.06	2.01	1.97	1.93	1.90	1.88	1.85	1.83	1.81	1.79
60	4.00	3.15	2.76	2.53	2.37	2.25	2.17	2.10	2.04	1.99	1.95	1.92	1.89	1.86	1.84	1.82	1.80	1.78
65	3.99	3.14	2.75	2.51	2.36	2.24	2.15	2.08	2.03	1.98	1.94	1.90	1.87	1.85	1.82	1.80	1.78	1.76
70	3.98	3.13	2.74	2.50	2.35	2.23	2.14	2.07	2.02	1.97	1.93	1.89	1.86	1.84	1.81	1.79	1.77	1.75
80	3.96	3.11	2.73	2.49	2.33	2.21	2.13	2.06	2.00	1.95	1.91	1.88	1.84	1.82	1.79	1.77	1.75	1.73
90	3.95	3.10	2.71	2.47	2.32	2.20	2.11	2.04	1.99	1.94	1.90	1.86	1.83	1.80	1.78	1.76	1.74	1.72
100	3.94	3.09	2.70	2.46	2.31	2.19	2.10	2.03	1.97	1.93	1.89	1.85	1.82	1.79	1.77	1.75	1.73	1.71
125	3.92	3.07	2.68	2.44	2.29	2.17	2.08	2.01	1.96	1.91	1.87	1.83	1.80	1.77	1.76	1.72	1.70	1.69
150	3.90	3.08	2.66	2.43	2.27	2.16	2.07	2.00	1.94	1.89	1.85	1.82	1.79	1.76	1.73	1.71	1.69	1.67
200	3.89	3.04	2.65	2.42	2.26	2.14	2.06	1.98	1.93	1.88	1.84	1.80	1.77	1.74	1.72	1.69	1.67	1.65
300	3.87	3.03	2.63	2.40	2.24	2.13	2.04	1.97	1.91	1.86	1.82	1.78	1.75	1.72	1.70	1.68	1.66	1.64
500	3.86	3.01	2.62	2.39	2.23	2.12	2.03	1.96	1.90	1.85	1.81	1.77	1.74	1.71	1.69	1.66	1.64	1.62
1000	3.85	3.00	2.61	2.38	2.22	2.11	2.02	1.95	1.89	1.84	1.80	1.76	1.73	1.70	1.68	1.65	1.63	1.61
∞	3.84	3.00	2.60	2.37	2.21	2.10	2.01	1.94	1.88	1.83	1.79	1.75	1.72	1.69	1.67	1.64	1.62	1.60

Degrees of freedom for the denominator (ν_2)

19	20	22	24	26	28	30	35	40	45	50	60	80	100	200	500	∞	
1.84	1.83	1.80	1.78	1.76	1.74	1.73	1.70	1.68	1.66	1.65	1.62	1.59	1.57	1.53	1.51	1.49	42
1.83	1.81	1.79	1.77	1.75	1.73	1.72	1.69	1.67	1.65	1.63	1.61	1.58	1.56	1.52	1.49	1.48	44
1.82	1.80	1.78	1.76	1.74	1.72	1.71	1.68	1.65	1.64	1.62	1.60	1.57	1.55	1.51	1.48	1.46	46
1.81	1.79	1.77	1.75	1.73	1.71	1.70	1.67	1.64	1.62	1.61	1.59	1.56	1.54	1.49	1.47	1.45	48
1.80	1.78	1.76	1.74	1.72	1.70	1.69	1.66	1.63	1.61	1.60	1.58	1.54	1.52	1.48	1.46	1.44	50
1.78	1.76	1.74	1.72	1.70	1.68	1.67	1.64	1.61	1.59	1.58	1.55	1.52	1.50	1.46	1.43	1.41	55
1.76	1.75	1.72	1.70	1.68	1.66	1.65	1.62	1.59	1.57	1.56	1.53	1.50	1.48	1.44	1.41	1.39	60
1.75	1.73	1.71	1.69	1.67	1.65	1.63	1.60	1.58	1.56	1.54	1.52	1.49	1.48	1.42	1.39	1.37	65
1.74	1.72	1.70	1.67	1.65	1.64	1.62	1.59	1.57	1.55	1.53	1.50	1.47	1.45	1.40	1.37	1.35	70
1.72	1.70	1.68	1.65	1.63	1.62	1.60	1.57	1.54	1.52	1.51	1.48	1.45	1.43	1.38	1.35	1.32	80
1.70	1.69	1.66	1.64	1.62	1.60	1.59	1.55	1.53	1.51	1.49	1.46	1.43	1.41	1.36	1.32	1.30	90
1.69	1.68	1.65	1.63	1.61	1.59	1.57	1.54	1.52	1.49	1.48	1.45	1.41	1.39	1.34	1.31	1.28	100
1.67	1.65	1.63	1.60	1.58	1.57	1.55	1.52	1.49	1.47	1.45	1.42	1.39	1.36	1.31	1.27	1.25	125
1.66	1.64	1.61	1.59	1.57	1.55	1.53	1.50	1.48	1.45	1.44	1.41	1.37	1.34	1.29	1.23	1.22	150
1.64	1.62	1.60	1.57	1.55	1.53	1.52	1.48	1.46	1.43	1.41	1.39	1.35	1.32	1.26	1.22	1.19	200
1.62	1.61	1.58	1.55	1.53	1.51	1.50	1.46	1.43	1.41	1.39	1.38	1.32	1.30	1.23	1.19	1.15	300
1.61	1.50	1.56	1.54	1.52	1.50	1.48	1.45	1.42	1.40	1.38	1.34	1.30	1.28	1.21	1.16	1.11	500
1.60	1.58	1.55	1.53	1.51	1.49	1.47	1.44	1.41	1.38	1.36	1.33	1.29	1.26	1.19	1.13	1.08	1000
1.59	1.57	1.54	1.52	1.50	1.48	1.46	1.42	1.39	1.37	1.35	1.32	1.27	1.24	1.17	1.11	1.00	∞

Degrees of freedom for the numerator (ν_1)

Degrees of freedom for the denominator (ν_2)

Appendix 4

Tabulation of the values of $F_{0.025;\nu_1;\nu_2}$ versus ν_1 and v_2

						Degrees of freedom for the numerator (ν_1)												
	1	*2*	*3*	*4*	*5*	*6*	*7*	*8*	*9*	*10*	*11*	*12*	*13*	*14*	*15*	*16*	*17*	*18*
1	648	800	864	900	922	937	948	957	963	969	975	977	980	983	985	987	989	990
2	38.5	39.0	39.2	29.2	39.3	39.3	39.4	39.4	39.4	39.4	39.4	39.4	39.4	39.4	39.4	39.4	39.4	39.4
3	17.4	16.0	15.4	15.1	14.9	14.7	14.6	14.5	14.5	14.4	14.4	14.3	14.3	14.3	14.3	14.2	14.2	14.2
4	12.2	10.6	9.98	9.60	9.36	9.20	9.07	8.98	8.90	8.84	8.79	8.75	8.79	8.69	8.66	8.64	8.62	8.60
5	10.0	8.43	7.76	7.39	7.15	6.98	6.85	6.76	6.68	6.62	6.57	6.52	6.49	6.46	6.43	6.41	6.39	6.37
6	8.81	7.28	6.00	6.23	5.90	5.82	5.70	5.60	5.52	5.48	5.41	5.37	5.33	5.30	5.27	5.25	5.23	5.21
7	8.07	6.54	5.89	5.52	5.29	5.12	4.99	4.90	4.82	4.76	4.71	4.67	4.63	4.60	4.57	4.54	4.52	4.50
8	7.57	6.06	5.42	5.05	4.62	4.65	4.53	4.43	4.66	4.30	4.24	4.20	4.16	4.13	4.10	4.06	4.05	4.03
9	7.21	5.71	5.08	4.72	4.48	4.32	4.20	4.10	4.03	3.96	3.91	3.87	3.83	3.80	3.77	3.74	3.72	3.70
10	6.94	5.46	4.83	4.47	4.24	4.07	3.95	3.85	3.78	3.72	3.86	3.67	3.59	3.55	3.52	3.50	3.47	3.45
11	6.72	5.26	4.63	4.28	4.04	3.88	3.76	3.66	3.59	3.53	3.47	3.43	3.39	3.36	3.33	3.30	3.28	3.26
12	6.55	5.10	4.47	4.12	3.89	3.73	3.61	3.51	3.44	3.37	3.32	3.28	3.24	3.21	3.18	3.15	3.13	3.11
13	6.41	4.97	4.35	4.00	3.77	3.70	3.48	3.39	3.31	3.24	3.20	3.14	3.12	3.08	3.05	3.03	3.00	2.98
14	6.30	4.86	4.24	3.89	3.66	3.50	3.38	3.29	3.21	3.15	3.09	3.05	3.01	2.98	2.95	2.92	2.90	2.88
15	6.20	4.76	4.15	3.80	3.58	3.41	3.29	3.20	3.12	3.06	3.01	2.96	2.92	2.89	2.86	2.84	2.81	2.79
16	6.12	4.69	4.08	3.73	3.50	3.34	3.22	3.12	3.05	2.99	2.93	2.89	2.85	2.82	2.79	2.76	2.74	2.72
17	6.04	4.62	4.01	3.66	3.44	3.28	3.16	3.06	2.98	2.92	2.87	2.82	2.79	2.75	2.72	2.70	2.67	2.65
18	5.98	4.56	3.95	3.61	3.38	3.22	3.10	3.01	2.93	2.87	2.81	2.77	2.73	2.70	2.67	2.64	2.62	2.60
19	5.92	4.51	3.90	3.56	3.33	3.17	3.05	2.96	2.88	2.82	2.76	2.72	2.68	2.65	2.62	2.59	2.57	2.55
20	5.87	4.48	3.86	3.51	3.29	3.13	3.01	2.91	2.84	2.77	2.72	2.68	2.64	2.60	2.57	2.55	2.52	2.50
21	5.83	4.42	3.82	3.48	3.25	3.09	2.97	2.87	2.80	2.73	2.68	2.64	2.60	2.56	2.53	2.51	2.48	2.46
22	5.79	4.38	3.78	3.44	3.22	3.05	2.93	2.84	2.76	2.70	2.65	2.60	2.56	2.53	2.50	2.47	2.45	2.43
23	5.75	4.35	3.75	3.41	3.18	3.02	2.90	2.81	2.73	2.67	2.62	2.57	2.53	2.50	2.47	2.46	2.42	2.39
24	5.72	4.32	3.72	3.38	3.15	2.99	2.87	2.78	2.70	2.64	2.59	2.54	2.50	2.47	2.44	2.41	2.39	2.36
25	5.69	4.29	3.69	3.35	3.13	2.97	2.85	2.75	2.68	2.61	2.56	2.51	2.48	2.44	2.41	2.38	2.36	2.34
26	5.66	4.27	3.67	3.33	3.10	2.94	2.82	2.73	2.65	2.59	2.54	2.49	2.45	2.42	2.39	2.36	2.34	2.31
27	5.63	4.24	3.65	3.31	3.08	2.92	2.80	2.71	2.63	2.57	2.51	2.47	2.43	2.39	2.36	2.34	2.31	2.29
28	5.61	4.22	3.63	3.29	3.06	2.90	2.78	2.69	2.61	2.55	2.49	2.45	2.41	2.37	2.34	2.32	2.29	2.27
29	5.59	4.20	3.61	3.27	3.04	2.88	2.76	2.67	2.59	2.53	2.48	2.43	2.39	2.36	2.32	2.30	2.27	2.25
30	5.57	4.18	3.59	3.25	3.03	2.87	2.75	2.65	2.57	2.51	2.46	2.41	2.37	2.34	2.31	2.28	2.26	2.23
32	5.53	4.15	3.56	3.22	3.00	2.84	2.72	2.62	2.54	2.48	2.43	2.38	2.34	2.31	2.28	2.25	2.22	2.20
34	5.50	4.12	3.53	3.19	2.97	2.81	2.69	2.59	2.52	2.45	2.40	2.35	2.31	2.28	2.25	2.22	2.19	2.17
36	5.47	4.09	3.51	3.17	2.94	2.79	2.56	2.57	2.49	2.43	2.37	2.33	2.29	2.25	2.22	2.20	2.17	2.15
38	5.45	4.07	3.48	3.15	2.92	2.76	2.64	2.55	2.47	2.41	2.35	2.31	2.27	2.23	2.20	2.17	2.15	2.13
40	5.42	4.05	3.46	3.13	2.90	2.74	2.62	2.53	2.45	2.39	2.33	2.29	2.25	2.21	2.18	2.15	2.13	2.11

Degrees of freedom for the denominator (v_2)

						Degrees of freedom for the numerator (ν_1)											
19	*20*	*22*	*24*	*26*	*28*	*30*	*35*	*40*	*45*	*50*	*60*	*80*	*100*	*200*	*500*	∞	
992	923	995	997	999	1000	1001	1004	1005	1007	1008	1010	1012	1013	1016	1017	1018	1
23.4	30.4	29.5	39.5	39.5	39.5	39.5	39.5	39.5	39.5	39.5	39.5	39.5	39.5	39.5	39.5	39.5	2
14.2	14.2	14.1	14.1	14.1	14.1	14.1	14.1	10.0	14.0	14.0	14.0	14.0	14.0	13.9	13.9	13.9	3
8.53	8.56	8.53	8.51	8.49	8.48	8.48	8.44	8.41	8.39	8.38	8.36	8.33	8.32	8.29	8.27	8.26	4
6.35	6.33	6.30	6.28	6.26	6.24	6.23	6.20	6.18	6.16	6.14	6.12	6.10	6.06	6.05	6.03	6.02	5
5.19	5.17	5.14	5.12	5.10	5.08	5.07	5.04	5.01	4.99	4.96	4.96	4.93	4.92	4.88	4.86	4.85	6
4.48	4.47	4.44	4.42	4.92	4.38	4.38	4.33	4.31	4.29	4.28	4.25	4.23	4.21	4.18	4.16	4.14	7
4.02	4.00	3.97	3.95	3.93	3.91	3.90	3.86	3.84	3.82	3.81	3.78	3.76	3.74	3.70	3.68	3.67	8
3.68	3.67	3.64	3.61	3.59	3.58	3.56	3.53	3.51	3.49	3.47	3.45	3.42	3.40	3.37	3.35	3.33	9
3.44	3.42	3.39	3.37	3.34	3.33	3.31	3.28	3.26	3.24	3.23	3.20	3.17	3.15	3.12	3.09	3.08	10
3.24	3.23	3.20	3.17	3.15	3.13	3.12	3.09	3.06	3.04	3.03	3.00	2.97	2.96	2.92	2.90	2.88	11
3.09	3.07	3.04	3.02	3.00	2.98	2.96	2.93	2.91	2.89	2.87	2.85	2.82	2.80	2.76	2.74	2.72	12
2.96	2.95	2.92	2.89	2.87	2.85	2.84	2.80	2.78	2.76	2.74	2.72	2.69	2.67	2.63	2.61	2.60	13
2.86	2.84	2.81	2.79	2.77	2.75	2.73	2.70	2.67	2.65	2.64	2.61	2.58	2.56	2.53	2.50	2.49	14
2.77	2.76	2.73	2.70	2.68	2.66	2.64	2.61	2.55	2.56	2.55	2.52	2.49	2.47	2.44	2.41	2.40	15
2.70	2.68	2.65	2.63	2.60	2.58	2.57	2.53	2.51	2.49	2.47	2.45	2.42	2.40	2.36	2.33	2.32	16
2.63	2.62	2.59	2.56	2.54	2.52	2.50	2.47	2.44	2.42	2.41	2.38	2.35	2.33	2.29	2.28	2.25	17
2.58	2.56	2.53	2.50	2.48	2.46	2.44	2.41	2.38	2.36	2.35	2.32	2.29	2.27	2.23	2.20	2.19	18
2.53	2.51	2.48	2.45	2.43	2.41	2.39	2.36	2.33	2.31	2.30	2.27	2.24	2.22	2.18	2.15	2.13	19
2.43	2.46	2.43	2.41	2.39	2.37	2.35	2.31	2.29	2.27	2.25	2.22	2.19	2.17	2.13	2.10	2.09	20
2.44	2.42	2.39	2.37	2.34	2.33	2.31	2.27	2.25	2.23	2.21	2.18	2.15	2.13	2.09	2.06	2.04	21
2.41	2.39	2.36	2.33	2.31	2.29	2.27	2.24	2.21	2.19	2.17	2.14	2.11	2.09	2.05	2.02	2.00	22
2.37	2.36	2.33	2.30	2.28	2.26	2.24	2.20	2.18	2.15	2.14	2.11	2.08	2.06	2.01	1.99	1.97	23
2.35	2.33	2.30	2.27	2.25	2.23	2.21	2.17	2.15	2.12	2.11	2.08	2.05	2.02	1.98	1.95	1.94	24
2.32	2.30	2.27	2.24	2.22	2.20	2.18	2.13	2.12	2.10	2.08	2.05	2.02	2.00	1.95	1.92	1.91	25
2.29	2.28	2.24	2.22	2.19	2.17	2.16	2.12	2.09	2.07	2.05	2.03	1.99	1.97	1.92	1.90	1.88	26
2.27	2.25	2.22	2.19	2.17	2.15	2.13	2.10	2.07	2.05	2.03	2.00	1.97	1.94	1.90	1.87	1.85	27
2.25	2.23	2.20	2.17	2.15	2.13	2.11	2.08	2.05	2.03	2.01	1.98	1.94	1.92	1.88	1.85	1.83	28
2.23	2.21	2.18	2.15	2.13	2.11	2.09	2.06	2.03	2.01	1.99	1.96	1.92	1.90	1.86	1.83	1.81	29
2.21	2.20	2.16	2.14	2.11	2.08	2.07	2.04	2.01	1.99	1.97	1.94	1.90	1.88	1.84	1.81	1.79	30
2.18	2.16	2.13	2.10	2.08	2.06	2.04	2.00	1.98	1.95	1.93	1.91	1.87	1.85	1.80	1.77	1.75	32
2.15	2.13	2.10	2.07	2.05	2.03	2.01	1.97	1.95	1.92	1.90	1.88	1.84	1.82	1.77	1.74	1.72	34
2.12	2.11	2.08	2.05	2.03	2.00	1.99	1.95	1.92	1.90	1.88	1.85	1.81	1.79	1.74	1.71	1.69	36
2.11	2.09	2.05	2.03	2.00	1.98	1.96	1.93	1.90	1.87	1.85	1.82	1.79	1.76	1.71	1.68	1.66	38
2.08	2.07	2.03	2.01	1.98	1.96	1.94	1.90	1.88	1.85	1.83	1.80	1.76	1.74	1.69	1.66	1.64	40

Degrees of freedom for the denominator (ν_2)

Appendix 4

		1	2	3	4	5	6	7	8	9	10	11	12	13	14	15	16	17	18
		\multicolumn{18}{c}{*Degrees of freedom for the numerator (ν_1)*}

Degrees of freedom for the numerator (ν_1)

Degrees of freedom for the denominator (ν_2)	1	2	3	4	5	6	7	8	9	10	11	12	13	14	15	16	17	18
42	5.40	4.03	3.45	3.11	2.89	2.73	2.61	2.51	2.44	2.37	2.32	2.27	2.23	2.20	2.16	2.14	2.11	2.09
44	5.39	4.02	3.43	3.09	2.87	2.71	2.59	2.50	2.42	2.36	2.30	2.26	2.21	2.18	2.15	2.12	2.10	2.07
46	5.37	4.00	3.42	3.08	2.88	2.70	2.58	2.48	2.41	2.34	2.29	2.24	2.20	2.17	2.13	2.11	2.08	2.06
48	5.35	3.99	3.40	3.07	2.84	2.69	2.57	2.47	2.39	2.33	2.27	2.23	2.19	2.15	2.12	2.09	2.07	2.05
50	3.34	3.98	3.39	3.06	2.83	2.67	2.55	2.46	2.38	2.32	2.26	2.22	2.18	2.14	2.11	2.06	2.06	2.03
55	5.31	3.95	3.36	3.03	2.81	2.65	2.53	2.43	2.36	2.29	2.24	2.19	2.15	2.11	2.08	2.05	2.03	2.01
60	5.29	3.93	3.34	3.01	2.79	2.63	2.51	2.41	2.33	2.27	2.22	2.17	2.13	2.09	2.06	2.03	2.01	1.98
65	5.27	3.91	3.32	2.99	2.77	2.61	2.49	2.39	2.32	2.25	2.20	2.15	2.11	2.07	2.04	2.01	1.99	1.97
70	5.25	3.89	3.31	2.98	2.75	2.60	2.48	2.38	2.30	2.24	2.18	2.14	2.10	2.06	2.03	2.00	1.97	1.95
80	5.22	3.86	3.28	2.95	2.73	2.57	2.45	2.36	2.28	2.21	2.16	2.11	2.07	2.03	2.00	1.97	1.95	1.93
90	5.20	3.84	3.27	2.93	2.71	2.55	2.43	2.34	2.26	2.19	2.14	2.09	2.05	2.02	1.98	1.95	1.93	1.91
100	5.18	3.83	3.25	2.92	2.70	2.54	2.42	2.32	2.24	2.18	2.12	2.08	2.04	2.00	1.97	1.94	1.91	1.89
125	5.15	3.80	3.22	2.89	2.67	2.51	2.39	2.30	2.22	2.14	2.10	2.05	2.01	1.97	1.94	1.91	1.89	1.86
150	5.13	3.78	3.20	2.87	2.65	2.49	2.37	2.28	2.20	2.13	2.08	2.03	1.99	1.95	1.92	1.89	1.87	1.84
200	5.10	3.76	3.18	2.85	2.63	2.47	2.35	2.26	2.18	2.11	2.06	2.01	1.97	1.93	1.90	1.87	1.84	1.82
300	5.08	3.74	3.16	2.83	2.61	2.45	2.33	2.23	2.16	2.09	2.04	1.99	1.95	1.91	1.88	1.85	1.82	1.80
500	5.05	3.72	3.14	2.81	2.59	2.43	2.31	2.22	2.14	2.07	2.02	1.97	1.93	1.89	1.86	1.83	1.80	1.78
1000	5.04	3.70	3.13	2.80	2.58	2.42	2.30	2.20	2.13	2.06	2.01	1.96	1.92	1.88	1.85	1.82	1.79	1.77
∞	5.02	3.69	3.12	2.79	2.57	2.41	2.29	2.19	2.11	2.05	1.99	1.94	1.90	1.87	1.83	1.80	1.78	1.75

					Degrees of freedom for the numerator (ν_1)												
19	20	22	24	25	28	30	35	40	45	50	60	80	100	200	500	∞	
2.07	2.05	2.02	1.99	1.96	1.94	1.92	1.89	1.86	1.83	1.81	1.78	1.74	1.72	1.67	1.64	1.62	42
2.05	2.03	2.00	1.97	1.95	1.93	1.91	1.87	1.84	1.82	1.80	1.77	1.73	1.70	1.65	1.62	1.60	44
2.04	2.02	1.99	1.96	1.93	1.91	1.89	1.85	1.82	1.80	1.78	1.75	1.71	1.69	1.63	1.60	1.58	46
2.03	2.01	1.97	1.94	1.92	1.90	1.88	1.84	1.81	1.79	1.77	1.73	1.69	1.67	1.62	1.58	1.56	48
2.01	1.99	1.96	1.93	1.91	1.88	1.87	1.83	1.80	1.77	1.75	1.72	1.68	1.66	1.59	1.57	1.55	50
1.99	1.97	1.93	1.90	1.88	1.86	1.84	1.80	1.77	1.74	1.72	1.69	1.65	1.62	1.57	1.54	1.51	55
1.96	1.94	1.91	1.88	1.86	1.83	1.82	1.78	1.74	1.72	1.70	1.67	1.62	1.60	1.54	1.51	1.48	60
1.95	1.93	1.89	1.86	1.84	1.82	1.80	1.76	1.72	1.70	1.68	1.65	1.60	1.58	1.52	1.48	1.46	65
1.93	1.91	1.88	1.85	1.82	1.80	1.78	1.74	1.71	1.68	1.66	1.63	1.58	1.56	1.50	1.46	1.44	70
1.90	1.88	1.85	1.82	1.79	1.77	1.75	1.71	1.68	1.65	1.63	1.60	1.55	1.53	1.47	1.43	1.40	80
1.88	1.86	1.83	1.80	1.77	1.75	1.73	1.69	1.66	1.63	1.61	1.58	1.53	1.50	1.44	1.40	1.37	90
1.87	1.85	1.81	1.78	1.76	1.74	1.71	1.67	1.64	1.61	1.59	1.56	1.51	1.48	1.42	1.38	1.35	100
1.84	1.82	1.79	1.75	1.73	1.71	1.68	1.64	1.61	1.58	1.56	1.52	1.48	1.45	1.38	1.34	1.30	125
1.82	1.80	1.77	1.74	1.71	1.69	1.67	1.62	1.58	1.56	1.54	1.50	1.45	1.42	1.35	1.31	1.27	150
1.80	1.78	1.74	1.71	1.68	1.66	1.64	1.60	1.56	1.53	1.51	1.47	1.42	1.39	1.32	1.27	1.23	200
1.77	1.75	1.72	1.69	1.66	1.64	1.62	1.57	1.54	1.51	1.48	1.45	1.39	1.36	1.28	1.23	1.18	200
1.78	1.74	1.70	1.67	1.64	1.62	1.60	1.55	1.51	1.49	1.46	1.42	1.37	1.34	1.25	1.19	1.14	500
1.74	1.72	1.69	1.65	1.63	1.60	1.58	1.54	1.50	1.47	1.44	1.41	1.35	1.32	1.23	1.16	1.09	1000
1.73	1.71	1.67	1.64	1.61	1.59	1.57	1.52	1.48	1.45	1.43	1.39	1.33	1.30	1.21	1.13	1.00	∞

Degrees of freedom for the denominator (ν_2)

Appendix 4

Tabulation of the values of $F_{0.01;\nu_1;\nu_2}$ versus ν_1 and ν_2

						Degrees of freedom for the numerator (ν_1)													
	1	*2*	*3*	*4*	*5*	*6*	*7*	*8*	*9*	*10*	*11*	*12*	*13*	*14*	*15*	*16*	*17*	*18*

Multiply the numbers of the first row ($\nu_2 = 1$) by 10

ν_2	1	2	3	4	5	6	7	8	9	10	11	12	13	14	15	16	17	18
1	405	500	540	563	576	596	598	598	602	606	608	611	613	614	616	617	618	619
2	93.5	99.0	99.3	99.3	99.3	99.3	99.4	99.4	99.4	99.4	99.4	99.4	99.4	99.4	99.4	99.4	99.4	99.4
3	34.1	30.8	30.5	28.7	28.2	27.9	27.7	27.5	27.3	27.2	27.1	27.1	27.0	26.9	26.9	25.8	26.8	26.8
4	21.2	18.0	16.7	16.0	15.5	15.2	15.0	14.8	14.7	14.5	14.4	14.4	14.3	14.2	14.2	14.2	14.1	14.1
5	16.8	13.2	12.1	11.4	11.0	10.7	10.5	10.3	10.2	10.1	9.06	9.89	9.82	9.77	9.72	9.68	9.64	9.61
6	13.7	10.9	9.78	9.15	8.75	8.47	8.28	8.10	7.98	7.87	7.79	7.72	7.66	7.60	7.56	7.52	7.48	7.45
7	12.2	9.55	8.45	7.85	7.46	7.19	6.99	6.94	6.72	6.62	6.54	6.47	6.41	6.36	5.31	6.27	6.24	6.21
8	11.3	8.65	7.89	7.01	6.63	6.37	6.18	6.03	5.91	5.81	5.73	5.67	5.61	5.56	5.52	5.48	5.44	5.41
9	10.6	8.02	6.99	6.42	6.06	5.80	5.61	5.47	5.35	5.26	5.13	5.11	5.05	5.00	4.96	4.92	4.89	4.86
10	10.0	7.56	6.55	5.99	5.64	5.39	5.20	5.06	4.94	4.85	4.77	4.71	4.65	4.60	4.56	4.52	4.49	4.46
11	9.65	7.21	6.22	5.67	5.32	5.07	4.89	4.74	4.63	4.54	4.46	4.40	4.34	4.39	4.25	4.21	4.18	4.15
12	9.33	6.93	5.95	5.41	5.06	4.82	4.64	4.50	4.30	4.30	4.22	4.16	4.10	4.05	4.01	3.97	3.94	3.91
13	9.07	6.70	5.74	5.21	4.86	4.62	4.44	4.30	4.19	4.10	4.02	3.96	3.91	3.86	3.82	3.78	3.75	3.72
14	8.86	6.51	5.58	5.04	4.70	4.46	4.28	4.14	4.03	3.94	3.88	3.80	3.75	3.70	3.66	3.62	3.50	3.56
15	8.68	6.26	5.42	4.89	4.56	4.32	4.14	4.00	3.89	3.80	3.72	3.67	3.61	3.56	3.52	3.49	3.45	3.42
16	8.53	6.22	5.29	4.77	4.44	4.20	4.03	3.89	3.78	3.69	3.62	3.55	3.50	3.45	3.41	3.37	3.34	3.31
17	8.60	6.11	5.18	4.67	4.34	4.10	3.93	3.79	3.68	3.59	3.52	3.46	3.40	3.35	3.31	3.27	3.24	3.21
18	8.20	6.01	5.09	4.58	4.25	4.01	3.84	3.71	3.60	3.51	3.43	3.37	3.32	3.27	3.22	3.19	3.16	3.13
19	8.18	5.93	5.01	4.50	4.17	3.94	3.77	3.68	3.52	3.43	3.36	3.30	3.24	3.19	3.15	3.12	3.08	3.05
20	8.10	5.85	4.94	4.43	4.10	3.87	3.70	3.56	3.46	3.37	3.29	3.23	3.18	3.13	3.09	3.05	3.02	2.99
21	8.02	5.78	4.87	4.37	4.04	3.81	3.64	3.51	3.40	3.31	3.24	3.17	3.12	3.07	3.03	2.99	2.96	2.93
22	7.95	5.72	4.82	4.31	3.99	3.76	3.59	3.45	3.35	3.26	3.18	3.12	3.07	3.02	2.98	2.94	2.91	2.88
23	7.86	5.66	4.76	4.26	3.94	3.71	3.54	3.41	3.30	3.21	3.14	3.07	3.02	2.97	2.93	2.89	2.86	2.83
24	7.82	5.61	4.72	4.22	3.90	3.67	3.50	3.36	3.26	3.17	3.09	3.03	2.98	2.93	2.89	2.85	2.83	2.79
25	7.77	5.57	4.68	4.18	3.86	3.63	3.46	3.32	3.22	3.13	3.06	2.99	2.94	2.89	2.85	2.81	2.78	2.75
26	7.72	5.53	4.64	4.14	3.82	3.59	3.42	3.29	3.18	3.09	3.02	2.96	2.90	2.86	2.82	2.78	2.74	2.72
27	7.66	5.49	4.60	4.11	3.78	3.56	3.39	3.26	3.15	3.06	2.99	2.93	2.87	2.82	2.78	2.75	2.71	2.68
28	7.64	5.45	4.57	4.07	3.75	3.53	3.36	3.23	3.12	3.03	2.96	2.90	2.84	2.79	2.75	2.72	2.68	2.65
29	7.60	5.42	4.54	4.04	3.73	3.50	3.33	3.20	3.09	3.00	2.93	2.87	2.81	2.77	2.73	2.69	2.66	2.63
30	7.56	5.39	4.51	4.03	3.70	3.47	3.30	3.17	3.07	2.98	2.91	2.84	2.79	2.74	2.70	2.66	2.63	2.60
32	7.50	5.34	4.46	3.97	3.65	3.43	3.26	3.13	3.02	2.93	2.86	2.80	2.74	2.70	2.66	2.62	2.58	2.55
34	7.44	5.29	4.42	3.93	3.61	3.39	3.23	3.09	2.96	2.89	2.82	2.76	2.70	2.66	2.62	2.58	2.55	2.51
36	7.40	5.25	4.36	3.89	3.57	3.35	3.18	3.05	2.95	2.86	2.79	2.72	2.67	2.62	2.58	2.54	2.51	2.48
38	7.35	5.21	4.34	3.86	3.54	3.32	3.15	3.02	2.92	2.83	2.75	2.69	2.64	2.59	2.55	2.51	2.48	2.45
40	7.31	5.18	4.31	3.83	3.51	3.29	3.12	2.99	2.90	2.80	2.73	2.66	2.61	2.56	2.52	2.48	2.45	2.42

Degrees of freedom for the denominator (ν_2)

Degrees of freedom for the numerator (ν_1)																	
19	*20*	*22*	*24*	*26*	*28*	*30*	*35*	*40*	*45*	*50*	*60*	*80*	*100*	*200*	*500*	∞	
Multiply the numbers of the first row ($\nu_2 = 1$) by 10																	
620	*621*	*623*	*623*	*624*	*625*	*626*	*628*	*629*	*630*	*630*	*631*	*633*	*633*	*635*	*636*	*637*	1
99.4	99.4	99.5	99.5	99.5	99.5	99.5	99.5	99.5	99.5	99.5	99.5	99.5	99.5	99.5	99.5	99.5	2
26.7	26.7	26.6	26.6	26.6	26.5	26.5	26.5	26.4	26.4	26.4	26.3	26.3	26.2	26.2	26.1	26.1	3
14.0	14.0	14.0	13.9	13.9	13.9	13.8	13.8	13.7	13.7	13.7	13.7	13.6	13.6	13.5	13.5	13.5	4
9.58	9.55	9.51	9.47	9.43	9.40	9.38	9.33	9.39	9.36	9.24	9.20	9.16	9.13	9.08	9.04	9.02	5
7.42	7.40	7.35	7.31	7.28	7.25	7.23	7.18	7.14	7.11	7.09	7.06	7.01	6.99	6.93	6.90	6.88	6
6.18	6.16	6.11	6.07	6.04	6.02	5.99	5.94	5.91	5.88	5.88	5.82	5.78	5.75	5.70	5.67	5.65	7
5.38	5.28	5.32	5.28	5.25	5.22	5.20	5.15	5.12	5.09	5.07	5.03	4.99	4.96	4.91	4.88	4.86	8
4.53	4.81	4.77	4.73	4.70	4.67	4.65	4.60	4.57	4.54	4.52	4.48	4.44	4.42	4.36	4.33	4.31	9
4.43	4.41	4.36	4.33	4.30	4.27	4.25	4.20	4.17	4.14	4.12	4.08	4.04	4.01	3.96	3.93	3.91	10
4.12	4.10	4.06	4.02	3.99	3.95	3.94	3.89	3.88	3.83	3.81	3.78	3.73	3.71	3.66	3.62	3.60	11
3.88	3.86	3.82	3.78	3.75	3.72	3.70	3.65	3.62	3.59	3.57	3.54	3.49	3.47	3.41	3.38	3.36	12
3.69	3.66	3.62	3.59	3.56	3.53	3.51	3.46	3.43	3.40	3.38	3.34	3.30	3.27	3.22	3.19	3.17	13
3.53	3.51	3.46	3.43	3.40	3.37	3.35	3.30	3.27	3.24	3.22	3.18	3.14	3.11	3.06	3.03	3.00	14
3.40	3.37	3.33	3.29	3.26	3.24	3.21	3.17	3.13	3.10	3.08	3.05	3.00	2.98	2.92	2.89	2.87	15
3.28	3.26	3.22	3.18	3.15	3.12	3.10	3.05	3.02	2.99	2.97	2.93	2.89	2.88	2.81	2.78	2.75	16
3.18	3.16	3.12	3.08	3.05	3.03	3.00	2.96	2.92	2.89	2.87	2.83	2.79	2.76	2.71	2.68	2.68	17
3.10	3.08	3.03	3.00	2.97	2.94	2.92	2.87	2.84	2.81	2.78	2.75	2.70	2.68	2.62	2.59	2.57	18
3.03	3.00	2.96	2.92	2.89	2.87	2.84	2.80	2.76	2.73	2.71	2.67	2.63	2.60	2.55	2.51	2.49	19
2.93	2.94	2.90	2.88	2.83	2.80	2.78	2.73	2.69	2.67	2.64	2.61	2.56	2.54	2.48	2.44	2.42	20
2.90	2.88	2.84	2.80	2.77	2.74	2.72	2.67	2.64	2.61	2.58	2.55	2.50	2.46	2.42	2.38	2.36	21
2.85	2.83	2.78	2.75	2.72	2.69	2.67	2.62	2.58	2.55	2.53	2.50	2.45	2.42	2.36	2.33	2.31	22
2.80	2.78	2.74	2.70	2.67	2.64	2.62	2.57	2.54	2.51	2.48	2.45	2.40	2.37	2.32	2.28	2.26	23
2.76	2.74	2.70	2.66	2.63	2.60	2.58	2.53	2.49	2.46	2.44	2.40	2.36	2.33	2.27	2.24	2.21	24
2.72	2.70	2.66	2.62	2.50	2.56	2.54	2.49	2.45	2.42	2.40	2.36	2.32	2.29	2.23	2.19	2.17	25
2.69	2.66	2.62	2.58	2.55	2.53	2.50	2.45	2.42	2.39	2.36	2.33	2.28	2.25	2.19	2.16	2.13	26
2.66	2.63	2.59	2.55	2.52	2.49	2.47	2.42	2.38	2.35	2.33	2.29	2.25	2.22	2.16	2.12	2.10	27
2.63	2.60	2.56	2.52	2.49	2.46	2.44	2.39	2.35	2.32	2.30	2.26	2.22	2.19	2.13	2.09	2.06	28
2.60	2.57	2.53	2.49	2.46	2.44	2.41	2.36	2.33	2.30	2.27	2.23	2.19	2.16	2.10	2.06	2.03	29
2.57	2.55	2.51	2.47	2.44	2.41	2.39	2.34	2.30	2.27	2.24	2.21	2.16	2.13	2.07	2.03	2.01	30
2.53	2.50	2.46	2.42	2.39	2.36	2.34	2.29	2.25	2.22	2.20	2.16	2.11	2.08	2.02	1.98	1.96	32
2.49	2.46	2.42	2.38	2.35	2.32	2.30	2.25	2.21	2.18	2.16	2.12	2.07	2.04	1.98	1.94	1.91	34
2.45	2.43	2.38	2.35	2.32	2.29	2.26	2.21	2.17	2.14	2.12	2.08	2.03	2.00	1.94	1.90	1.87	36
2.42	2.40	2.35	2.32	2.28	2.26	2.23	2.18	2.14	2.11	2.09	2.05	2.00	1.97	1.90	1.86	1.84	38
2.39	2.37	2.33	2.39	2.26	2.23	2.20	2.15	2.11	2.08	2.06	2.02	1.97	1.94	1.87	1.83	1.80	40

Degrees of freedom for the denominator (ν_2)

	Degrees of freedom for the numerator (ν_1)																	
	1	2	3	4	5	6	7	8	9	10	11	12	13	14	15	16	17	18

Multiply the numbers of the first row ($\nu_2 = 1$) by 10

Degrees of freedom for the denominator (ν_2):

ν_2	1	2	3	4	5	6	7	8	9	10	11	12	13	14	15	16	17	18
42	7.28	5.15	4.29	3.80	3.49	3.27	3.10	2.97	2.86	2.78	2.70	2.64	2.59	2.54	2.50	2.46	2.43	2.40
44	7.25	5.12	4.26	3.78	3.47	3.24	3.08	2.95	2.84	2.75	2.68	2.62	2.56	2.52	2.47	2.44	2.40	2.37
46	7.22	5.10	4.24	3.76	3.44	3.22	3.06	2.93	2.82	2.73	2.66	2.60	2.54	2.50	2.45	2.42	2.38	2.35
48	7.19	5.08	4.22	3.74	3.43	3.20	3.04	2.91	2.80	2.72	2.64	2.58	2.53	2.48	2.44	2.40	2.37	2.33
50	7.17	5.06	4.20	3.72	3.41	3.19	3.02	2.89	2.79	2.70	2.63	2.56	2.51	2.46	2.42	2.38	2.35	2.32
55	7.12	5.01	4.16	3.68	3.37	3.15	2.98	2.85	2.75	2.66	2.59	2.53	2.47	2.42	2.38	2.34	2.31	2.28
60	7.08	4.98	4.12	3.65	3.34	3.12	2.95	2.82	2.72	2.63	2.56	2.50	2.44	2.39	2.35	2.31	2.28	2.24
65	7.04	4.95	4.10	3.62	3.31	3.09	2.93	2.80	2.69	2.61	2.53	2.47	2.42	2.37	2.33	2.29	2.26	2.23
70	7.01	4.92	4.08	3.60	3.29	3.07	2.91	2.78	2.67	2.59	2.51	2.45	2.40	2.24	2.31	2.27	2.23	2.20
80	6.98	4.88	4.04	3.56	3.26	3.04	2.87	2.74	2.64	2.55	2.48	2.42	2.36	2.21	2.27	2.23	2.20	2.17
90	6.93	4.85	4.01	3.54	3.23	3.01	2.84	2.72	2.61	2.52	2.45	2.39	2.33	2.29	2.24	2.21	2.17	2.14
100	6.90	4.83	3.96	3.51	3.21	2.99	2.82	2.69	2.59	2.50	2.43	2.37	2.31	2.26	2.22	2.19	2.15	2.12
125	6.84	4.78	3.94	3.47	3.17	2.95	2.79	2.66	2.55	2.47	2.39	2.33	2.28	2.23	2.19	2.15	2.11	2.08
150	6.81	4.75	3.92	3.45	3.14	2.92	2.76	2.63	2.53	2.44	2.37	2.31	2.25	2.20	2.16	2.12	2.09	2.06
200	6.76	4.71	3.88	3.41	3.11	2.89	2.73	2.60	2.50	2.41	2.34	2.27	2.22	2.17	2.13	2.09	2.06	2.02
300	6.72	4.68	3.85	3.38	3.08	2.86	2.70	2.57	2.47	2.36	2.31	2.24	2.19	2.14	2.10	2.06	2.03	1.99
500	6.69	4.65	3.82	3.36	3.05	2.84	2.68	2.55	2.44	2.36	2.28	2.22	2.17	2.12	2.07	2.04	2.00	1.97
1000	6.66	4.63	3.80	3.34	3.04	2.82	2.66	2.53	2.43	2.34	2.27	2.20	2.15	2.10	2.06	2.02	1.98	1.95
∞	6.66	4.61	3.78	3.32	3.02	2.80	2.64	2.51	2.41	2.32	2.25	2.18	2.13	2.08	2.04	2.00	1.97	1.00

Degrees of freedom for the numerator (ν_1)																	

19	20	22	24	26	28	30	35	40	45	50	60	80	100	200	500	∞	
							Multiply the numbers of the first row ($\nu_2 = 1$) by 10										
2.37	2.34	2.30	2.26	2.23	2.20	2.18	2.13	2.09	2.08	2.03	1.99	1.94	1.91	1.85	1.80	1.78	42
2.35	2.32	2.28	2.24	2.21	2.18	2.15	2.10	2.06	2.03	2.01	1.97	1.92	1.89	1.82	1.78	1.75	44
2.33	2.30	2.26	2.22	2.19	2.16	2.13	2.08	2.04	2.01	1.99	1.95	1.90	1.86	1.80	1.75	1.73	46
2.31	2.28	2.24	2.20	2.17	2.14	2.12	2.06	2.02	1.99	1.97	1.93	1.88	1.84	1.78	1.73	1.70	48
2.29	2.27	2.22	2.18	2.15	2.12	2.10	2.05	2.01	1.97	1.95	1.91	1.86	1.82	1.76	1.71	1.68	50
2.25	2.23	2.18	2.15	2.11	2.08	2.06	2.01	1.97	1.92	1.91	1.87	1.81	1.78	1.71	1.67	1.64	55
2.22	2.20	2.15	2.12	2.08	2.05	2.03	1.98	1.94	1.90	1.88	1.84	1.78	1.75	1.68	1.63	1.60	60
2.20	2.17	2.13	2.09	2.06	2.03	2.00	1.95	1.91	1.88	1.85	1.81	1.75	1.72	1.65	1.60	1.57	65
2.18	2.15	2.11	2.07	2.03	2.01	1.98	1.93	1.89	1.85	1.83	1.78	1.73	1.70	1.62	1.57	1.54	70
2.14	2.12	2.07	2.03	2.00	1.97	1.94	1.89	1.85	1.81	1.79	1.75	1.69	1.66	1.58	1.53	1.49	80
2.11	2.09	2.04	2.00	1.97	1.94	1.92	1.86	1.82	1.79	1.76	1.72	1.66	1.62	1.54	1.49	1.46	90
2.09	2.07	2.02	1.98	1.94	1.92	1.89	1.84	1.80	1.76	1.73	1.69	1.63	1.60	1.52	1.47	1.43	100
2.05	2.03	1.98	1.94	1.91	1.88	1.85	1.80	1.79	1.72	1.69	1.65	1.59	1.55	1.47	1.41	1.37	125
2.03	2.00	1.96	1.92	1.88	1.85	1.83	1.77	1.73	1.69	1.66	1.63	1.56	1.52	1.43	1.35	1.33	150
2.00	1.97	1.93	1.89	1.84	1.82	1.79	1.74	1.69	1.66	1.63	1.58	1.52	1.48	1.39	1.33	1.28	200
1.97	1.94	1.89	1.85	1.82	1.79	1.76	1.71	1.68	1.62	1.58	1.55	1.48	1.44	1.35	1.28	1.22	300
1.94	1.92	1.87	1.83	1.79	1.76	1.74	1.68	1.63	1.60	1.55	1.53	1.45	1.41	1.31	1.23	1.16	500
1.92	1.90	1.85	1.81	1.77	1.74	1.72	1.66	1.61	1.57	1.54	1.50	1.43	1.38	1.28	1.19	1.11	1000
1.90	1.86	1.83	1.79	1.76	1.72	1.70	1.64	1.58	1.55	1.52	1.47	1.40	1.36	1.25	1.15	1.00	∞

Degrees of freedom for the denominator (ν_2)

Tabulation of the values of $F_{0.005;\nu_1;\nu_2}$ versus ν_1 and ν_2

	Degrees of freedom for the numerator (ν_1)																	
	1	*2*	*3*	*4*	*5*	*6*	*7*	*8*	*9*	*10*	*11*	*12*	*13*	*14*	*15*	*16*	*17*	*18*

Multiply the numbers of the first row ($\nu_2 = 1$) by 100

ν_2	1	2	3	4	5	6	7	8	9	10	11	12	13	14	15	16	17	18
1	162	200	216	225	231	234	237	239	241	242	243	244	245	246	246	247	247	248
2	198	199	199	199	199	199	199	199	199	199	199	199	199	199	199	199	199	199
3	55.6	49.8	47.5	46.2	45.4	44.3	44.4	44.1	43.9	43.7	43.5	43.4	43.3	43.2	43.1	43.0	42.9	42.9
4	31.2	26.3	24.3	23.2	22.5	22.0	21.6	21.4	21.1	21.0	20.8	20.7	20.6	20.5	20.4	20.4	20.3	20.2
5	22.8	18.2	16.5	15.6	14.3	14.5	14.2	14.0	13.8	13.6	13.5	13.4	13.3	12.2	12.1	12.1	13.0	13.0
6	18.6	14.5	12.9	13.0	11.5	11.1	10.8	10.6	10.4	10.2	10.1	10.0	9.95	9.88	9.81	9.76	9.71	9.08
7	16.2	12.4	10.9	10.0	9.52	9.16	8.89	8.68	8.51	8.38	8.27	8.18	8.10	8.03	7.97	7.93	7.87	7.83
8	14.7	11.0	9.60	8.81	8.30	7.95	7.69	7.50	7.34	7.21	7.10	7.01	6.94	6.37	6.34	6.76	6.72	6.68
9	13.6	10.1	8.72	7.96	7.47	7.13	6.88	6.69	6.54	6.42	6.31	6.23	6.15	6.09	6.03	5.98	5.94	5.90
10	12.8	9.43	8.06	7.34	6.87	6.54	6.30	6.12	5.97	5.85	5.75	5.66	5.50	5.53	5.47	5.42	5.38	5.34
11	12.2	8.91	7.60	6.88	6.42	6.10	5.86	5.68	5.54	5.42	5.32	5.24	5.16	5.10	5.05	5.00	4.96	4.92
12	11.8	8.51	7.23	6.52	6.07	5.76	5.52	5.35	5.20	5.09	4.99	4.91	4.84	4.77	4.72	4.67	4.63	4.50
13	11.4	8.19	6.93	6.23	5.79	5.48	5.25	5.08	4.94	4.82	4.72	4.54	4.57	4.54	4.46	4.41	4.37	4.23
14	11.1	7.92	6.68	6.00	5.56	5.26	5.03	4.86	4.72	4.60	4.51	4.43	4.36	4.30	4.25	4.20	4.18	4.12
15	10.8	7.70	6.48	5.80	5.37	5.07	4.85	4.67	4.54	4.42	4.33	4.25	4.18	4.12	4.07	4.02	3.93	3.95
16	10.6	7.51	6.30	5.64	5.21	4.91	4.69	4.52	4.48	4.27	4.18	4.10	4.03	3.97	3.92	3.87	3.83	3.80
17	10.4	7.35	6.16	5.50	5.07	4.78	4.56	4.39	4.25	4.14	4.05	3.97	3.90	3.84	3.79	3.75	3.71	3.67
18	10.2	7.21	6.03	5.37	4.98	4.66	4.44	4.28	4.14	4.03	3.94	3.86	3.79	3.73	3.68	3.64	3.60	3.56
19	10.1	7.09	5.92	5.27	4.85	4.56	4.34	4.18	4.04	3.93	3.84	3.76	3.70	3.64	3.59	3.54	3.50	3.46
20	9.94	6.99	5.82	5.17	4.76	4.47	4.28	4.09	3.96	3.85	3.76	3.68	3.61	3.55	3.50	3.48	3.42	3.28
21	9.82	6.89	5.73	5.09	4.39	4.63	4.18	4.01	3.88	3.77	3.68	3.60	3.54	3.48	3.43	3.38	3.34	3.31
22	9.73	6.81	5.65	5.02	4.61	4.32	4.11	3.94	3.81	3.70	3.61	3.54	3.47	3.41	3.36	3.31	3.27	3.24
23	9.63	6.72	5.58	4.95	4.54	4.26	4.05	3.85	3.75	3.64	3.55	3.47	3.41	3.35	3.30	3.25	3.21	3.18
24	9.55	6.66	5.52	4.89	4.49	4.20	3.99	3.83	3.69	3.59	3.50	3.42	3.35	3.30	3.25	3.20	3.16	3.12
25	9.48	6.60	5.46	4.84	4.43	4.15	3.94	3.78	3.64	3.54	3.45	3.37	3.30	3.25	3.20	3.15	3.11	3.08
26	9.41	6.54	5.41	4.79	4.36	4.10	3.89	3.73	3.60	3.49	3.40	3.33	3.26	3.20	3.15	3.17	3.07	3.03
27	9.34	6.49	5.36	4.74	3.44	4.06	3.85	3.69	3.56	3.45	3.36	3.28	3.22	3.16	3.11	3.07	3.03	2.99
28	9.28	6.44	5.32	4.70	4.30	4.02	3.81	3.65	3.52	3.41	3.32	3.25	3.18	3.12	3.07	2.08	2.99	2.95
29	9.23	6.40	5.26	4.66	4.26	3.98	3.77	3.61	3.48	3.38	3.29	3.21	3.15	3.09	3.04	2.09	2.95	2.92
30	9.18	6.35	5.24	4.62	4.23	3.95	3.74	3.58	3.45	3.34	3.25	3.18	3.11	3.06	3.01	2.98	2.92	2.89
32	9.09	6.28	5.17	4.56	4.17	3.89	3.68	3.52	3.39	3.29	3.20	3.12	3.06	3.00	2.95	2.90	2.86	2.83
34	9.01	6.22	5.11	4.50	4.11	3.84	3.63	3.47	3.34	3.24	3.15	3.07	3.01	2.95	2.90	2.85	2.81	2.78
36	8.94	6.15	5.06	4.46	4.06	3.79	3.58	3.42	3.30	3.19	3.10	3.03	2.96	2.90	2.85	2.81	2.77	2.73
38	8.88	6.11	5.02	4.41	4.02	3.75	3.54	3.39	3.25	3.15	3.06	2.99	2.92	2.87	2.82	2.77	2.73	2.70
40	8.83	6.07	4.98	4.37	3.99	3.71	3.51	3.35	3.22	3.12	3.03	2.95	2.90	2.83	2.78	2.74	2.70	2.66

Degrees of freedom for the numerator (ν_1)																		
19	*20*	*22*	*24*	*26*	*28*	*30*	*35*	*40*	*45*	*50*	*60*	*80*	*100*	*200*	*500*	*∞*	

Multiply the numbers of the first row ($\nu_2 = 1$) by 100

																	ν_2
248	*248*	*249*	*249*	*250*	*250*	*250*	*251*	*251*	*252*	*252*	*253*	*253*	*253*	*254*	*254*	*255*	1
199	199	199	199	199	199	199	199	199	199	199	199	199	199	199	200	200	2
42.8	42.8	42.7	42.6	42.6	42.5	42.5	42.4	42.3	42.3	42.2	42.1	42.1	42.0	41.9	41.9	41.8	3
20.2	20.2	20.1	20.0	20.0	19.9	19.9	19.8	19.8	19.7	19.7	19.6	19.5	19.5	19.4	19.4	19.3	4
12.9	12.9	12.8	12.8	12.7	12.6	12.7	12.6	12.5	12.5	12.5	12.4	12.3	12.3	12.2	12.2	12.1	5
9.62	9.59	9.53	9.47	9.43	9.39	9.36	9.29	9.24	9.20	9.17	9.12	9.06	9.03	8.95	8.91	8.83	6
7.79	7.75	7.69	7.64	7.60	7.57	7.53	7.47	7.42	7.38	7.25	7.21	7.25	7.22	7.15	7.10	7.08	7
6.64	6.61	6.55	6.50	6.46	6.43	6.40	6.33	6.29	6.25	6.22	6.18	6.12	6.09	6.02	5.98	5.95	8
5.86	5.83	5.78	5.73	5.69	5.65	5.62	5.56	5.52	5.48	5.45	5.41	5.36	5.32	5.26	5.21	5.19	9
5.30	5.27	5.22	5.17	5.13	5.10	5.07	5.01	4.97	4.93	4.90	4.98	4.80	4.77	4.71	4.67	4.64	10
4.89	4.86	4.80	4.76	4.72	4.68	4.65	4.60	4.55	4.52	4.49	4.44	4.39	4.36	4.20	4.25	4.23	11
4.56	4.53	4.48	4.43	4.39	4.36	4.33	4.27	4.23	4.19	4.17	4.12	4.07	4.04	3.97	3.93	3.90	12
4.30	4.27	4.22	4.17	4.13	4.10	4.07	4.01	3.97	3.94	3.91	3.87	3.81	3.78	3.71	3.67	3.65	13
4.09	4.08	4.01	3.96	3.92	3.89	3.86	3.80	3.76	3.73	3.70	3.66	3.60	3.57	3.50	3.46	3.44	14
3.91	3.88	3.83	3.79	3.75	3.72	3.69	3.63	3.58	3.55	3.52	3.43	3.43	3.30	3.33	3.29	3.28	15
3.76	3.73	3.68	3.64	3.60	3.57	3.54	3.48	3.44	3.40	3.37	3.33	3.28	3.25	3.18	3.14	3.11	16
3.64	3.61	3.56	3.51	3.47	3.44	3.41	3.35	3.31	3.28	3.25	3.21	3.15	3.12	3.05	3.01	2.98	17
3.53	3.50	3.45	3.40	3.36	3.33	3.33	3.25	3.20	3.17	3.14	3.10	3.04	3.01	2.94	2.90	2.87	18
3.43	3.40	3.35	3.31	3.27	3.24	3.21	3.15	3.11	3.07	3.04	3.00	2.95	2.91	2.85	2.80	2.78	19
3.35	3.32	3.27	3.22	3.18	3.15	3.12	3.07	3.02	2.99	2.96	2.92	2.86	2.83	2.76	2.72	2.69	20
3.27	3.24	3.19	3.15	3.11	3.08	3.05	2.99	2.95	2.91	2.88	2.84	2.78	2.75	2.68	2.64	2.61	21
3.20	3.18	3.12	3.08	3.04	3.01	2.98	2.92	2.88	2.84	2.82	2.77	2.72	2.69	2.62	2.57	2.55	22
3.15	3.12	3.06	3.02	2.98	2.95	2.92	2.86	2.82	2.78	2.76	2.71	2.66	2.62	2.56	2.51	2.48	23
3.09	3.06	3.01	2.97	2.93	2.90	2.87	2.81	2.77	2.73	2.70	2.68	2.60	2.57	2.50	2.46	2.43	24
3.04	3.01	2.96	2.92	2.88	2.85	2.82	2.76	2.72	2.68	2.65	2.61	2.55	2.52	2.45	2.41	2.38	25
3.00	2.97	2.92	2.87	2.83	2.80	2.77	2.72	2.67	2.64	2.61	2.56	2.51	2.47	2.40	2.36	2.33	26
2.96	2.93	2.88	2.83	2.79	2.76	2.73	2.67	2.63	2.58	2.57	2.52	2.47	2.43	2.36	2.32	2.29	27
2.92	2.89	2.84	2.79	2.76	2.72	2.69	2.64	2.59	2.56	2.53	2.48	2.43	2.39	2.32	2.28	2.25	28
2.88	2.86	2.80	2.76	2.72	2.69	2.66	2.60	2.56	2.52	2.49	2.45	2.39	2.36	2.28	2.24	2.21	29
2.85	2.82	2.77	2.73	2.69	2.66	2.63	2.57	2.52	2.49	2.46	2.42	2.36	2.32	2.25	2.21	2.18	30
2.80	2.77	2.71	2.67	2.63	2.60	2.57	2.51	2.47	2.43	2.40	2.36	2.30	2.26	2.19	2.15	2.11	32
2.75	2.72	2.66	2.62	2.58	2.55	2.52	2.46	2.42	2.38	2.35	2.30	2.25	2.21	2.14	2.09	2.06	34
2.70	2.67	2.62	2.58	2.54	2.50	2.48	2.42	2.37	2.33	2.30	2.26	2.20	2.17	2.09	2.04	2.01	36
2.66	2.63	2.58	2.54	2.50	2.47	2.44	2.38	2.33	2.29	2.27	2.22	2.16	2.12	2.05	2.00	1.97	38
2.63	2.60	2.55	2.50	2.46	2.43	2.40	2.34	2.30	2.28	2.23	2.18	2.12	2.09	2.01	1.96	1.93	40

Degrees of freedom for the denominator (ν_2)

		1	2	3	4	5	6	7	8	9	10	11	12	13	14	15	16	17	18

Degrees of freedom for the numerator (ν_1)

Multiply the numbers of the first row ($\nu_2 = 1$) by 100

ν_2	1	2	3	4	5	6	7	8	9	10	11	12	13	14	15	16	17	18
42	8.78	6.03	4.94	4.34	3.95	3.68	3.48	3.32	3.19	3.09	3.00	2.92	2.86	2.80	2.75	2.71	2.67	2.63
44	8.74	5.99	4.91	4.31	3.92	3.65	3.45	3.29	3.16	3.06	2.97	2.89	2.83	2.77	2.72	2.68	2.64	2.60
46	8.70	5.96	4.88	4.28	3.90	3.62	3.42	3.26	3.14	3.03	2.94	2.87	2.80	2.75	2.70	2.65	2.61	2.58
48	8.66	5.93	4.85	4.25	3.87	3.60	3.40	3.24	3.11	3.01	2.92	2.85	2.78	2.72	2.67	2.63	2.59	2.55
50	8.63	5.90	4.83	4.23	3.85	3.58	3.38	3.22	3.09	2.99	2.90	2.82	2.76	2.70	2.65	2.61	2.57	2.53
55	8.55	5.84	4.77	4.18	3.80	3.53	3.33	3.17	3.05	2.94	2.85	2.78	2.71	2.66	2.61	2.56	2.52	2.49
60	8.49	5.80	4.73	4.14	3.76	3.49	3.29	3.12	3.01	2.90	2.82	2.74	2.68	2.62	2.57	2.53	2.49	2.45
65	8.44	5.75	4.68	4.11	3.73	3.45	3.26	3.10	2.98	2.87	2.79	2.71	2.63	2.59	2.54	2.49	2.45	2.42
70	8.40	5.72	4.65	4.08	3.70	3.43	3.23	3.08	2.95	2.85	2.76	2.68	2.62	2.56	2.51	2.47	2.43	2.39
80	8.33	5.67	4.61	4.03	3.65	3.39	3.19	3.03	2.91	2.80	2.72	2.64	2.58	2.52	2.47	2.43	2.39	2.35
90	8.28	5.62	4.57	3.99	3.62	3.35	3.15	3.00	2.87	2.77	2.68	2.61	2.54	2.49	2.44	2.39	2.35	2.32
100	8.24	5.59	4.54	3.96	3.59	3.33	3.13	2.97	2.85	2.74	2.66	2.58	2.52	2.46	2.41	2.37	2.33	2.29
125	8.17	5.53	4.49	3.91	3.54	3.28	3.08	2.93	2.80	2.70	2.61	2.54	2.47	2.42	2.37	2.32	2.28	2.24
150	8.12	5.49	4.43	3.88	3.51	3.25	3.05	2.89	2.77	2.67	2.58	2.51	2.44	2.38	2.33	2.29	2.25	2.21
200	8.06	5.44	4.41	3.84	3.47	3.21	3.01	2.85	2.73	2.63	2.54	2.47	2.40	2.35	2.30	2.26	2.21	2.18
300	8.00	5.39	4.37	3.80	3.43	3.17	2.97	2.81	2.69	2.59	2.51	2.43	2.37	2.31	2.26	2.21	2.17	2.14
500	7.95	5.36	4.33	3.76	3.40	3.14	2.94	2.79	2.66	2.56	2.48	2.40	2.34	2.28	2.23	2.19	2.14	2.11
1000	7.92	5.33	4.31	3.74	3.37	3.11	2.92	2.77	2.64	2.54	2.45	2.38	2.32	2.28	2.21	2.16	2.12	2.09
∞	7.88	5.30	4.28	3.72	3.35	3.09	2.90	2.74	2.62	2.52	2.43	2.36	2.29	2.24	2.19	2.14	2.10	2.08

Degrees of freedom for the denominator (ν_2)

							Degrees of freedom for the numerator (ν_1)										
19	20	22	24	26	28	30	35	40	45	50	60	80	100	200	500	∞	

Multiply the numbers of the first row ($\nu_2 = 1$) by 100

19	20	22	24	26	28	30	35	40	45	50	60	80	100	200	500	∞	
2.60	2.57	2.52	2.47	2.43	2.40	2.37	2.31	2.26	2.23	2.20	2.15	2.09	2.06	1.98	1.93	1.90	42
2.57	2.54	2.49	2.44	2.40	2.37	2.34	2.28	2.24	2.20	2.17	2.12	2.06	2.03	1.95	1.90	1.87	44
2.54	2.51	2.46	2.42	2.38	2.34	2.32	2.26	2.21	2.17	2.14	2.10	2.04	2.00	1.92	1.87	1.84	46
2.52	2.49	2.44	2.39	2.36	2.32	2.29	2.23	2.19	2.15	2.12	2.07	2.01	1.97	1.89	1.84	1.81	48
2.50	2.47	2.42	2.37	2.33	2.30	2.27	2.21	2.16	2.13	2.10	2.05	1.99	1.95	1.87	1.82	1.79	50
2.45	2.42	2.37	2.33	2.29	2.26	2.23	2.16	2.12	2.08	2.05	2.00	1.94	1.90	1.82	1.77	1.73	55
2.42	2.39	2.33	2.29	2.24	2.22	2.19	2.13	2.08	2.04	2.01	1.98	1.90	1.86	1.78	1.73	1.69	60
2.39	2.36	2.30	2.26	2.22	2.19	2.16	2.09	2.05	2.01	1.98	1.93	1.87	1.83	1.74	1.69	1.65	65
2.36	2.33	2.28	2.23	2.19	2.16	2.13	2.07	2.02	1.98	1.93	1.90	1.84	1.80	1.71	1.65	1.62	70
2.32	2.29	2.23	2.19	2.15	2.11	2.08	2.02	1.97	1.93	1.90	1.85	1.79	1.75	1.66	1.60	1.56	80
2.28	2.25	2.20	2.15	2.12	2.08	2.05	1.99	1.94	1.90	1.87	1.82	1.75	1.71	1.62	1.56	1.52	90
2.26	2.23	2.17	2.13	2.09	2.05	2.02	1.96	1.91	1.87	1.84	1.79	1.72	1.68	1.59	1.53	1.49	100
2.21	2.18	2.13	2.08	2.04	2.01	1.98	1.91	1.86	1.82	1.79	1.74	1.67	1.63	1.53	1.47	1.42	125
2.18	2.15	2.10	2.05	2.01	1.98	1.94	1.88	1.83	1.79	1.76	1.70	1.63	1.59	1.49	1.42	1.37	150
2.14	2.11	2.06	2.01	1.97	1.94	1.91	1.84	1.79	1.75	1.71	1.66	1.58	1.54	1.44	1.37	1.31	200
2.10	2.07	2.02	1.97	1.93	1.90	1.87	1.80	1.75	1.71	1.67	1.61	1.54	1.50	1.39	1.31	1.25	300
2.07	2.04	1.99	1.94	1.90	1.87	1.84	1.77	1.72	1.67	1.64	1.58	1.51	1.46	1.35	1.26	1.18	500
2.05	2.02	1.97	1.92	1.88	1.84	1.81	1.75	1.69	1.65	1.61	1.56	1.48	1.43	1.31	1.22	1.12	1000
2.03	2.00	1.95	1.90	1.86	1.82	1.79	1.72	1.67	1.63	1.59	1.53	1.45	1.40	1.28	1.17	1.00	∞

Degrees of freedom for the denominator (ν_2)

Appendix 4

Tabulation of the values of $F_{0.001;\nu_1;\nu_2}$ versus ν_1 and ν_2

	Degrees of freedom for the numerator (ν_1)																	
	1	2	3	4	5	6	7	8	9	10	15	20	30	50	100	200	500	∞
	Multiply the numbers of the first row ($\nu_2 = 1$) by 1000																	
1	405	500	540	562	576	586	593	598	602	606	616	621	626	630	633	635	636	637
2	998	999	999	999	999	999	999	999	999	999	999	999	999	999	999	999	999	999
3	168	148	141	137	135	133	132	131	130	129	127	126	125	125	124	124	124	124
4	74.1	61.2	56.2	53.4	51.7	50.5	49.7	49.0	48.5	48.0	46.8	46.1	45.4	44.9	44.5	44.3	44.1	44.0
5	47.0	36.6	33.2	31.1	29.8	28.8	28.2	27.6	27.2	26.9	25.9	25.4	24.9	24.4	24.1	23.9	23.8	23.8
6	35.5	27.0	23.7	21.9	20.8	20.0	19.5	19.0	18.7	18.4	17.6	17.1	16.7	16.3	16.0	15.9	15.8	15.8
7	29.2	21.7	18.8	17.2	16.2	15.5	15.0	14.6	14.3	14.1	13.3	12.9	12.5	12.2	11.9	11.8	11.7	11.7
8	25.4	18.5	15.8	14.4	13.5	12.9	12.4	12.0	11.8	11.5	10.8	10.5	10.1	9.80	9.57	9.46	9.39	9.34
9	22.9	16.4	13.9	12.6	11.7	11.1	10.7	10.4	10.1	9.89	9.24	8.90	8.55	8.26	8.04	7.93	7.86	7.81
10	21.0	14.9	12.6	11.3	10.5	9.92	9.52	9.20	8.96	8.75	8.13	7.80	7.47	7.19	6.98	6.87	6.81	6.76
11	19.7	13.8	11.6	10.4	9.58	9.05	8.66	8.35	8.12	7.92	7.32	7.01	6.68	6.41	6.21	6.10	6.04	6.00
12	18.6	13.0	10.8	9.63	8.89	8.38	8.00	7.71	7.48	7.29	6.71	6.40	6.09	5.83	5.63	5.52	5.46	5.42
13	17.8	12.3	10.2	9.07	8.35	7.86	7.49	7.21	6.98	6.80	6.23	5.93	5.62	5.37	5.17	5.07	5.01	4.97
14	17.1	11.8	9.73	8.62	7.92	7.43	7.08	6.80	6.58	6.40	5.85	5.56	5.25	5.00	4.80	4.70	4.64	4.60
15	16.6	11.3	9.34	8.25	7.57	7.09	6.74	6.47	6.26	6.08	5.53	5.25	4.95	4.70	4.51	4.41	4.35	4.31
16	16.1	11.0	9.00	7.94	7.27	6.81	6.46	6.19	5.98	5.81	5.27	4.99	4.70	4.45	4.26	4.16	4.10	4.06
17	15.7	10.7	8.73	7.68	7.02	6.56	6.22	5.96	5.75	5.58	5.05	4.78	4.48	4.24	4.05	3.95	3.89	3.85
18	15.4	10.4	8.49	7.46	6.81	6.35	6.02	5.76	5.56	5.39	4.87	4.59	4.30	4.06	3.87	3.77	3.71	3.67
19	15.1	10.2	8.28	7.26	6.61	6.18	5.84	5.59	5.39	5.22	4.70	4.43	4.14	3.90	3.71	3.61	3.55	3.51
20	14.8	9.95	8.10	7.10	6.46	6.02	5.69	5.44	5.24	5.08	4.56	4.29	4.01	3.77	3.58	3.48	3.42	3.38
22	14.4	9.61	7.80	6.81	6.19	5.76	5.44	5.19	4.99	4.83	4.32	4.06	3.77	3.53	3.34	3.25	3.19	3.15
24	14.0	9.34	7.55	6.59	5.98	5.55	5.23	4.99	4.80	4.64	4.14	3.87	3.59	3.35	3.16	3.07	3.01	2.97
26	13.7	9.12	7.36	6.41	5.80	5.38	5.07	4.83	4.64	4.48	3.99	3.72	3.45	3.20	3.01	2.92	2.86	2.82
28	13.5	8.93	7.19	6.25	5.66	5.24	4.93	4.69	4.50	4.35	3.86	3.60	3.32	3.08	2.89	2.79	2.73	2.70
30	13.3	8.77	7.05	6.12	5.53	5.12	4.82	4.58	4.39	4.24	3.75	3.49	3.22	2.98	2.79	2.69	2.63	2.59
40	12.6	8.25	6.60	5.70	5.13	4.73	4.43	4.21	4.02	3.87	3.40	3.15	2.87	2.64	2.44	2.34	2.28	2.23
50	12.2	7.95	6.34	5.46	4.90	4.51	4.22	4.00	3.83	3.67	3.20	2.95	2.68	2.44	2.24	2.14	2.07	2.03
60	12.0	7.76	6.17	5.31	4.76	4.37	4.09	3.87	3.69	3.54	3.08	2.83	2.56	2.31	2.11	2.01	1.93	1.89
80	11.7	7.54	5.97	5.13	4.58	4.21	3.92	3.70	3.53	3.39	2.93	2.68	2.40	2.16	1.95	1.84	1.77	1.72
100	11.5	7.41	5.85	5.01	4.48	4.11	3.83	3.61	3.44	3.30	2.84	2.59	2.32	2.07	1.87	1.75	1.68	1.62
200	11.2	7.15	5.64	4.81	4.29	3.92	3.65	3.43	3.26	3.12	2.67	2.42	2.15	1.90	1.68	1.55	1.46	1.39
500	11.0	7.01	5.51	4.69	4.18	3.82	3.54	3.33	3.16	3.02	2.58	2.33	2.05	1.80	1.57	1.43	1.32	1.23
∞	10.8	6.91	5.42	4.62	4.10	3.74	3.47	3.27	3.10	2.96	2.51	2.27	1.99	1.73	1.49	1.34	1.21	1.00

Degrees of freedom for the denominator (ν_2)

Appendix 5.   Kolmogorov–Smirnov Tables

Critical values, $d_\alpha(n)^a$, of the maximum absolute difference between sample $F_n(x)$ and population $F(x)$ cumulative distribution.

Number of trials, n	Level of significance, α			
	0.10	0.05	0.02	0.01
1	0.95000	0.97500	0.99000	0.99500
2	0.77639	0.84189	0.90000	0.92929
3	0.63604	0.70760	0.78456	0.82900
4	0.56522	0.62394	0.68887	0.73424
5	0.50945	0.56328	0.62718	0.66853
6	0.46799	0.51926	0.57741	0.61661
7	0.43607	0.48342	0.53844	0.57581
8	0.40962	0.45427	0.50654	0.54179
9	0.38746	0.43001	0.47960	0.51332
10	0.36866	0.40925	0.45662	0.48893
11	0.35242	0.39122	0.43670	0.46770
12	0.33815	0.37543	0.41918	0.44905
13	0.32549	0.36143	0.40362	0.43247
14	0.31417	0.34890	0.38970	0.41762
15	0.30397	0.33760	0.37713	0.40420
16	0.29472	0.32733	0.36571	0.39201
17	0.28627	0.31796	0.35528	0.38086
18	0.27851	0.30936	0.34569	0.37062
19	0.27136	0.30143	0.33685	0.36117
20	0.26473	0.29408	0.32866	0.35241
21	0.25858	0.28724	0.32104	0.34427
22	0.25283	0.28087	0.31394	0.33666
23	0.24746	0.27490	0.30728	0.32954
24	0.24242	0.26931	0.30104	0.32286
25	0.23768	0.26404	0.29516	0.31657
26	0.23320	0.25907	0.28962	0.31064
27	0.22898	0.25438	0.28438	0.30502
28	0.22497	0.24993	0.27942	0.29971
29	0.22117	0.24571	0.27471	0.29466
30	0.21756	0.24170	0.27023	0.28987
31	0.21412	0.23788	0.26596	0.28530
32	0.21085	0.23424	0.26189	0.28094
33	0.20771	0.23076	0.25801	0.27677
34	0.20472	0.22743	0.25429	0.27279
35	0.20185	0.22425	0.26073	0.26897

Appendix 5

Critical values, $d_\alpha(n)^a$, of the maximum absolute difference between sample $F_n(x)$ and population $F(x)$ cumulative distribution.

Number of trials, n	Level of significance, α			
	0.10	0.05	0.02	0.01
36	0.19910	0.22119	0.24732	0.26532
37	0.19646	0.21826	0.24404	0.26180
38	0.19392	0.21544	0.24089	0.25843
39	0.19148	0.21273	0.23786	0.25518
40^b	0.18913	0.21012	0.23494	0.25205

[a]Values of $d_\alpha(n)$ such that $p(\max)|F^n(x) - F(x)|d_\alpha(n) = \alpha$.

[b]$N > 40 \approx \dfrac{1.22}{N^{1/2}}, \dfrac{1.36}{N^{1/2}}, \dfrac{1.51}{N^{1/2}}$ and $\dfrac{1.63}{N^{1/2}}$ for the four levels of significance.

Appendix 6.   Rank Tables (Median, 5%, 95%)

MEDIAN RANKS

$j\backslash n$	SAMPLE SIZE									
	1	2	3	4	5	6	7	8	9	10
1	50.000	29.289	20.630	15.910	12.945	10.910	9.428	8.300	7.412	6.697
2		70.711	50.000	38.573	31.381	26.445	22.849	20.113	17.962	16.226
3			79.370	61.427	50.000	42.141	36.412	32.052	28.624	25.857
4				84.090	68.619	57.859	50.000	44.015	39.308	35.510
5					87.055	73.555	63.588	55.984	50.000	45.169
6						89.090	77.151	67.948	60.691	54.831
7							90.572	79.887	71.376	64.490
8								91.700	82.038	74.142
9									92.587	83.774
10										93.303

MEDIAN RANKS

$j\backslash n$	SAMPLE SIZE									
	11	12	13	14	15	16	17	18	19	20
1	6.107	5.613	5.192	4.830	4.516	4.240	3.995	3.778	3.582	3.406
2	14.796	13.598	12.579	11.702	10.940	10.270	9.678	9.151	8.677	8.251
3	23.578	21.669	20.045	18.647	17.432	16.365	15.422	14.581	13.827	13.147
4	32.380	29.758	27.528	25.608	23.939	22.474	21.178	20.024	18.988	18.055
5	41.189	37.853	35.016	32.575	30.452	28.589	26.940	25.471	24.154	22.967
6	50.000	45.951	42.508	39.544	36.967	34.605	32.704	30.921	29.322	27.880
7	58.811	54.049	50.000	46.515	43.483	40.823	38.469	36.371	34.491	32.795
8	67.620	62.147	57.492	53.485	50.000	46.941	44.234	41.823	39.660	37.710
9	76.421	70.242	64.984	60.456	56.517	53.059	50.000	47.274	44.830	42.626
10	85.204	78.331	72.472	67.425	63.033	59.177	55.766	52.726	50.000	47.542
11	93.893	86.402	79.955	74.392	69.548	65.295	61.531	58.177	55.170	52.458
12		94.387	87.421	81.353	76.061	71.411	67.296	63.629	60.340	57.374
13			94.808	88.298	82.568	77.525	73.060	69.079	65.509	62.289
14				95.169	89.060	83.635	78.821	74.529	70.678	67.205
15					95.484	89.730	84.578	79.976	75.846	72.119
16						95.760	90.322	85.419	81.011	77.033
17							96.005	90.849	86.173	81.945
18								96.222	91.322	86.853
19									96.418	91.749
20										96.594

MEDIAN RANKS

$j\backslash n$	SAMPLE SIZE									
	21	22	23	24	25	26	27	28	29	30
1	3.247	3.101	2.969	2.847	2.734	2.631	2.534	2.445	2.362	2.284
2	7.864	7.512	7.191	6.895	6.623	6.372	6.139	5.922	5.720	5.532
3	12.531	11.970	11.458	10.987	10.553	10.153	9.781	9.436	9.114	8.814
4	17.209	16.439	15.734	15.088	14.492	13.942	13.432	12.958	12.517	12.104
5	21.890	20.911	20.015	19.192	18.435	17.735	17.086	16.483	15.922	15.397
6	26.574	25.384	24.297	23.299	22.379	21.529	20.742	20.010	19.328	18.691
7	31.258	29.859	28.580	27.406	26.324	25.325	24.398	23.537	22.735	21.986
8	35.943	34.334	32.863	31.513	30.269	29.120	28.055	27.065	26.143	25.281
9	40.629	38.810	37.147	35.621	34.215	32.916	31.712	30.593	29.550	28.576
10	45.314	43.286	41.431	39.729	38.161	36.712	35.370	34.121	32.958	31.872
11	50.000	47.762	45.716	43.837	42.107	40.509	39.027	37.650	36.367	35.168
12	54.686	52.238	50.000	47.946	46.054	44.305	42.685	41.178	39.775	38.464
13	59.371	56.714	54.284	52.054	50.000	48.102	46.342	44.707	43.183	41.760
14	64.057	61.190	58.568	56.162	53.946	51.898	50.000	48.236	46.592	45.056
15	68.742	65.665	62.853	60.271	57.892	55.695	53.658	51.764	50.000	48.352
16	73.426	70.141	67.137	64.379	61.839	59.491	57.315	55.293	53.408	51.648
17	78.109	74.616	71.420	68.487	65.785	63.287	60.973	58.821	56.817	54.944
18	82.791	70.089	75.703	72.594	69.730	67.084	64.630	62.350	60.225	58.240
19	87.469	83.561	79.985	76.701	73.676	70.880	68.288	65.878	63.633	61.536
20	92.136	88.030	84.266	80.808	77.621	74.675	71.945	69.407	67.041	64.852
21	96.753	92.488	88.542	84.912	81.565	78.471	75.602	72.935	70.450	68.128
22		96.898	92.809	89.013	85.507	82.265	79.258	76.463	73.857	71.424
23			97.031	93.105	89.447	86.058	82.914	79.990	77.265	74.719
24				97.153	93.377	89.847	86.568	83.517	80.672	78.014
25					97.265	93.628	90.219	87.042	84.078	81.309
26						97.369	93.861	90.564	87.483	84.603
27							97.465	94.078	90.865	87.896
28								97.555	94.280	91.186
29									97.638	94.468
30										97.716

MEDIAN RANKS

$j\backslash n$	SAMPLE SIZE									
	31	32	33	34	35	36	37	38	39	40
1	2.211	2.143	2.078	2.018	1.961	1.907	1.856	1.807	1.762	1.718
2	5.355	5.190	5.034	4.887	4.749	4.618	4.495	4.377	4.266	4.160
3	8.533	8.269	8.021	7.787	7.567	7.359	7.162	6.975	6.798	6.629
4	11.716	11.355	11.015	10.694	10.391	10.105	9.835	9.578	9.335	9.103
5	14.905	14.445	14.011	13.603	13.218	12.855	12.510	12.184	11.874	11.580
6	18.094	17.535	17.009	16.514	16.046	15.605	15.187	14.791	14.415	14.057
7	21.284	20.626	20.007	19.425	18.875	18.355	17.864	17.398	16.956	16.535
8	24.474	23.717	23.006	22.336	21.704	21.107	20.541	20.005	19.497	19.013
9	27.664	26.809	26.005	25.247	24.533	23.858	23.219	22.613	22.038	21.492
10	30.855	29.901	29.004	28.159	27.362	26.609	25.897	25.221	24.580	23.971
11	34.046	32.993	32.003	31.071	30.192	29.361	28.575	27.829	27.122	26.449
12	37.236	36.085	35.003	33.983	33.022	32.113	31.253	30.437	29.664	28.928
13	40.427	39.177	38.002	36.895	35.851	34.865	33.931	33.046	32.206	31.407
14	43.618	42.269	41.001	39.807	38.681	37.616	36.609	35.654	34.748	33.886
15	46.809	45.362	44.001	42.720	41.511	40.368	39.287	38.262	37.290	36.365
16	50.000	48.454	47.000	45.632	44.340	43.120	41.965	40.871	39.832	38.844
17	53.191	51.546	50.000	48.544	47.170	45.872	44.644	43.479	42.374	41.323
18	56.382	54.638	52.999	51.456	50.000	48.624	47.322	46.087	44.916	43.802
19	59.573	57.731	55.999	54.368	52.830	51.376	50.000	48.696	47.458	46.281
20	62.763	60.823	58.998	57.280	55.660	54.128	52.678	51.304	50.000	48.760
21	65.954	63.915	61.998	60.193	58.489	56.830	55.356	53.913	52.542	51.239
22	69.145	67.007	64.997	63.105	61.319	59.632	58.035	56.521	55.084	53.719
23	72.335	70.099	67.997	66.017	64.149	62.383	60.713	59.129	57.626	56.198
24	75.526	73.191	70.996	68.929	66.973	65.135	63.391	61.738	60.168	58.677
25	78.716	76.283	73.995	71.841	69.808	67.837	66.069	64.346	62.710	61.156
26	81.906	79.374	76.994	74.752	72.637	70.639	68.747	66.954	65.252	63.635
27	85.094	82.465	79.993	77.664	75.467	73.391	71.425	69.562	67.794	66.114
28	88.282	85.555	82.991	80.575	78.296	76.142	74.103	72.171	70.336	68.593
29	91.467	88.644	85.989	83.486	81.125	78.899	76.781	74.779	72.878	71.072
30	94.645	91.731	88.985	86.397	83.954	81.645	79.459	77.387	75.420	73.550
31	97.789	94.810	91.979	89.306	86.782	84.395	82.136	79.994	77.962	76.029
32		97.857	94.966	92.213	89.608	87.145	84.813	82.602	80.503	78.508
33			97.921	95.113	92.433	89.894	87.490	85.209	83.044	80.986
34				97.982	95.251	92.641	90.165	87.816	85.585	83.465
35					98.039	95.382	92.838	90.422	88.126	85.943
36						98.093	95.505	93.025	90.665	88.420
37							98.144	95.622	93.202	90.897
38								98.192	95.734	93.371
39									98.238	95.839
40										98.282

Appendix 6

MEDIAN RANKS

					SAMPLE SIZE					
$j\backslash n$	41	42	43	44	45	46	47	48	49	50
1	1.676	1.637	1.599	1.563	1.528	1.495	1.464	1.434	1.405	1.377
2	4.060	3.964	3.872	3.785	3.702	3.622	3.545	3.472	3.402	3.334
3	6.649	6.316	6.170	6.031	5.898	5.771	5.649	5.532	5.420	5.312
4	8.883	8.673	8.473	8.282	8.099	7.925	7.757	7.597	7.443	7.295
5	11.300	11.033	10.778	10.535	10.303	10.080	9.867	9.663	9.467	9.279
6	13.717	13.393	13.084	12.789	12.507	12.237	11.979	11.731	11.493	11.265
7	16.135	15.754	15.391	15.043	14.712	14.394	14.090	13.799	13.519	13.250
8	18.554	18.115	17.697	17.298	16.917	16.551	16.202	15.867	15.545	15.236
9	20.972	20.477	20.004	19.553	19.122	18.709	18.314	17.935	17.571	17.222
10	23.391	22.838	22.311	21.808	21.327	20.867	20.426	20.003	19.598	19.209
11	25.810	25.200	24.618	24.063	23.532	23.025	22.538	22.072	21.625	21.195
12	28.228	27.562	26.926	26.318	25.738	25.182	24.650	24.140	23.651	23.181
13	30.647	29.924	29.233	28.574	27.943	27.340	26.763	26.209	25.678	25.168
14	33.066	32.285	31.540	30.829	30.149	29.498	28.875	28.278	27.705	27.154
15	35.485	34.647	33.848	33.084	32.355	31.656	30.988	30.347	29.731	29.141
16	37.905	37.009	36.155	35.340	34.560	33.814	33.100	32.415	31.758	31.127
17	40.324	39.371	38.463	37.595	36.766	35.972	35.212	34.484	33.785	33.114
18	42.743	41.733	40.770	39.851	38.972	38.130	37.325	36.553	35.822	35.100
19	45.162	44.095	43.078	42.106	41.177	40.289	39.437	38.622	37.839	37.087
20	47.581	46.457	45.385	44.361	43.383	42.447	41.550	40.690	39.866	39.074
21	50.000	48.819	47.692	46.617	45.589	44.605	43.662	42.759	41.892	41.060
22	52.419	51.181	50.000	48.872	47.794	46.763	45.775	44.825	43.919	43.047
23	54.838	53.543	52.307	51.128	50.000	48.921	47.887	46.897	45.946	45.033
24	57.257	55.905	54.615	53.383	52.206	51.079	50.000	48.966	47.973	47.020
25	59.676	58.267	56.922	55.639	54.411	53.237	52.112	51.034	50.000	49.007
26	62.095	60.629	59.230	57.894	56.617	55.395	54.225	53.103	52.027	50.993
27	64.514	62.991	61.537	60.149	58.823	57.553	56.337	55.172	54.054	52.980
28	66.933	65.353	63.845	62.405	61.028	59.711	58.450	57.241	56.081	54.966
29	69.352	67.714	66.152	64.660	63.234	61.869	60.562	59.310	58.107	56.953
30	71.771	70.076	68.459	66.916	65.440	64.027	62.675	61.378	60.134	58.940
31	74.190	72.438	70.767	69.171	67.645	66.186	64.767	63.447	62.161	60.926
32	76.609	74.800	73.074	71.426	69.851	68.344	66.900	65.516	64.188	62.913
33	79.028	77.162	75.381	73.681	72.056	70.502	69.012	67.585	66.215	64.899
34	81.446	79.523	77.689	75.937	74.262	72.660	71.125	69.653	68.242	66.886
35	83.865	81.885	79.996	78.192	76.467	74.817	73.237	71.722	70.268	68.873
36	86.283	84.246	82.303	80.447	78.673	76.975	75.349	73.791	72.295	70.859
37	88.700	86.607	84.609	82.702	80.878	79.133	77.462	75.859	74.322	72.546
38	91.117	88.967	86.916	84.956	83.083	81.291	79.574	77.928	76.349	74.832
39	93.531	91.327	89.222	87.211	85.283	83.448	81.686	79.997	78.375	76.819
40	95.940	93.684	91.527	89.465	87.493	85.606	83.798	82.065	80.402	78.805
41	98.324	96.036	93.830	91.718	89.697	87.763	85.910	84.133	82.428	80.791
42		98.363	96.127	93.969	91.900	89.920	88.021	86.201	84.455	82.778
43			98.401	96.215	94.102	92.075	90.132	88.269	86.481	84.764
44				98.437	96.298	94.229	92.243	90.337	88.507	86.750
45					98.471	96.378	94.351	92.403	90.532	88.735
46						98.504	96.455	94.468	92.557	90.721
47							98.536	96.528	94.580	92.705
48								98.566	96.598	94.688
49									98.595	96.666
50										98.623

5 PER CENT RANKS

$j\backslash n$	1	2	3	4	SAMPLE SIZE 5	6	7	8	9	10
1	5.000	2.532	1.695	1.274	1.021	0.851	0.730	0.639	0.568	0.512
2		22.361	13.535	9.761	7.644	6.285	5.337	4.639	4.102	3.677
3			36.840	24.860	18.925	15.316	12.876	11.111	9.775	8.726
4				47.237	34.259	27.134	22.532	19.290	16.875	15.003
5					54.928	41.820	34.126	28.924	25.137	22.244
6						60.696	47.930	40.031	34.494	30.354
7							65.184	52.932	45.036	39.338
8								68.766	57.086	49.310
9									71.687	60.584
10										74.113

5 PER CENT RANKS

$j\backslash n$	11	12	13	14	15	16	17	18	19	20
1	0.465	0.426	0.394	0.366	0.341	0.320	0.301	0.285	0.270	0.256
2	3.332	3.046	2.805	2.600	2.423	2.268	2.132	2.011	1.903	1.806
3	7.882	7.187	6.605	6.110	5.685	5.315	4.990	4.702	4.446	4.217
4	13.507	12.285	11.267	10.405	9.666	9.025	8.464	7.969	7.529	7.135
5	19.958	18.102	16.566	15.272	14.166	13.211	12.377	11.643	10.991	10.408
6	27.125	24.530	22.395	20.607	19.086	17.777	16.636	15.634	14.747	13.955
7	34.981	31.524	28.705	26.358	24.373	22.669	21.191	19.895	18.750	17.731
8	43.563	39.086	35.480	32.503	29.999	27.860	26.011	24.396	22.972	21.707
9	52.991	47.267	42.738	39.041	35.956	33.337	31.083	29.120	27.395	25.865
10	63.564	56.189	50.535	45.999	42.256	39.101	36.401	34.060	32.009	30.195
11	76.160	66.132	58.990	53.434	48.925	45.165	41.970	39.215	36.811	34.693
12		77.908	68.366	61.461	56.022	51.560	47.808	44.595	41.806	39.358
13			79.418	70.327	63.656	58.343	53.945	50.217	47.003	44.197
14				80.736	72.060	65.617	60.436	56.112	52.420	49.218
15					81.896	73.604	67.381	62.332	58.088	54.442
16						82.925	74.988	68.974	64.057	59.897
17							83.843	76.234	70.420	65.634
18								84.668	77.363	71.738
19									85.413	78.389
20										86.089

5 PER CENT RANKS

					SAMPLE SIZE					
$j\backslash n$	21	22	23	24	25	26	27	28	29	30
1	0.244	0.233	0.223	0.213	0.205	0.197	0.190	0.183	0.177	0.171
2	1.719	1.640	1.567	1.501	1.440	1.384	1.332	1.284	1.239	1.198
3	4.010	3.822	3.651	3.495	3.352	3.220	3.098	2.985	2.879	2.781
4	6.781	6.460	6.167	5.901	5.656	5.431	5.223	5.031	4.852	4.685
5	9.884	9.411	8.981	8.588	8.229	7.899	7.594	7.311	7.049	6.806
6	13.245	12.603	12.021	11.491	11.006	10.560	10.148	9.768	9.415	9.087
7	16.818	15.994	15.248	14.569	13.947	13.377	12.852	12.367	11.917	11.499
8	20.575	19.556	18.634	17.796	17.030	16.328	15.682	15.085	14.532	14.018
9	24.499	23.272	22.164	21.157	20.238	19.396	18.622	17.908	17.246	16.633
10	28.580	27.131	25.824	24.639	23.559	22.570	21.662	20.824	20.050	19.331
11	32.811	31.126	29.609	28.236	26.985	25.842	24.793	23.827	22.934	22.106
12	37.190	35.254	33.515	31.942	30.513	29.508	28.012	26.911	25.894	24.953
13	41.720	39.516	37.539	35.756	34.139	32.664	31.314	30.072	28.927	27.867
14	46.406	43.913	41.684	39.678	37.862	36.209	34.697	33.309	32.030	30.846
15	51.261	48.454	45.954	43.711	41.684	39.842	38.161	36.620	35.200	33.889
16	56.302	53.151	50.356	47.858	45.607	43.566	41.707	40.004	38.439	36.995
17	61.559	58.020	54.902	52.127	49.636	47.384	45.336	43.464	41.746	40.163
18	67.079	63.091	59.610	56.531	53.779	51.300	49.052	47.002	45.123	43.394
19	72.945	68.409	64.507	61.086	58.048	55.323	52.861	50.621	48.573	46.691
20	79.327	74.053	69.636	65.819	62.459	59.465	56.770	54.327	52.099	50.056
21	86.705	80.188	75.075	70.773	67.039	63.740	60.790	58.127	55.706	53.493
22		87.269	80.980	76.020	71.828	68.176	64.936	62.033	59.403	57.007
23			87.788	81.711	76.896	72.810	69.237	66.060	63.200	60.605
24				88.265	82.388	77.711	73.726	70.231	67.113	64.299
25					88.707	83.017	78.470	74.583	71.168	68.103
26						89.117	83.603	79.179	75.386	72.038
27							89.498	84.149	79.844	76.140
28								89.853	84.661	80.467
29									90.185	85.140
30										90.497

5 PER CENT RANKS

$j \backslash n$	31	32	33	34	35	36	37	38	39	40
					SAMPLE SIZE					
1	0.165	0.160	0.155	0.151	0.146	0.142	0.138	0.135	0.131	0.128
2	1.158	1.122	1.086	1.055	1.025	0.996	0.969	0.943	0.919	0.896
3	2.690	2.604	2.524	2.448	2.377	2.310	2.246	2.186	2.129	2.075
4	4.530	4.384	4.246	4.120	3.999	3.885	3.778	3.676	3.580	3.488
5	6.578	6.365	6.166	5.978	5.802	5.636	5.479	5.331	5.190	5.057
6	8.781	8.495	8.227	7.976	7.739	7.516	7.306	7.107	6.919	6.740
7	11.109	10.745	10.404	10.084	9.783	9.499	9.232	8.979	8.740	8.513
8	13.540	13.093	12.675	12.283	11.914	11.567	11.240	10.931	10.638	10.361
9	16.061	15.528	15.029	14.561	14.122	13.708	13.318	12.950	12.601	12.271
10	18.662	18.038	17.455	16.909	16.396	15.913	15.458	15.028	14.622	14.237
11	21.336	20.618	19.948	19.319	18.730	18.175	17.653	17.160	16.694	16.252
12	24.077	23.262	22.501	21.788	21.119	20.491	19.898	19.340	18.812	18.312
13	26.883	25.966	25.111	24.310	23.560	22.855	22.191	21.565	20.973	20.413
14	29.749	28.727	27.775	26.884	26.049	25.265	24.527	23.832	23.175	22.553
15	32.674	31.544	30.491	29.507	28.585	27.719	26.905	26.138	25.414	24.729
16	35.657	34.415	33.258	32.177	31.165	30.216	29.324	28.483	27.690	26.940
17	38.698	37.339	36.074	34.894	33.789	32.754	31.781	30.865	30.001	29.185
18	41.797	40.317	38.940	37.657	36.457	35.332	34.276	33.283	32.346	31.461
19	44.956	43.349	41.656	40.466	39.167	37.951	36.809	35.736	34.725	33.770
20	48.175	46.436	44.823	43.321	41.920	40.609	39.380	38.224	37.136	36.109
21	51.458	49.581	47.841	46.225	44.717	43.309	41.988	40.748	39.581	38.480
22	54.810	52.786	50.914	49.177	47.560	46.049	44.634	43.307	42.058	40.881
23	58.234	56.055	54.344	52.181	50.448	48.832	47.320	45.902	44.569	43.314
24	64.739	59.314	57.235	55.239	53.385	51.658	50.045	48.534	47.114	45.778
25	65.336	62.810	60.493	58.355	56.374	54.532	52.812	51.204	49.694	48.275
26	69.036	66.313	63.824	61.534	59.416	57.454	55.624	53.914	52.311	50.805
27	72.563	69.916	67.237	64.754	62.523	60.429	58.483	56.666	54.966	53.370
28	76.650	73.640	70.748	68.113	65.695	63.483	61.392	59.463	57.661	55.972
29	81.054	77.518	74.375	71.535	68.944	66.561	64.357	62.309	60.399	58.612
30	85.591	81.606	76.150	75.069	72.282	69.732	67.384	65.209	63.185	61.294
31	90.789	86.015	82.127	78.747	75.728	72.990	70.482	68.168	66.021	64.021
32		91.063	86.415	82.619	79.312	76.352	73.663	71.196	68.916	66.797
33			91.322	86.793	83.085	79.848	76.946	74.304	71.876	69.629
34				91.566	87.150	83.526	80.357	77.510	74.915	72.525
35					91.797	87.488	83.946	80.841	78.048	75.497
36						92.015	87.809	84.344	81.302	78.560
37							92.222	88.115	84.723	81.741
38								92.419	88.405	85.085
39									92.606	88.681
40										92.784

5 PER CENT RANKS

					SAMPLE SIZE					
$j\backslash n$	41	42	43	44	45	46	47	48	49	50
1	0.125	0.122	0.119	0.116	0.114	0.111	0.109	0.107	0.105	0.102
2	0.874	0.853	0.833	0.814	0.795	0.778	0.761	0.745	0.730	0.715
3	2.024	1.975	1.928	1.884	1.842	1.801	1.762	1.725	1.689	1.655
4	3.402	3.319	3.240	3.165	3.093	3.025	2.959	2.897	2.836	2.779
5	4.930	4.810	4.695	4.586	4.481	4.382	4.286	4.195	4.108	4.024
6	6.570	6.409	6.256	6.109	5.969	5.836	5.708	5.586	5.469	5.357
7	8.298	8.093	7.898	7.713	7.536	7.366	7.205	7.050	6.902	6.760
8	10.097	9.847	9.609	9.382	9.166	8.959	8.762	8.573	8.392	8.218
9	11.958	11.660	11.377	11.107	10.850	10.605	10.370	10.146	9.931	9.725
10	13.872	13.525	13.195	12.881	12.582	12.296	12.023	11.762	11.512	11.272
11	15.833	15.436	15.058	14.698	14.355	14.028	13.715	13.416	13.130	12.856
12	17.838	17.389	16.961	16.554	16.166	15.796	15.443	15.105	14.782	14.472
13	19.883	19.379	18.901	18.445	18.012	17.598	17.203	16.825	16.464	16.117
14	21.964	21.406	20.875	20.370	19.889	19.430	18.993	18.574	18.174	17.790
15	24.081	23.466	22.881	22.326	21.796	21.292	20.810	20.350	19.910	19.488
16	26.230	25.557	24.918	24.311	23.732	23.180	22.654	22.151	21.671	21.210
17	28.412	27.679	26.984	26.323	25.694	25.095	24.523	23.977	23.455	22.955
18	30.624	29.831	29.078	28.363	27.683	27.034	26.416	25.825	25.261	24.721
19	32.867	32.011	31.200	30.429	29.696	28.997	28.331	27.696	27.088	26.507
20	35.138	34.219	33.348	32.520	31.733	30.984	30.269	29.588	28.936	28.313
21	37.440	36.455	35.522	34.636	33.794	32.993	32.229	31.500	30.804	30.138
22	39.770	38.719	37.722	36.777	35.879	35.025	34.210	33.434	32.692	31.980
23	42.129	41.009	39.949	38.943	37.987	37.078	36.212	35.387	34.599	33.845
24	44.518	43.328	42.201	41.133	40.118	39.154	38.235	37.360	36.524	35.726
25	46.937	45.674	44.480	43.347	42.273	41.251	40.279	39.353	38.469	37.625
26	49.388	48.050	46.785	45.587	44.451	43.371	42.344	41.366	40.432	39.541
27	51.869	50.454	49.117	47.852	46.652	45.513	44.430	43.398	42.415	41.476
28	54.385	52.889	51.478	50.143	48.878	47.678	46.537	45.451	44.416	43.428
29	56.935	55.356	53.868	52.461	51.129	49.866	48.666	47.524	46.436	45.399
30	59.522	57.857	56.288	54.807	53.406	52.078	50.817	49.618	48.477	47.388
31	62.149	60.393	58.741	57.183	55.710	54.315	52.991	51.734	50.537	49.396
32	64.820	62.968	61.228	59.590	58.042	56.578	55.190	53.871	52.617	51.423
33	67.539	65.585	63.753	62.029	60.404	58.868	57.413	56.032	54.720	53.470
34	70.311	68.248	66.318	64.505	62.798	61.187	59.662	58.217	56.844	55.538
35	73.146	70.963	68.927	67.020	65.227	63.537	61.940	60.427	58.991	57.627
36	76.053	73.738	71.587	69.578	67.694	65.921	64.247	62.664	61.164	59.738
37	79.049	76.584	74.306	72.185	70.203	68.341	66.587	64.931	63.362	61.874
38	82.160	79.517	77.093	74.849	72.759	70.805	68.963	67.228	65.589	64.034
39	85.429	82.561	79.964	77.580	75.370	73.309	71.378	69.561	67.846	66.222
40	88.945	85.759	82.944	80.392	78.046	75.870	73.838	71.932	70.136	68.440
41	92.954	89.196	86.073	83.310	80.802	78.494	76.350	74.347	72.465	70.691
42		93.116	89.437	86.374	83.661	81.196	78.924	76.812	74.836	72.978
43			93.270	89.666	86.662	83.998	81.573	79.337	77.256	75.306
44				93.418	89.887	86.939	84.321	81.936	79.734	77.683
45					93.560	90.098	87.204	84.631	82.285	80.117
46						93.695	90.300	87.459	84.929	82.621
47							93.825	90.494	87.703	85.216
48								93.950	90.681	87.939
49									94.069	90.860
50										94.184

95 PER CENT RANKS

					SAMPLE SIZE					
$j\backslash n$	1	2	3	4	5	6	7	8	9	10
1	95.000	77.639	63.160	52.713	45.072	39.304	34.816	31.234	28.313	25.887
2		97.468	86.465	75.139	65.741	58.180	52.070	47.068	42.914	39.416
3			98.305	90.239	81.075	72.866	65.874	59.969	54.964	50.690
4				98.726	92.356	84.684	77.468	71.076	65.506	60.662
5					98.979	93.715	87.124	80.710	74.863	69.646
6						99.149	94.662	88.889	83.125	77.756
7							99.270	95.361	90.225	84.997
8								99.361	95.898	91.274
9									99.432	96.323
10										99.488

95 PER CENT RANKS

					SAMPLE SIZE					
$j\backslash n$	11	12	13	14	15	16	17	18	19	20
1	23.840	22.092	20.582	19.264	18.104	17.075	16.157	15.332	14.587	13.911
2	36.436	33.868	31.634	29.673	27.940	26.396	25.012	23.766	22.637	21.611
3	47.009	43.811	41.010	38.539	36.344	34.383	32.619	31.026	29.580	28.262
4	56.437	52.733	49.465	46.566	43.978	41.657	39.564	37.668	35.943	34.366
5	65.019	60.914	57.262	54.000	51.075	48.440	46.055	43.888	41.912	40.103
6	72.875	68.476	64.520	60.928	57.744	54.835	52.192	49.783	47.580	45.558
7	80.042	75.470	71.295	67.497	64.043	60.899	58.029	55.404	52.997	50.782
8	86.492	81.898	77.604	73.641	70.001	66.663	63.599	60.784	58.194	55.803
9	92.118	87.715	83.434	79.393	75.627	72.140	68.917	65.940	63.188	60.641
10	96.668	92.813	88.733	84.728	80.913	77.331	73.989	70.880	67.991	65.307
11	99.535	96.954	93.395	89.595	85.834	82.223	78.809	75.604	72.605	69.805
12		99.573	97.195	93.890	90.334	86.789	83.364	80.105	77.028	74.135
13			99.606	97.400	94.315	90.975	87.623	84.366	81.250	78.293
14				99.634	97.577	94.685	91.535	88.357	85.253	82.269
15					99.659	97.732	95.010	92.030	89.009	86.045
16						99.680	97.868	95.297	92.471	89.592
17							99.699	97.989	95.553	92.865
18								99.715	98.097	95.783
19									99.730	98.193
20										99.744

Appendix 6

95 PER CENT RANKS

$j \backslash n$	SAMPLE SIZE 21	22	23	24	25	26	27	28	29	30
1	13.295	12.731	12.212	11.735	11.293	10.883	10.502	10.147	9.814	9.503
2	20.673	19.812	19.020	18.289	17.612	16.983	16.397	15.851	15.339	14.860
3	27.055	25.947	24.925	23.980	23.104	22.289	21.530	20.821	20.156	19.533
4	32.921	31.591	30.364	29.227	28.172	27.190	26.274	25.417	24.614	23.860
5	38.441	36.909	35.193	34.181	32.961	31.824	30.763	29.769	28.837	27.962
6	43.698	41.980	40.390	38.914	37.541	36.260	35.062	33.940	32.887	31.897
7	48.739	46.849	45.097	43.469	41.952	40.535	39.210	37.967	36.800	35.701
8	53.954	51.546	49.643	47.873	46.221	44.677	43.230	41.873	40.597	39.395
9	58.280	56.087	54.046	52.142	50.364	48.700	47.139	45.673	44.294	42.993
10	62.810	60.484	58.315	56.289	54.393	52.616	50.948	49.379	47.901	46.507
11	67.189	64.746	62.461	60.321	58.316	56.434	54.664	52.998	51.427	49.944
12	71.420	68.874	66.485	64.244	62.138	60.158	58.293	56.536	54.877	53.309
13	75.501	72.869	70.391	68.058	65.861	63.791	61.839	59.996	58.254	56.605
14	79.425	76.728	74.176	71.764	69.487	67.336	65.303	63.380	61.561	59.837
15	83.182	80.444	77.836	75.361	73.015	70.792	68.686	66.691	64.799	63.005
16	86.755	84.006	81.366	78.843	76.441	74.158	71.988	69.927	67.970	66.111
17	90.116	87.397	84.752	82.204	79.762	77.430	75.207	73.089	71.073	69.154
18	93.219	90.589	87.978	85.431	82.970	80.604	78.338	76.173	74.106	72.133
19	95.990	93.540	91.019	88.509	86.052	83.672	81.378	79.176	77.066	75.047
20	98.281	96.178	93.832	91.411	88.994	86.623	84.318	82.092	79.950	77.894
21	99.756	98.360	96.348	94.099	91.771	89.440	87.148	84.915	82.753	80.669
22		99.767	98.433	96.505	94.344	92.101	89.851	87.633	85.468	83.367
23			99.777	98.499	96.648	94.569	92.406	90.232	88.083	85.981
24				99.786	98.560	96.780	94.777	92.689	90.584	88.501
25					99.795	98.616	96.902	94.969	92.951	90.913
26						99.803	98.668	97.015	95.148	93.194
27							99.810	98.716	97.120	95.314
28								99.817	98.761	97.218
29									99.823	98.802
30										99.829

95 PER CENT RANKS

$j \backslash n$	31	32	33	34	35	36	37	38	39	40
					SAMPLE SIZE					
1	9.211	8.937	8.678	8.434	8.203	7.985	7.778	7.581	7.394	7.216
2	14.409	13.985	13.585	13.207	12.850	12.512	12.191	11.885	11.595	11.319
3	18.946	18.394	17.873	17.381	16.915	16.474	16.054	15.656	15.277	14.915
4	23.150	22.482	21.850	21.253	20.688	20.152	19.643	19.159	18.698	18.259
5	27.137	26.360	25.625	24.931	24.272	23.648	23.054	22.490	21.952	21.440
6	30.964	30.084	29.252	28.465	27.718	27.010	26.337	25.696	25.085	24.503
7	34.665	33.687	32.763	31.887	31.056	30.268	29.518	28.804	28.124	27.475
8	38.261	37.190	36.176	35.216	34.305	33.439	32.616	31.832	31.084	30.371
9	41.766	40.606	39.507	38.466	37.477	36.537	35.643	34.791	33.979	33.203
10	45.190	43.945	42.765	41.645	40.582	39.571	38.608	37.691	36.815	35.979
11	48.542	47.214	45.956	44.761	43.626	42.546	41.517	40.537	39.601	38.706
12	51.825	50.419	49.086	47.819	46.615	45.468	44.376	43.334	42.339	41.388
13	55.044	53.564	52.159	50.823	49.552	48.341	47.187	46.086	45.034	44.028
14	58.203	56.651	55.177	53.775	52.440	51.168	49.955	48.796	47.689	46.630
15	61.302	59.683	58.144	56.678	55.282	53.951	52.680	51.466	50.305	49.195
16	64.343	62.661	61.060	59.534	58.080	56.691	55.366	54.098	52.886	51.725
17	67.326	65.585	63.926	62.343	60.833	59.391	58.012	56.693	55.431	54.222
18	70.251	68.456	66.742	65.106	63.543	62.049	60.620	59.252	57.942	56.686
19	73.117	71.272	69.509	67.823	66.210	64.668	63.190	61.776	60.419	59.119
20	75.922	74.034	72.225	70.493	68.835	67.246	65.723	64.264	62.864	61.520
21	78.664	76.738	74.889	73.116	71.415	69.784	68.219	66.717	65.275	63.891
22	81.338	79.382	77.499	75.689	73.951	72.280	70.676	69.135	67.654	66.230
23	83.939	81.961	80.052	78.212	76.440	74.735	73.094	71.517	69.999	68.539
24	86.460	84.472	82.545	80.680	78.881	77.145	75.473	73.862	72.310	70.815
25	88.891	86.907	84.971	83.091	81.270	79.509	77.809	76.168	74.586	73.060
26	91.219	89.255	87.325	85.439	83.604	81.825	80.101	78.435	76.825	75.270
27	93.422	91.505	89.596	87.717	85.878	84.087	82.347	80.660	79.027	77.447
28	95.470	93.635	91.772	89.916	88.086	86.292	84.542	82.840	81.188	79.587
29	97.310	95.615	93.834	92.024	90.217	88.433	86.682	84.972	83.306	81.688
30	98.841	97.396	95.752	94.021	92.261	90.501	88.760	87.050	85.378	83.746
31	99.835	98.878	97.476	95.880	94.198	92.483	90.768	89.069	87.399	85.763
32		99.840	98.912	97.552	96.001	94.364	92.694	91.021	89.362	87.729
33			99.845	98.945	97.623	96.114	94.521	92.893	91.260	89.639
34				99.849	98.975	97.690	96.222	94.669	93.081	91.487
35					99.854	99.004	97.754	96.324	94.810	93.260
36						99.858	99.031	97.814	96.420	94.943
37							99.861	99.057	97.871	96.511
38								99.865	99.081	97.925
39									99.869	99.104
40										99.872

95 PER CENT RANKS

$j\backslash n$	41	42	43	44	45	46	47	48	49	50
					SAMPLE SIZE					
1	7.046	6.884	6.730	6.582	6.440	6.305	6.175	6.050	5.931	5.816
2	11.055	10.804	10.563	10.334	10.113	9.902	9.700	9.506	9.319	9.140
3	14.571	14.241	13.927	13.626	13.338	13.061	12.796	12.541	12.297	12.061
4	17.840	17.439	17.056	16.690	16.339	16.002	15.679	15.369	15.071	14.784
5	20.951	20.483	20.036	19.608	19.198	18.804	18.427	18.064	17.715	17.379
6	23.947	23.416	22.907	22.420	21.954	21.506	21.076	20.663	20.266	19.883
7	26.854	26.262	25.694	25.151	24.630	24.130	23.650	23.188	22.744	22.317
8	29.689	29.037	28.413	27.814	27.241	26.691	26.162	25.623	25.164	24.694
9	32.461	31.752	31.073	30.422	29.797	29.198	28.622	28.068	27.535	27.022
10	35.180	34.415	33.682	32.980	32.306	31.659	31.037	30.439	29.864	29.309
11	37.851	37.032	36.247	35.495	34.773	34.079	33.413	32.772	32.154	31.560
12	40.478	39.607	38.772	37.971	37.202	36.463	35.753	35.069	34.411	33.778
13	43.065	42.143	41.259	40.410	39.596	38.813	38.060	37.336	36.698	35.933
14	45.615	44.644	43.712	42.817	41.958	41.132	40.338	39.573	38.836	38.126
15	48.131	47.110	46.132	45.193	44.290	43.422	42.587	41.783	41.008	40.262
16	50.612	49.546	48.522	47.539	46.594	45.665	44.810	43.968	43.156	42.373
17	53.062	51.950	50.883	49.857	48.871	47.922	47.009	46.129	45.280	44.462
18	55.482	54.326	53.215	52.148	51.122	50.134	49.183	48.266	47.382	46.530
19	57.871	56.672	55.520	54.413	53.348	52.322	51.334	50.382	49.463	48.577
20	60.230	58.991	57.799	56.653	55.549	54.487	53.463	52.476	51.523	50.604
21	62.560	61.281	60.051	58.867	57.727	56.629	55.570	54.549	53.563	52.612
22	64.861	63.545	62.273	61.057	59.882	58.749	57.556	56.602	55.584	54.801
23	67.133	65.781	64.478	63.223	62.013	60.846	59.721	58.634	57.585	56.572
24	69.376	67.989	66.652	65.363	64.121	62.922	61.765	60.647	59.568	58.524
25	71.588	70.169	68.800	67.480	66.205	64.975	63.787	62.640	61.531	60.459
26	73.769	72.320	70.922	69.571	68.267	67.007	65.790	64.613	63.476	62.375
27	75.919	74.443	73.016	71.637	70.304	69.016	67.771	66.566	65.401	64.274
28	78.035	76.534	75.082	73.677	72.317	71.002	69.730	68.500	67.308	66.155
29	80.117	78.594	77.119	75.689	74.306	72.966	71.668	70.412	69.196	68.017
30	82.162	80.621	79.125	77.674	76.268	74.905	73.584	72.304	71.064	69.862
31	84.166	82.611	81.099	79.630	78.203	76.819	75.477	74.175	72.912	71.687
32	86.128	84.564	83.039	81.554	80.111	78.708	77.346	76.023	74.739	73.493
33	88.042	86.475	84.942	83.446	81.988	80.569	79.190	77.848	76.545	75.279
34	89.903	88.340	86.805	85.302	83.834	82.402	81.007	79.650	78.329	77.045
35	91.702	90.153	88.623	87.119	85.645	84.204	82.797	81.426	80.090	78.790
36	93.430	91.907	90.391	88.892	87.418	85.972	84.557	83.175	81.826	80.511
37	95.070	93.591	92.102	90.618	89.150	87.704	86.285	84.895	83.536	82.210
38	96.598	95.190	93.744	92.287	90.834	89.395	87.977	86.584	85.218	83.882
39	97.976	96.681	95.305	93.891	92.464	91.041	89.630	88.238	86.870	85.528
40	99.126	98.025	96.760	95.414	94.030	92.633	91.238	89.854	88.488	87.144
41	99.875	99.147	98.071	96.835	95.518	94.164	92.795	91.427	90.069	88.728
42		99.878	99.167	98.116	96.907	95.618	94.291	92.950	91.608	90.275
43			99.881	99.186	98.158	96.975	95.714	94.414	93.098	91.781
44				99.883	99.205	98.199	97.041	95.805	94.531	93.240
45					99.886	99.222	98.238	97.103	95.892	94.643
46						99.889	99.239	98.275	97.163	95.976
47							99.891	99.255	98.311	97.221
48								99.893	99.270	98.345
49									99.895	99.285
50										99.897

Appendix 7.   Matrix Algebra Revision

The solution of the second-order matrix

$$\begin{vmatrix} a_1 & b_1 \\ a_2 & b_2 \end{vmatrix}$$

is $a_1 b_2 - a_2 b_1$.

The solution of the third-order matrix

$$\begin{vmatrix} a_1 & b_1 & c_1 \\ a_2 & b_2 & c_2 \\ a_3 & b_3 & c_3 \end{vmatrix}$$

is

$$a_1 \begin{vmatrix} b_2 & c_2 \\ b_3 & c_3 \end{vmatrix} - b_1 \begin{vmatrix} a_2 & c_2 \\ a_3 & c_3 \end{vmatrix} + c_1 \begin{vmatrix} a_2 & b_2 \\ a_3 & b_3 \end{vmatrix}$$

$$= a_1 b_2 c_3 + a_2 b_3 c_1 + a_3 b_1 c_2 - a_1 b_3 c_2 - a_2 b_1 c_3 - a_3 b_2 c_1$$

$a_1, a_2, \ldots, c_2, c_3$ are called the *elements* of the matrix.
The lower order matrix associated with a_1, i.e.

$$\begin{vmatrix} b_2 & c_2 \\ b_3 & c_3 \end{vmatrix}$$

is called the *cofactor* of a_1, denoted A_1. Similarly,

$$- \begin{vmatrix} a_2 & c_2 \\ a_3 & c_3 \end{vmatrix}$$

is the cofactor of b_1, denoted B_1, and so on. Thus the solution of the third order matrix is

$$a_1 A_1 + b_1 B_1 + c_1 C_1$$

and so on for higher orders.

To get the cofactor signs correct, think of the element positions in the matrix as having positive and negative signs associated with them, as follows:

$$
\begin{vmatrix}
+ & - & + & - & \cdot & \cdot & \cdot \\
- & + & - & + & \cdot & \cdot & \cdot \\
+ & - & + & - & \cdot & \cdot & \cdot \\
\cdot & \cdot & \cdot & \cdot & & & \\
\cdot & \cdot & \cdot & \cdot & & & \\
\cdot & \cdot & \cdot & \cdot & & &
\end{vmatrix}
$$

Matrix multiplication

$$
\begin{vmatrix} a_1 & b_1 \\ a_2 & b_2 \end{vmatrix} \times \begin{vmatrix} A_1 & B_1 \\ A_2 & B_2 \end{vmatrix}
$$

$$
= \begin{vmatrix} (a_1 A_1 + b_1 A_2) & (a_1 B_1 + b_1 B_2) \\ (a_2 A_1 + b_2 A_2) & (a_2 B_1 + b_2 B_2) \end{vmatrix}
$$

Appendix 8. Failure Reporting, Analysis and Corrective Action System (FRACAS)

Failure Reporting Form				
Report No.				
System		**Serial No.**	**Date** / /	**Time** :
Location/customer			**Run time**	**h:** min
Test/operation	**Tick**	1 2 3 4 Schedule 1 Schedule 2 Schedule 3 Calibration Unscheduled Other		
Describe (Conditions, special features)				
Result/failure				
Repair action				

Parts replaced

Item	1	2	3	4	5
Part No. Serial No. Mod strike					

Time started:	**Time finished:**	**Time worked:** h min
Work done by:	**Checked by:**	

Analysis
Previous occurrences   Trend   Cause/s   Comments
(Report Nos.)

Corrective action
Recommended

Agreed

Tested

Accepted

	Approved	**Date** / /

ANALYSES

1. Listing of failures:

Failure Report No.	System	Sub-system	Assembly	Part

Output selectable by system, sub-system, assembly, part. Also by time period and other appropriate feature, e.g. location, modification status.

2. Pareto analysis of top 10–20 failure modes, selectable as above (Chapter 12). See Note 1.
3. MTBF for system, sub-system, assembly. Show number of failures for each MTBF value. Selectable as above.
4. Trend analyses, selectable as above. See Chapter 12.
5. Probability/hazard plots and derivation of distribution parameters. Selected as above. See chapter 3.

Notes

1. The Pareto analysis should form the basis for the critical items list (page 336). Failure reporting for critical items should be amplified as necessary to aid investigation, and special reporting, e.g. by phone, fax or e-mail, could be used.
2. Calendar time could be used instead of run time when appropriate. Then trend analysis, probability plots, and MTBF would be calculated on a calendar time basis. This is acceptable if run time is quite closely correlated with calendar time and if run time data are not easily obtainable, e.g. if the equipment does not have run time indicators. Using calendar time can be easier and cheaper, since it is then not necessary to obtain the run time at each failure and total (fleet) run time. Only the startup date for each unit and the total number of units in use need be ascertained.
3. The FRACAS should be made simple to operate and to understand.
4. Input data should be sufficiently detailed to be meaningful to users. Coding of data such as causes of failure, either by the person filling in the job report or by people at the data centre, can lead to ambiguity and errors.
5. Data input directly into portable computers, rather than onto paper, can greatly improve the quality and speed of data collection and analysis.

Appendix 9. Reliability, Maintainability (and Safety) Plan Example

**RELIABILITY AND MAINTAINABILITY (AND SAFETY) PLAN
SUPER SYSTEM**

CONTENTS

Appendix 1   RAMS Work Plans
Appendix 2   RAMS Deliverables

References

1.   RELIABILITY, MAINTAINABILITY (AND SAFETY) PLAN OVERVIEW

1.1   Introduction

This Plan describes the organisation and responsibilities for the reliability, maintainability (and safety) (RAMS) engineering tasks that will be integrated into the design, development, production and in-service support activities for the Super system project. It also describes the RAMS tasks that will be undertaken.

The RAMS requirements form part of the overall performance requirements for the system, as described in Reference 1.

The RAMS tasks will managed and performed in compliance with the requirements and guidelines in Reference 2.

During the design and development stage, the Company will ensure that their system and equipment suppliers and subcontractors fully understand and comply with the RAMS requirements and with the RAMS engineering tasks specified to them.

In order to achieve these requirements the design, development test, and production philosophy will be for the creation of intrinsically robust, failure-free designs, including the design of all production processes, and followed by stringent production quality assurance and improvement. The failure-free design (FFD) philosophy of hardware and of processes will ensure that all stresses, variations, and other potential or actual causes of failure will be identified and corrected, by the adoption of an integrated, concurrent approach to design, development, and production control. The primary objective of the reliability programme will be to ensure that designs are inherently robust in relation to manufacturing processes and to the environmental conditions of storage, maintenance and operation, throughout the life of the system. To this end, all design analyses and tests will be directed towards identifying and eliminating causes of failure. Particular features of the RAMS programme in this respect will be:

—Prediction and measurement of reliability will be performed as described, but these activities will be treated as secondary to the primary objective of creating an inherently failure-free design.
—The effects of variation of environmental conditions, parameter values and manufacturing processes will be assessed by analysis and by the use of statistically designed tests, including Taguchi methods, to ensure that all designs are robust in relation to all sources of variation over the life of the system.
—The methods of Highly Accelerated Life Testing (HALT) and Highly Accelerated Stress Screening (HASS) will be applied to development and production testing. The objective of these tests will be to force failures by applying high stresses, so that designs and processes can be optimized.

—The reliability test programme will be fully integrated with the overall development test programme. A common failure reporting and corrective action system (FRACAS) will be applied. All failures will be fully investigated and corrective action taken to prevent recurrence. The test programme will include sub-system and system level tests, statistical experiments to assess variation, environmental tests, trials, as well as tests specifically designed to stimulate reliability growth. Reliability growth in development will be monitored in terms of problems discovered and corrected. All test and failure data and corrective action information will be reported, with assessments of reliability achieved and expected in relation to development programme objectives and the in-service requirements.

—By the end of the development programme the Company will have demonstrated that the system and subsystem designs are inherently capable of being produced and tested, and of withstanding the storage, operating and maintenance environments without failure during the in-service life. All relevant failures which occur during development testing will have been corrected by changes to designs or processes, and the effectiveness of the corrective action will have been proved.

—All subcontractors will be required to undertake reliability programmes based upon the same philosophy and methods. The results of their analyses and tests will be closely monitored to ensure a common approach, and to ensure that design improvements and corrective action is implemented promptly and effectively. Subcontractors will be selected on the basis of their proven excellence in the technologies involved, and they will be motivated to adopt the same philosophy for RAMS.

1.2 Reliability, availability, maintainability and safety (RAMS) requirements

The RAMS requirements for the project, related to a service life of ____ are:

—System reliability requirement: not more than N failures per ____, causing ____.
—System maintainability requirement: repairs to be performed in not more than __ minutes for ____% of failures.
—(Safety)

Where responsibility for design and development of a sub- system or component is let by the company to a subcontractor the RAMS requirements for that system, sub-system or equipment will be fully specified by the Project RAMS Engineer in accordance with the relevant Company Procedure.

1.3 RAMS tasks

To ensure achievement and assurance of the RAMS requirements, comprehensive reliability, maintainability (and safety) engineering tasks will be applied. Parts 2 and 3 detail the RAMS engineering tasks that will be undertaken to satisfy the particular requirements in the specification. The tasks will be applied where relevant, from the initial design phase through to, and including, the in-service warranty period.

The RAMS tasks that will be applied at appropriate stages of design and development are:

—Quality function deployment (QFD) to identify and prioritize key design and process requirements.
—Reliability modelling of the system and sub-systems using reliability block diagrams.
—Reliability predictions and apportionment for sub-systems and components, as appropriate.
—Failure analysis by failure mode, effect and criticality analysis (FMECA) and fault tree analysis (FTA).
—Testing, with the emphasis on HALT.
—Failure reporting, analysis and corrective action (FRACAS).
—Maintainability analysis and demonstration.
—System safety analysis by the application of hazard identification and hazard analysis techniques.
—RAMS reviews, in which compliance with RAMS tasks will be audited.
—Production quality assurance and improvement methods.

The Work Plan for these tasks is shown in Appendix 1, and the list of RAMS deliverables is in Appendix 2.

2. RELIABILITY & MAINTAINABILITY ENGINEERING TASKS

2.1 Reliability modelling

RBDs will be constructed following the guidelines given in . . . ., and will be updated to reflect the state of the design. RBD interfaces will be agreed with and controlled by the Project RAMS Engineer.

2.2 Reliability prediction and apportionment

Reliability prediction will be carried out, covering all areas of design following the guidelines set out in . . . . . The methods and data sources will be declared to the Customer.

The reliability prediction process will be started at the commencement of the project, and will be updated to take account of design changes. Predictions will be used to identify high risk components and sub-systems, and for updating reliability apportionments.

2.3 Failure modes, effects and criticality analysis

FMECA will be carried out following the guidelines given in . . . . FMECA software will be used to create and record the analysis. Wherever appropriate computer-aided design (CAD) models and data will be used as inputs and for analysis.

The relevant Design, Quality Assurance and Safety staff will be responsible for formal response to the analyses regarding preventive actions, compensating factors or the effects on safety.

2.4   Fault tree analysis

FTA will be undertaken on equipment designs where safety-related Top Events have been identified. To enable FTA to be undertaken Top Events will be defined and described.

FTA software will be used to perform and record the analysis.

2.5   Reliability testing

An integrated test programme will be conducted, as described in the Project Test Plan (Reference 3). The main features of the test approach will be:

(HALT)

(Taguchi)

2.6   Failure reporting and RAMS monitoring

During the design and development phase of the project the Company will operate a failure reporting, analysis and corrective action system (FRACAS) as described in the Company Quality Manual.

Failure reporting action will be taken on all failures that occur on hardware and software used on the Company and subcontractor tests and trials undertaken during the design and development phase of the project. Failures during production testing will be reported and managed in accordance with the Company Quality Manual.

Reliability and maintainability achievement will be monitored during all development testing.

2.7   Production reliability assurance

The Quality Assurance Plan describes the methods that will be applied prior to and during production to ensure that production systems will achieve the reliability requirements. The QA activities will be integrated with the reliability activities wherever appropriate.

Particular features of the QA programme to ensure reliability will be:

—FMECA will be used for the derivation of production functional test and inspection methods. Production FMECA will also be performed.

—Statistical Process Control (SPC) will be applied to all manufacturing processes in which variation can affect yield and reliability. SPC limits will be based where relevant on the results of analysis and test of development hardware, particularly when statistical experiments have been conducted to optimize product and process designs.

—Production stress screening methods (HASS) will be developed as part of the development test programme. HASS will be applied to all production hardware, at sub-system and system level, and will be tailored to provide the optimum screens for the items concerned. HASS profiles and durations will be continuously monitored and modified during production, to ensure the most cost-effective approach. The HASS will provide assurance that all production hardware is function and capable of withstanding the storage and use environments.

—All failures occurring at any test or inspection stage will be investigated, with the objective of preventing recurrence. The objective of the failure reporting and corrective action system (FRACAS), in conjunction with monitoring of SPC, will be to generate continuous improvement of all processes.

—All subcontractors will be required to work to the same philosophy of continuous improvement. Their performance will be closely monitored, and they will be assisted where necessary.

2.7   *Maintainability analysis and demonstration*

Maintainability will be analysed during design to ensure that the requirements are achieved, and will be measured during development. Preventive maintenance tasks will be optimized using the reliability centred maintenance (RCM) method.

2.9   *In-service RAMS monitoring (ISRM)*

The Company will maintain a FRACAS for all systems in service, to monitor RAMS achievement, to ensure that all reliability requirements are achieved, and that any shortcomings or failures are promptly investigated and corrected.

3.   SAFETY ENGINEERING TASKS

3.1   *Preliminary hazard analysis*

The Company and subcontractors will apply preliminary hazard analysis (PHA) to all areas of design responsibility in accordance with the Company RAMS Manual and Reference . . . .

The output from the PHA will consist of documented hazards associated with the system.

3.2   *System and sub-system hazard analysis*

System and sub-system hazard analysis (SHA) will be carried out to identify hazards associated with the system and sub-system design, which may not have been identified in the PHA, including component fault modes, critical human error inputs, and hazards resulting from interfaces within the equipment. The techniques that will be applied will be HAZOPS, FMECA, FTA, and event tree analysis (ETA), as appropriate. They will be applied to all areas of design responsibility in accordance with the Company RAMS Manual.

The output from the SHA will detail system and sub-system hazards, their severity and probability values, together with recommendations for actions necessary to eliminate them, or to control the risk to a level that is agreed to be as low as practicable (ALARP).

3.3   Hazard tracking (Hazard log)

Hazard Tracking will be applied to all conditions which could possibly produce a Catastrophic (Severity Level 1) effect or Hazardous (Severity Level 2) effect, as defined in Reference . . . .

Hazards will be tracked from the point of identification until the hazard is eliminated or the associated risk is reduced to a level agreed with the Customer as being acceptable. The Hazard Tracking system will be maintained after design work is complete and throughout the warranty period.

A Hazard Log will be maintained, containing as a minimum:

—Description of each hazard, by nature, cause, and effect.
—Severity rating.
—Status of actions to resolve.
—Traceability of resolution to the point of risk acceptance.

The Company will ensure that the hazard tracking system is correctly maintained, and will make the records available to the Customer for audit and review.

4.   PROJECT RAMS ENGINEERING MANAGEMENT AND REPORTING

4.1   Responsibilities

For the effective management of RAMS engineering, formal management procedures and guidelines on analysis techniques will be applied for all tasks outlined in the RAMS Plan. The management procedures are contained within the relevant Company Procedures.

The application of the RAMS Plan will be the responsibility of the Project Reliability (and Safety) Engineer on behalf of the Project Manager.

Subcontractors will be required to prepare RAMS Plans that comply with the Project RAMS Plan, as appropriate to the sub-systems.

4.2   RAMS reviews

The RAMS Review is the formal audit of the RAMS engineering tasks undertaken by the Company and implemented as detailed in the RAMS Work Plan.

RAMS Reviews will be conducted in accordance with the relevant Company Procedure, on all areas of design for which the Company and subcontractors are responsible. The responsibility for conducting RAMS Reviews is assigned to the Company Project Reliability (and Safety) Engineer.

On completion of a RAMS Review, a RAMS Progress Report will be raised by the Company, to highlight areas of non-conformance or risk, and to advise on the extent to which the RAMS requirements are expected to be achieved.

References

1. Super System Specification.
2. (Detailed method guidelines/descriptions: relevant company procedures, standards, etc.).
3. Super System Test Plan.

Appendices

1. RAMS Work Plans.
2. RAMS Deliverables.

Index